Discrete Geometry and Symmetry

Discrete Geometry and Symmetry

Edited by Isla Harvey

www.statesacademicpress.com

States Academic Press,
109 South 5th Street,
Brooklyn, NY 11249, USA

Visit us on the World Wide Web at:
www.statesacademicpress.com

ISBN: 978-1-63989-743-8

Cataloging-in-Publication Data

Discrete geometry and symmetry / edited by Isla Harvey.
p. cm.
Includes bibliographical references and index.
ISBN 978-1-63989-743-8
1. Discrete geometry. 2. Symmetry (Mathematics). 3. Combinatorial geometry.
4. Discrete mathematics. 5. Group theory. I. Harvey, Isla.

QA640.7 .D57 2023
516.11--dc23

Table of Contents

Preface

Every book is a source of knowledge and this one is no exception. The idea that led to the conceptualization of this book was the fact that the world is advancing rapidly; which makes it crucial to document the progress in every field. I am aware that a lot of data is already available, yet, there is a lot more to learn. Hence, I accepted the responsibility of editing this book and contributing my knowledge to the community.

Discrete geometry refers to a subfield of geometry that examines discrete and highly symmetric structures in geometry and how discrete geometric objects might be constructed. It is closely connected to computational geometry as well as other mathematical subjects including combinatorial geometry and topology, combinatorial optimization, geometric graph theory, and finite geometry. Discrete geometry focuses on the problems that aim to characterize specific discrete arrangements of geometric objects having high symmetry. Many of the techniques used to investigate discrete arrangements are based on a wide range of symetrizations. Discrete geometry is used in numerous theoretical disciplines including algebraic geometry and topology, geometry of numbers, mathematical physics, functional analysis, geometric measure theory, calculus of variations, and group theory. This book is compiled in such a manner, that it will provide an in-depth knowledge about the theory and applications of discrete geometry and symmetry. It is a resource guide for experts as well as students.

While editing this book, I had multiple visions for it. Then I finally narrowed down to make every chapter a sole standing text explaining a particular topic, so that they can be used independently. However, the umbrella subject sinews them into a common theme. This makes the book a unique platform of knowledge.

I would like to give the major credit of this book to the experts from every corner of the world, who took the time to share their expertise with us. Also, I owe the completion of this book to the never-ending support of my family, who supported me throughout the project.

Editor

1

Icosahedral Polyhedra from D_6 Lattice and Danzer's *ABCK* Tiling

Abeer Al-Siyabi [1], Nazife Ozdes Koca [1,*] and Mehmet Koca [2,†]

[1] Department of Physics, College of Science, Sultan Qaboos University, P.O. Box 36, Al-Khoud, Muscat 123, Oman; s21168@student.squ.edu.om
[2] Department of Physics, Cukurova University, Adana 1380, Turkey; mehmetkocaphysics@gmail.com
* Correspondence: nazife@squ.edu.om
† Retired.

Abstract: It is well known that the point group of the root lattice D_6 admits the icosahedral group as a maximal subgroup. The generators of the icosahedral group H_3, its roots, and weights are determined in terms of those of D_6. Platonic and Archimedean solids possessing icosahedral symmetry have been obtained by projections of the sets of lattice vectors of D_6 determined by a pair of integers (m_1, m_2) in most cases, either both even or both odd. Vertices of the Danzer's *ABCK* tetrahedra are determined as the fundamental weights of H_3, and it is shown that the inflation of the tiles can be obtained as projections of the lattice vectors characterized by the pair of integers, which are linear combinations of the integers (m_1, m_2) with coefficients from the Fibonacci sequence. Tiling procedure both for the *ABCK* tetrahedral and the <*ABCK*> octahedral tilings in 3D space with icosahedral symmetry H_3, and those related transformations in 6D space with D_6 symmetry are specified by determining the rotations and translations in 3D and the corresponding group elements in D_6. The tetrahedron *K* constitutes the fundamental region of the icosahedral group and generates the rhombic triacontahedron upon the group action. Properties of "*K*-polyhedron", "*B*-polyhedron", and "*C*-polyhedron" generated by the icosahedral group have been discussed.

Keywords: lattices; Coxeter–Weyl groups; icosahedral group; projections of polytopes; polyhedra; aperiodic tilings; quasicrystals

---

1. Introduction

Quasicrystallography as an emerging science attracts the interests of many scientists varying from the fields of material science, chemistry, and physics. For a review, see, for instance, the references [1–3]. It is mathematically intriguing, as it requires the aperiodic tiling of the space by some prototiles. There have been several approaches to describe the aperiodicity of the quasicrystallographic space such as the set theoretic techniques, cut-and-project scheme of the higher dimensional lattices, and the intuitive approaches such as the Penrose-like tilings of the space. For a review of these techniques, we propose the reference [4].

There have been two major approaches for the aperiodic tiling of the 3D space with local icosahedral symmetry. One of them is the Socolar–Steinhardt tiles [5] consisting of acute rhombohedron with golden rhombic faces, Bilinski rhombic dodecahedron, rhombic icosahedron, and rhombic triacontahedron. The latter three are constructed with two Ammann tiles of acute and obtuse rhombohedra. Later, it was proved that [6,7] they can be constructed by the Danzer's *ABCK* tetrahedral tiles [8,9], and recently, Hann–Socolar–Steinhardt [10] proposed a model of tiling scheme with decorated Ammann tiles. A detailed account of the Danzer *ABCK* tetrahedral tilings can be found in [11] and in page 231 of the reference [4] where the substitution matrix, its eigenvalues, and the corresponding eigenvectors are studied. The right and left eigenvectors of the substitution matrix corresponding to the Perron–Frobenius eigenvalue are well known and will not be repeated here.

Ammann rhombohedral and Danzer *ABCK* tetrahedral tilings are intimately related with the projection of six-dimensional cubic lattice and the root and weight lattices of D_6, the point symmetry group of which is of order $2^5 6!$. See for a review the paper "Modelling of quasicrystals" by Kramer [12] and references therein. Similar work has also been carried out in the reference [13]. Kramer and Andrle [14] have investigated Danzer tiles from the wavelet point of view and their relations with the lattice D_6.

In what follows, we point out that the icosahedral symmetry requires a subset of the root lattice D_6 characterized by a pair of integers (m_1, m_2) with $m_1 + m_2 = even$, which are the coefficients of the orthogonal set of vectors $l_i, (i = 1, 2, \ldots, 6)$. Our approach is different than the cut and project scheme of lattice D_6, as will be seen in the sequel. The paper consists of two major parts; the first part deals with the determination of Platonic and Archimedean icosahedral polyhedra by projection of the fundamental weights of the root lattice D_6 into 3D space, and the second part employs the technique to determine the images of the Danzer tiles in D_6. Inflation of the Danzer tiles are related to a redefinition of the pair of integers (m_1, m_2) by the Fibonacci sequence. Embeddings of basic tiles in the inflated ones require translations and rotations in 3D space, where the corresponding transformations in 6D space can be easily determined by the technique we have introduced. This technique, which restricts the lattice D_6 to its subset, has not been discussed elsewhere.

The paper is organized as follows. In Section 2, we introduce the root lattice D_6, its icosahedral subgroup, and decomposition of its weights in terms of the weights of the icosahedral group H_3 leading to the Archimedean polyhedra projected from the D_6 lattice. It turns out that the lattice vectors to be projected are determined by a pair of integers (m_1, m_2). In Section 3, we introduce the *ABCK* tetrahedral Danzer tiles in terms of the fundamental weights of the icosahedral group H_3. Tiling by inflation with τ^n where $\tau = \frac{1+\sqrt{5}}{2}$ and $n \in \mathbb{Z}$ is studied in H_3 space by prescribing the appropriate rotation and translation operators. The corresponding group elements of D_6 are determined noting that the pair of integers (m_1, m_2) can be expressed as the linear combinations of similar integers with coefficients from Fibonacci sequence. Section 4 is devoted for conclusive remarks.

2. Projection of D_6 Lattice under the Icosahedral Group H_3 and the Archimedean Polyhedra

We will use the Coxeter diagrams of D_6 and H_3 to introduce the basic concepts of the root systems, weights, and the projection technique. A vector of the D_6 lattice can be written as a linear combination of the simple roots with integer coefficients:

$$\lambda = \sum_{i=1}^{6} n_i \alpha_i = \sum_{i=1}^{6} m_i l_i, \ \sum_{i=1}^{6} m_i = even, \ n_i, m_i \in \mathbb{Z}. \tag{1}$$

Here, α_i are the simple roots of D_6 defined in terms of the orthonormal set of vectors as $\alpha_i = l_i - l_{i+1}$, $i = 1, \ldots, 5$ and $\alpha_6 = l_5 + l_6$, and the reflection generators of D_6 act as $r_i : l_i \longleftrightarrow l_{i+1}$ and $r_6 : l_5 \longleftrightarrow -l_6$. The generators of H_3 can be defined [13] as $R_1 = r_1 r_5$, $R_2 = r_2 r_4$ and $R_3 = r_3 r_6$, where the Coxeter element, for example, can be taken as $R = R_1 R_2 R_3$. The weights of D_6 are given by

$$\begin{aligned} &\omega_1 = l_1, \ \omega_2 = l_1 + l_2, \ \omega_3 = l_1 + l_2 + l_3, \ \omega_4 = l_1 + l_2 + l_3 + l_4, \\ &= \tfrac{1}{2}(l_1 + l_2 + l_3 + l_4 + l_5 - l_6), \ \omega_6 = \tfrac{1}{2}(l_1 + l_2 + l_3 + l_4 + l_5 + l_6). \end{aligned} \tag{2}$$

The Voronoi cell of the root lattice D_6 is the dual polytope of the root polytope of ω_2 [15] and determined as the union of the orbits of the weights ω_1, ω_5 and ω_6, which correspond to the holes of the root lattice [16]. If the roots and weights of H_3 are defined in two complementary 3D spaces $E_{\parallel}$ and $E_{\perp}$ as $\beta_i(\acute{\beta}_i)$ and $v_i(\acute{v}_i)$, $(i = 1, 2, 3)$, respectively, then they can be expressed in terms of the roots and weights of D_6 as

$$\begin{aligned}\beta_1 &= \tfrac{1}{\sqrt{2+\tau}}(\alpha_1+\tau\alpha_5),\ v_1 = \tfrac{1}{\sqrt{2+\tau}}(\omega_1+\tau\omega_5),\\ \beta_2 &= \tfrac{1}{\sqrt{2+\tau}}(\alpha_2+\tau\alpha_4),\ v_2 = \tfrac{1}{\sqrt{2+\tau}}(\omega_2+\tau\omega_4),\\ \beta_3 &= \tfrac{1}{\sqrt{2+\tau}}(\alpha_6+\tau\alpha_3),\ v_3 = \tfrac{1}{\sqrt{2+\tau}}(\omega_6+\tau\omega_3).\end{aligned} \tag{3}$$

For the complementary 3D space, replace β_i by $\acute{\beta}_i$, v_i by $\acute{v}_i$ in (3), and $\tau = \frac{1+\sqrt{5}}{2}$ by its algebraic conjugate $\sigma = \frac{1-\sqrt{5}}{2} = -\tau^{-1}$. Consequently, Cartan matrix of D_6 (Gram matrix in lattice terminology) and its inverse block diagonalize as

$$\begin{gathered} C = \begin{bmatrix} 2 & -1 & 0 & 0 & 0 & 0\\ -1 & 2 & -1 & 0 & 0 & 0\\ 0 & -1 & 2 & -1 & 0 & 0\\ 0 & 0 & -1 & 2 & -1 & -1\\ 0 & 0 & 0 & -1 & 2 & 0\\ 0 & 0 & 0 & -1 & 0 & 2 \end{bmatrix} \rightarrow \begin{bmatrix} 2 & -1 & 0 & 0 & 0 & 0\\ -1 & 2 & -\tau & 0 & 0 & 0\\ 0 & -\tau & 2 & 0 & 0 & 0\\ 0 & 0 & 0 & 2 & -1 & 0\\ 0 & 0 & 0 & -1 & 2 & -\sigma\\ 0 & 0 & 0 & 0 & -\sigma & 2 \end{bmatrix}\\ C^{-1} = \frac{1}{2}\begin{bmatrix} 2 & 2 & 2 & 2 & 1 & 1\\ 2 & 4 & 4 & 4 & 2 & 2\\ 2 & 4 & 6 & 6 & 3 & 3\\ 2 & 4 & 6 & 8 & 4 & 4\\ 1 & 2 & 3 & 4 & 3 & 2\\ 1 & 2 & 3 & 4 & 2 & 3 \end{bmatrix} \rightarrow \frac{1}{2}\begin{bmatrix} 2+\tau & 2\tau^2 & \tau^3 & 0 & 0 & 0\\ 2\tau^2 & 4\tau^2 & 2\tau^3 & 0 & 0 & 0\\ \tau^3 & 2\tau^3 & 3\tau^2 & 0 & 0 & 0\\ 0 & 0 & 0 & 2+\sigma & 2\sigma^2 & \sigma^3\\ 0 & 0 & 0 & 2\sigma^2 & 4\sigma^2 & 2\sigma^3\\ 0 & 0 & 0 & \sigma^3 & 2\sigma^3 & 3\sigma^2 \end{bmatrix} \end{gathered} \tag{4}$$

Figure 1 shows how one can illustrate the decomposition of the 6D space as the direct sum of two complementary 3D spaces where the corresponding nodes $\alpha_i, (i = 1, 2, \ldots, 6)$, $\beta_i, (i = 1, 2, 3)$ represent the corresponding simple roots, and $\acute{\beta}_i$ stands for β_i in the complementary space.

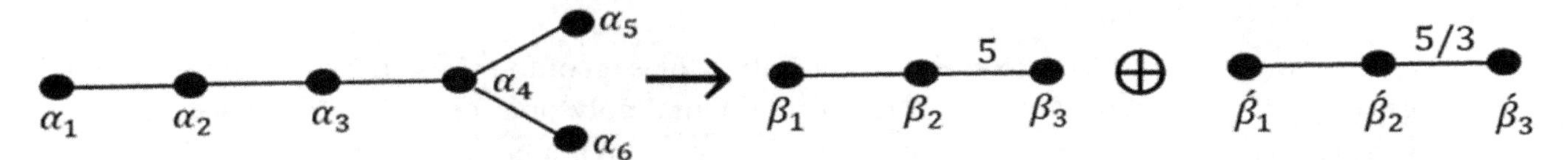

Figure 1. Coxeter–Dynkin diagrams of D_6 and H_3 illustrating the symbolic projection.

For a choice of the simple roots of H_3 as $\beta_1 = \left(\sqrt{2}, 0, 0\right)$, $\beta_2 = -\frac{1}{\sqrt{2}}(1,\sigma,\tau)$, $\beta_3 = \left(0, 0, \sqrt{2}\right)$ and the similar expressions for the roots in the complementary space, we can express the components of the set of vectors l_i $(i = 1, 2, \ldots, 6)$ as

$$\begin{bmatrix} l_1\\ l_2\\ l_3\\ l_4\\ l_5\\ l_6 \end{bmatrix} = \frac{1}{\sqrt{2(2+\tau)}}\begin{bmatrix} 1 & \tau & 0 & \tau & -1 & 0\\ -1 & \tau & 0 & -\tau & -1 & 0\\ 0 & 1 & \tau & 0 & \tau & -1\\ 0 & 1 & -\tau & 0 & \tau & 1\\ \tau & 0 & 1 & -1 & 0 & \tau\\ -\tau & 0 & 1 & 1 & 0 & \tau \end{bmatrix} \tag{5}$$

The first three and last three components project the vectors l_i into $E_\parallel$ and $E_\perp$ spaces, respectively. On the other hand, the weights are represented by the vectors $v_1 = \frac{1}{\sqrt{2}}(1,\tau,0)$, $v_2 = \sqrt{2}(0,\tau,0)$, $v_3 = \frac{1}{\sqrt{2}}\left(0,\tau^2,1\right)$. The generators of H_3 in the space $E_\parallel$ read

$$R_1 = \begin{bmatrix} -1 & 0 & 0\\ 0 & 1 & 0\\ 0 & 0 & 1 \end{bmatrix},\quad R_2 = \frac{1}{2}\begin{bmatrix} 1 & -\sigma & -\tau\\ -\sigma & \tau & 1\\ -\tau & 1 & \sigma \end{bmatrix},\quad R_3 = \begin{bmatrix} 1 & 0 & 0\\ 0 & 1 & 0\\ 0 & 0 & -1 \end{bmatrix}, \tag{6}$$

satisfying the generating relations

$$R_1{}^2 = R_2{}^2 = R_3{}^2 = (R_1R_3)^2 = (R_1R_2)^3 = (R_2R_3)^5 = 1 \tag{7}$$

Their representations in the $E_\perp$ space follows from (6) by algebraic conjugation. The orbits of the weights v_i under the icosahedral group H_3 can be obtained by applying the group elements on the weight vectors as

$$\begin{aligned}\frac{(v_1)_{h_3}}{\sqrt{2}} &= \tfrac{1}{2}\{(\pm1,\pm\tau,0),(\pm\tau,0,\pm1),(0,\pm1,\pm\tau)\},\\ \frac{(\tau^{-1}\,v_2)_{h_3}}{2\sqrt{2}} &= \tfrac{1}{2}\Big\{(\pm1,0,0),(0,0,\pm1),(0,\pm1,0),\tfrac{1}{2}(\pm1,\pm\sigma,\pm\tau),\tfrac{1}{2}(\pm\sigma,\pm\tau,\pm1),\tfrac{1}{2}(\pm\tau,\pm1,\pm\sigma)\Big\},\\ \frac{(\tau^{-1}v_3)_{h_3}}{\sqrt{2}} &= \tfrac{1}{2}\{(\pm1,\pm1,\pm1),(0,\pm\tau,\pm\sigma),(\pm\tau,\pm\sigma,0),(\pm\sigma,0,\pm\tau)\}\end{aligned} \tag{8}$$

where the notation $(v_i)_{h_3}$ is introduced for the set of vectors generated by the action of the icosahedral group on the weight v_i.

The sets of vectors in (8) represent the vertices of an icosahedron, an icosidodecahedron, and a dodecahedron, respectively, in the $E_{||}$ space. The weights v_1, v_2, and v_3 also denote the five-fold, two-fold, and three-fold symmetry axes of the icosahedral group. The union of the orbits of $\frac{(v_1)_{h_3}}{\sqrt{2}}$ and $\frac{(\tau^{-1}v_3)_{h_3}}{\sqrt{2}}$ constitute the vertices of a rhombic triacontahedron.

It is obvious that the Coxeter group D_6 is symmetric under the algebraic conjugation, and this is more apparent in the characteristic equation of the Coxeter element of D_6 given by

$$\left(\lambda^3+\sigma\lambda^2+\sigma\lambda+1\right)\left(\lambda^3+\tau\lambda^2+\tau\lambda+1\right)=0, \tag{9}$$

whose eigenvalues lead to the Coxeter exponents of D_6. The first bracket is the characteristic polynomial of the Coxeter element of the matrices in (6) describing it in the $E_{||}$ space, and the second bracket describes it in the $E_\perp$ space [17].

Therefore, projection of D_6 into either space is the violation of the algebraic conjugation. It would be interesting to discuss the projections of the fundamental polytopes of D_6 into 3D space possessing the icosahedral symmetry. It is beyond the scope of the present paper; however, we may discuss a few interesting cases.

The orbit generated by the weight ω_1 is a polytope with 12 vertices called cross polytope and represents an icosahedron when projected into either space. The orbit of weight ω_2 constitutes the "root polytope" of D_6 with 60 vertices, which project into 3D space as two icosidodecahedra with 30 vertices each, the ratio of radii of the circumspheres is τ. The dual of the root polytope is the union of the three polytopes generated by ω_1, ω_5 and ω_6, which constitute the Voronoi cell of the lattice D_6, as mentioned earlier and projects into an icosahedron and two rhombic triacontahedra. Actually, they consist of three icosahedra with the ratio of radii $1,\tau,\tau^2$ and two dodecahedra with the radii in proportion to τ. The orbit generated by the weight vector ω_3 is a polytope with 160 vertices and constitutes the Voronoi cell of the weight lattice $D_6{}^*$. It projects into two dodecahedra and two polyhedra with 60 vertices each. Voronoi cells can be used as windows for the cut and projects scheme; however, we prefer the direct projection of the root lattice as described in what follows.

A general root vector can be decomposed in terms of the weights $v_i(\acute{v}_i)$ as

$$\begin{aligned}&m_1l_1+m_2l_2+m_3l_3+m_4l_4+m_5l_5+m_6l_6\\ &\quad=\tfrac{1}{\sqrt{2+\tau}}[(m_1-m_2+\tau m_5-\tau m_6)v_1+(m_2-m_3+\tau m_4-\tau m_5)v_2\\ &\quad+(m_5+m_6+\tau m_3-\tau m_4)v_3]+(v_i\to\acute{v}_i,\ \tau\to\sigma).\end{aligned} \tag{10}$$

Projection of an arbitrary vector of D_6 into the space $E_{||}$, or more thoroughly, onto a particular weight vector, for example, onto the weight v_1 is given by

$$[m_1l_1+m_2(l_2+l_3+l_4+l_5-l_6)]_{||}\equiv[(m_1-m_2)\,\omega_1+2m_2\,\omega_5]_{||}=c(\frac{v_1}{\sqrt{2}}) \tag{11}$$

And represents an icosahedron where $c \equiv c(m_1, m_2) = \sqrt{\frac{2}{2+\tau}}(m_1 - m_2 + 2m_2\tau)$ is an overall scale factor. The subscript $\|$ means the projection into the space $E_{\|}$. The expression in (11) shows that $m_1 + m_2 = even$, implying that the pair of integers (m_1, m_2) are either both even or both odd. We will see that not only icosahedron but also dodecahedron and the icosahedral Archimedean polyhedra can be obtained by relations similar to (11). The Platonic and Archimedean polyhedra with icosahedral symmetry are listed in Table 1 as projections of D_6 lattice vectors determined by the pair of integers (m_1, m_2). Table 1 shows that only a certain subset of vectors of D_6 project onto regular icosahedral polyhedra. We will see in the next section that the Danzer tiling also restricts the D_6 lattice in a domain, where the vectors are determined by the pair of integers (m_1, m_2). These quantities studied here are potentially applicable to many graph indices [18].

Table 1. Platonic and Archimedean icosahedral polyhedra projected from D_6 (vertices are orbits under the icosahedral group).

Vector of the Polyhedron in H_3 $c \equiv \sqrt{\frac{2}{2+\tau}}(m_1 - m_2 + 2m_2\tau)$		**Corresponding Vector in D_6** $m_1, m_2 \in \mathbb{Z}$
Icosahedron $c\frac{v_1}{\sqrt{2}} = \frac{c}{2}(1, \tau, 0)$		$m_1 l_1 + m_2(l_2 + l_3 + l_4 + l_5 - l_6) = (m_1 - m_2)\omega_1 + 2m_2\omega_5$ $m_1,\ m_2 \in 2\mathbb{Z}$ or $2\mathbb{Z}+1$
Dodecahedron $c\frac{v_3}{\sqrt{2}} = \frac{c}{2}(0, \tau^2, 1)$		$\frac{1}{2}[(m_1 + 3m_2)(l_1 + l_2 + l_3) + (m_1 - m_2)(l_4 + l_5 + l_6)] =$ $(m_1 - m_2)\omega_6 + 2m_2\omega_3$ $m_1,\ m_2 \in 2\mathbb{Z}$ or $2\mathbb{Z}+1$
Icosidodecahedron $c\frac{v_2}{\sqrt{2}} = c(0, \tau, 0)$		$(m_1 + m_2)(l_1 + l_2) + 2m_2(l_3 + l_4) =$ $(m_1 - m_2)\omega_2 + 2m_2\omega_4$ $m_1,\ m_2 \in 2\mathbb{Z}$ or $2\mathbb{Z}+1$
Truncated Icosahedron $c\frac{(v_1+v_2)}{\sqrt{2}} = \frac{c}{2}(1,\ 3\tau, 0)$		$(2m_1 + m_2)l_1 + (m_1 + 2m_2)l_2 +$ $m_2(3l_3 + 3l_4 + l_5 - l_6) =$ $(m_1 - m_2)(\omega_1 + \omega_2) + 2m_2(\omega_4 + \omega_5)$ $m_1,\ m_2 \in 2\mathbb{Z}$ or $2\mathbb{Z}+1$
Small Rhombicosidodecahedron $c\frac{(v_1+v_3)}{\sqrt{2}} = \frac{c}{2}(1,\ 2\tau+1,\ 1)$		$\frac{1}{2}[(3m_1 + 3m_2)l_1 + (m_1 + 5m_2)(l_2 + l_3)$ $+(m_1 + m_2)(l_4 + l_5) + (m_1 - 3m_2)l_6]$ $= (m_1 - m_2)(\omega_1 + \omega_6) + 2m_2(\omega_3 + \omega_5)$
Truncated Dodecahedron $c\frac{(v_2+v_3)}{\sqrt{2}} = \frac{c}{2}(0,\ 3\tau+1,\ 1)$		$\frac{1}{2}[(3m_1 + 5m_2)(l_1 + l_2) + (m_1 + 7m_2)l_3 + (m_1 + 3m_2)l_4$ $+(m_1 - m_2)(l_5 + l_6)]$ $= (m_1 - m_2)(\omega_2 + \omega_6) + 2m_2(\omega_3 + \omega_4)$
Great Rhombicosidodecahedron $c\frac{(v_1+v_2+v_3)}{\sqrt{2}} = \frac{c}{2}(1,\ 4\tau+1,\ 1)$		$\frac{1}{2}[5(m_1 + m_2)l_1 + (3m_1 + 7m_2)l_2 + (m_1 + 9m_2)l_3$ $+(m_1 + 5m_2)l_4 + (m_1 + m_2)l_5 + (m_1 - 3m_2)l_6]$ $= (m_1 - m_2)(\omega_1 + \omega_2 + \omega_6) + 2m_2(\omega_3 + \omega_4 + \omega_5)$

3. Danzer's *ABCK* Tiles and D_6 Lattice

We introduce the *ABCK* tiles with their coordinates in Figure 2 as well as their images in lattice D_6. For a fixed pair of integers $(m_1, m_2) \neq (0,\ 0)$, let us define the image of K by vertices $D_1(m_1, m_2)$, $D_2(m_1, m_2)$, $D_3(m_1, m_2)$, and $D_0\ (0,0)$ in lattice D_6 and its projection in 3D space, where $D_0\ (0,0)$ represents the origin. They are given as

$$\begin{aligned} D_1(m_1, m_2) &= \tfrac{cv_1}{\sqrt{2}} = \left[m_1 l_1 + m_2(l_2 + l_3 + l_4 + l_5 - l_6)\right]_{\|} = \left[(m_1 - m_2)\omega_1 + 2m_2\omega_5\right]_{\|} \\ D_2(m_1, m_2) &= \tfrac{cv_2}{2\sqrt{2}} = \tfrac{1}{2}\left[(m_1 + m_2)(l_1 + l_2) + 2m_2(l_3 + l_4)\right]_{\|} = \tfrac{1}{2}\left[(m_1 - m_2)\omega_2 + 2m_2\omega_4\right]_{\|} \\ D_3(m_1, m_2) &= \tfrac{c\tau^{-1}v_3}{\sqrt{2}} = \tfrac{1}{2}\left[(m_1 + m_2)(l_1 + l_2 + l_3) + (-m_1 + 3m_2)(l_4 + l_5 + l_6)\right]_{\|} \\ &= \left[2m_2\omega_3 + (-m_1 + 3m_2)\omega_6\right]_{\|} \end{aligned} \tag{12}$$

which follows from Table 1 with redefinitions of the pair of integers (m_1, m_2). With removal of the notation $\|$, they represent the vectors in 6D space. A set of scaled vertices defined by $\acute{D}_i \equiv \frac{D_i(m_1,\ m_2)}{c}$ are used for the vertices of the Danzer's tetrahedra in Figure 2.

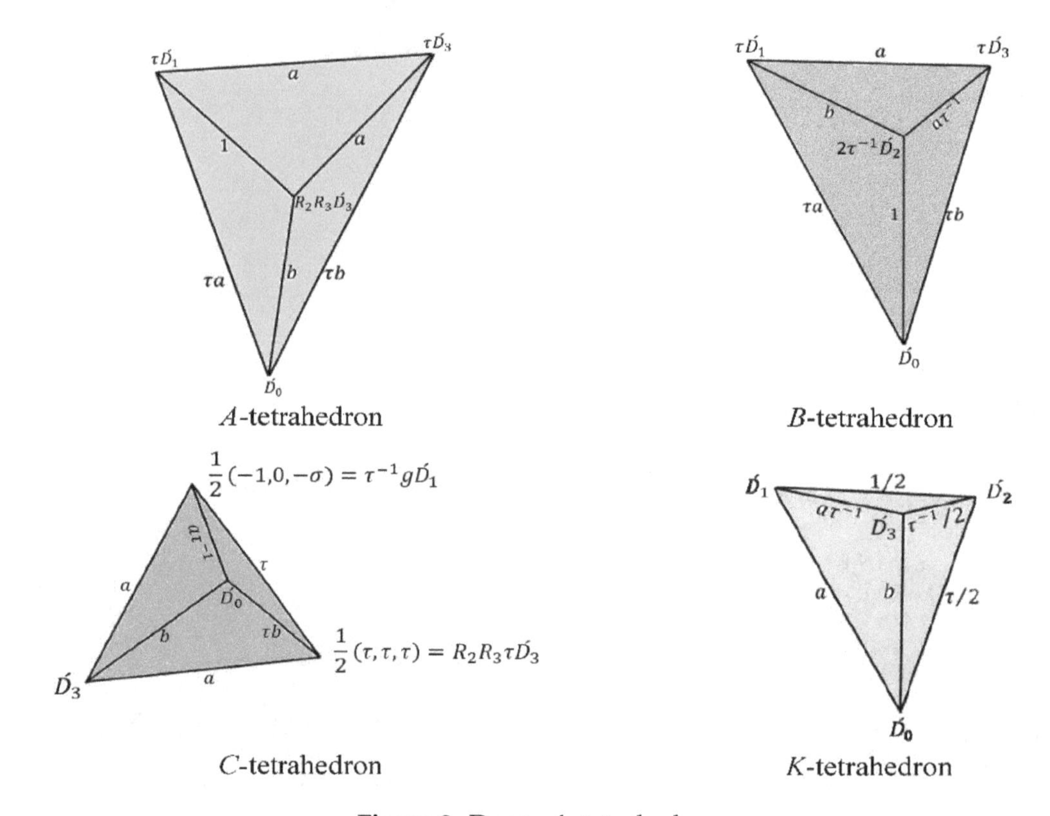

Figure 2. Danzer's tetrahedra.

It is instructive to give a brief introduction to Danzer's tetrahedra before we go into details. The edge lengths of *ABC* tetrahedra are related to the weights of the icosahedral group H_3 whose edge lengths are given by $a = \frac{\sqrt{2+\tau}}{2} = \frac{\|v_1\|}{\sqrt{2}}$, $b = \frac{\sqrt{3}}{2} = \frac{\|\tau^{-1}v_3\|}{\sqrt{2}}$, $1 = \frac{\|\tau^{-1}v_2\|}{\sqrt{2}}$ and their multiples by τ and τ^{-1}, where $a, b, 1$ are the original edge lengths introduced by Danzer. However, the tetrahedron K has edge lengths also involving $\frac{1}{2}$, $\frac{\tau}{2}$, $\frac{\tau^{-1}}{2}$. As we noted in Section 2, the vertices of a rhombic triacontahedron consist of the union of the orbits $\frac{(v_1)_{h_3}}{\sqrt{2}}$ and $\frac{(\tau^{-1}v_3)_{h_3}}{\sqrt{2}}$. One of its cells is a pyramid based on a golden rhombus with vertices

$$\frac{v_1}{\sqrt{2}} = \frac{1}{2}(1,\tau,0),\ \frac{\tau^{-1}v_3}{\sqrt{2}} = \frac{1}{2}(0,\tau,-\sigma),\ R_1\frac{v_1}{\sqrt{2}} = \frac{1}{2}(-1,\tau,0),\ R_3\frac{\tau^{-1}v_3}{\sqrt{2}} = \frac{1}{2}(0,\tau,\sigma), \tag{13}$$

And the apex is at the origin. The coordinate of the intersection of the diagonals of the rhombus is the vector $\frac{v_2}{2\sqrt{2}} = \frac{1}{2}(0, \tau, 0)$, and its magnitude is the in-radius of the rhombic triacontahedron. Therefore, the weights $\frac{v_1}{\sqrt{2}}, \frac{\tau^{-1}v_3}{\sqrt{2}}, \frac{v_2}{2\sqrt{2}}$ and the origin can be taken as the vertices of the tetrahedron K. As such, it is the fundamental region of the icosahedral group from which the rhombic triacontahedron is generated [19]. The octahedra generated by these tetrahedra denoted by $\langle B \rangle$, $\langle C \rangle$and $\langle A \rangle$ comprise four copies of each obtained by a group of order four generated by two commuting generators R_1 and R_3 or their conjugate groups. The octahedron $\langle K \rangle$ consists of $8K$ generated by a group of order eight consisting of three commuting generators R_1, R_3 as mentioned earlier, and the third generator R_0 is an affine reflection [20] with respect to the golden rhombic face induced by the affine Coxeter group $\widetilde{D}_6$.

One can dissect the octahedron $\langle K \rangle$ into three non-equivalent pyramids with rhombic bases by cutting along the lines orthogonal to three planes $v_1 - v_2, v_2 - v_3, v_3 - v_1$. One of the pyramids is generated by the tetrahedron K upon the actions of the group generated by the reflections R_1 and R_3. If we call $R_1 K = \acute{K}$ as the mirror image of K the others can be taken as the tetrahedra obtained from K and $R_1 K$ by a rotation of 180° around the axis v_2. A mirror image of $4K = (2K + 2\acute{K})$ with respect to the rhombic plane (corresponding to an affine reflection) complements it up to an octahedron of $8K$, as we mentioned above. In addition to the vertices in (13) the octahedron $8K$ also includes the vertices $(0, 0, 0)$ and $(0, \tau, 0)$. Octahedral tiles are depicted in Figure 3. Dissected pyramids constituting the octahedron $\langle K \rangle$, as depicted in Figure 4, have bases, two of which with bases of golden rhombuses with edge lengths (heights) $\tau^{-1}a\left(\frac{\tau}{2}\right)$, $a\left(\frac{\tau^{-1}}{2}\right)$, and the third is the one with $b\left(\frac{1}{2}\right)$. The rhombic triacontahedron consists of $60K + 60\acute{K} = 30\left(2K + 2\acute{K}\right)$, where $\left(2K + 2\acute{K}\right)$ form a cell of pyramid based on a golden rhombus of edge $\tau^{-1}a$ and height $\frac{\tau}{2}$. Faces of the rhombic triacontahedron are the rhombuses orthogonal to one of 30 vertices of the icosidodecahedron given in (8).

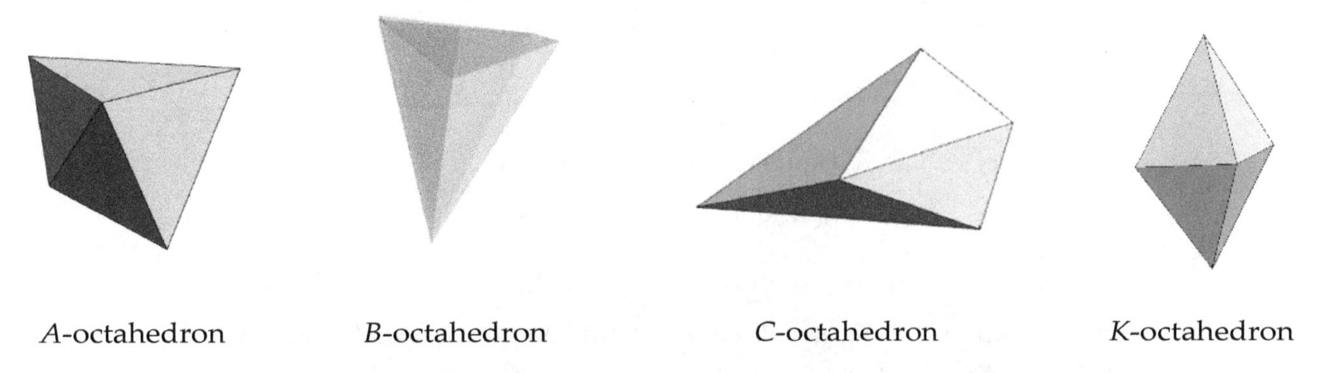

A-octahedron B-octahedron C-octahedron K-octahedron

Figure 3. The octahedra generated by the $ABCK$ tetrahedra.

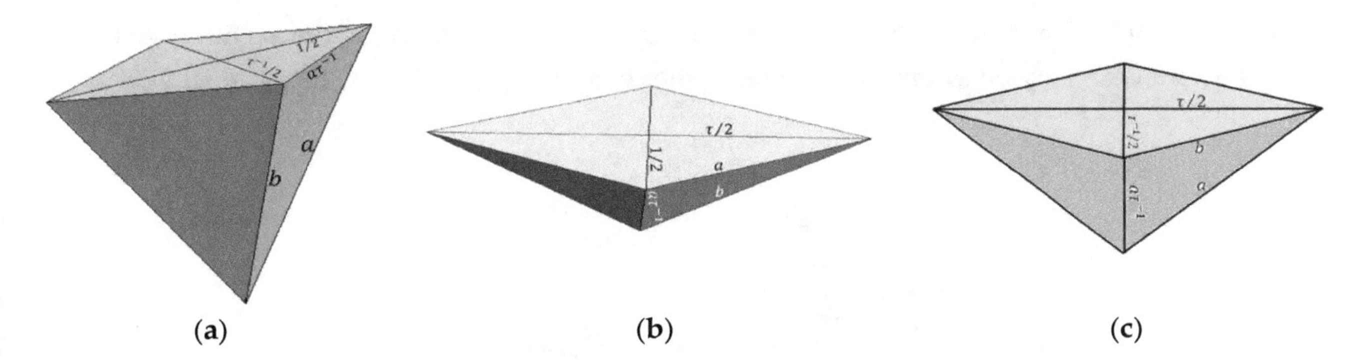

(a) **(b)** **(c)**

Figure 4. The three pyramids (**a–c**) composed of 4K tetrahedra.

When the $4K$ of Figure 4a is rotated by the icosahedral group, it generates the rhombic triacontahedron as shown in Figure 5.

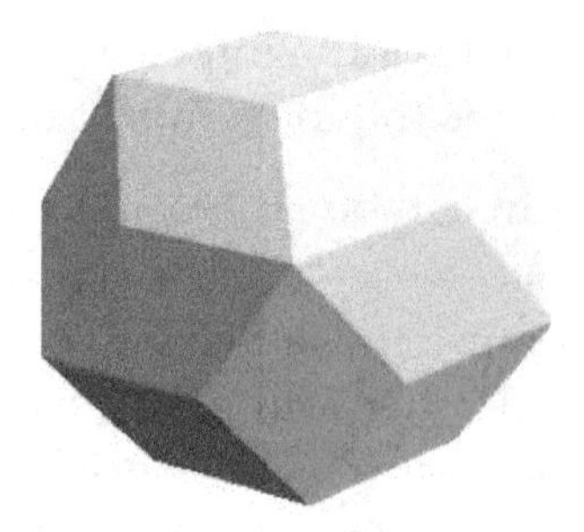

Figure 5. A view of rhombic triacontahedron.

The octahedron $\langle B \rangle$ is a non-convex polyhedron whose vertices can be taken as $\acute{D}_1$, $\tau R_1 \acute{D}_1$, $\tau \acute{D}_3$, $\tau R_3 \acute{D}_3, 2\tau^{-1}\acute{D}_2$, and the sixth vertex is the origin $\acute{D}_0$. It consists of four triangular faces with edges $a, \tau a, \tau b$ and four triangular faces with edges $\tau^{-1}a, a, b$. Full action of the icosahedral group on the tetrahedron B would generate the "B-polyhedron" consisting of $60B + 60\acute{B}$ where $\acute{B} = R_1 B$ is the mirror image of B. It has 62 vertices (12 like$\frac{(\tau v_1)_{h_3}}{\sqrt{2}}$, 30 like$\frac{(\tau^{-1}v_2)_{h_3}}{\sqrt{2}}$, 20 like$\frac{(v_3)_{h_3}}{\sqrt{2}}$), 180 edges, and 120 faces consisting of faces with triangles of edges with $\tau^{-1}a, a, b$. The face transitive "B-polyhedron" is depicted in Figure 6 showing three-fold and five-fold axes simultaneously.

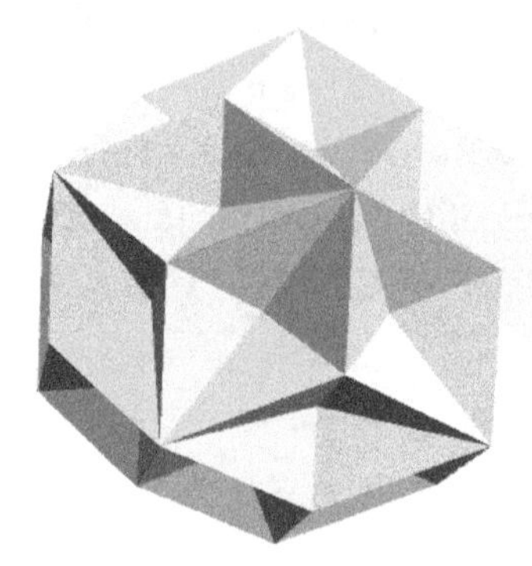

Figure 6. The "B-polyhedron".

The octahedron $\langle C \rangle$ is a convex polyhedron, which can be represented by the vertices

$$\frac{1}{2}(\tau, 0, 1),\ \frac{1}{2}(\tau, \tau, \tau),\ \frac{1}{2}(1, \tau, 0), \frac{1}{2}\left(\tau^2, 1, 0\right), \frac{1}{2}\left(\tau^2, \tau, 1\right),\ (0, 0, 0). \tag{14}$$

This is obtained from that of Figure 2 by a translation and inversion. It consists of four triangular faces with edges $a, b, \tau b$ and four triangular faces with edges $\tau^{-1}a, a$ and b.

The "C-polyhedron" is a non-convex polyhedron with 62 vertices (12 like $\frac{(v_1)_{h_3}}{\sqrt{2}}$, 30 like $\frac{(v_2)_{h_3}}{\sqrt{2}}$, 20 like $\frac{(v_3)_{h_3}}{\sqrt{2}}$), 180 edges, and 120 faces consisting of only one type of triangular face with edge lengths $\tau^{-1}a, a$ and b as can be seen in Figure 7 (see also the reference [4]). It is also face transitive same as "B-polyhedron".

Figure 7. The "C-polyhedron".

The octahedron $\langle A \rangle$ is a non-convex polyhedron, which can be represented by the vertices

$$\frac{1}{2}(-\tau,0,1),\frac{1}{2}(1,-\tau,0),\ \frac{1}{2}(1,1,1),\ \frac{1}{2}(\sigma,0,-\tau),\frac{1}{2}(\sigma,-\tau,1),\ (0,0,0), \tag{15}$$

as shown in Figure 3.

Now, we discuss how the tiles are generated by inflation with an inflation factor τ. First of all, let us recall that the projection of subset of the D_6 vectors specified by the pair of integers (m_1, m_2) are some linear combinations of the weights v_i, $(i = 1, 2, 3)$ with an overall factor c. It is easy to find out the vector of D_6 corresponding to the inflated vertex of any $ABCK$ tetrahedron by noting that $c(\tau^n v_i) = c(\acute{m}_1,\acute{m}_2)v_i$. Here, we use $\tau^n = F_{n-1} + F_n\tau$, and we define

$$\acute{m}_1 \equiv m_1F_{n-1} + \frac{1}{2}(m_1+5m_2)F_n,\quad \acute{m}_2 \equiv m_2F_{n-1} + \frac{1}{2}(m_1+m_2)F_n, \tag{16}$$

where F_n represents the Fibonacci sequence satisfying

$$F_{n+1} = F_n + F_{n-1},\ F_{-n} = (-1)^{n+1}F_n,\ F_0 = 0, F_1 = 1. \tag{17}$$

It follows from (16) that $\acute{m}_1 + \acute{m}_2$ = even if $m_1 + m_2$ = even, otherwise $\acute{m}_1,\ \acute{m}_2 \in \mathbb{Z}$. This proves that the pair of integers $(\acute{m}_1,\acute{m}_2)$ obtained by inflation of the vertices of the Danzer's tetrahedra remain in the subset of D_6 lattice. We conclude that the inflated vectors by τ^n in (12) can be obtained by replacing m_1 by $\acute{m}_1$ and m_2 by $\acute{m}_2$. For example, radii of the icosahedra projected by D_6 vectors

$$2\omega_1, 2\omega_5,\ 2\omega_1 + 2\omega_5, 2\omega_1 + 4\omega_5 \tag{18}$$

are in proportion to $1, \tau,\ \tau^2$ and τ^3, respectively. It will not be difficult to obtain the D_6 image of any general vector in the 3D space in the form of $c(\tau^p,\ \tau^q,\ \tau^r)$ where $p, q, r \in \mathbb{Z}$. Now we discuss the inflation of each tile one by one.

3.1. Construction of $\tau K = B + K$

The vertices of τK is shown in Figure 8 where the origin coincides with one of the vertices of B as shown in Figure 2 and the other vertices of K and B are depicted. A transformation is needed to translate K to its new position. The face of K opposite to the vertex $\acute{D}_2$ having a normal vector $\frac{1}{2}(-1,-\sigma,-\tau)$ outward should match with the face of B opposite to the vertex $\acute{D}_0$ with a normal vector $\frac{1}{2}(\sigma,\tau,-1)$ inward. For this reason, we perform a rotation of K given by

$$g_K = \begin{bmatrix} 0 & -1 & 0 \\ 0 & 0 & -1 \\ 1 & 0 & 0 \end{bmatrix} \tag{19}$$

matching the normal of these two faces. A translation by the vector $\frac{1}{2}\left(\tau,\tau^2,0\right)$ will locate K in its proper place in τK. This is the simplest case where a rotation and a translation would do the work. The corresponding rotation and translation in D_6 can be calculated easily, and the results are illustrated in Table 2.

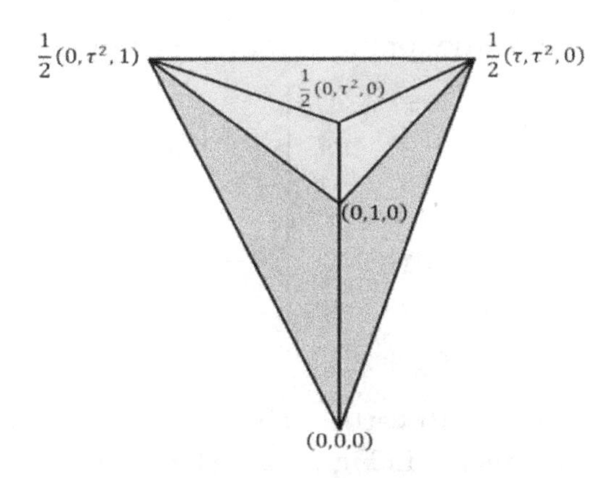

Figure 8. Vertices of τK.

Table 2. Rotation and translation in H_3 and D_6.

H_3		D_6
g_K	←rotation→	$g_K:(1\bar{3}6)(24\bar{5})$
$t_K=\frac{1}{2}(\tau,\tau^2,0)$	←translation→	$\frac{1}{2}[(m_1+5m_2)l_1+(m_1+m_2)(l_2+l_3+l_4+l_5-l_6)]$

The rotation in D_6 is represented as the permutations of components of the vectors in the l_i basis.

Rhombic triacontahedron generated by τK now consists of a "B-Polyhedron" centered at the origin and 30 pyramids of $4K$ with golden rhombic bases of edge lengths a and heights $\frac{\tau^{-1}}{2}$ occupy the 30 inward gaps of "B-Polyhedron".

3.2. Construction of $\tau B = C + 4K + B_1 + B_2$

The first step is to rotate B by a matrix

$$g_B=\frac{1}{2}\begin{bmatrix} -\sigma & \tau & -1 \\ -\tau & 1 & -\sigma \\ 1 & -\sigma & \tau \end{bmatrix} \tag{20}$$

and then inflate by τ to obtain the vertices of $\tau g_B B$ as shown in Figure 9.

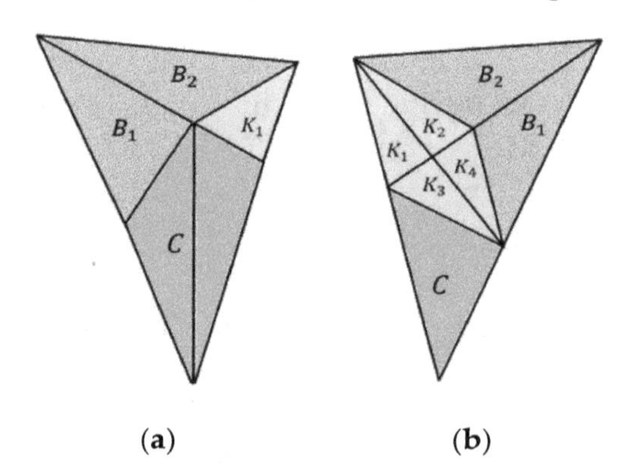

Figure 9. Views from $\tau g_B B$. (**a**) Top view of $\tau g_B B$; (**b**) bottom view of $\tau g_B B$.

Representation of $\tau g_B B$ is chosen to coincide the origin with the $\acute{D}_0$ vertex of tetrahedron C. It is obvious to see that the bottom view illustrates that the $4K$ is the pyramid based on the rhombus with edge length (height), $b\left(\frac{1}{2}\right)$ as shown in Figure 4c. The $4K$ takes its position in Figure 9 by a rotation followed by a translation given by

$$\text{rotation}:\ g_{4K}=\frac{1}{2}\begin{bmatrix} 1 & -\sigma & -\tau \\ -\sigma & \tau & 1 \\ -\tau & 1 & \sigma \end{bmatrix},\ \text{translation}:\ t_{4K}=\frac{1}{2}(\tau,0,1). \tag{21}$$

To translate B_1 and B_2, we follow the rotation and translation sequences as

$$\begin{aligned} &\text{rotation}:\ g_{B_1}=\tfrac{1}{2}\begin{bmatrix} \sigma & -\tau & 1 \\ -\tau & 1 & -\sigma \\ -1 & \sigma & -\tau \end{bmatrix},\ \text{translation}:\ t_B=\tfrac{1}{2}\left(\tau^3,0,\ \tau^2\right), \\ &\text{rotation}:\ g_{B_2}=\tfrac{1}{2}\begin{bmatrix} -\sigma & -\tau & 1 \\ \tau & 1 & -\sigma \\ 1 & \sigma & -\tau \end{bmatrix},\ \text{translation}:\ t_B=\tfrac{1}{2}\left(\tau^3,0,\ \tau^2\right). \end{aligned} \tag{22}$$

3.3. Construction of $\tau C = K_1 + K_2 + C_1 + C_2 + A$

We inflate by τ the tetrahedron C with vertices shown in Figure 10. Top view of τC where one of the vertices of K_1 is at the origin is depicted in Figure 11. The vertices of the constituting tetrahedra are given as follows:

$$\begin{aligned}
K_1 &: \left\{(0,0,0),\ \tfrac{1}{2}(1,1,1),\ \tfrac{1}{2}(\tau,0,1),\ \tfrac{1}{4}\left(\tau^2,\tau,1\right)\right\};\\
K_2 &: \left\{\tfrac{1}{2}(1,1,1),\ \tfrac{1}{2}(\tau,0,1),\ \tfrac{1}{4}\left(\tau^2,\tau,1\right),\ \tfrac{1}{2}\left(\tau^2,\tau,1\right)\right\};\\
C_1 &: \left\{\tfrac{1}{2}(1,1,1),\ \tfrac{1}{2}(\tau,0,1),\ \tfrac{1}{2}\left(\tau^2,\tau,1\right),\ \tfrac{1}{2}\left(\tau^2,\tau^2,\tau^2\right)\right\};\\
C_2 &: \left\{\tfrac{1}{2}(\tau,0,1),\ \tfrac{1}{2}\left(\tau^2,\tau,1\right),\ \tfrac{1}{2}\left(\tau^2,0,\tau\right),\ \tfrac{1}{2}\left(\tau^2,\tau^2,\tau^2\right)\right\};\\
A &: \left\{\tfrac{1}{2}\left(\tau^2,\tau,1\right),\ \tfrac{1}{2}\left(\tau^2,0,\tau\right),\ \tfrac{1}{2}\left(\tau^2,\tau^2,\tau^2\right),\ \tfrac{1}{2}\left(\tau^3,\tau^2,\tau\right)\right\}.
\end{aligned} \tag{23}$$

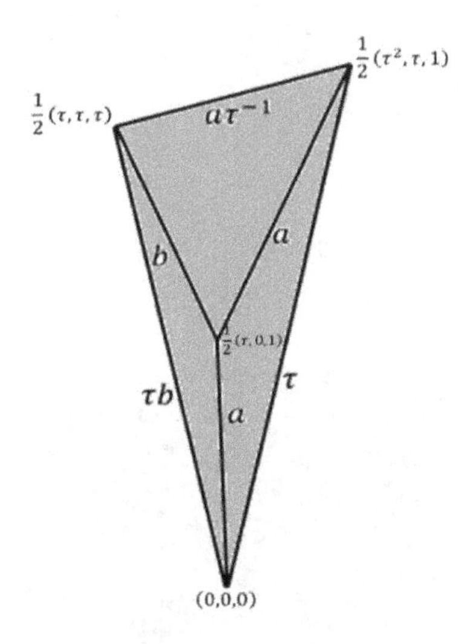

Figure 10. C-tetrahedron obtained from that of Figure 2 by translation and inversion.

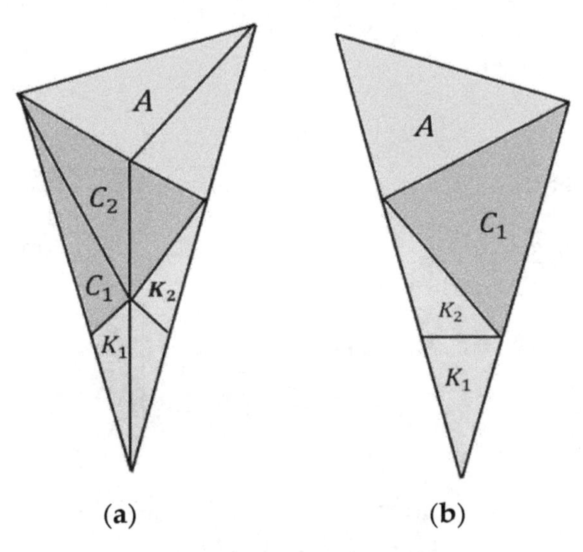

Figure 11. Views from τC. (**a**) Top view of τC; (**b**) bottom view of τC.

The vertices of K_1 can be obtained from K by a rotation $K_1 = g_{K_1}K$, where

$$g_{K_1} = \frac{1}{2}\begin{bmatrix} -\sigma & \tau & -1 \\ -\tau & 1 & -\sigma \\ 1 & -\sigma & \tau \end{bmatrix}, \quad \left(g_{K_1}\right)^5 = 1. \tag{24}$$

The transformation to obtain K_2 is a rotation g_{K_2} followed by a translation of $t_{K_2} = \frac{1}{2}\left(\tau^2,\tau,1\right)$, where

$$g_{K_2} = \frac{1}{2}\begin{bmatrix} -\sigma & -\tau & -1 \\ -\tau & -1 & -\sigma \\ 1 & \sigma & \tau \end{bmatrix}, \quad \left(g_{K_2}\right)^{10} = 1 \tag{25}$$

To obtain the coordinates of $C_1$and C_2, first translate C by $-\frac{1}{2}(\tau,\tau,\tau)$ and rotate by g_{C_1} and g_{C_2}, respectively, followed by the translation $t_{C_1} = \frac{1}{2}\left(\tau^2,\tau^2,\tau^2\right)$, where

$$g_{C_1} = \frac{1}{2}\begin{bmatrix} \sigma & \tau & 1 \\ \tau & 1 & \sigma \\ 1 & \sigma & \tau \end{bmatrix}, \quad g_{C_2} = \frac{1}{2}\begin{bmatrix} 1 & -\sigma & -\tau \\ -\sigma & \tau & 1 \\ \tau & -1 & -\sigma \end{bmatrix} \tag{26}$$

Similarly, vertices of tetrahedron A are rotated by

$$g_A = \frac{1}{2}\begin{bmatrix} 1 & -\sigma & -\tau \\ -\sigma & \tau & 1 \\ -\tau & 1 & \sigma \end{bmatrix} \tag{27}$$

followed by a translation $t_A = \frac{1}{2}\left(\tau^2, 0, \tau\right)$.

3.4. *Construction of $\tau A = 3B + 2C + 6K$*

It can also be written as $\tau A = C + K_1 + K_2 + B + \tau B$, where τB is already studied. A top view of τA is depicted in Figure 12 with the vertices of τB given in Figure 9.

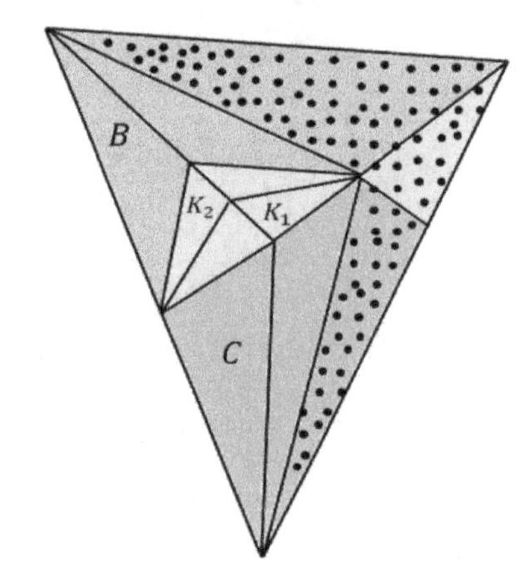

Figure 12. Top view of τA (bottom view is the same as bottom view of τB- dotted region).

The vertices of C $\{(0,0,0), \frac{1}{2}(\tau, 0, 1), \frac{1}{2}\left(\tau^2, 1, 0\right), \frac{1}{2}\left(\tau^2, \tau, 1\right)\}$ can be obtained from those in Figure 2 by a translation $\frac{1}{2}(-\tau, -\tau, -\tau)$ followed by a rotation

$$g_c = \frac{1}{2}\begin{bmatrix} -1 & \sigma & -\tau \\ \sigma & -\tau & 1 \\ -\tau & 1 & -\sigma \end{bmatrix} \tag{28}$$

Vertices of $K_1$and K_2 are given by

$$\begin{aligned} K_1 &: \left\{\tfrac{1}{2}(\tau, 0, 1), \tfrac{1}{2}\left(\tau^2, 1, 0\right), \tfrac{1}{2}\left(\tau^2, \tau, 1\right), \tfrac{1}{4}(3\tau + 1, \tau, 1)\right\}, \\ K_2 &: \left\{\tfrac{1}{2}(\tau, 0, 1), \tfrac{1}{2}\left(\tau^2, \tau, 1\right), \tfrac{1}{4}(3\tau + 1, \tau, 1), \tfrac{1}{2}(2\tau, -\sigma, 1)\right\}. \end{aligned} \tag{29}$$

To obtain K_1 and K_2 with these vertices, first rotate K by g_{K_1} and g_{K_2} given by the matrices

$$g_{K_1} = \frac{1}{2}\begin{bmatrix} \sigma & \tau & -1 \\ \tau & 1 & -\sigma \\ 1 & \sigma & -\tau \end{bmatrix}, \quad g_{K_2} = \frac{1}{2}\begin{bmatrix} \sigma & \tau & 1 \\ \tau & 1 & \sigma \\ 1 & \sigma & \tau \end{bmatrix}, \tag{30}$$

and translate both by the vector $t_{K_1} = \frac{1}{2}(\tau, 0, 1)$. The vertices of B in Figure 12 can be obtained by a rotation then a translation given, respectively, by

$$g_B = \frac{1}{2}\begin{bmatrix} -\tau & -1 & \sigma \\ -1 & -\sigma & \tau \\ -\sigma & -\tau & 1 \end{bmatrix}, \quad t_B = \frac{1}{2}\left(\tau^3, 0, \tau^2\right). \tag{31}$$

All these procedures are described in Table 3, which shows the rotations and translations both in $E_{||}$ and D_6 spaces.

Table 3. Rotation and translation in H_3 and D_6 (see the text for the definition of rotations in $E_{\parallel}$ space).

H_3		D_6
	$\tau K = B + K$	
g_K	←rotation→	$(1\bar{3}6)(24\bar{5})$
$t_K = \frac{c}{2}(\tau, \tau^2, 0)$	←translation→	$\frac{1}{2}[(m_1 + 5m_2)l_1 + (m_1 + m_2)(l_2 + l_3 + l_4 + l_5 - l_6)]$
	$\tau B = C + 4K + B_1 + B_2$	
g_B	←rotation→	$(3)(126\bar{4}5)$
g_{4K}	←rotation→	(1)(23)(45)(6)
$t_{4K} = \frac{c}{2}(\tau, 0, 1)$	←translation→	$m_1 l_5 + m_2(l_1 - l_2 + l_3 - l_4 - l_6)$
g_{B_1}	←rotation→	$(1643\bar{5})(2)$
g_{B_2}	←rotation→	$(15\overline{215}2)$, $(3\overline{643}64)$
$t_B = \frac{c}{2}(\tau^3, 0, \tau^2)$	←translation→	$\frac{1}{2}[(3m_1 + 5m_2)l_5 + (m_1 + 3m_2)(l_1 - l_2 + l_3 - l_4 - l_6)]$
	$\tau C = K_1 + K_2 + C_1 + C_2 + A$	
g_{K_1}	←rotation→	$(5\bar{4}621)(3)$
	(identity translation)	
g_{K_2}	←rotation→	$(1\bar{1})(63\bar{4}5\overline{263}4\bar{5}2)$
$t_K = \frac{c}{2}(\tau^2, \tau, 1)$	←translation→	$(m_1 + m_2)(l_1 + l_5) + 2m_2)(l_3 - l_6)$
$t_C = -\frac{c}{2}(\tau, \tau, \tau)$	←translation→	$-\frac{1}{2}[(m_1 + 3m_2)(l_1 + l_3 + l_5) + (m_1 - m_2)(l_2 - l_4 - l_6)]$
g_{C_1}	←rotation→	$(1)(4)(2\bar{6})(35)$
g_{C_2}	←rotation→	$(1)(35\bar{6}42)$
$t_C = \frac{c}{2}(\tau^2, \tau^2, \tau^2)$	←translation→	$(m_1 + 2m_2)(l_1 + l_3 + l_5) + m_2(l_2 - l_4 - l_6)$
g_A	←rotation→	(1)(6)(23)(45)
$t_A = \frac{c}{2}(\tau^2, 0, \tau)$	←translation→	$\frac{1}{2}[(m_1 + 5m_2)l_5 + (m_1 + m_2)(l_1 - l_2 + l_3 - l_4 - l_6)]$
	$\tau A = C + K_1 + K_2 + B + \tau B$	
$t_C = -\frac{c}{2}(\tau, \tau, \tau)$	←translation→	$-\frac{1}{2}[(m_1 + 3m_2)(l_1 + l_3 + l_5) + (m_1 - m_2)(l_2 - l_4 - l_6)]$
g_C	←rotation→	$(1\bar{1})(2\bar{4})(36)(5\bar{5})$
g_{K_1}	←rotation→	$(1)(2543\bar{6})$
g_{K_2}	←rotation→	$(1)(4)(2\bar{6})(35)$
$t_K = \frac{c}{2}(\tau, 0, 1)$	←translation→	$m_1 l_5 + m_2(l_1 - l_2 + l_3 - l_4 - l_6)$
g_B	←rotation→	$(65\overline{1651})$, $(\overline{423}423)$
$t_B = \frac{c}{2}(\tau^3, 0, \tau^2)$	←translation→	$\frac{1}{2}[(3m_1 + 5m_2)l_5 + (m_1 + 3m_2)(l_1 - l_2 + l_3 - l_4 - l_6)]$

4. Discussions

The six-dimensional reducible representation of the icosahedral subgroup of the point group of D_6 is decomposed into the direct sum of its two three-dimensional representations described also by the direct sum of two graphs of icosahedral group. We have shown that the subset of the D_6 lattice characterized by a pair of integers (m_1, m_2) with $m_1 + m_2 =$ even projects onto the Platonic and Archimedean polyhedra possessing icosahedral symmetry, which are determined as the orbits of the fundamental weights v_i or their multiples by τ^n. The edge lengths of the Danzer tiles are related to the weights of the icosahedral group H_3 and via Table 1 to the weights of D_6, a group theoretical property, which has not been discussed elsewhere. It turns out that the tetrahedron K constitutes the fundamental region of the icosahedral group H_3 as being the cell of the rhombic triacontahedron. Images of the Danzer tiles and their inflations are determined in D_6 by employing translations and icosahedral rotations. This picture gives a one-to-one correspondence between the translations–rotations of 3D space and 6D space.

Faces of the Danzer tiles are all parallel to the faces of the rhombic triacontahedron; in other words, they are all orthogonal to the two-fold axes. Since the faces of the Ammann rhombohedral tiles are orthogonal to the two-fold axes of the icosahedral group it is this common feature that any tilings obtained from the Ammann tiles either with decoration or in the form of the Socolar–Steinhardt model can also be obtained by the Danzer tiles. These geometrical properties have not been studied from the group theoretical point of view, a subject which is beyond the present work but definitely deserves to be studied. Note that the inflation introduces a cyclic permutation among the rhombic triacontahedron, "*B*-polyhedron", and "*C*-polyhedron", as they occur alternatively centered at the origin.

Another novel feature of the paper is to show that *ABCK* tiles and their inflations are directly related to the transformations in the subset of D_6 lattice characterized by integers (m_1, m_2), leading to an alternative projection technique different from the cut and project scheme. Details of the composition of the *ABCK* tiles into a structure with long-range quasiperiodic order follows from the inflation matrix (see for instance [4]).

Author Contributions: Conceptualization, M.K., N.O.K., A.A.-S.; methodology, M.K., A.A.-S., N.O.K.; software, A.A.-S., N.O.K.; validation, M.K., A.A.-S., N.O.K.; formal analysis, A.A.-S., N.O.K.; investigation, A.A.-S., M.K., N.O.K.; writing-original draft preparation, M.K., A.A.-S., N.O.K.; writing-review and editing, M.K., N.O.K., A.A.-S.; visualization, A.A.-S.; N.O.K.; project administration, M.K.; supervision, M.K., N.O.K., A.A.-S. All authors have read and agreed to the published version of the manuscript.

Acknowledgments: We would like to thank Ramazan Koc for his contributions to Figures 6 and 7.

References

1. Di Vincenzo, D.; Steinhardt, P.J. *Quasicrystals: The State of the Art*; World Scientific Publishers: Singapore, 1991.
2. Janot, C. *Quasicrystals: A Primer*; Oxford University Press: Oxford, UK, 1993.
3. Senechal, M. *Quasicrystals and Geometry*; Cambridge University Press: Cambridge, UK, 1995.
4. Baake, M.; Grimm, U. *Aperiodic Order*; A Mathematical Invitation; Cambridge University Press: Cambridge, UK, 2013; Volume 1.
5. Socolar, J.E.S.; Steinhardt, P.J. Quasicrystals. II. Unit-cell configurations. *Phys. Rev. B* **1986**, *34*, 617–647. [CrossRef] [PubMed]
6. Danzer, L.; Papadopolos, D.; Talis, A. Full equivalence between Socolar's tilings and the (A, B, C, K)-tilings leading to a rather natural decoration. *J. Mod. Phys.* **1993**, *7*, 1379–1386. [CrossRef]
7. Roth, J. The equivalence of two face-centered icosahedral tilings with respect to local derivability. *J. Phys. A Math. Gen.* **1993**, *26*, 1455–1461. [CrossRef]
8. Danzer, L. Three-dimensional analogs of the planar Penrose tilings and quasicrystals. *Discrete Math.* **1989**, *76*, 1–7. [CrossRef]
9. Katz, A. *Some Local Properties of the Three-Dimensional Tilings, in Introduction of the Mathematics of Quasicrystals*; Jaric, M.V., Ed.; Academic Press, Inc.: New York, NY, USA, 1989; Volume 2, pp. 147–182.
10. Hann, C.T.; Socolar, J.E.S.; Steinhardt, P.J. Local growth of icosahedral quasicrystalline tilings. *Phys. Rev. B* **2016**, *94*, 014113. [CrossRef]
11. Available online: https://www.math.uni-bielefeld.de/~{}frettloe/papers/ikosa.pdf (accessed on 1 October 2020).
12. Kramer, P. Modelling of quasicrystals. *Phys. Scr.* **1993**, *1993*, 343–348. [CrossRef]
13. Koca, M.; Koca, N.; Koc, R. Group-theoretical analysis of aperiodic tilings from projections of higher-dimensional lattices Bn. *Acta Cryst. A* **2015**, *71*, 175–185. [CrossRef] [PubMed]
14. Kramer, P.; Andrle, M. Inflation and wavelets for the icosahedral Danzer tiling. *J. Phys. A Math. Gen.* **2004**, *37*, 3443–3457. [CrossRef]
15. Koca, M.; Koca, N.; Al-Siyabi, A.; Koc, R. Explicit construction of the Voronoi and Delaunay cells of W (An) and W (Dn) lattices and their facets. *Acta. Cryst. A* **2018**, *74*, 499–511. [CrossRef] [PubMed]

16. Conway, J.H.; Sloane, N.J.A. *Sphere Packings, Lattices and Groups*, 3rd ed.; Springer New York Inc.: New York, NY, USA, 1999.
17. Koca, M.; Koc, R.; Al-Barwani, M. on crystallographic Coxeter group H4 in E8. *J. Phys. A Math. Gen.* **2001**, *34*, 11201–11213. [CrossRef]
18. Alhevaz, A.; Baghipur, M.; Shang, Y. On generalized distance Gaussian Estrada index of graphs. *Symmetry* **2019**, *11*, 1276. [CrossRef]
19. Coxeter, H.S.M. *Regular Polytopes*, 3rd ed.; Dover Publications: Mineola, NY, USA, 1973.
20. Humphreys, J.E. *Reflection Groups and Coxeter Groups*; Cambridge University Press: Cambridge, UK, 1992.

2

On Degree-based Topological Indices of Symmetric Chemical Structures

Jia-Bao Liu [1], Haidar Ali [2], Muhammad Kashif Shafiq [2,*] and Usman Munir [2]

1 School of Mathematics and Physics, Anhui Jianzhu University, Hefei 230601, China; liujiabaoad@163.com
2 Department of Mathematics, Government College University, Faisalabad 38000, Pakistan; haidarali@gcuf.edu.pk (H.A.); mianusman24120@gmail.com (U.M.)
* Correspondence: kashif4v@gmail.com

Abstract: A Topological index also known as connectivity index is a type of a molecular descriptor that is calculated based on the molecular graph of a chemical compound. Topological indices are numerical parameters of a graph which characterize its topology and are usually graph invariant. In QSAR/QSPR study, physico-chemical properties and topological indices such as Randić, atom-bond connectivity (ABC) and geometric-arithmetic (GA) index are used to predict the bioactivity of chemical compounds. Graph theory has found a considerable use in this area of research. In this paper, we study HDCN1(m,n) and HDCN2(m,n) of dimension m, n and derive analytical closed results of general Randić index $R_\alpha(\mathcal{G})$ for different values of α. We also compute the general first Zagreb, ABC, GA, ABC_4 and GA_5 indices for these Hex derived cage networks for the first time and give closed formulas of these degree-based indices.

Keywords: general randić index; atom-bond connectivity (ABC) index; geometric-arithmetic (GA) index; Hex-Derived Cage networks; $HDCN1(m,n)$, $HDCN2(m,n)$

---

1. Introduction

A graph is formed by vertices and edges connecting the vertices. A network is a connected simple graph having no multiple edges and loops. A topological index is a function Top : $\sum \rightarrow \mathbb{R}$ where $\mathbb{R}$ is the set of real numbers and $\sum$ is the finite simple graph with property that $\text{Top}(G_1) = Top(G_2)$ if G_1 and G_2 are isomorphic. A topological index is a numerical value associated with chemical constitution for correlation of chemical structure with various physical properties, chemical reactivity or biological activity. Many tools, such as topological indices has provided by graph theory to the chemists. *Cheminformatics* is new subject which is a combination of chemistry, mathematics and information science. It studies Quantitative structure-activity (QSAR) and structure-property (QSPR) relationships that are used to predict the biological activities and properties of chemical compounds. In the QSAR /QSPR study, physico-chemical properties and topological indices such as Wiener index, Szeged index, Randić index, Zagreb indices and ABC index are used to predict bioactivity of the chemical compounds. "In terms of graph theory, the structural formula of a chemical compound represents the molecular graph, in which vertices are represents to atoms and edges as chemical bonds". A molecular descriptor is a numeric number, which represents the properties of a chemical graph. Basically, a molecular descriptor and topological descriptor are different from each other. A molecular descriptor represents the underlying chemical graph but a topological descriptor are the representation of physico-chemical properties of underlying chemical graph in addition to show the whole structure. Topological indices have many applications in the field of nanobiotechnology and QSAR/QSPR study. Topological indices were firstly introduced by Wiener [1], he named the resulting index as path number while he was working on boiling point of Paraffin. Later on, it renamed as Wiener index [2]. Consider "n" Hex-Derived networks $(HDN1(1,1))$,$(HDN1(2,2))$ and so on to $(HDN1(m,n))$. Connect every boundary vertices of $(HDN1(1,1))$ to its mirror image vertices in $(HDN1(2,2))$ by an edge and so on to $(HDN1(m,n))$. As a result, we found a graph, which is called Hex-Derived Cage networks with "n" layers. In this article, the notations which we used take from the books [3,4].

In this article, Graph $(\mathcal{G})$ is considered to be a graph with vertex set $V(\mathcal{G})$ and edge set $E(\mathcal{G})$, the $d(a)$ is the degree of vertex $a \in V(\mathcal{G})$ and $S(a) = \sum_{b \in N_{\mathcal{G}}(a)} d(b)$ where $N_{\mathcal{G}}(a) = \{b \in V(\mathcal{G}) \mid ab \in E(\mathcal{G})\}$.

Let $\mathcal{G}$ be a graph. Then the Wiener index is written as

$$W(\mathcal{G}) = \frac{1}{2} \sum_{(a,b)} d(a,b) \tag{1}$$

The Randić index [5] is the oldest degree-based topological index invented by Milan Randić, denoted as $R_{-\frac{1}{2}}(\mathcal{G})$ and defined as

$$R_{-\frac{1}{2}}(\mathcal{G}) = \sum_{ab \in E(\mathcal{G})} \frac{1}{\sqrt{d(a)d(b)}} \tag{2}$$

$R_{\alpha}(\mathcal{G})$ is a general Randić index and it is defined as

$$R_{\alpha}(\mathcal{G}) = \sum_{ab \in E(\mathcal{G})} (d(a)d(b))^{\alpha} \quad \text{for} \quad \alpha \in \mathbb{R} \tag{3}$$

A topological index which has a great importance was introduced by Ivan Gutman and *Trinajstić* is Zagreb index and defined as

$$M_1(\mathcal{G}) = \sum_{ab \in E(\mathcal{G})} (d(a) + d(b)) \tag{4}$$

Estrada et al. in [6] invented a very famous degree-based topological index ABC and defined as

$$ABC(\mathcal{G}) = \sum_{ab \in E(\mathcal{G})} \sqrt{\frac{d(a) + d(b) - 2}{d(a)d(b)}} \tag{5}$$

GA index is also a very famous connectivity topological descriptor, which invented by Vukičević et al. [7] and denoted as

$$GA(\mathcal{G}) = \sum_{ab \in E(\mathcal{G})} \frac{2\sqrt{d(a)d(b)}}{(d(a) + d(b))} \tag{6}$$

ABC_4 and GA_5 indices find only if we find the edge partition of interconnection networks each edge in the graphs depend on sum of the degrees of end vertices. ABC_4 index invented by Ghorbani et al. [8] and written as

$$ABC_4(\mathcal{G}) = \sum_{ab \in E(\mathcal{G})} \sqrt{\frac{S(a) + S(b) - 2}{S(a)S(b)}} \tag{7}$$

The latest version of index is GA_5 invented by Graovac et al. [9] and defined as

$$GA_5(\mathcal{G}) = \sum_{ab \in E(\mathcal{G})} \frac{2\sqrt{S(a)S(b)}}{(S(a) + S(b))} \tag{8}$$

For any graph G for $\alpha = 1$, the general Randić index is second Zagreb index.

2. Main Results

Hex-Derived Cage networks $HDCN1(m,n)$ (show in Figure 1) and $HDCN2(m,n)$ (show in Figure 2) give closed formulas of that indices, we study the general Randić, first Zagreb, ABC, GA, ABC_4 and GA_5 indices of certain graphs in [10]. These days there is a broad research activity on ABC and GA indices and their variants, for additionally investigation of topological indices of different families see, [1,11–23].

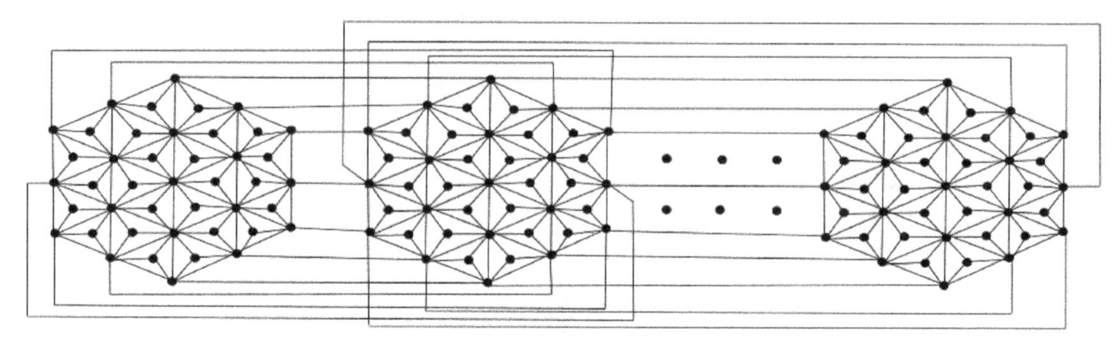

Figure 1. Hex-Derived Network ($HDCN1(3,n)$).

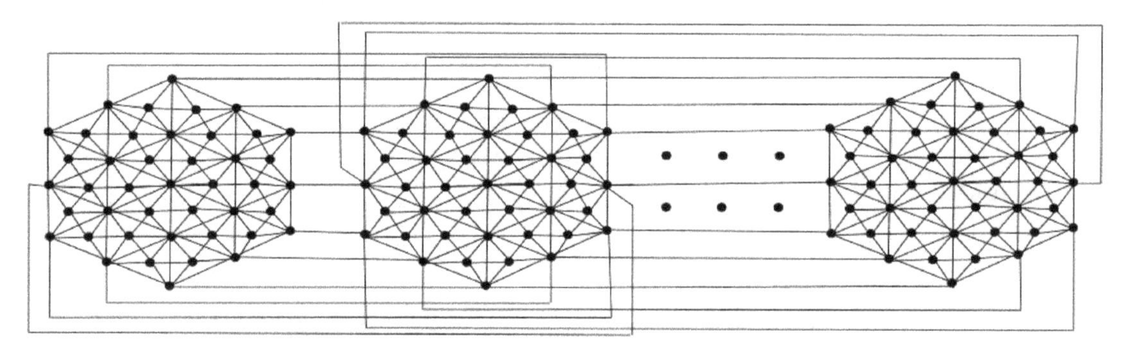

Figure 2. Hex-Derived Network ($HDCN2(3,n)$).

2.1. Results for Hex-Derived Cage Networks

We compute specific degree-based topological indices of Hex-Derived Cage networks. In this paper, we calculate Randić index $R_\alpha(\mathcal{G})$ with $\alpha = 1, -1, \frac{1}{2}, -\frac{1}{2}$, M_1, ABC, GA, ABC_4 and GA_5 for Hex-Derived Cage networks $HDCN1(m,n)$ and $HDCN2(m,n)$.

Theorem 1. *Let $\mathcal{G}_1 \cong HDCN1(m,n)$ be the Hex-Derived Cage network, then its general Randić index is equal to*

$$R_\alpha(\mathcal{G}_1) = \begin{cases} 18(108n^3 - 219n^2 + 25n + 91), & \alpha = 1; \\ 6(36n^3 + 3(7\sqrt{3} - 34)n^2 + (4\sqrt{21} + 6\sqrt{7} + \\ 28\sqrt{6} - 84\sqrt{3} + 12\sqrt{2} + 35)n + 2\sqrt{42} - \\ 8\sqrt{21} - 12\sqrt{7} - 56\sqrt{6} + 100\sqrt{3} + 39), & \alpha = \frac{1}{2}; \\ \frac{11907n^3 - 17003n^2 + 12343n - 3051}{21168}, & \alpha = -1; \\ \frac{15n^3}{4} + (\frac{8}{\sqrt{3}} - \frac{125}{12})n^2 + (\frac{4}{\sqrt{7}} + \\ 4\sqrt{6} - \frac{32}{\sqrt{3}} + \sqrt{2} + 5\sqrt{\frac{3}{7}} + \frac{109}{21})n - \\ \frac{8}{\sqrt{7}} - 8\sqrt{6} + \frac{38}{\sqrt{3}} + 3\sqrt{2} + 2\sqrt{\frac{6}{7}} - \\ 10\sqrt{\frac{3}{7}} + \frac{13}{14}, & \alpha = -\frac{1}{2}. \end{cases}$$

Proof. Let $\mathcal{G}_1$ be the Hex-Derived Cage network $(HDCN1(m,n))$ where $m = n \geq 5$. The edge set of $HDCN1(m,n)$ are divided into seventeen partitions based on the degree of end vertices shows in Table 1. Thus from Equation (3), is follows that

$$R_\alpha(\mathcal{G}_1) = \sum_{ab\in E(\mathcal{G})} (d(a)d(b))^\alpha$$

For $\alpha = 1$

$$R_1(\mathcal{G}_1) = \sum_{j=1}^{17} \sum_{ab\in E_j(\mathcal{G})} deg(u) \cdot deg(v)$$

By using the edge partition given in Table 1, we have
$R_1(\mathcal{G}_1)=18|E_1(\mathcal{G}_1)| + 21|E_2(\mathcal{G}_1)| + 24|E_3(\mathcal{G}_1)| + 27|E_4(\mathcal{G}_1)| + 36|E_5(\mathcal{G}_1)| + 42|E_6(\mathcal{G}_1)| + 48|E_7(\mathcal{G}_1)| + 72|E_8(\mathcal{G}_1)| + 49|E_9(\mathcal{G}_1)| + 63|E_{10}(\mathcal{G}_1)| + 84|E_{11}(\mathcal{G}_1)| + 64|E_{12}(\mathcal{G}_1)| + 72|E_{13}(\mathcal{G}_1)| + 96|E_{14}(\mathcal{G}_1)| + 81|E_{15}(\mathcal{G}_1)| + 108|E_{16}(\mathcal{G}_1)| + 144|E_{17}(\mathcal{G}_1)|$

After simplification, we have

$$R_1(\mathcal{G}_1) = 18(108n^3 - 219n^2 + 25n + 91)$$

For $\alpha = \frac{1}{2}$

$$R_{\frac{1}{2}}(\mathcal{G}_1) = \sum_{j=1}^{17} \sum_{ab\in E_j(\mathcal{G})} \sqrt{d(a) \cdot d(b)}$$

Using the edge partition from Table 1, we have
$R_{\frac{1}{2}}(\mathcal{G}_1)=3\sqrt{2}|E_1(\mathcal{G}_1)| + \sqrt{21}|E_2(\mathcal{G}_1)| + 2\sqrt{6}|E_3(\mathcal{G}_1)| + 3\sqrt{3}|E_4(\mathcal{G}_1)| + 6|E_5(\mathcal{G}_1)| + \sqrt{42}|E_6(\mathcal{G}_1)| + 4\sqrt{3}|E_7(\mathcal{G}_1)| + 6\sqrt{2}|E_8(\mathcal{G}_1)| + 7|E_9(\mathcal{G}_1)| + 3\sqrt{7}|E_{10}(\mathcal{G}_1)| + 2\sqrt{21}|E_{11}(\mathcal{G}_1)| + 8|E_{12}(\mathcal{G}_1)| + 6\sqrt{2}|E_{13}(\mathcal{G}_1)| + 4\sqrt{6}|E_{14}(\mathcal{G}_1)| + 9|E_{15}(\mathcal{G}_1)| + 6\sqrt{3}|E_{16}(\mathcal{G}_1)| + 12|E_{17}(\mathcal{G}_1)|$

After simplification, we have
$R_{\frac{1}{2}}(\mathcal{G}_1)=6(36n^3 + 3(7\sqrt{3} - 34)n^2 + (4\sqrt{21} + 6\sqrt{7} + 28\sqrt{6} - 84\sqrt{3} + 12\sqrt{2} + 35)n + 2\sqrt{42} - 8\sqrt{21} - 12\sqrt{7} - 56\sqrt{6} + 100\sqrt{3} + 39)$

For $\alpha = -1$

$$R_{-1}(\mathcal{G}_1) = \sum_{j=1}^{17} \sum_{ab\in E_j(\mathcal{G})} \frac{1}{d(a) \cdot d(b)}$$

$R_{-1}(\mathcal{G}_1)=\frac{1}{18}|E_1(\mathcal{G}_1)| + \frac{1}{21}|E_2(\mathcal{G}_1)| + \frac{1}{24}|E_3(\mathcal{G}_1)| + \frac{1}{27}|E_4(\mathcal{G}_1)| + \frac{1}{36}|E_5(\mathcal{G}_1)| + \frac{1}{42}|E_6(\mathcal{G}_1)| + \frac{1}{48}|E_7(\mathcal{G}_1)| + \frac{1}{72}|E_8(\mathcal{G}_1)| + \frac{1}{49}|E_9(\mathcal{G}_1)| + \frac{1}{63}|E_{10}(\mathcal{G}_1)| + \frac{1}{84}|E_{11}(\mathcal{G}_1)| + \frac{1}{64}|E_{12}(\mathcal{G}_1)| + \frac{1}{72}|E_{13}(\mathcal{G}_1)| + \frac{1}{96}|E_{14}(\mathcal{G}_1)| + \frac{1}{81}|E_{15}(\mathcal{G}_1)| + \frac{1}{108}|E_{16}(\mathcal{G}_1)| + \frac{1}{144}|E_{17}(\mathcal{G}_1)|$

After simplification, we have

$$R_{-1}(\mathcal{G}_1) = \frac{11907n^3 - 17003n^2 + 12343n - 3051}{21168}$$

For $\alpha = -\frac{1}{2}$

$$R_{-\frac{1}{2}}(\mathcal{G}_1) = \sum_{j=1}^{17} \sum_{ab\in E_j(\mathcal{G})} \frac{1}{\sqrt{d(a) \cdot d(b)}}$$

$R_{-\frac{1}{2}}(\mathcal{G}_1) = \frac{\sqrt{2}}{6}|E_1(\mathcal{G}_1)| + \frac{\sqrt{21}}{21}|E_2(\mathcal{G}_1)| + \frac{\sqrt{6}}{12}|E_3(\mathcal{G}_1)| + \frac{\sqrt{3}}{9}|E_4(\mathcal{G}_1)| + \frac{1}{6}|E_5(\mathcal{G}_1)| + \frac{\sqrt{42}}{42}|E_6(\mathcal{G}_1)| + \frac{\sqrt{3}}{12}|E_7(\mathcal{G}_1)| + \frac{\sqrt{2}}{12}|E_8(\mathcal{G}_1)| + \frac{1}{7}|E_9(\mathcal{G}_1)| + \frac{\sqrt{7}}{21}|E_{10}(\mathcal{G}_1)| + \frac{\sqrt{21}}{42}|E_{11}(\mathcal{G}_1)| + \frac{1}{8}|E_{12}(\mathcal{G}_1)| + \frac{\sqrt{2}}{12}|E_{13}(\mathcal{G}_1)| + \frac{\sqrt{6}}{24}|E_{14}(\mathcal{G}_1)| + \frac{1}{9}|E_{15}(\mathcal{G}_1)| + \frac{\sqrt{3}}{18}|E_{16}(\mathcal{G}_1)| + \frac{1}{12}|E_{17}(\mathcal{G}_1)|$

After simplification, we have

$R_{-\frac{1}{2}}(\mathcal{G}_1)=\frac{15n^3}{4} + (\frac{8}{\sqrt{3}} - \frac{125}{12})n^2 + (\frac{4}{\sqrt{7}} + 4\sqrt{6} - \frac{32}{\sqrt{3}} + \sqrt{2} + 5\sqrt{\frac{3}{7}} + \frac{109}{21})n - \frac{8}{\sqrt{7}} - 8\sqrt{6} + \frac{38}{\sqrt{3}} + 3\sqrt{2} + 2\sqrt{\frac{6}{7}} - 10\sqrt{\frac{3}{7}} + \frac{13}{14}$ □

In the below theorem, we calculate the Zagreb index of $\mathcal{G}_1$(m,n).

Theorem 2. *The first Zagreb index of hex-derived cage network $HDCN1(m,n)$ is equal to*

$$M_1(\mathcal{G}_1) = 18(27n^3 - 51n^2 + 10n + 14)$$

Proof. With the help of Table 1, we calculate the Zagreb index as

$$M_1(\mathcal{G}_1) = \sum_{ab\in E(\mathcal{G})} (d(a)+d(b)) = \sum_{j=1}^{17} \sum_{ab\in E_j(\mathcal{G})} (d(a)+d(b))$$

$M_1(\mathcal{G}_1) = 9|E_1(\mathcal{G}_1)| + 10|E_2(\mathcal{G}_1)| + 11|E_3(\mathcal{G}_1)| + 12|E_4(\mathcal{G}_1)| + 15|E_5(\mathcal{G}_1)| + 13|E_6(\mathcal{G}_1)| + 14|E_7(\mathcal{G}_1)| + 18|E_8(\mathcal{G}_1)| + 14|E_9(\mathcal{G}_1)| + 16|E_{10}(\mathcal{G}_1)| + 19|E_{11}(\mathcal{G}_1)| + 16|E_{12}(\mathcal{G}_1)| + 17|E_{13}(\mathcal{G}_1)| + 20|E_{14}(\mathcal{G}_1)| + 18|E_{15}(\mathcal{G}_1)| + 21|E_{16}(\mathcal{G}_1)| + 24|E_{17}(\mathcal{G}_1)|$

After some calculations, we get

$$M_1(\mathcal{G}_1) = 18(27n^3 - 51n^2 + 10n + 14)$$

□

Table 1. Edge partition of Hex-Derived Cage network ($HDCN1$) based on degrees of end vertices of each edge.

(d_u, d_v) where $ab \in E(\mathcal{G}_1)$	Number of Edges	(d_u, d_v) where $ab \in E(\mathcal{G}_1)$	Number of Edges
$E_1 = (3,6)$	24	$E_{10} = (7,7)$	$6n - 18$
$E_2 = (3,7)$	$2(6n - 12)$	$E_{11} = (7,9)$	$2(6n - 12)$
$E_3 = (3,8)$	$6(6n - 12)$	$E_{12} = (7,12)$	$6n - 12$
$E_4 = (3,9)$	$18n^2 - 72n + 72$	$E_{13} = (8,8)$	$2(6n - 18)$
$E_5 = (3,12)$	$18n^3 - 54n^2 + 42n$	$E_{14} = (8,9)$	$2(6n - 12)$
$E_6 = (6,7)$	12	$E_{15} = (8,12)$	$4(6n - 12)$
$E_7 = (6,8)$	24	$E_{16} = (9,9)$	$12n^2 - 60n + 72$
$E_8 = (6,12)$	12	$E_{17} = (9,12)$	$12n^2 - 48n + 48$
$E_9 = (12,12)$	$9n^3 - 33n^2 + 30n$		

In the next theorem, we calculate the ABC, GA, ABC_4 and GA_5 indices of Hex-Derived Cage network $HDCN1(m,n)$.

Theorem 3. *Let HDCN1(m,n) be Hex-Derived Cage network, then we have*

- $ABC(\mathcal{G}_1) = \frac{3}{4}(4\sqrt{13} + \sqrt{22})n^3 + \frac{1}{12}(8\sqrt{57} + 24\sqrt{30} - 33\sqrt{22} - 108\sqrt{13} + 64)n^2 + (-\frac{80}{3} + 8\sqrt{\frac{6}{7}} + 4\sqrt{2} + \frac{54\sqrt{3}}{7} + 3\sqrt{\frac{7}{2}} + 5\sqrt{\frac{11}{2}} + 9\sqrt{6} - 8\sqrt{\frac{19}{3}} + \sqrt{\frac{51}{7}} + 7\sqrt{13} - 7\sqrt{30})n + 44 - 16\sqrt{\frac{6}{7}} - 4\sqrt{2} - \frac{120\sqrt{3}}{7} - \sqrt{\frac{7}{2}} - 18\sqrt{6} + 8\sqrt{\frac{19}{3}} - 2\sqrt{\frac{51}{7}} + 2\sqrt{\frac{66}{7}} + 6\sqrt{30}.$

- $GA(\mathcal{G}_1)) = \frac{117n^3}{5} + \frac{3}{35}(185\sqrt{3} - 749)n^2 + (\frac{108}{5} + \frac{144\sqrt{2}}{17} - \frac{444\sqrt{3}}{7} + \frac{1248\sqrt{6}}{55} + \frac{9\sqrt{7}}{2} + \frac{348\sqrt{21}}{95})n + \frac{24\sqrt{42}}{13} - \frac{696\sqrt{21}}{95} - 9\sqrt{7} - \frac{2496\sqrt{6}}{55} + \frac{540\sqrt{3}}{7} + \frac{120\sqrt{2}}{17} + 18.$

- $ABC_4(\mathcal{G}_1) = \frac{6}{5}\sqrt{\frac{38}{7}}(n-4)^2 + 2\sqrt{\frac{69}{79}}(n-2) + \frac{1}{3}\sqrt{\frac{62}{5}}n(3n^2 - 15n + 19) + \frac{1}{15}\sqrt{\frac{89}{2}}n(3n^2 - 17n + 24) + \sqrt{\frac{177}{14}}(n^2 - 5n + 6) + 2\sqrt{\frac{115}{77}}(n^2 - 5n + 6) + 2\sqrt{\frac{86}{105}}(n^2 - 5n + 6) + 2\sqrt{\frac{2}{5}}(n^2 - 5n + 6) + \frac{1}{7}\sqrt{\frac{83}{2}}(n^2 - 6n + 8) + \frac{8}{23}\sqrt{34}(n^2 - 9n + 20) + 2\sqrt{\frac{113}{79}}(n-2) + \sqrt{\frac{334}{395}}(n-2) + 4\sqrt{\frac{30}{79}}(n-2) + 6\sqrt{\frac{15}{79}}(n-2) + 6\sqrt{\frac{58}{41}}(n-3) + 4\sqrt{\frac{34}{41}}(n-3) + 12\sqrt{\frac{7}{41}}(n-3) + 18\sqrt{\frac{6}{287}}(n-3) + \frac{20(n-4)^2}{\sqrt{253}} + 2\sqrt{\frac{194}{115}}(n-4)^2 + 2\sqrt{\frac{151}{161}}(n-4)^2 + \frac{144(n-4)}{\sqrt{5293}} + \frac{48(n-4)}{\sqrt{85}} + \frac{3}{5}\sqrt{\frac{254}{79}}(n-4) + \frac{54}{41}\sqrt{2}(n-4) + 2\sqrt{\frac{114}{67}}(n-4) + 2\sqrt{\frac{447}{469}}(n-4) + 3\sqrt{\frac{22}{29}}(n-4) + 4\sqrt{\frac{6}{23}}(n-4) + 12\sqrt{\frac{6}{29}}(n-4) + 6\sqrt{\frac{93}{469}}(n-4) + 12\sqrt{\frac{138}{1189}}(n-4) + 30\sqrt{\frac{2}{119}}(n-4) + 18\sqrt{\frac{5}{391}}(n-4) + 28\sqrt{\frac{6}{737}}(n-4) + \frac{3}{17}\sqrt{134}(n-5) + \frac{6}{29}\sqrt{114}(n-5) + \frac{24}{67}\sqrt{33}(n-5) + \frac{21}{25}\sqrt{2}(n-5) + \frac{206\sqrt{194}}{385} + 4\frac{\sqrt{\frac{678}{11}}}{7} + \frac{6\sqrt{41}}{7} + \frac{24}{\sqrt{29}} + 2\sqrt{\frac{110}{19}} + 3\sqrt{\frac{43}{14}} + 3\sqrt{\frac{53}{19}} + 4\sqrt{\frac{138}{77}} + 4\sqrt{\frac{94}{55}} + 4\sqrt{\frac{786}{737}} + 4\sqrt{\frac{78}{79}} + 4\sqrt{\frac{74}{77}} + 2\sqrt{\frac{82}{95}} + 6\sqrt{\frac{190}{287}} + \frac{24\sqrt{\frac{22}{41}}}{7} + 6\sqrt{\frac{65}{133}} + \frac{60\sqrt{\frac{3}{7}}}{7} + 12\sqrt{\frac{10}{41}} + 6\sqrt{\frac{115}{779}} + 12\sqrt{\frac{10}{77}} + 12\sqrt{\frac{34}{287}} + 12\sqrt{\frac{2}{17}} + 12\sqrt{\frac{78}{779}} + 24\sqrt{\frac{5}{91}} + 60\sqrt{\frac{2}{247}} + 36\sqrt{\frac{2}{553}}.$

- $GA_5(\mathcal{G}_1) = \frac{40}{13}\sqrt{14}(n-4)^2 + \frac{48}{163}\sqrt{1659}(n-2) + \frac{12}{7}\sqrt{10}n(3n^2 - 15n + 19) + \frac{24}{29}\sqrt{210}(n^2 - 5n + 6) + \frac{16}{13}\sqrt{77}(n^2 - 5n + 6) + \frac{12}{19}\sqrt{70}(n^2 - 5n + 6) + \frac{18}{5}\sqrt{21}(n^2 - 5n + 6) + 9n^3 - 33n^2 - 66n + \frac{3}{14}\sqrt{2607}(n-2) + \frac{36}{169}\sqrt{790}(n-2) + \frac{48}{107}\sqrt{553}(n-2) + \frac{144}{115}\sqrt{79}(n-2) + \frac{24}{55}\sqrt{574}(n-3) + \frac{72}{43}\sqrt{205}(n-3) + \frac{216}{59}\sqrt{82}(n-3) + \frac{64}{19}\sqrt{41}(n-3) + \frac{6}{17}\sqrt{253}(n-4)^2 + \frac{8}{11}\sqrt{230}(n-4)^2 + \frac{16}{17}\sqrt{161}(n-4)^2 + \frac{12}{73}\sqrt{5293}(n-4) + \frac{3}{17}\sqrt{4623}(n-4) + \frac{6}{25}\sqrt{2211}(n-4) + \frac{24}{97}\sqrt{2010}(n-4) + \frac{48}{151}\sqrt{1407}(n-4) + \frac{24}{35}\sqrt{1189}(n-4) + \frac{48}{137}\sqrt{1173}(n-4) + \frac{8}{21}\sqrt{986}(n-4) + \frac{48}{101}\sqrt{561}(n-4) + \frac{48}{49}\sqrt{510}(n-4) + \frac{48}{95}\sqrt{469}(n-4) + \frac{48}{43}\sqrt{406}(n-4) + \frac{24}{19}\sqrt{357}(n-4) + \frac{20}{43}\sqrt{158}(n-4) + \frac{40}{39}\sqrt{134}(n-4) + \frac{32}{15}\sqrt{29}(n-4) + 12(n-4) + 36(n-5) + \frac{48\sqrt{5214}}{145} + \frac{48\sqrt{4422}}{133} + \frac{48\sqrt{1558}}{79} + \frac{48\sqrt{1122}}{67} + \frac{48\sqrt{1066}}{67} + \frac{16\sqrt{779}}{39} + \frac{96\sqrt{574}}{97} + \frac{16\sqrt{494}}{17} + \frac{48\sqrt{462}}{47} + \frac{32\sqrt{287}}{23} + \frac{16\sqrt{266}}{11} + \frac{96\sqrt{231}}{61} + \frac{32\sqrt{203}}{19} + \frac{72\sqrt{190}}{83} + \frac{48\sqrt{154}}{25} + \frac{96\sqrt{91}}{41} + \frac{21\sqrt{79}}{16} + \frac{336\sqrt{66}}{115} + 3\sqrt{55} + \frac{28\sqrt{41}}{15} + \frac{32\sqrt{38}}{9} + \frac{36\sqrt{19}}{7} + \frac{192\sqrt{7}}{11} + \frac{1336\sqrt{2}}{33} + 258.$

Proof. From Table 1 we calculate the $ABC(\mathcal{G}_1)$ as

$$ABC(\mathcal{G}_1) = \sum_{ab \in E(\mathcal{G})} \sqrt{\frac{d(a)+d(b)-2}{d(a)\cdot d(b)}} = \sum_{j=1}^{17} \sum_{ab \in E_j(\mathcal{G})} \sqrt{\frac{d(a)+d(b)-2}{d(a)\cdot d(b)}}$$

$ABC(\mathcal{G}_1) = \frac{\sqrt{14}}{6}|E_1(\mathcal{G}_1)| + \frac{2\sqrt{42}}{21}|E_2(\mathcal{G}_1)| + \frac{\sqrt{6}}{4}|E_3(\mathcal{G}_1)| + \frac{\sqrt{30}}{9}|E_4(\mathcal{G}_1)| + \frac{\sqrt{13}}{6}|E_5(\mathcal{G}_1)| + \frac{\sqrt{462}}{42}|E_6(\mathcal{G}_1)| + \frac{1}{2}|E_7(\mathcal{G}_1)| + \frac{\sqrt{2}}{3}|E_8(\mathcal{G}_1)| + \frac{2\sqrt{3}}{7}|E_9(\mathcal{G}_1)| + \frac{\sqrt{2}}{3}|E_{10}(\mathcal{G}_1)| + \frac{\sqrt{357}}{42}|E_{11}(\mathcal{G}_1)| + \frac{\sqrt{14}}{8}|E_{12}(\mathcal{G}_1)| + \frac{\sqrt{30}}{12}|E_{13}(\mathcal{G}_1)| + \frac{\sqrt{3}}{4}|E_{14}(\mathcal{G}_1)| + \frac{4}{9}|E_{15}(\mathcal{G}_1)| + \frac{\sqrt{57}}{18}|E_{16}(\mathcal{G}_1)| + \frac{\sqrt{22}}{12}|E_{17}(\mathcal{G}_1)|.$

After simplification, we have

$ABC(\mathcal{G}_1) = \frac{3}{4}(4\sqrt{13} + \sqrt{22})n^3 + \frac{1}{12}(8\sqrt{57} + 24\sqrt{30} - 33\sqrt{22} - 108\sqrt{13} + 64)n^2 + (-\frac{80}{3} + 8\sqrt{\frac{6}{7}} + 4\sqrt{2} + \frac{54\sqrt{3}}{7} + 3\sqrt{\frac{7}{2}} + 5\sqrt{\frac{11}{2}} + 9\sqrt{6} - 8\sqrt{\frac{19}{3}} + \sqrt{\frac{51}{7}} + 7\sqrt{13} - 7\sqrt{30})n + 44 - 16\sqrt{\frac{6}{7}} - 4\sqrt{2} - \frac{120\sqrt{3}}{7} - \sqrt{\frac{7}{2}} - 18\sqrt{6} + 8\sqrt{\frac{19}{3}} - 2\sqrt{\frac{51}{7}} + 2\sqrt{\frac{66}{7}} + 6\sqrt{30}.$

Now we calculate GA from Equation (6) as

$$GA(\mathcal{G}_1) = \sum_{ab \in E(\mathcal{G})} \frac{2\sqrt{d(a)d(b)}}{(d(a)+d(b))} = \sum_{j=1}^{17} \sum_{ab \in E_j(\mathcal{G})} \frac{2\sqrt{d(a)d(b)}}{(d(a)+d(b))}$$

From Table 1 calculate $GA(\mathcal{G}_1)$ as

$GA(\mathcal{G}_1)=\frac{2\sqrt{2}}{3}|E_1(\mathcal{G}_1)| + \frac{\sqrt{21}}{5}|E_2(\mathcal{G}_1)| + \frac{4\sqrt{6}}{11}|E_3(\mathcal{G}_1)| + \frac{\sqrt{3}}{2}|E_4(\mathcal{G}_1)| + \frac{4}{5}|E_5(\mathcal{G}_1)| + \frac{2\sqrt{42}}{13}|E_6(\mathcal{G}_1)| + \frac{4\sqrt{3}}{7}|E_7(\mathcal{G}_1)| + \frac{2\sqrt{2}}{3}|E_8(\mathcal{G}_1)| + 1|E_9(\mathcal{G}_1)| + \frac{3\sqrt{7}}{8}|E_{10}(\mathcal{G}_1)| + \frac{4\sqrt{21}}{19}|E_{11}(\mathcal{G}_1)| + 1|E_{12}(\mathcal{G}_1)| + \frac{12\sqrt{2}}{17}|E_{13}(\mathcal{G}_1)| + \frac{2\sqrt{6}}{5}|E_{14}(\mathcal{G}_1)| + 1|E_{15}(\mathcal{G}_1)| + \frac{4\sqrt{3}}{7}|E_{16}(\mathcal{G}_1)| + 1|E_{17}(\mathcal{G}_1)|.$

After simplification, we have

$GA(\mathcal{G}_1) =\frac{117n^3}{5} + \frac{3}{35}(185\sqrt{3} - 749)n^2 + (\frac{108}{5} + \frac{144\sqrt{2}}{17} - \frac{444\sqrt{3}}{7} + \frac{1248\sqrt{6}}{55} + \frac{9\sqrt{7}}{2} + \frac{348\sqrt{21}}{95})n + \frac{24\sqrt{42}}{13} - \frac{696\sqrt{21}}{95} - 9\sqrt{7} - \frac{2496\sqrt{6}}{55} + \frac{540\sqrt{3}}{7} + \frac{120\sqrt{2}}{17} + 18.$

If we consider an edge partition based on degree sum of neighbors of end vertices; then the edge set $E(HDCN1(m,n))$ are divided into sixtynine edge partition $E_j(HDCN1(m,n)), 18 \leq j \leq 86$ shows in Table 2.

From Equation (7), we have

$$ABC_4(\mathcal{G}_1) = \sum_{ab\in E(\mathcal{G})}\sqrt{\frac{S(a)+S(b)-2}{S(a)S(b)}} = \sum_{j=18}^{86}\sum_{ab\in E_j(\mathcal{G})}\sqrt{\frac{S(a)+S(b)-2}{S(a)S(b)}}.$$

From Table 2 we use edge partition, we get

$ABC_4(\mathcal{G}_1) = \frac{2\sqrt{1066}}{67}|E_{18}(\mathcal{G}_1)| + \frac{\sqrt{1456}}{41}|E_{19}(\mathcal{G}_1)| + \frac{\sqrt{1976}}{51}|E_{20}(\mathcal{G}_1)| + \frac{2\sqrt{1372}}{77}|E_{21}(\mathcal{G}_1)| + \frac{\sqrt{1400}}{39}|E_{22}(\mathcal{G}_1)| + \frac{\sqrt{1568}}{42}|E_{23}(\mathcal{G}_1)| + \frac{\sqrt{1624}}{43}|E_{24}(\mathcal{G}_1)| + \frac{\sqrt{1848}}{47}|E_{25}(\mathcal{G}_1)| + \frac{2\sqrt{1876}}{95}|E_{26}(\mathcal{G}_1)| + \frac{2\sqrt{2212}}{107}|E_{27}(\mathcal{G}_1)| + \frac{\sqrt{2296}}{55}|E_{28}(\mathcal{G}_1)| + \frac{\sqrt{1980}}{48}|E_{29}(\mathcal{G}_1)| + \frac{2\sqrt{2010}}{97}|E_{30}(\mathcal{G}_1)| + \frac{\sqrt{2040}}{\sqrt{49}}|E_{31}(\mathcal{G}_1)| + \frac{2\sqrt{2070}}{99}|E_{32}(\mathcal{G}_1)| + \frac{\sqrt{2520}}{57}|E_{33}(\mathcal{G}_1)| + \frac{\sqrt{1792}}{44}|E_{34}(\mathcal{G}_1)| + \frac{\sqrt{1856}}{45}|E_{35}(\mathcal{G}_1)| + \frac{\sqrt{2432}}{54}|E_{36}(\mathcal{G}_1)| + \frac{\sqrt{2624}}{57}|E_{37}(\mathcal{G}_1)| + \frac{2\sqrt{2178}}{99}|E_{38}(\mathcal{G}_1)| + \frac{\sqrt{2211}}{50}|E_{39}(\mathcal{G}_1)| + \frac{2\sqrt{2244}}{101}|E_{40}(\mathcal{G}_1)| + \frac{\sqrt{2277}}{51}|E_{41}(\mathcal{G}_1)| + \frac{\sqrt{2607}}{3}|E_{56}(\mathcal{G}_1)| + \frac{2\sqrt{2772}}{117}|E_{43}(\mathcal{G}_1)| + \frac{\sqrt{2736}}{56}|E_{44}(\mathcal{G}_1)| + \frac{2\sqrt{2844}}{115}|E_{45}(\mathcal{G}_1)| + \frac{\sqrt{2952}}{59}|E_{46}(\mathcal{G}_1)| + \frac{\sqrt{3024}}{60}|E_{47}(\mathcal{G}_1)| + \frac{\sqrt{3240}}{63}|E_{48}(\mathcal{G}_1)| + \frac{\sqrt{2009}}{45}|E_{49}(\mathcal{G}_1)| + \frac{2\sqrt{2296}}{97}|E_{50}(\mathcal{G}_1)| + \frac{2\sqrt{3116}}{117}|E_{51}(\mathcal{G}_1)| + \frac{2\sqrt{2450}}{99}|E_{52}(\mathcal{G}_1)| + \frac{2\sqrt{3234}}{115}|E_{53}(\mathcal{G}_1)| + \frac{\sqrt{3871}}{64}|E_{54}(\mathcal{G}_1)| + 1|E_{55}(\mathcal{G}_1)| + \frac{2\sqrt{3350}}{117}|E_{56}(\mathcal{G}_1)| + \frac{2\sqrt{3950}}{129}|E_{57}(\mathcal{G}_1)| + \frac{\sqrt{3248}}{57}|E_{58}(\mathcal{G}_1)| + \frac{\sqrt{3696}}{56}|E_{59}(\mathcal{G}_1)| + \frac{\sqrt{4256}}{66}|E_{60}(\mathcal{G}_1)| + \frac{\sqrt{4292}}{69}|E_{61}(\mathcal{G}_1)| + \frac{\sqrt{3364}}{58}|E_{62}(\mathcal{G}_1)| + \frac{\sqrt{3944}}{63}|E_{63}(\mathcal{G}_1)| + \frac{\sqrt{4756}}{70}|E_{64}(\mathcal{G}_1)| + \frac{2\sqrt{4422}}{133}|E_{65}(\mathcal{G}_1)| + \frac{\sqrt{4488}}{67}|E_{66}(\mathcal{G}_1)| + \frac{2\sqrt{5214}}{145}|E_{67}(\mathcal{G}_1)| + \frac{\sqrt{5544}}{75}|E_{68}(\mathcal{G}_1)| + \frac{\sqrt{4489}}{67}|E_{69}(\mathcal{G}_1)| + \frac{\sqrt{4623}}{68}|E_{70}(\mathcal{G}_1)| + \frac{\sqrt{5293}}{73}|E_{71}(\mathcal{G}_1)| + \frac{2\sqrt{5628}}{151}|E_{72}(\mathcal{G}_1)| + \frac{\sqrt{4624}}{68}|E_{73}(\mathcal{G}_1)| + \frac{2\sqrt{4692}}{137}|E_{74}(\mathcal{G}_1)| + \frac{\sqrt{5712}}{76}|E_{75}(\mathcal{G}_1)| + \frac{\sqrt{4761}}{69}|E_{76}(\mathcal{G}_1)| + \frac{2\sqrt{5796}}{153}|E_{77}(\mathcal{G}_1)| + \frac{\sqrt{6232}}{79}|E_{78}(\mathcal{G}_1)| + \frac{\sqrt{6840}}{83}|E_{79}|(\mathcal{G}_1) + \frac{2\sqrt{6636}}{163}|E_{80}(\mathcal{G}_1)| + \frac{2\sqrt{7110}}{169}|E_{81}(\mathcal{G}_1)| + \frac{\sqrt{4724}}{82}|E_{82}(\mathcal{G}_1)| + \frac{\sqrt{7380}}{86}|E_{83}(\mathcal{G}_1)| + \frac{\sqrt{7056}}{84}|E_{84}(\mathcal{G}_1)| + \frac{\sqrt{7560}}{87}|E_{85}(\mathcal{G}_1)| + 1|E_{86}(\mathcal{G}_1)|.$

After simplification, we get

$ABC_4(\mathcal{G}_1) =\frac{6}{5}\sqrt{\frac{38}{7}}(n-4)^2 + 2\sqrt{\frac{69}{79}}(n-2) + \frac{1}{3}\sqrt{\frac{62}{5}}n(3n^2 - 15n + 19) + \frac{1}{15}\sqrt{\frac{89}{2}}n(3n^2 - 17n + 24) + \sqrt{\frac{177}{14}}(n^2 - 5n + 6) + 2\sqrt{\frac{115}{77}}(n^2 - 5n + 6) + 2\sqrt{\frac{86}{105}}(n^2 - 5n + 6) + 2\sqrt{\frac{2}{5}}(n^2 - 5n + 6) + \frac{1}{7}\sqrt{\frac{83}{2}}(n^2 - 6n + 8) + \frac{8}{23}\sqrt{34}(n^2 - 9n + 20) + 2\sqrt{\frac{113}{79}}(n-2) + \sqrt{\frac{334}{395}}(n-2) + 4\sqrt{\frac{30}{79}}(n-2) + 6\sqrt{\frac{15}{79}}(n-2) + 6\sqrt{\frac{58}{41}}(n-3) + 4\sqrt{\frac{34}{41}}(n-3) + 12\sqrt{\frac{7}{41}}(n-3) + 18\sqrt{\frac{6}{287}}(n-3) + \frac{20(n-4)^2}{\sqrt{253}} + 2\sqrt{\frac{194}{115}}(n-4)^2 + 2\sqrt{\frac{151}{161}}(n-4)^2 + \frac{144(n-4)}{\sqrt{5293}} + \frac{48(n-4)}{\sqrt{85}} + \frac{3}{5}\sqrt{\frac{254}{79}}(n-4) + \frac{54}{41}\sqrt{2}(n-4) + 2\sqrt{\frac{114}{67}}(n-4) + 2\sqrt{\frac{447}{469}}(n-4) + 3\sqrt{\frac{22}{29}}(n-4) + 4\sqrt{\frac{6}{23}}(n-4) + 12\sqrt{\frac{6}{29}}(n-4) + 6\sqrt{\frac{93}{469}}(n-4) + 12\sqrt{\frac{138}{1189}}(n-4) + 30\sqrt{\frac{2}{119}}(n-4) + 18\sqrt{\frac{5}{391}}(n-4) + 28\sqrt{\frac{6}{737}}(n-4) + \frac{3}{17}\sqrt{134}(n-5) + \frac{6}{29}\sqrt{114}(n-5) + \frac{24}{67}\sqrt{33}(n-5) + \frac{21}{25}\sqrt{2}(n-5) + \frac{206\sqrt{194}}{385} + 4\frac{\sqrt{\frac{678}{11}}}{7} + \frac{6\sqrt{41}}{7} + \frac{24}{\sqrt{29}} + 2\sqrt{\frac{110}{19}} + 3\sqrt{\frac{43}{14}} + 3\sqrt{\frac{53}{19}} + 4\sqrt{\frac{138}{77}} + 4\sqrt{\frac{94}{55}} + 4\sqrt{\frac{786}{737}} + 4\sqrt{\frac{78}{79}} + 4\sqrt{\frac{74}{77}} + 2\sqrt{\frac{82}{95}} + 6\sqrt{\frac{190}{287}} + \frac{24\sqrt{\frac{22}{41}}}{7} + 6\sqrt{\frac{65}{133}} + \frac{60\sqrt{\frac{3}{7}}}{7} + 12\sqrt{\frac{10}{41}} + 6\sqrt{\frac{115}{779}} + 12\sqrt{\frac{10}{77}} + 12\sqrt{\frac{34}{287}} + 12\sqrt{\frac{2}{17}} + 12\sqrt{\frac{78}{779}} + 24\sqrt{\frac{5}{91}} + 60\sqrt{\frac{2}{247}} + 36\sqrt{\frac{2}{553}}$

Now we find $GA_5(\mathcal{G}_1)$ as

$$GA_5(\mathcal{G}_1) = \sum_{ab \in E(\mathcal{G})} \frac{2\sqrt{S(a)S(b)}}{(S(a)+S(b))} = \sum_{j=18}^{86} \sum_{ab \in E_j(\mathcal{G})} \frac{2\sqrt{S(a)S(b)}}{(S(a)+S(b))}.$$

Using the edge partition from Table 2, we get

$$GA_5(\mathcal{G}_1) = \sqrt{\tfrac{65}{1066}}|E_{18}(\mathcal{G}_1)| + \sqrt{\tfrac{5}{91}}|E_{19}(\mathcal{G}_1)| + \tfrac{5}{\sqrt{494}}|E_{20}(\mathcal{G}_1)| + \sqrt{\tfrac{75}{1372}}|E_{21}(\mathcal{G}_1)| + \sqrt{\tfrac{19}{350}}|E_{22}(\mathcal{G}_1)| + \sqrt{\tfrac{41}{784}}|E_{23}(\mathcal{G}_1)| + \sqrt{\tfrac{3}{58}}|E_{24}(\mathcal{G}_1)| + \sqrt{\tfrac{23}{462}}|E_{25}(\mathcal{G}_1)| + \sqrt{\tfrac{93}{1876}}|E_{26}(\mathcal{G}_1)| + \sqrt{\tfrac{105}{2212}}|E_{27}(\mathcal{G}_1)| + \sqrt{\tfrac{27}{574}}|E_{28}(\mathcal{G}_1)| + \sqrt{\tfrac{47}{990}}|E_{29}(\mathcal{G}_1)| + \sqrt{\tfrac{19}{402}}|E_{30}(\mathcal{G}_1)| + \sqrt{\tfrac{4}{85}}|E_{31}(\mathcal{G}_1)| + \sqrt{\tfrac{97}{2070}}|E_{32}(\mathcal{G}_1)| + \sqrt{\tfrac{14}{315}}|E_{33}(\mathcal{G}_1)| + \sqrt{\tfrac{43}{896}}|E_{34}(\mathcal{G}_1)| + \sqrt{\tfrac{11}{232}}|E_{35}(\mathcal{G}_1)| + \sqrt{\tfrac{53}{1216}}|E_{36}(\mathcal{G}_1)| + \sqrt{\tfrac{7}{164}}|E_{37}(\mathcal{G}_1)| + \sqrt{\tfrac{97}{2178}}|E_{38}(\mathcal{G}_1)| + \sqrt{\tfrac{98}{2211}}|E_{39}(\mathcal{G}_1)| + \sqrt{\tfrac{9}{204}}|E_{40}(\mathcal{G}_1)| + \sqrt{\tfrac{10}{2277}}|E_{41}(\mathcal{G}_1)| + \sqrt{\tfrac{110}{2607}}|E_{56}(\mathcal{G}_1)| + \sqrt{\tfrac{115}{2772}}|E_{43}(\mathcal{G}_1)| + \sqrt{\tfrac{55}{1368}}|E_{44}(\mathcal{G}_1)| + \sqrt{\tfrac{113}{2844}}|E_{45}(\mathcal{G}_1)| + \sqrt{\tfrac{29}{738}}|E_{46}(\mathcal{G}_1)| + \sqrt{\tfrac{59}{1512}}|E_{47}(\mathcal{G}_1)| + \sqrt{\tfrac{31}{810}}|E_{48}(\mathcal{G}_1)| + \sqrt{\tfrac{88}{2009}}|E_{49}(\mathcal{G}_1)| + \sqrt{\tfrac{95}{2296}}|E_{50}(\mathcal{G}_1)| + \sqrt{\tfrac{115}{3116}}|E_{51}(\mathcal{G}_1)| + \sqrt{\tfrac{97}{2450}}|E_{52}(\mathcal{G}_1)| + \sqrt{\tfrac{113}{3234}}|E_{53}(\mathcal{G}_1)| + \sqrt{\tfrac{126}{3871}}|E_{54}(\mathcal{G}_1)| + \tfrac{\sqrt{98}}{50}|E_{55}(\mathcal{G}_1)| + \sqrt{\tfrac{23}{670}}|E_{56}(\mathcal{G}_1)| + \sqrt{\tfrac{127}{3950}}|E_{57}(\mathcal{G}_1)| + \sqrt{\tfrac{7}{203}}|E_{58}(\mathcal{G}_1)| + \sqrt{\tfrac{5}{154}}|E_{59}(\mathcal{G}_1)| + \sqrt{\tfrac{65}{2128}}|E_{60}(\mathcal{G}_1)| + \sqrt{\tfrac{17}{574}}|E_{61}(\mathcal{G}_1)| + \sqrt{\tfrac{57}{1682}}|E_{62}(\mathcal{G}_1)| + \sqrt{\tfrac{31}{952}}|E_{63}(\mathcal{G}_1)| + \sqrt{\tfrac{69}{2378}}|E_{64}(\mathcal{G}_1)| + \sqrt{\tfrac{131}{4422}}|E_{65}(\mathcal{G}_1)| + \sqrt{\tfrac{33}{1122}}|E_{66}(\mathcal{G}_1)| + \sqrt{\tfrac{143}{5214}}|E_{67}(\mathcal{G}_1)| + \sqrt{\tfrac{37}{1344}}|E_{68}(\mathcal{G}_1)| + \sqrt{\tfrac{132}{4489}}|E_{69}(\mathcal{G}_1)| + \sqrt{\tfrac{134}{4623}}|E_{70}(\mathcal{G}_1)| + \sqrt{\tfrac{144}{5293}}|E_{71}(\mathcal{G}_1)| + \sqrt{\tfrac{149}{5628}}|E_{72}(\mathcal{G}_1)| + \sqrt{\tfrac{67}{2312}}|E_{73}(\mathcal{G}_1)| + \sqrt{\tfrac{135}{4692}}|E_{74}(\mathcal{G}_1)| + \sqrt{\tfrac{75}{4856}}|E_{75}(\mathcal{G}_1)| + \sqrt{\tfrac{136}{4761}}|E_{76}(\mathcal{G}_1)| + \sqrt{\tfrac{151}{5796}}|E_{77}(\mathcal{G}_1)| + \sqrt{\tfrac{159}{6232}}|E_{78}(\mathcal{G}_1)| + \sqrt{\tfrac{41}{1710}}|E_{79}|(\mathcal{G}_1) + \sqrt{\tfrac{161}{6384}}|E_{80}(\mathcal{G}_1)| + \sqrt{\tfrac{167}{7110}}|E_{81}(\mathcal{G}_1)| + \sqrt{\tfrac{81}{6724}}|E_{82}(\mathcal{G}_1)| + \sqrt{\tfrac{17}{738}}|E_{83}(\mathcal{G}_1)| + \sqrt{\tfrac{83}{3528}}|E_{84}(\mathcal{G}_1)| + \sqrt{\tfrac{86}{3785}}|E_{85}(\mathcal{G}_1)| + \sqrt{\tfrac{89}{450}}|E_{86}(\mathcal{G}_1)|.$$

After simplification, we get

$$GA_5(\mathcal{G}_1) = \tfrac{40}{13}\sqrt{14}(n-4)^2 + \tfrac{48}{163}\sqrt{1659}(n-2) + \tfrac{12}{7}\sqrt{10}n(3n^2-15n+19) + \tfrac{24}{29}\sqrt{210}(n^2-5n+6) + \tfrac{16}{13}\sqrt{77}(n^2-5n+6) + \tfrac{12}{19}\sqrt{70}(n^2-5n+6) + \tfrac{18}{5}\sqrt{21}(n^2-5n+6) + 9n^3 - 33n^2 - 66n + \tfrac{3}{14}\sqrt{2607}(n-2) + \tfrac{36}{169}\sqrt{790}(n-2) + \tfrac{48}{107}\sqrt{553}(n-2) + \tfrac{144}{115}\sqrt{79}(n-2) + \tfrac{24}{55}\sqrt{574}(n-3) + \tfrac{72}{43}\sqrt{205}(n-3) + \tfrac{216}{59}\sqrt{82}(n-3) + \tfrac{64}{19}\sqrt{41}(n-3) + \tfrac{6}{17}\sqrt{253}(n-4)^2 + \tfrac{8}{11}\sqrt{230}(n-4)^2 + \tfrac{16}{17}\sqrt{161}(n-4)^2 + \tfrac{12}{73}\sqrt{5293}(n-4) + \tfrac{3}{17}\sqrt{4623}(n-4) + \tfrac{6}{25}\sqrt{2211}(n-4) + \tfrac{24}{97}\sqrt{2010}(n-4) + \tfrac{48}{151}\sqrt{1407}(n-4) + \tfrac{24}{35}\sqrt{1189}(n-4) + \tfrac{48}{137}\sqrt{1173}(n-4) + \tfrac{8}{21}\sqrt{986}(n-4) + \tfrac{48}{101}\sqrt{561}(n-4) + \tfrac{48}{49}\sqrt{510}(n-4) + \tfrac{48}{95}\sqrt{469}(n-4) + \tfrac{48}{43}\sqrt{406}(n-4) + \tfrac{24}{19}\sqrt{357}(n-4) + \tfrac{20}{43}\sqrt{158}(n-4) + \tfrac{40}{39}\sqrt{134}(n-4) + \tfrac{32}{15}\sqrt{29}(n-4) + 12(n-4) + 36(n-5) + \frac{48\sqrt{5214}}{145} + \frac{48\sqrt{4422}}{133} + \frac{48\sqrt{1558}}{79} + \frac{48\sqrt{1122}}{67} + \frac{48\sqrt{1066}}{67} + \frac{16\sqrt{779}}{39} + \frac{96\sqrt{574}}{97} + \frac{16\sqrt{494}}{17} + \frac{48\sqrt{462}}{47} + \frac{32\sqrt{287}}{23} + \frac{16\sqrt{266}}{11} + \frac{96\sqrt{231}}{61} + \frac{32\sqrt{203}}{19} + \frac{72\sqrt{190}}{83} + \frac{48\sqrt{154}}{25} + \frac{96\sqrt{91}}{41} + \frac{21\sqrt{79}}{16} + \frac{336\sqrt{66}}{115} + 3\sqrt{55} + \frac{28\sqrt{41}}{15} + \frac{32\sqrt{38}}{9} + \frac{36\sqrt{19}}{7} + \frac{192\sqrt{7}}{11} + \frac{1336\sqrt{2}}{33} + 258 \quad \square$$

Table 2. Edge partition of Hex-Derived Cage network ($HDCN1$) based on degrees of end vertices of each edge.

(S_u, S_v) where $ab \in E(\mathcal{G}_1)$	Number of Edges	(S_u, S_v) where $ab \in E(\mathcal{G}_1)$	Number of Edges
$E_{18} = (26, 41)$	24	$E_{53} = (49, 66)$	24
$E_{19} = (26, 56)$	24	$E_{54} = (49, 79)$	12
$E_{20} = (26, 76)$	24	$E_{55} = (50, 50)$	$6n - 30$
$E_{21} = (28, 49)$	24	$E_{56} = (50, 67)$	$2(6n - 24)$
$E_{22} = (28, 50)$	$2(6n - 24)$	$E_{57} = (50, 79)$	$6n - 24$
$E_{23} = (28, 56)$	24	$E_{58} = (56, 58)$	24
$E_{24} = (28, 58)$	$4(6n - 24)$	$E_{59} = (56, 66)$	24
$E_{25} = (28, 66)$	24	$E_{60} = (56, 76)$	24
$E_{26} = (28, 67)$	$2(6n - 24)$	$E_{61} = (56, 82)$	24
$E_{27} = (28, 79)$	$2(6n - 12)$	$E_{62} = (58, 58)$	$2(6n - 30)$
$E_{28} = (28, 82)$	$2(6n - 18)$	$E_{63} = (58, 68)$	$2(6n - 24)$
$E_{29} = (30, 66)$	24	$E_{64} = (58, 82)$	$4(6n - 24)$
$E_{30} = (30, 67)$	$2(6n - 24)$	$E_{65} = (66, 67)$	24
$E_{31} = (30, 68)$	$4(6n - 24)$	$E_{66} = (66, 68)$	24
$E_{32} = (30, 69)$	$12n^2 - 96n + 192$	$E_{67} = (66, 79)$	24
$E_{33} = (30, 84)$	$6n^2 - 30n + 36$	$E_{68} = (66, 84)$	24
$E_{34} = (32, 56)$	24	$E_{69} = (67, 67)$	$2(6n - 30)$
$E_{35} = (32, 58)$	$2(6n - 24)$	$E_{70} = (67, 69)$	$2(6n - 24)$
$E_{36} = (32, 76)$	24	$E_{71} = (67, 79)$	$2(6n - 24)$
$E_{37} = (32, 82)$	$4(6n - 18)$	$E_{72} = (67, 84)$	$2(6n - 24)$
$E_{38} = (33, 66)$	24	$E_{73} = (68, 68)$	$2(6n - 30)$
$E_{39} = (33, 67)$	$2(6n - 24)$	$E_{74} = (68, 69)$	$2(6n - 24)$
$E_{40} = (33, 68)$	$2(6n - 24)$	$E_{75} = (68, 84)$	$4(6n - 24)$
$E_{41} = (33, 69)$	$6n^2 - 48n + 96$	$E_{76} = (69, 69)$	$12n^2 - 108n + 240$
$E_{42} = (33, 79)$	$2(6n - 12)$	$E_{77} = (69, 84)$	$12n^2 - 96n + 192$
$E_{43} = (33, 84)$	$12n^2 - 60n + 72$	$E_{78} = (76, 82)$	24
$E_{44} = (36, 76)$	24	$E_{79} = (76, 90)$	12
$E_{45} = (36, 79)$	$2(6n - 12)$	$E_{80} = (79, 84)$	$2(6n - 12)$
$E_{46} = (36, 82)$	$6(6n - 18)$	$E_{81} = (79, 90)$	$6n - 12$
$E_{47} = (36, 84)$	$18n^2 - 90n + 108$	$E_{82} = (82, 82)$	$2(6n - 24)$
$E_{48} = (36, 90)$	$18n^3 - 90n^2 + 114n$	$E_{83} = (82, 90)$	$4(6n - 18)$
$E_{49} = (41, 49)$	12	$E_{84} = (84, 84)$	$6n^2 - 36n + 48$
$E_{50} = (41, 56)$	24	$E_{85} = (84, 90)$	$12n^2 - 60n + 72$
$E_{51} = (41, 76)$	12	$E_{86} = (90, 90)$	$9n^3 - 51n^2 + 72n$
$E_{52} = (49, 50)$	12		

2.2. *Results for Hex-Derived Cage Network (HDCN2(m,n))*

In this portion, we find some degree-based topological indices for Hex-Derived Cage network ($HDCN2(m,n)$). We calculate the general Randić index $R_\alpha(\mathcal{G})$ with $\alpha = \{1, -1, \frac{1}{2}, -\frac{1}{2}\}$, ABC, GA' ABC_4 and GA_5 in the the below theorems for ($HDCN2(m,n)$).

Theorem 4. *Let $G_2 \cong HDCN2(m,n)$ be the Hex-Derived Cage network, then its general Randić index is equal to*

$$R_\alpha(\mathcal{G}_2) = \begin{cases} 6(486n^3 - 1068n^2 + 312n + 293), & \alpha = 1; \\ 6(9(2\sqrt{2}+3)n^3 + (2\sqrt{30}+2\sqrt{15}+3\sqrt{6}+6\sqrt{5}+ \\ 12\sqrt{3} - 60\sqrt{2} - 81)n^2 + 2(\sqrt{35} - 2\sqrt{30} + \sqrt{21}- \\ \sqrt{15} + 4\sqrt{10} + 3\sqrt{7} + 2\sqrt{6} - 12\sqrt{5} - 20\sqrt{3} \\ +30\sqrt{2} + 14)n + 2\sqrt{42} - 4\sqrt{35} + 4\sqrt{30} - 4\sqrt{21}- \\ 16\sqrt{10} - 12\sqrt{7} - 20\sqrt{6} + 24\sqrt{5} + 48\sqrt{3}- \\ 12\sqrt{2} + 39), & \alpha = \frac{1}{2}; \\ \frac{297675n^3 - 445655n^2 + 282283n - 40155}{529200}, & \alpha = -1; \\ \frac{3}{4}(2\sqrt{2}+3)n^3 + (\frac{4}{\sqrt{5}} + \frac{2}{\sqrt{3}} - 5\sqrt{2} + 2\sqrt{\frac{6}{5}} + \sqrt{\frac{2}{3}}+ \\ \sqrt{\frac{3}{5}} - \frac{83}{12})n^2 + (\frac{461}{105}) + 6\sqrt{\frac{2}{5}} + \sqrt{\frac{3}{7}} - \sqrt{\frac{3}{5}} - \sqrt{\frac{2}{3}}- \\ 4\sqrt{\frac{6}{5}} + 5\sqrt{2} - \frac{5}{\sqrt{3}} - \frac{16}{\sqrt{5}} + \frac{4}{\sqrt{7}} + \frac{12}{\sqrt{35}})n - \frac{24}{\sqrt{35}}- \\ \frac{8}{\sqrt{7}} + \frac{16}{\sqrt{5}} + \frac{8}{\sqrt{3}} - \sqrt{2} + 4\sqrt{\frac{6}{5}} + 2\sqrt{\frac{6}{7}} - 2\sqrt{\frac{2}{3}}- \\ 2\sqrt{\frac{3}{7}} - 12\sqrt{\frac{2}{5}} + \frac{13}{14}, & \alpha = -\frac{1}{2}. \end{cases}$$

Proof. Let $\mathcal{G}_2$ be the Hex-Derived Cage network $(HDCN2(m,n))$ where $m = n \geq 5$. The edge set of $HDCN2(m,n)$ is divided into twenty partitions based on the degree of end vertices. Table 3 shows these edge partition of $HDCN2(m,n)$.

$$R_\alpha(\mathcal{G}_2) = \sum_{ab \in E(\mathcal{G})} (d(a)d(b))^\alpha$$

For $\alpha = 1$

$$R_1(\mathcal{G}_2) = \sum_{j=1}^{20} \sum_{ab \in E_j(\mathcal{G})} deg(u) \cdot deg(v)$$

Using the edge partition from Table 3, we get
$R_1(\mathcal{G}_2) = 25|E_1(\mathcal{G}_2)| + 30|E_2(\mathcal{G}_2)| + 35|E_3(\mathcal{G}_2)| + 40|E_4(\mathcal{G}_2)| + 45|E_5(\mathcal{G}_2)| + 60|E_6(\mathcal{G}_2)| + 36|E_7(\mathcal{G}_2)| + 42|E_8(\mathcal{G}_2)| + 48|E_9(\mathcal{G}_2)| + 54|E_{10}(\mathcal{G}_2)| + 72|E_{11}(\mathcal{G}_2)| + 49|E_{12}(\mathcal{G}_2)| + 63|E_{13}(\mathcal{G}_2)| + 84|E_{14}(\mathcal{G}_2)| + 64|E_{15}(\mathcal{G}_2)| + 72|E_{16}(\mathcal{G}_2)| + 96|E_{17}(\mathcal{G}_2)| + 81|E_{18}(\mathcal{G}_2)| + 108|E_{19}(\mathcal{G}_2)| + 144|E_{20}(\mathcal{G}_2)|$
After simplification, we get

$$R_1(\mathcal{G}_2) = 6(486n^3 - 1068n^2 + 312n + 293)$$

For $\alpha = \frac{1}{2}$

$$R_{\frac{1}{2}}(\mathcal{G}_2) = \sum_{j=1}^{20} \sum_{ab \in E_j(\mathcal{G})} \sqrt{d(a) \cdot d(b)}$$

Using edge partition from Table 3, we get
$R_{\frac{1}{2}}(\mathcal{G}_2) = 5|E_1(\mathcal{G}_2)| + \sqrt{30}|E_2(\mathcal{G}_2)| + \sqrt{35}|E_3(\mathcal{G}_2)| + 2\sqrt{10}|E_4(\mathcal{G}_2)| + 3\sqrt{5}|E_5(\mathcal{G}_2)| + 2\sqrt{15}|E_6(\mathcal{G}_2)| + 6|E_7(\mathcal{G}_2)| + \sqrt{42}|E_8(\mathcal{G}_2)| + 4\sqrt{3}|E_9(\mathcal{G}_2)| + 3\sqrt{6}|E_{10}(\mathcal{G}_2)| + 6\sqrt{2}|E_{11}(\mathcal{G}_2)| + 7|E_{12}(\mathcal{G}_2)| + 3\sqrt{7}|E_{13}(\mathcal{G}_2)| + 2\sqrt{21}|E_{14}(\mathcal{G}_2)| + 8|E_{15}(\mathcal{G}_2)| + 6\sqrt{2}|E_{16}(\mathcal{G}_2)| + 4\sqrt{6}|E_{17}(\mathcal{G}_2)| + 9|E_{18}(\mathcal{G}_2)| + 6\sqrt{3}|E_{19}(\mathcal{G}_2)| + 12|E_{20}(\mathcal{G}_2)|$
After simplification, we get
$R_{\frac{1}{2}}(\mathcal{G}_2) = 6(9(2\sqrt{2}+3)n^3 + (2\sqrt{30} + 2\sqrt{15} + 3\sqrt{6} + 6\sqrt{5} + 12\sqrt{3} - 60\sqrt{2} - 81)n^2 + 2(\sqrt{35} - 2\sqrt{30} + \sqrt{21} - \sqrt{15} + 4\sqrt{10} + 3\sqrt{7} + 2\sqrt{6} - 12\sqrt{5} - 20\sqrt{3} + 30\sqrt{2} + 14)n + 2\sqrt{42} - 4\sqrt{35} + 4\sqrt{30} - 4\sqrt{21} - 16\sqrt{10} - 12\sqrt{7} - 20\sqrt{6} + 24\sqrt{5} + 48\sqrt{3} - 12\sqrt{2} + 39)$

For $\alpha = -1$

$$R_{-1}(\mathcal{G}_2) = \sum_{j=1}^{20} \sum_{ab \in E_j(\mathcal{G})} \frac{1}{d(a) \cdot d(b)}$$

$R_{-1}(\mathcal{G}_2) = \frac{1}{25}|E_1(\mathcal{G}_2)| + \frac{1}{30}|E_2(\mathcal{G}_2)| + \frac{1}{35}|E_3(\mathcal{G}_2)| + \frac{1}{40}|E_4(\mathcal{G}_2)| + \frac{1}{45}|E_5(\mathcal{G}_2)| + \frac{1}{60}|E_6(\mathcal{G}_2)| + \frac{1}{36}|E_7(\mathcal{G}_2)| + \frac{1}{42}|E_8(\mathcal{G}_2)| + \frac{1}{48}|E_9(\mathcal{G}_2)| + \frac{1}{54}|E_{10}(\mathcal{G}_2)| + \frac{1}{72}|E_{11}(\mathcal{G}_2)| + \frac{1}{49}|E_{12}(\mathcal{G}_2)| + \frac{1}{63}|E_{13}(\mathcal{G}_2)| + \frac{1}{84}|E_{14}(\mathcal{G}_2)| + \frac{1}{64}|E_{15}(\mathcal{G}_2)| + \frac{1}{72}|E_{16}(\mathcal{G}_2)| + \frac{1}{96}|E_{17}(\mathcal{G}_2)| + \frac{1}{81}|E_{18}(\mathcal{G}_2)| + \frac{1}{108}|E_{19}(\mathcal{G}_2)| + \frac{1}{144}|E_{20}(\mathcal{G}_2)|$

After simplification, we get

$$R_{-1}(\mathcal{G}_2) = \frac{297675n^3 - 445655n^2 + 282283n - 40155}{529200}$$

For $\alpha = -\frac{1}{2}$

$$R_{-\frac{1}{2}}(\mathcal{G}_2) = \sum_{j=1}^{20} \sum_{ab \in E_j(\mathcal{G})} \frac{1}{\sqrt{d(a) \cdot d(b)}}$$

$R_{-\frac{1}{2}}(\mathcal{G}_2) = \frac{1}{5}|E_1(\mathcal{G}_2)| + \frac{1}{\sqrt{30}}|E_2(\mathcal{G}_2)| + \frac{1}{\sqrt{35}}|E_3(\mathcal{G}_2)| + \frac{1}{2\sqrt{10}}|E_4(\mathcal{G}_2)| + \frac{1}{3\sqrt{5}}|E_5(\mathcal{G}_2)| + \frac{1}{2\sqrt{15}}|E_6(\mathcal{G}_2)| + \frac{1}{6}|E_7(\mathcal{G}_2)| + \frac{1}{\sqrt{42}}|E_8(\mathcal{G}_2)| + \frac{1}{4\sqrt{3}}|E_9(\mathcal{G}_2)| + \frac{1}{3\sqrt{6}}|E_{10}(\mathcal{G}_2)| + \frac{1}{6\sqrt{2}}|E_{11}(\mathcal{G}_2)| + \frac{1}{7}|E_{12}(\mathcal{G}_2)| + \frac{1}{3\sqrt{7}}|E_{13}(\mathcal{G}_2)| + \frac{1}{2\sqrt{21}}|E_{14}(\mathcal{G}_2)| + \frac{1}{8}|E_{15}(\mathcal{G}_2)| + \frac{1}{6\sqrt{2}}|E_{16}(\mathcal{G}_2)| + \frac{1}{4\sqrt{6}}|E_{17}(\mathcal{G}_2)| + \frac{1}{9}|E_{18}(\mathcal{G}_2)| + \frac{1}{6\sqrt{3}}|E_{19}(\mathcal{G}_2)| + \frac{1}{12}|E_{20}(\mathcal{G}_2)|$

After simplification, we get

$R_{-\frac{1}{2}}(\mathcal{G}_2) = \frac{3}{4}(2\sqrt{2}+3)n^3 + (\frac{4}{\sqrt{5}} + \frac{2}{\sqrt{3}} - 5\sqrt{2} + 2\sqrt{\frac{6}{5}} + \sqrt{\frac{2}{3}} + \sqrt{\frac{3}{5}} - \frac{83}{12})n^2 + (\frac{461}{105}) + 6\sqrt{\frac{2}{5}} + \sqrt{\frac{3}{7}} - \sqrt{\frac{3}{5}} - \sqrt{\frac{2}{3}} - 4\sqrt{\frac{6}{5}} + 5\sqrt{2} - \frac{5}{\sqrt{3}} - \frac{16}{\sqrt{5}} + \frac{4}{\sqrt{7}} + \frac{12}{\sqrt{35}})n - \frac{24}{\sqrt{35}} - \frac{8}{\sqrt{7}} + \frac{16}{\sqrt{5}} + \frac{8}{\sqrt{3}} - \sqrt{2} + 4\sqrt{\frac{6}{5}} + 2\sqrt{\frac{6}{7}} - 2\sqrt{\frac{2}{3}} - 2\sqrt{\frac{3}{7}} - 12\sqrt{\frac{2}{5}} + \frac{13}{14}$ □

In this theorem, we find the first Zagreb index for hex-derived cage network $\mathcal{G}_2$.

Theorem 5. *For Hex-Derived Cage Network ($\mathcal{G}_2$), the first Zagreb index is equal to*

$$M_1(\mathcal{G}_2) = 12(54n^3 - 109n^2 + 34n + 21)$$

Proof. Let $\mathcal{G}_2$ be the Hex-Derived Cage Network ($\mathcal{G}_2$). Using the edge partition from Table 3, we have

$$M_1(\mathcal{G}_2) = \sum_{ab \in E(\mathcal{G})} (d(a) + d(b)) = \sum_{j=1}^{20} \sum_{ab \in E_j(\mathcal{G})} (d(a) + d(b))$$

$M_1(\mathcal{G}_2) = 10|E_1(\mathcal{G}_2)| + 11|E_2(\mathcal{G}_2)| + 12|E_3(\mathcal{G}_2)| + 13|E_4(\mathcal{G}_2)| + 14|E_5(\mathcal{G}_2)| + 17|E_6(\mathcal{G}_2)| + 12|E_7(\mathcal{G}_2)| + 13|E_8(\mathcal{G}_2)| + 14|E_9(\mathcal{G}_2)| + 15|E_{10}(\mathcal{G}_2)| + 18|E_{11}(\mathcal{G}_2)| + 14|E_{12}(\mathcal{G}_2)| + 16|E_{13}(\mathcal{G}_2)| + 19|E_{14}(\mathcal{G}_2)| + 16|E_{15}(\mathcal{G}_2)| + 17|E_{16}(\mathcal{G}_2)| + 20|E_{17}(\mathcal{G}_2)| + 18|E_{18}(\mathcal{G}_2)| + 21|E_{19}(\mathcal{G}_2)| + 24|E_{20}(\mathcal{G}_2)|$.

After simplification, we get

$$M_1(\mathcal{G}_2) = 12(54n^3 - 109n^2 + 34n + 21)$$

□

In below theorem, we calculate the ABC, GA, ABC_4 and GA_5 indices of Hex-Derived Cage Network $\mathcal{G}_2$.

Theorem 6. *Let $\mathcal{G}_2$ be the Hex-Derived Cage Network for every positive integer $m = n \geq 5$; then we have*

- $ABC(\mathcal{G}_2) = 2\sqrt{2}(3n^3 - 10n^2 + 8n + 2) + \frac{1}{2}\sqrt{\frac{11}{2}}n(3n^2 - 11n + 10) + \sqrt{\frac{5}{2}}n(3n^2 - 11n + 10) + 6\sqrt{\frac{6}{5}}(n^2 - 2n + 2) + \frac{16}{3}(n^2 - 5n + 6) + 3n(n-1) + \frac{12\sqrt{2}n}{5} + 6n + \sqrt{\frac{26}{3}}(n-2)^2 + 2\sqrt{\frac{19}{3}}(n-2)^2 + 8\sqrt{\frac{3}{5}}(n-2)^2 + \sqrt{30}(n-2) + \sqrt{\frac{51}{7}}(n-2) + 6\sqrt{\frac{22}{5}}(n-2) + 6\sqrt{3}(n-2) + 4\sqrt{2}(n-2) + 12\sqrt{\frac{2}{7}}(n-2) + 3\sqrt{\frac{7}{2}}(n-3) + \frac{12}{7}\sqrt{3}(n-3) + 2\sqrt{\frac{66}{7}}$

- $GA(\mathcal{G}_2) = 4\sqrt{2}(3n^3 - 10n^2 + 8n + 2) + \frac{24}{11}\sqrt{30}(n^2 - 2n + 2) + 18n^3 - 54n^2 + \frac{24}{17}\sqrt{15}(n-1)n + \frac{48\sqrt{3}n}{7} + 12n + \frac{12}{5}\sqrt{6}(n-2)^2 + \frac{36}{7}\sqrt{5}(n-2)^2 + \frac{48}{7\sqrt{3}}(n-2)^2 + 2\sqrt{35}(n-2) + \frac{24}{19}\sqrt{21}(n-2) + \frac{96}{13}\sqrt{10}(n-2) + \frac{9}{2}\sqrt{7}(n-2) + \frac{48}{5}\sqrt{6}(n-2) + \frac{144}{17}\sqrt{2}(n-2) + 12(n-3) + \frac{24\sqrt{42}}{13} + 54.$

- $ABC_4(\mathcal{G}_2) = \frac{1}{18}\sqrt{\frac{107}{2}}n(3n^2 - 17n + 24) + \frac{1}{9}\sqrt{\frac{53}{2}}(3n^3 - 13n^2 + 16n - 10) + \frac{4}{9}\sqrt{5}n(3n^2 - 15n + 19) + \frac{2}{7}\sqrt{\frac{534}{7}}(n^2 - 5n + 6) + 3\sqrt{\frac{102}{101}}(n^2 - 5n + 6) + 2\sqrt{\frac{69}{101}}(n^2 - 5n + 6) + \frac{24}{7}\sqrt{\frac{37}{101}}(n^2 - 5n + 6) + 3\sqrt{\frac{94}{707}}(n^2 - 5n + 6) + \frac{60}{101}\sqrt{2}(n^2 - 6n + 8) + \frac{15}{19}\sqrt{6}(n^2 - 9n + 20) + 7\sqrt{\frac{2}{95}}(3n - 8) + \frac{1}{7}\sqrt{\frac{202}{3}}(n-2)^2 + \frac{4}{7}\sqrt{\frac{258}{13}}(n-2) + \frac{4}{13}\sqrt{19}(n-2) + \frac{5}{3}\sqrt{2}(n-2) + \frac{12}{7}\sqrt{\frac{142}{95}}(n-2) + \sqrt{\frac{67}{95}}(n-2) + 12\sqrt{\frac{194}{9595}}(n-2) + 24\sqrt{\frac{11}{1235}}(n-2) + \sqrt{\frac{129}{5}}(n-3) + 2\sqrt{\frac{302}{33}}(n-3) + \frac{4}{3}\sqrt{\frac{205}{33}}(n-3) + \sqrt{\frac{274}{55}}(n-3) + 2\sqrt{\frac{145}{33}}(n-3) + \frac{3}{7}\sqrt{\frac{123}{19}}(n-4)^2 + 2\sqrt{\frac{174}{133}}(n-4)^2 + 30\sqrt{\frac{7}{1919}}(n-4)^2 + \frac{4}{35}\sqrt{366}(n-4) + \frac{2}{5}\sqrt{\frac{447}{19}}(n-4) + \frac{6}{5}\sqrt{\frac{206}{13}}(n-4) + \frac{2}{3}\sqrt{14}(n-4) + 4\sqrt{\frac{46}{35}}(n-4) + 2\sqrt{\frac{42}{37}}(n-4) + \frac{12}{5}\sqrt{\frac{46}{65}}(n-4) + \frac{24}{5}\sqrt{\frac{58}{101}}(n-4) + 3\sqrt{\frac{37}{65}}(n-4) + 6\sqrt{\frac{2}{13}}(n-4) + 6\sqrt{\frac{38}{259}}(n-4) + 6\sqrt{\frac{2}{19}}(n-4) + 6\sqrt{\frac{334}{3515}}(n-4) + 6\sqrt{\frac{346}{3737}}(n-4) + \frac{66}{7}\sqrt{\frac{2}{37}}(n-4) + 72\sqrt{\frac{2}{715}}(n-4) + \frac{56}{33}(n-4) + \frac{6}{37}\sqrt{146}(n-5) + \frac{8}{25}\sqrt{37}(n-5) + \frac{96}{65}\sqrt{2}(n-5) + \frac{32}{\sqrt{37}} + 2\sqrt{\frac{249}{37}} + \frac{8\sqrt{6}}{5} + 2\sqrt{\frac{202}{35}} + 2\sqrt{\frac{109}{21}} + \frac{8\sqrt{\frac{106}{35}}}{3} + \sqrt{\frac{66}{23}} + \frac{183\sqrt{2}}{37} + 4\sqrt{\frac{678}{511}} + 2\sqrt{\frac{70}{53}} + \frac{48\sqrt{\frac{30}{73}}}{7} + 8\sqrt{\frac{14}{37}} + 8\sqrt{\frac{330}{949}} + 8\sqrt{\frac{134}{511}} + 8\sqrt{\frac{6}{23}} + 12\sqrt{\frac{127}{851}} + 6\sqrt{\frac{3}{23}} + \frac{32\sqrt{\frac{10}{77}}}{3} + 12\sqrt{\frac{290}{2701}} + 12\sqrt{\frac{17}{161}} + 12\sqrt{\frac{21}{253}} + 24\sqrt{\frac{30}{689}} + 24\sqrt{\frac{2}{65}} + 12\sqrt{\frac{146}{5035}} + 24\sqrt{\frac{166}{6935}} + 16\sqrt{\frac{6}{265}} + 48\sqrt{\frac{31}{3869}} + 48\sqrt{\frac{43}{7373}}.$

- $GA_5(\mathcal{G}_2) = 3n(3n^2 - 17n + 24) + 4\sqrt{2}n(3n^2 - 15n + 19) + \frac{12}{143}\sqrt{4242}(n^2 - 5n + 6) + \frac{28}{25}\sqrt{101}(n^2 - 5n + 6) + \frac{24}{13}\sqrt{42}(n^2 - 5n + 6) + \frac{2424}{209}(n^2 - 5n + 6) + \frac{36}{149}\sqrt{570}(3n - 8) + 9n^3 - 21n^2 - 90n + \frac{252}{103}\sqrt{6}(n-2)^2 + \frac{6}{49}\sqrt{9595}(n-2) + \frac{12}{67}\sqrt{3705}(n-2) + \frac{72}{203}\sqrt{285}(n-2) + \frac{7}{6}\sqrt{95}(n-2) + \frac{21}{11}\sqrt{39}(n-2) + \frac{144}{17}\sqrt{2}(n-2) + \frac{144}{139}\sqrt{110}(n-3) + \frac{72}{17}\sqrt{66}(n-3) + \frac{9120\sqrt{33}(n-3)}{1127} + \frac{48}{11}\sqrt{30}(n-3) + \frac{16}{59}\sqrt{1919}(n-4)^2 + \frac{24}{59}\sqrt{798}(n-4)^2 + \frac{168}{125}\sqrt{19}(n-4)^2 + \frac{24}{175}\sqrt{7474}(n-4) + \frac{24}{169}\sqrt{7030}(n-4) + \frac{24}{113}\sqrt{2886}(n-4) + \frac{8}{25}\sqrt{1406}(n-4) + \frac{12}{29}\sqrt{777}(n-4) + \frac{36}{41}\sqrt{715}(n-4) + \frac{15}{11}\sqrt{303}(n-4) + \frac{1350}{791}\sqrt{195}(n-4) + \frac{9}{8}\sqrt{111}(n-4) + \frac{56}{41}\sqrt{74}(n-4) + \frac{240}{151}\sqrt{57}(n-4) + \frac{1496}{217}\sqrt{26}(n-4) + \frac{80}{13}\sqrt{14}(n-4) + \frac{210}{31}\sqrt{3}(n-4) + 12(n-4) + 36(n-5) + \frac{8\sqrt{7373}}{29} + \frac{2\sqrt{6935}}{7} + \frac{16\sqrt{5402}}{49} + \frac{6\sqrt{5035}}{37} + \frac{8\sqrt{3869}}{21} + \frac{48\sqrt{3066}}{115} + \frac{3\sqrt{2847}}{7} + \frac{12\sqrt{2067}}{23} + \frac{32\sqrt{851}}{43} + \frac{18\sqrt{511}}{17} + \frac{9\sqrt{455}}{8} + \frac{72\sqrt{318}}{107} + \frac{36\sqrt{265}}{49} + \frac{36\sqrt{259}}{25} + \frac{288\sqrt{253}}{191} + \frac{60\sqrt{219}}{37} + \frac{72\sqrt{185}}{41} + \frac{288\sqrt{161}}{155} + \frac{144\sqrt{138}}{73} + \frac{144\sqrt{115}}{137} + \frac{192\sqrt{111}}{85} + \frac{8\sqrt{77}}{3} + \frac{168\sqrt{73}}{61} + \frac{288\sqrt{70}}{103} + \frac{732\sqrt{69}}{175} + 4\sqrt{35} + \frac{192\sqrt{21}}{37} + 258.$

Proof. Using the edge partition from Table 3, we find ABC as

$$ABC(\mathcal{G}_2) = \sum_{ab \in E(\mathcal{G})} \sqrt{\frac{d(a) + d(b) - 2}{d(a) \cdot d(b)}} = \sum_{j=1}^{20} \sum_{ab \in E_j(\mathcal{G})} \sqrt{\frac{d(a) + d(b) - 2}{d(a) \cdot d(b)}}$$

$ABC(\mathcal{G}_2) = \frac{2\sqrt{2}}{5}|E_1(\mathcal{G}_2)| + \sqrt{\frac{3}{10}}|E_2(\mathcal{G}_2)| + \sqrt{\frac{2}{7}}|E_3(\mathcal{G}_2)| + \frac{\sqrt{11}}{2\sqrt{10}}|E_4(\mathcal{G}_2)| + \frac{2\sqrt{3}}{3\sqrt{5}}|E_5(\mathcal{G}_2)| + \frac{1}{2}|E_6(\mathcal{G}_2)| + \sqrt{\frac{5}{18}}|E_7(\mathcal{G}_2)| + \sqrt{\frac{11}{42}}|E_8(\mathcal{G}_2)| + \frac{1}{2}|E_9(\mathcal{G}_2)| + \sqrt{\frac{13}{54}}|E_{10}(\mathcal{G}_2)| + \frac{\sqrt{2}}{3}|E_{11}(\mathcal{G}_2)| + \frac{2\sqrt{3}}{7}|E_{12}(\mathcal{G}_2)| +$

$\sqrt{\frac{14}{63}}|E_{13}(\mathcal{G}_2)| + \sqrt{\frac{17}{84}}|E_{14}(\mathcal{G}_2)| + \sqrt{\frac{7}{32}}|E_{15}(\mathcal{G}_2)| + \sqrt{\frac{5}{24}}|E_{16}(\mathcal{G}_2)| + \sqrt{\frac{3}{16}}|E_{17}(\mathcal{G}_2)| + \sqrt{\frac{16}{81}}|E_{18}(\mathcal{G}_2)| + \sqrt{\frac{19}{108}}|E_{19}(\mathcal{G}_2)| + \sqrt{\frac{11}{72}}|E_{20}(\mathcal{G}_2)|$.

After simplification, we get

$ABC(\mathcal{G}_2)=2\sqrt{2}(3n^3 - 10n^2 + 8n + 2) + \frac{1}{2}\sqrt{\frac{11}{2}}n(3n^2 - 11n + 10) + \sqrt{\frac{5}{2}}n(3n^2 - 11n + 10) + 6\sqrt{\frac{6}{5}}(n^2 - 2n + 2) + \frac{16}{3}(n^2 - 5n + 6) + 3n(n-1) + \frac{12\sqrt{2}n}{5} + 6n + \sqrt{\frac{26}{3}}(n-2)^2 + 2\sqrt{\frac{19}{3}}(n-2)^2 + 8\sqrt{\frac{3}{5}}(n-2)^2 + \sqrt{30}(n-2) + \sqrt{\frac{51}{7}}(n-2) + 6\sqrt{\frac{22}{5}}(n-2) + 6\sqrt{3}(n-2) + 4\sqrt{2}(n-2) + 12\sqrt{\frac{2}{7}}(n-2) + 3\sqrt{\frac{7}{2}}(n-3) + \frac{12}{7}\sqrt{3}(n-3) + 2\sqrt{\frac{66}{7}}$

Using the edge partition from Table 3, we find GA as

$GA(\mathcal{G}_2) = 1|E_1(\mathcal{G}_2)| + \frac{2\sqrt{30}}{11}|E_2(\mathcal{G}_2)| + \frac{\sqrt{35}}{6}|E_3(\mathcal{G}_2)| + \frac{4\sqrt{10}}{13}|E_4(\mathcal{G}_2)| + \frac{3\sqrt{5}}{7}|E_5(\mathcal{G}_2)| + \frac{4\sqrt{15}}{17}|E_6(\mathcal{G}_2)| + 1|E_7(\mathcal{G}_2)| + \frac{2\sqrt{42}}{13}|E_8(\mathcal{G}_2)| + \frac{2\sqrt{12}}{7}|E_9(\mathcal{G}_2)| + \frac{2\sqrt{54}}{15}|E_{10}(\mathcal{G}_2)| + \frac{2\sqrt{2}}{3}|E_{11}(\mathcal{G}_2)| + 1|E_{12}(\mathcal{G}_2)| + \frac{3\sqrt{7}}{8}|E_{13}(\mathcal{G}_2)| + \frac{4\sqrt{21}}{19}|E_{14}(\mathcal{G}_2)| + 1|E_{15}(\mathcal{G}_2)| + \frac{12\sqrt{2}}{17}|E_{16}(\mathcal{G}_2)| + \frac{\sqrt{23}}{5}|E_{17}(\mathcal{G}_2)| + 1|E_{18}(\mathcal{G}_2)| + \frac{2\sqrt{3}}{7}|E_{19}(\mathcal{G}_2)| + 1|E_{20}(\mathcal{G}_2)|$.

After simplification, we get

$GA(\mathcal{G}_2) = 4\sqrt{2}(3n^3 - 10n^2 + 8n + 2) + \frac{24}{11}\sqrt{30}(n^2 - 2n + 2) + 18n^3 - 54n^2 + \frac{24}{17}\sqrt{15}(n-1)n + \frac{48\sqrt{3}n}{7} + 12n + \frac{12}{5}\sqrt{6}(n-2)^2 + \frac{36}{7}\sqrt{5}(n-2)^2 + \frac{48}{7\sqrt{3}}(n-2)^2 + 2\sqrt{35}(n-2) + \frac{24}{19}\sqrt{21}(n-2) + \frac{96}{13}\sqrt{10}(n-2) + \frac{9}{2}\sqrt{7}(n-2) + \frac{48}{5}\sqrt{6}(n-2) + \frac{144}{17}\sqrt{2}(n-2) + 12(n-3) + \frac{24\sqrt{42}}{13} + 54.$

If we suppose an edge partition based on degree sum of neighbors of end vertices, then the edge set $E(\mathcal{G}_2)$ can be divided into seventy six edge partition $E_j(\mathcal{G}_2)$, $21 \le j \le 96$. Table 4 shows these edge partitions.

From Equation (7), we get

$$ABC_4(\mathcal{G}_2) = \sum_{ab \in E(\mathcal{G})} \sqrt{\frac{S(a) + S(b) - 2}{S(a)S(b)}} = \sum_{j=21}^{96} \sum_{ab \in E_j(\mathcal{G})} \sqrt{\frac{S(a) + S(b) - 2}{S(a)S(b)}}.$$

Using the edge partition from Table 3, we get

$ABC_4(\mathcal{G}_2) = \sqrt{\frac{72}{1369}}|E_{21}(\mathcal{G}_2)| + \frac{4}{\sqrt{333}}|E_{22}(\mathcal{G}_2)| + \sqrt{\frac{83}{1776}}|E_{23}(\mathcal{G}_2)| + \sqrt{\frac{98}{2321}}|E_{24}(\mathcal{G}_2)| + \sqrt{\frac{127}{3404}}|E_{25}(\mathcal{G}_2)| + \sqrt{\frac{74}{1443}}|E_{26}(\mathcal{G}_2)| + \sqrt{\frac{86}{1911}}|E_{27}(\mathcal{G}_2)| + \sqrt{\frac{90}{2067}}|E_{28}(\mathcal{G}_2)| + \sqrt{\frac{42}{2106}}|E_{29}(\mathcal{G}_2)| + \sqrt{\frac{110}{2847}}|E_{30}(\mathcal{G}_2)| + \sqrt{\frac{111}{2886}}|E_{31}(\mathcal{G}_2)| + \sqrt{\frac{44}{1235}}|E_{32}(\mathcal{G}_2)| + \sqrt{\frac{43}{940}}|E_{33}(\mathcal{G}_2)| + \sqrt{\frac{101}{2520}}|E_{34}(\mathcal{G}_2)| + \sqrt{\frac{103}{2600}}|E_{35}(\mathcal{G}_2)| + \sqrt{\frac{137}{3960}}|E_{36}(\mathcal{G}_2)| + \sqrt{\frac{89}{2058}}|E_{37}(\mathcal{G}_2)| + \sqrt{\frac{113}{3066}}|E_{38}(\mathcal{G}_2)| + \sqrt{\frac{17}{1554}}|E_{39}(\mathcal{G}_2)| + \sqrt{\frac{115}{3150}}|E_{40}(\mathcal{G}_2)| + \sqrt{\frac{29}{798}}|E_{41}(\mathcal{G}_2)| + \sqrt{\frac{141}{4242}}|E_{42}(\mathcal{G}_2)| + \sqrt{\frac{96}{2385}}|E_{43}(\mathcal{G}_2)| + \sqrt{\frac{106}{2835}}|E_{44}(\mathcal{G}_2)| + \sqrt{\frac{27}{558}}|E_{45}(\mathcal{G}_2)| + \frac{5}{\sqrt{648}}|E_{46}(\mathcal{G}_2)| + \sqrt{\frac{109}{3024}}|E_{47}(\mathcal{G}_2)| + \sqrt{\frac{111}{3120}}|E_{48}(\mathcal{G}_2)| + \sqrt{\frac{69}{228}}|E_{49}(\mathcal{G}_2)| + \sqrt{\frac{145}{4752}}|E_{50}(\mathcal{G}_2)| + \sqrt{\frac{101}{2646}}|E_{51}(\mathcal{G}_2)| + \sqrt{\frac{120}{3577}}|E_{52}(\mathcal{G}_2)| + \frac{11}{\sqrt{3626}}|E_{53}(\mathcal{G}_2)| + \sqrt{\frac{122}{3675}}|E_{54}(\mathcal{G}_2)| + \sqrt{\frac{123}{3724}}|E_{55}(\mathcal{G}_2)| + \sqrt{\frac{142}{4655}}|E_{56}(\mathcal{G}_2)| + \sqrt{\frac{148}{4949}}|E_{57}(\mathcal{G}_2)| + \sqrt{\frac{105}{2862}}|E_{58}(\mathcal{G}_2)| + \sqrt{\frac{124}{3869}}|E_{59}(\mathcal{G}_2)| + \sqrt{\frac{146}{5035}}|E_{60}(\mathcal{G}_2)| + \sqrt{\frac{53}{1458}}|E_{61}(\mathcal{G}_2)| + \sqrt{\frac{63}{1998}}|E_{62}(\mathcal{G}_2)| + \sqrt{\frac{105}{621}}|E_{63}(\mathcal{G}_2)| + \sqrt{\frac{147}{5130}}|E_{64}(\mathcal{G}_2)| + \sqrt{\frac{151}{5346}}|E_{65}(\mathcal{G}_2)| + \sqrt{\frac{153}{5454}}|E_{66}(\mathcal{G}_2)| + \sqrt{\frac{20}{729}}|E_{67}(\mathcal{G}_2)| + \sqrt{\frac{126}{4095}}|E_{68}(\mathcal{G}_2)| + \sqrt{\frac{134}{4599}}|E_{69}(\mathcal{G}_2)| + \sqrt{\frac{153}{5796}}|E_{70}(\mathcal{G}_2)| + \sqrt{\frac{160}{6237}}|E_{71}(\mathcal{G}_2)| + \sqrt{\frac{128}{4225}}|E_{72}(\mathcal{G}_2)| + \sqrt{\frac{46}{1625}}|E_{73}(\mathcal{G}_2)| + \sqrt{\frac{54}{2145}}|E_{74}(\mathcal{G}_2)| + \sqrt{\frac{145}{5402}}|E_{75}(\mathcal{G}_2)| + \sqrt{\frac{116}{5475}}|E_{76}(\mathcal{G}_2)| + \sqrt{\frac{166}{6935}}|E_{77}(\mathcal{G}_2)| + \sqrt{\frac{172}{7373}}|E_{78}(\mathcal{G}_2)| + \sqrt{\frac{73}{2738}}|E_{79}(\mathcal{G}_2)| + \sqrt{\frac{37}{1406}}|E_{80}(\mathcal{G}_2)| + \sqrt{\frac{167}{7030}}|E_{81}(\mathcal{G}_2)| + \sqrt{\frac{173}{7474}}|E_{82}(\mathcal{G}_2)| + \sqrt{\frac{148}{5625}}|E_{83}(\mathcal{G}_2)| + \sqrt{\frac{149}{5700}}|E_{84}(\mathcal{G}_2)| + \sqrt{\frac{58}{2525}}|E_{85}(\mathcal{G}_2)| + \sqrt{\frac{75}{2888}}|E_{86}(\mathcal{G}_2)| + \sqrt{\frac{175}{7676}}|E_{87}(\mathcal{G}_2)| + \sqrt{\frac{189}{9108}}|E_{88}(\mathcal{G}_2)| + \sqrt{\frac{99}{4948}}|E_{89}(\mathcal{G}_2)| + \sqrt{\frac{194}{9595}}|E_{90}(\mathcal{G}_2)| +$

$\sqrt{\frac{201}{10260}}|E_{91}(\mathcal{G}_2)| + \frac{14}{\sqrt{9801}}|E_{92}(\mathcal{G}_2)| + \sqrt{\frac{205}{10692}}|E_{93}(\mathcal{G}_2)| + \sqrt{\frac{200}{10201}}|E_{94}(\mathcal{G}_2)| + \sqrt{\frac{207}{10908}}|E_{95}(\mathcal{G}_2)| + \sqrt{\frac{3}{162}}|E_{96}(\mathcal{G}_2)|.$

After simplification, we have

$ABC_4(\mathcal{G}_2) = \frac{1}{18}\sqrt{\frac{107}{2}}n(3n^2 - 17n + 24) + \frac{1}{9}\sqrt{\frac{53}{2}}(3n^3 - 13n^2 + 16n - 10) + \frac{4}{9}\sqrt{5}n(3n^2 - 15n + 19) + \frac{2}{7}\sqrt{\frac{534}{7}}(n^2 - 5n + 6) + 3\sqrt{\frac{102}{101}}(n^2 - 5n + 6) + 2\sqrt{\frac{69}{101}}(n^2 - 5n + 6) + \frac{24}{7}\sqrt{\frac{37}{101}}(n^2 - 5n + 6) + 3\sqrt{\frac{94}{707}}(n^2 - 5n + 6) + \frac{60}{101}\sqrt{2}(n^2 - 6n + 8) + \frac{15}{19}\sqrt{6}(n^2 - 9n + 20) + 7\sqrt{\frac{2}{95}}(3n - 8) + \frac{1}{7}\sqrt{\frac{202}{3}}(n - 2)^2 + \frac{4}{7}\sqrt{\frac{258}{13}}(n - 2) + \frac{4}{13}\sqrt{19}(n - 2) + \frac{5}{3}\sqrt{2}(n - 2) + \frac{12}{7}\sqrt{\frac{142}{95}}(n - 2) + \sqrt{\frac{67}{95}}(n - 2) + 12\sqrt{\frac{194}{9595}}(n - 2) + 24\sqrt{\frac{11}{1235}}(n - 2) + \sqrt{\frac{129}{5}}(n - 3) + 2\sqrt{\frac{302}{33}}(n - 3) + \frac{4}{3}\sqrt{\frac{205}{33}}(n - 3) + \sqrt{\frac{274}{55}}(n - 3) + 2\sqrt{\frac{145}{33}}(n - 3) + \frac{3}{7}\sqrt{\frac{123}{19}}(n - 4)^2 + 2\sqrt{\frac{174}{133}}(n - 4)^2 + 30\sqrt{\frac{7}{1919}}(n - 4)^2 + \frac{4}{35}\sqrt{366}(n - 4) + \frac{2}{5}\sqrt{\frac{447}{19}}(n - 4) + \frac{6}{5}\sqrt{\frac{206}{13}}(n - 4) + \frac{2}{3}\sqrt{14}(n - 4) + 4\sqrt{\frac{46}{35}}(n - 4) + 2\sqrt{\frac{42}{37}}(n - 4) + \frac{12}{5}\sqrt{\frac{46}{65}}(n - 4) + \frac{24}{5}\sqrt{\frac{58}{101}}(n - 4) + 3\sqrt{\frac{37}{65}}(n - 4) + 6\sqrt{\frac{2}{13}}(n - 4) + 6\sqrt{\frac{38}{259}}(n - 4) + 6\sqrt{\frac{2}{19}}(n - 4) + 6\sqrt{\frac{334}{3515}}(n - 4) + 6\sqrt{\frac{346}{3737}}(n - 4) + \frac{66}{7}\sqrt{\frac{2}{37}}(n - 4) + 72\sqrt{\frac{2}{715}}(n - 4) + \frac{56}{33}(n - 4) + \frac{6}{37}\sqrt{146}(n - 5) + \frac{8}{25}\sqrt{37}(n - 5) + \frac{96}{65}\sqrt{2}(n - 5) + \frac{32}{\sqrt{37}} + 2\sqrt{\frac{249}{37}} + \frac{8\sqrt{6}}{5} + 2\sqrt{\frac{202}{35}} + 2\sqrt{\frac{109}{21}} + \frac{8\sqrt{\frac{106}{35}}}{3} + \sqrt{\frac{66}{23}} + \frac{183\sqrt{2}}{37} + 4\sqrt{\frac{678}{511}} + 2\sqrt{\frac{70}{53}} + \frac{48\sqrt{\frac{30}{73}}}{7} + 8\sqrt{\frac{14}{37}} + 8\sqrt{\frac{330}{949}} + 8\sqrt{\frac{134}{511}} + 8\sqrt{\frac{6}{23}} + 12\sqrt{\frac{127}{851}} + 6\sqrt{\frac{3}{23}} + \frac{32\sqrt{\frac{10}{77}}}{3} + 12\sqrt{\frac{290}{2701}} + 12\sqrt{\frac{17}{161}} + 12\sqrt{\frac{21}{253}} + 24\sqrt{\frac{30}{689}} + 24\sqrt{\frac{2}{65}} + 12\sqrt{\frac{146}{5035}} + 24\sqrt{\frac{166}{6935}} + 16\sqrt{\frac{6}{265}} + 48\sqrt{\frac{31}{3869}} + 48\sqrt{\frac{43}{7373}}$

From Equation (8), we get

$$GA_5(\mathcal{G}_2) = \sum_{ab \in E(\mathcal{G})} \frac{2\sqrt{S(a)S(b)}}{(S(a) + S(b))} = \sum_{j=21}^{96} \sum_{ab \in E_j(\mathcal{G})} \frac{2\sqrt{S(a)S(b)}}{(S(a)S(b))}.$$

Table 3. Edge partition of Hex-Derived Cage network ($HDCN2$) based on degrees of end vertices of each edge.

(d_u, d_v) where $ab \in E(\mathcal{G}_2)$	Number of Edges	(d_u, d_v) where $ab \in E(\mathcal{G}_2)$	Number of Edges
$E_1 = (5,5)$	$6n$	$E_{11} = (6,12)$	$18n^3 - 60n^2 + 48n + 12$
$E_2 = (5,6)$	$12n^2 - 24n + 24$	$E_{12} = (7,7)$	$6n - 18$
$E_3 = (5,7)$	$2(6n - 12)$	$E_{13} = (7,9)$	$2(6n - 12)$
$E_4 = (5,8)$	$4(6n - 12)$	$E_{14} = (7,12)$	$6n - 12$
$E_5 = (5,9)$	$12n^2 - 48n + 48$	$E_{15} = (8,8)$	$2(6n - 18)$
$E_6 = (5,12)$	$6n^2 - 6n$	$E_{16} = (8,9)$	$2(6n - 12)$
$E_7 = (6,6)$	$9n^3 - 33n^2 + 30n$	$E_{17} = (8,12)$	$4(6n - 12)$
$E_8 = (6,7)$	12	$E_{18} = (9,9)$	$12n^2 - 60n + 72$
$E_9 = (6,8)$	$12n$	$E_{19} = (9,12)$	$12n^2 - 48n + 48$
$E_{10} = (6,9)$	$6n^2 - 24n + 24$	$E_{20} = (12,12)$	$9n^3 - 33n^2 + 30n$

Using the edge partition from Table 4, we get

$GA_5(\mathcal{G}_2) = 1|E_{21}(\mathcal{G}_2)| + \frac{\sqrt{1665}}{41}|E_{22}(\mathcal{G}_2)| + \frac{2\sqrt{1776}}{85}|E_{23}(\mathcal{G}_2)| + \frac{\sqrt{2331}}{50}|E_{24}(\mathcal{G}_2)| + \frac{2\sqrt{3404}}{129}|E_{25}(\mathcal{G}_2)| + \frac{\sqrt{1443}}{38}|E_{26}(\mathcal{G}_2)| + \frac{\sqrt{1911}}{44}|E_{27}(\mathcal{G}_2)| + \frac{\sqrt{2067}}{46}|E_{28}(\mathcal{G}_2)| + \frac{2\sqrt{2106}}{93}|E_{29}(\mathcal{G}_2)| + \frac{\sqrt{2847}}{56}|E_{30}(\mathcal{G}_2)| + \frac{2\sqrt{2886}}{113}|E_{31}(\mathcal{G}_2)| + \frac{\sqrt{3705}}{67}|E_{32}(\mathcal{G}_2)| + \frac{\sqrt{1920}}{44}|E_{33}(\mathcal{G}_2)| + \frac{2\sqrt{2520}}{103}|E_{34}(\mathcal{G}_2)| + \frac{2\sqrt{2600}}{105}|E_{35}(\mathcal{G}_2)| + \frac{2\sqrt{3960}}{139}|E_{36}(\mathcal{G}_2)| + \frac{2\sqrt{2058}}{91}|E_{37}(\mathcal{G}_2)| + \frac{2\sqrt{3066}}{115}|E_{38}(\mathcal{G}_2)| + \frac{\sqrt{3108}}{58}|E_{39}(\mathcal{G}_2)| + \frac{2\sqrt{3150}}{117}|E_{40}(\mathcal{G}_2)| + \frac{\sqrt{3192}}{59}|E_{41}(\mathcal{G}_2)| + \frac{2\sqrt{4242}}{143}|E_{42}(\mathcal{G}_2)| + \frac{\sqrt{2385}}{49}|E_{43}(\mathcal{G}_2)| + \frac{\sqrt{2835}}{54}|E_{44}(\mathcal{G}_2)| + \frac{2\sqrt{4140}}{137}|E_{45}(\mathcal{G}_2)| + \frac{\sqrt{2592}}{51}|E_{46}(\mathcal{G}_2)| + \frac{2\sqrt{3024}}{111}|E_{47}(\mathcal{G}_2)| + \frac{2\sqrt{3120}}{113}|E_{48}(\mathcal{G}_2)| + \frac{\sqrt{4416}}{70}|E_{49}(\mathcal{G}_2)| + \frac{2\sqrt{4752}}{147}|E_{50}(\mathcal{G}_2)| + \frac{2\sqrt{2646}}{103}|E_{51}(\mathcal{G}_2)| + \frac{\sqrt{3577}}{61}|E_{52}(\mathcal{G}_2)| + \frac{2\sqrt{3626}}{123}|E_{53}(\mathcal{G}_2)| + \frac{\sqrt{3675}}{62}|E_{54}(\mathcal{G}_2)| + \frac{2\sqrt{3724}}{125}|E_{55}(\mathcal{G}_2)| + \frac{\sqrt{4655}}{72}|E_{56}(\mathcal{G}_2)| + \frac{\sqrt{4949}}{75}|E_{57}(\mathcal{G}_2)| + \frac{2\sqrt{2862}}{107}|E_{58}(\mathcal{G}_2)| + \frac{\sqrt{3869}}{63}|E_{59}(\mathcal{G}_2)| + \frac{\sqrt{5035}}{74}|E_{60}(\mathcal{G}_2)| + 1|E_{61}(\mathcal{G}_2)| +$

$\frac{\sqrt{3996}}{64}|E_{62}(\mathcal{G}_2)| + \frac{\sqrt{4968}}{73}|E_{63}(\mathcal{G}_2)| + \frac{2\sqrt{5130}}{149}|E_{64}(\mathcal{G}_2)| + \frac{2\sqrt{5346}}{153}|E_{65}(\mathcal{G}_2)| + \frac{2\sqrt{5454}}{155}|E_{66}(\mathcal{G}_2)| + \frac{\sqrt{5832}}{81}|E_{67}(\mathcal{G}_2)| + \frac{\sqrt{4095}}{64}|E_{68}(\mathcal{G}_2)| + \frac{\sqrt{4599}}{68}|E_{69}(\mathcal{G}_2)| + \frac{2\sqrt{5796}}{155}|E_{70}(\mathcal{G}_2)| + \frac{\sqrt{6237}}{81}|E_{71}(\mathcal{G}_2)| + 1|E_{72}(\mathcal{G}_2)| + \frac{\sqrt{4875}}{70}|E_{73}(\mathcal{G}_2)| + \frac{\sqrt{6435}}{82}|E_{74}(\mathcal{G}_2)| + \frac{2\sqrt{5402}}{147}|E_{75}(\mathcal{G}_2)| + \frac{\sqrt{5475}}{74}|E_{76}(\mathcal{G}_2)| + \frac{\sqrt{6935}}{84}|E_{77}(\mathcal{G}_2)| + \frac{\sqrt{7373}}{87}|E_{78}(\mathcal{G}_2)| + 1|E_{79}(\mathcal{G}_2)| + \frac{\sqrt{5624}}{75}|E_{80}(\mathcal{G}_2)| + \frac{2\sqrt{7030}}{169}|E_{81}(\mathcal{G}_2)| + \frac{2\sqrt{7474}}{175}|E_{82}(\mathcal{G}_2)| + 1|E_{83}(\mathcal{G}_2)| + \frac{2\sqrt{5700}}{151}|E_{84}(\mathcal{G}_2)| + \frac{\sqrt{7575}}{88}|E_{85}(\mathcal{G}_2)| + 1|E_{86}(\mathcal{G}_2)| + \frac{2\sqrt{7676}}{177}|E_{87}(\mathcal{G}_2)| + \frac{2\sqrt{9108}}{191}|E_{88}(\mathcal{G}_2)| + \frac{\sqrt{9936}}{100}|E_{89}(\mathcal{G}_2)| + \frac{\sqrt{9595}}{98}|E_{90}(\mathcal{G}_2)| + \frac{2\sqrt{10260}}{203}|E_{91}(\mathcal{G}_2)| + 1|E_{92}(\mathcal{G}_2)| + \frac{2\sqrt{10692}}{207}|E_{93}(\mathcal{G}_2)| + 1|E_{94}(\mathcal{G}_2)| + \frac{2\sqrt{10908}}{209}|E_{95}(\mathcal{G}_2)| + 1|E_{96}(\mathcal{G}_2)|$.

After simplification, we get

$GA_5(\mathcal{G}_2) = 3n(3n^2-17n+24) + 4\sqrt{2}n(3n^2-15n+19) + \frac{12}{143}\sqrt{4242}(n^2-5n+6) + \frac{28}{25}\sqrt{101}(n^2-5n+6) + \frac{24}{13}\sqrt{42}(n^2-5n+6) + \frac{2424}{209}(n^2-5n+6) + \frac{36}{149}\sqrt{570}(3n-8) + 9n^3 - 21n^2 - 90n + \frac{252}{103}\sqrt{6}(n-2)^2 + \frac{6}{49}\sqrt{9595}(n-2) + \frac{12}{67}\sqrt{3705}(n-2) + \frac{72}{203}\sqrt{285}(n-2) + \frac{7}{6}\sqrt{95}(n-2) + \frac{21}{11}\sqrt{39}(n-2) + \frac{144}{17}\sqrt{2}(n-2) + \frac{144}{139}\sqrt{110}(n-3) + \frac{72}{17}\sqrt{66}(n-3) + \frac{9120\sqrt{33}(n-3)}{1127} + \frac{48}{11}\sqrt{30}(n-3) + \frac{16}{59}\sqrt{1919}(n-4)^2 + \frac{24}{59}\sqrt{798}(n-4)^2 + \frac{168}{125}\sqrt{19}(n-4)^2 + \frac{24}{175}\sqrt{7474}(n-4) + \frac{24}{169}\sqrt{7030}(n-4) + \frac{24}{113}\sqrt{2886}(n-4) + \frac{8}{25}\sqrt{1406}(n-4) + \frac{12}{29}\sqrt{777}(n-4) + \frac{36}{41}\sqrt{715}(n-4) + \frac{15}{11}\sqrt{303}(n-4) + \frac{1350}{791}\sqrt{195}(n-4) + \frac{9}{8}\sqrt{111}(n-4) + \frac{56}{41}\sqrt{74}(n-4) + \frac{240}{151}\sqrt{57}(n-4) + \frac{1496}{217}\sqrt{26}(n-4) + \frac{80}{13}\sqrt{14}(n-4) + \frac{210}{31}\sqrt{3}(n-4) + 12(n-4) + 36(n-5) + \frac{8\sqrt{7373}}{29} + \frac{2\sqrt{6935}}{7} + \frac{16\sqrt{5402}}{49} + \frac{6\sqrt{5035}}{37} + \frac{8\sqrt{3869}}{21} + \frac{48\sqrt{3066}}{115} + \frac{3\sqrt{2847}}{7} + \frac{12\sqrt{2067}}{23} + \frac{32\sqrt{851}}{43} + \frac{18\sqrt{511}}{17} + \frac{9\sqrt{455}}{8} + \frac{72\sqrt{318}}{107} + \frac{36\sqrt{265}}{49} + \frac{36\sqrt{259}}{25} + \frac{288\sqrt{253}}{191} + \frac{60\sqrt{219}}{37} + \frac{72\sqrt{185}}{41} + \frac{288\sqrt{161}}{155} + \frac{144\sqrt{138}}{73} + \frac{144\sqrt{115}}{137} + \frac{192\sqrt{111}}{85} + \frac{8\sqrt{77}}{3} + \frac{168\sqrt{73}}{61} + \frac{288\sqrt{70}}{103} + \frac{732\sqrt{69}}{175} + 4\sqrt{35} + \frac{192\sqrt{21}}{37} + 258.$ □

The Comparison graphs for ABC, GA, ABC_4 and GA_5 in case of a Hex Derived Cage networks $HDCN1(m,n)$ and $HDCN2(m,n)$ of dimension m and n are shown in Figures 3 and 4 respectively.

Table 4. Edge partition of Hex-Derived Cage network ($HDCN2$) based on sum of degrees of end vertices of each edge.

(S_u, S_v) where $ab \in E(\mathcal{G}_2)$	Number of Edges	(S_u, S_v) where $ab \in E(\mathcal{G}_2)$	Number of Edges
$E_{21} = (37,37)$	12	$E_{59} = (53,73)$	24
$E_{22} = (37,45)$	24	$E_{60} = (53,95)$	12
$E_{23} = (37,48)$	24	$E_{61} = (54,54)$	$9n^3 - 39n^2 + 48n - 30$
$E_{24} = (37,63)$	24	$E_{62} = (54,74)$	$2(6n-24)$
$E_{25} = (37,92)$	24	$E_{63} = (54,92)$	24
$E_{26} = (37,39)$	$6n-12$	$E_{64} = (54,95)$	$3(6n-16)$
$E_{27} = (39,49)$	$2(6n-12)$	$E_{65} = (54,99)$	$6(6n-18)$
$E_{28} = (39,53)$	24	$E_{66} = (54,101)$	$18n^2 - 90n + 108$
$E_{29} = (39,54)$	$2(6n-24)$	$E_{67} = (54,108)$	$18n^3 - 90n^2 + 114n$
$E_{30} = (39,73)$	24	$E_{68} = (63,65)$	24
$E_{31} = (39,74)$	$2(6n-24)$	$E_{69} = (63,73)$	24
$E_{32} = (39,95)$	$2(6n-12)$	$E_{70} = (63,92)$	24
$E_{33} = (40,48)$	$4(6n-18)$	$E_{71} = (63,99)$	24
$E_{34} = (40,63)$	24	$E_{72} = (65,65)$	$2(6n-30)$
$E_{35} = (40,65)$	$4(6n-24)$	$E_{73} = (65,75)$	$2(6n-24)$
$E_{36} = (40,99)$	$2(6n-18)$	$E_{74} = (65,99)$	$4(6n-24)$
$E_{37} = (42,49)$	$12n^2 - 60n + 72$	$E_{75} = (73,74)$	24
$E_{38} = (42,73)$	24	$E_{76} = (73,75)$	24
$E_{39} = (42,74)$	$2(6n-24)$	$E_{77} = (73,95)$	24
$E_{40} = (42,75)$	$4(6n-24)$	$E_{78} = (73,101)$	24
$E_{41} = (42,76)$	$12n^2 - 96n + 192$	$E_{79} = (74,74)$	$2(6n-30)$
$E_{42} = (42,101)$	$6n^2 - 30n + 36$	$E_{80} = (74,76)$	$2(6n-24)$
$E_{43} = (45,53)$	12	$E_{81} = (74,95)$	$2(6n-24)$
$E_{44} = (45,63)$	24	$E_{82} = (74,101)$	$2(6n-24)$
$E_{45} = (45,92)$	12	$E_{83} = (75,75)$	$2(6n-30)$

Table 4. *Cont.*

(S_u, S_v) where $ab \in E(\mathcal{G}_2)$	Number of Edges	(S_u, S_v) where $ab \in E(\mathcal{G}_2)$	Number of Edges
$E_{46} = (48, 54)$	$2(6n - 12)$	$E_{84} = (75, 76)$	$2(6n - 24)$
$E_{47} = (48, 63)$	24	$E_{85} = (75, 101)$	$4(6n - 24)$
$E_{48} = (48, 65)$	$2(6n - 24)$	$E_{86} = (76, 76)$	$12n^2 - 108n + 240$
$E_{49} = (48, 92)$	24	$E_{87} = (76, 101)$	$12n^2 - 96n + 192$
$E_{50} = (48, 99)$	$4(6n - 18)$	$E_{88} = (92, 99)$	24
$E_{51} = (49, 54)$	$6n^2 - 24n + 24$	$E_{89} = (92, 108)$	12
$E_{52} = (49, 73)$	24	$E_{90} = (95, 101)$	$2(6n - 12)$
$E_{53} = (49, 74)$	$2(6n - 24)$	$E_{91} = (95, 108)$	$6n - 12$
$E_{54} = (49, 75)$	$2(6n - 24)$	$E_{92} = (99, 99)$	$2(6n - 24)$
$E_{55} = (49, 76)$	$6n^2 - 48n + 96$	$E_{93} = (99, 108)$	$4(6n - 18)$
$E_{56} = (49, 95)$	$2(6n - 12)$	$E_{94} = (101, 101)$	$6n^2 - 36n + 48$
$E_{57} = (49, 101)$	$12n^2 - 60n + 72$	$E_{95} = (101, 108)$	$12n^2 - 60n + 72$
$E_{58} = (53, 54)$	12	$E_{96} = (108, 108)$	$9n^3 - 51n^2 + 72n$

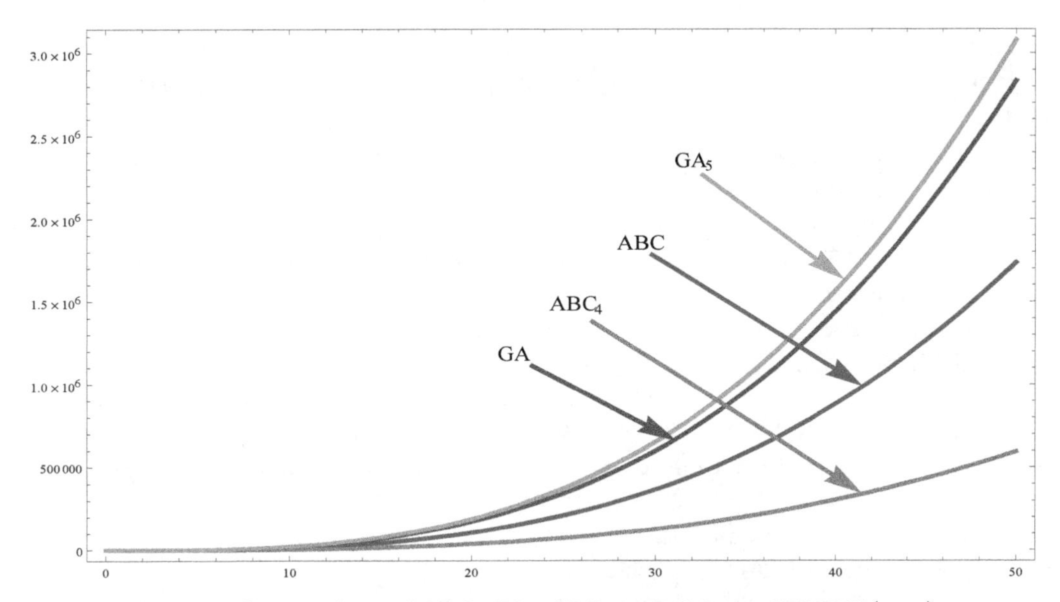

Figure 3. Comparison of ABC, GA, ABC_4 and GA_5 for $HDCN1(m, n)$.

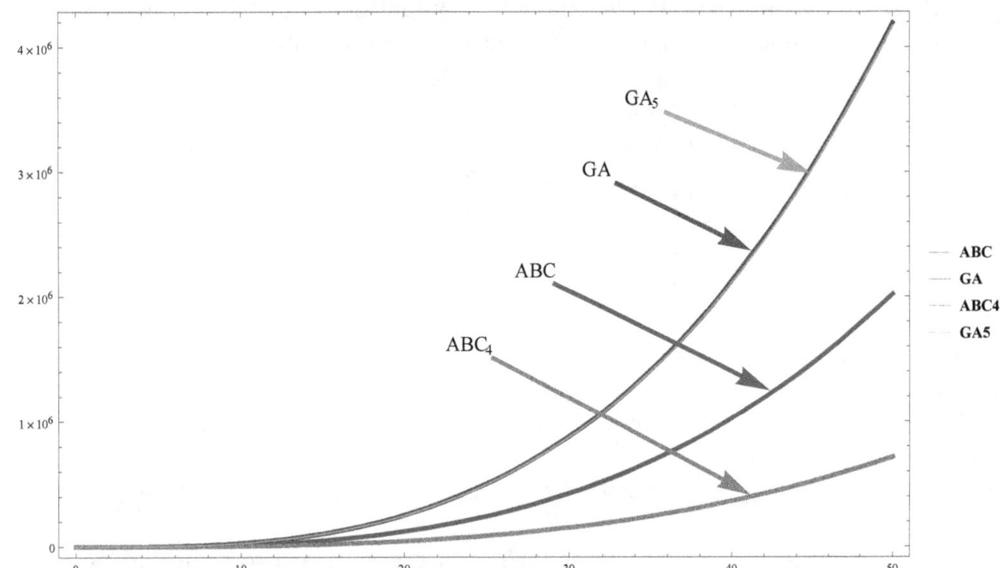

Figure 4. Comparison of ABC, GA, ABC_4 and GA_5 for $HDCN2(m, n)$.

3. Conclusions

In this paper, certain degree-based topological indices, namely the general Randić index, atomic-bond connectivity index (ABC), geometric-arithmetic index (GA) and first Zagreb index were studied for the first time and analytical closed formulas for $HDCN1(m,n)$ and $HDCN2(m,n)$ cage networks were determined which will help the people working in network science to understand and explore the underlying topologies of these networks.

For the future, we are interested in designing some new architectures/networks and then study their topological indices which will be quite helpful to understand their underlying topologies.

Author Contributions: Data curation, U.M.; Funding acquisition, J.-B.L.; Methodology, J.-B.L.; Software, U.M.; Supervision, M.K.S.; Writing—original draft, H.A.

Acknowledgments: The authors would like to thank all the respected reviewers for their suggestions and useful comments, which resulted in an improved version of this paper.

References

1. Wiener, H. Structural determination of paraffin boiling points. *J. Am. Chem. Soc.* **1947**, *69*, 17–20. [CrossRef] [PubMed]
2. Deza, M.; Fowler, P.W.; Rassat, A.; Rogers, K.M. Fullerenes as tiling of surfaces. *J. Chem. Inf. Comput. Sci.* **2000**, *40*, 550–558. [CrossRef] [PubMed]
3. Diudea, M.V.; Gutman, I.; Lorentz, J. *Molecular Topology*; Huntington: Columbus, OH, USA, 2001.
4. Gutman, I.; Polansky, O.E. *Mathematical Concepts in Organic Chemistry*; Springer: New York, NY, USA, 1986.
5. Randić, M. On Characterization of molecular branching. *J. Am. Chem. Soc.* **1975**, *97*, 6609–6615. [CrossRef]
6. Estrada, E.; Torres, L.; Rodríguez, L.; Gutman, I. An atom-bond connectivity index: Modelling the enthalpy of formation of alkanes. *Indian J. Chem.* **1998**, *37A*, 849–855.
7. Vukičević, D.; Furtula, B. Topological index based on the ratios of geometrical and arithmetical means of end-vertex degrees of edges. *J. Math. Chem.* **2009**, *46*, 1369–1376. [CrossRef]
8. Ghorbani, M.; Hosseinzadeh, M.A. Computing ABC_4 index of nanostar dendrimers. *Optoelectron. Adv. Mater. Rapid Commun.* **2010**, *4*, 1419–1422.
9. Graovac, A.; Ghorbani, M.; Hosseinzadeh, M.A. Computing fifth geometric-arithmetic index for nanostar dendrimers. *J. Math. Nanosci.* **2011**, *1*, 33–42.
10. Hayat, S.; Imran, M. Computation of topological indices of certain graphs. *Appl. Math. Comput.* **2014**, *240*, 213–228.
11. Hussain, Z.; Munir, M.; Rafique, S.; Min Kang, S. Topological Characterizations and Index-Analysis of New Degree-Based Descriptors of Honeycomb Networks. *Symmetry* **2018**, *10*, 478. [CrossRef]
12. Amic, D.; Beslo, D.; Lucic, B.; Nikolic, S.; Trinajstić, N. The vertex-connectivity index revisited. *J. Chem. Inf. Comput. Sci.* **1998**, *38*, 819–822. [CrossRef]
13. Bača, M.; Horváthová, J.; Mokrišová, M.; Suhányiovă, A. On topological indices of fullerenes. *Appl. Math. Comput.* **2015**, *251*, 154–161. [CrossRef]
14. Baig, A.Q.; Imran, M.; Ali, H. Computing Omega, Sadhana and PI polynomials of benzoid carbon nanotubes, *Optoelectron. Adv. Mater. Rapid Commun.* **2015**, *9*, 248–255.
15. Baig, A.Q.; Imran, M.; Ali, H. On Topological Indices of Poly Oxide, Poly Silicate, DOX and DSL Networks. *Can. J. Chem.* **2015**, *93*, 730–739. [CrossRef]
16. Caporossi, G.; Gutman, I.; Hansen, P.; Pavlovíc, L. Graphs with maximum connectivity index. *Comput. Biol. Chem.* **2003**, *27*, 85–90. [CrossRef]
17. Imran, M.; Baig, A.Q.; Ali, H. On topological properties of dominating David derived graphs. *Can. J. Chem.* **2016**, *94*, 137–148. [CrossRef]
18. Imran, M.; Baig, A.Q.; Ali, H. On molecular topological properties of hex-derived graphs. *J. Chemom.* **2016**, *30*, 121–129. [CrossRef]
19. Imran, M.; Baig, A.Q.; Ali, H.; Rehman, S.U. On topological properties of poly honeycomb graphs. *Period. Math. Hung.* **2016**, *73*, 100–119. [CrossRef]

20. Iranmanesh, A.; Zeraatkar, M. Computing GA index for some nanotubes. *Optoelectron. Adv. Mater. Rapid Commun.* **2010**, *4*, 1852–1855.
21. Lin, W.; Chen, J.; Chen, Q.; Gao, T.; Lin, X.; Cai, B. Fast computer search for trees with minimal ABC index based on tree degree sequences. *MATCH Commun. Math. Comput. Chem.* **2014**, *72*, 699–708.
22. Manuel, P.D.; Abd-El-Barr, M.I.; Rajasingh, I.; Rajan, B. An efficient representation of Benes networks and its applications. *J. Discret. Algorithms* **2008**, *6*, 11–19. [CrossRef]
23. Palacios, J.L. A resistive upper bound for the ABC index. *MATCH Commun. Math. Comput. Chem.* **2014**, *72*, 709–713.

3

Generalized Permanental Polynomials of Graphs

Shunyi Liu

School of Science, Chang'an University, Xi'an 710064, China; liu@chd.edu.cn

Abstract: The search for complete graph invariants is an important problem in graph theory and computer science. Two networks with a different structure can be distinguished from each other by complete graph invariants. In order to find a complete graph invariant, we introduce the generalized permanental polynomials of graphs. Let G be a graph with adjacency matrix $A(G)$ and degree matrix $D(G)$. The generalized permanental polynomial of G is defined by $P_G(x,\mu) = \mathrm{per}(xI - (A(G) - \mu D(G)))$. In this paper, we compute the generalized permanental polynomials for all graphs on at most 10 vertices, and we count the numbers of such graphs for which there is another graph with the same generalized permanental polynomial. The present data show that the generalized permanental polynomial is quite efficient for distinguishing graphs. Furthermore, we can write $P_G(x,\mu)$ in the coefficient form $\sum_{i=0}^{n} c_{\mu i}(G)x^{n-i}$ and obtain the combinatorial expressions for the first five coefficients $c_{\mu i}(G)$ $(i = 0, 1, \ldots, 4)$ of $P_G(x,\mu)$.

Keywords: generalized permanental polynomial; coefficient; co-permanental

---

1. Introduction

A graph invariant f is a function from the set of all graphs into any commutative ring, such that f has the same value for any two isomorphic graphs. Graph invariants can be used to check whether two graphs are not isomorphic. If a graph invariant f satisfies the condition that $f(G) = f(H)$ implies G and H are isomorphic, then f is called a complete graph invariant. The problem of finding complete graph invariants is closely related to the graph isomorphism problem. Up to now, no complete graph invariant for general graphs has been found. However, some complete graph invariants have been identified for special cases and graph classes (see, for example, [1]).

Graph polynomials are graph invariants whose values are polynomials, which have been developed for measuring the structural information of networks and for characterizing graphs [2]. Noy [3] surveyed results for determining graphs that can be characterized by graph polynomials. In a series of papers [1,4–6], Dehmer et al. studied highly discriminating descriptors to distinguish graphs (networks) based on graph polynomials. In [5], it was found that the graph invariants based on the zeros of permanental polynomials are quite efficient in distinguishing graphs. Balasubramanian and Parthasarathy [7,8] introduced the bivariate permanent polynomial of a graph and conjectured that this graph polynomial is a complete graph invariant. In [9], Liu gave counterexamples to the conjecture by a computer search.

In order to find almost complete graph invariants, we introduce a graph polynomial by employing graph matrices and the permanent of a square matrix. We will see that this graph polynomial turns out to be quite efficient when we use it to distinguish graphs (networks).

The permanent of an $n \times n$ matrix M with entries m_{ij} $(i, j = 1, 2, \ldots, n)$ is defined by

$$\mathrm{per}(M) = \sum_{\sigma} \prod_{i=1}^{n} m_{i\sigma(i)},$$

where the sum is over all permutations σ of $\{1,2,\ldots,n\}$. Valiant [10] proved that computing the permanent is #P-complete, even when restricted to (0,1)-matrices. The permanental polynomial of M, denoted by $\pi(M,x)$, is defined to be the permanent of the characteristic matrix of M; that is,

$$\pi(M,x) = \mathrm{per}(xI_n - M),$$

where I_n is the identity matrix of size n.

Let $G = (V(G), E(G))$ be a graph with adjacency matrix $A(G)$ and degree matrix $D(G)$. The Laplacian matrix and signless Laplacian matrix of G are defined by $L(G) = D(G) - A(G)$ and $Q(G) = D(G) + A(G)$, respectively. The ordinary permanental polynomial of a graph G is defined as the permanental polynomial of the adjacency matrix $A(G)$ of G (i.e., $\pi(A(G),x)$). We call $\pi(L(G),x)$ (respectively, $\pi(Q(G),x)$) the Laplacian (respectively, the signless Laplacian) permanental polynomial of G.

The permanental polynomial $\pi(A(G),x)$ of a graph G was first studied in mathematics by Merris et al. [11], and it was first studied in the chemical literature by Kasum et al. [12]. It was found that the coefficients and roots of $\pi(A(G),x)$ encode the structural information of a (chemical) graph G (see, e.g., [13,14]). Characterization of graphs by the permanental polynomial has been investigated, see [15–19]. The Laplacian permanental polynomial of a graph was first considered by Merris et al. [11], and the signless Laplacian permanental polynomial was first studied by Faria [20]. For more on permanental polynomials of graphs, we refer the reader to the survey [21].

We consider a bivariate graph polynomial of a graph G on n vertices, defined by

$$P_G(x,\mu) = \mathrm{per}(xI_n - (A(G) - \mu D(G))).$$

It is easy to see that $P_G(x,\mu)$ generalizes some well-known permanental polynomials of a graph G. For example, the ordinary permanental polynomial of G is $P_G(x,0)$, the Laplacian permanental polynomial of G is $(-1)^{|V(G)|}P_G(-x,1)$, and the signless Laplacian permanental polynomial of G is $P_G(x,-1)$. We call $P_G(x,\mu)$ the generalized permanental polynomial of G.

We can write the generalized permanental polynomial $P_G(x,\mu)$ in the coefficient form

$$P_G(x,\mu) = \sum_{i=0}^{n} c_{\mu i}(G)x^{n-i}.$$

The general problem is to achieve a better understanding of the coefficients of $P_G(x,\mu)$. For any graph polynomial, it is interesting to determine its ability to characterize or distinguish graphs. A natural question is how well the generalized permanental polynomial distinguishes graphs.

The rest of the paper is organized as follows. In Section 2, we obtain the combinatorial expressions for the first five coefficients $c_{\mu 0}$, $c_{\mu 1}$, $c_{\mu 2}$, $c_{\mu 3}$, and $c_{\mu 4}$ of $P_G(x,\mu)$, and we compute the first five coefficients of $P_G(x,\mu)$ for some specific graphs. In Section 3, we compute the generalized permanental polynomials for all graphs on at most 10 vertices, and we count the numbers of such graphs for which there is another graph with the same generalized permanental polynomial. The presented data shows that the generalized permanental polynomial is quite efficient in distinguishing graphs. It may serve as a powerful tool for dealing with graph isomorphisms.

2. Coefficients

In Section 2.1, we obtain a general relation between the generalized and the ordinary permanental polynomials of graphs. Explicit expressions for the first five coefficients of the generalized permanental polynomial are given in Section 2.2. As an application, we obtain the explicit expressions for the first five coefficients of the generalized permanental polynomials of some specific graphs in Section 2.3.

2.1. Relation between the Generalized and the Ordinary Permanental Polynomials

First, we present two properties of the permanent.

Lemma 1. *Let A, B, and C be three $n \times n$ matrices. If A, B, and C differ only in the rth row (or column), and the rth row (or column) of C is the sum of the rth rows (or columns) of A and B, then* $\mathrm{per}(C) = \mathrm{per}(A) + \mathrm{per}(B)$.

Lemma 2. *Let $M = (m_{ij})$ be an $n \times n$ matrix. Then, for any $i \in \{1, 2, \ldots, n\}$,*

$$\mathrm{per}(M) = \sum_{j=1}^{n} m_{ij}\,\mathrm{per}(M(i,j)),$$

where $M(i,j)$ denotes the matrix obtained by deleting the ith row and jth column from M.

Since Lemmas 1 and 2 can be easily verified using the definition of the permanent, the proofs are omitted.

We need the following notations. Let $G = (V(G), E(G))$ be a graph with vertex set $V(G) = \{v_1, v_2, \ldots, v_n\}$ and edge set $E(G)$. Let $d_i = d_G(v_i)$ be the degree of v_i in G. The degree matrix $D(G)$ of G is the diagonal matrix whose (i,i)th entry is $d_G(v_i)$. Let $v_{r_1}, v_{r_2}, \ldots, v_{r_k}$ be k distinct vertices of G. Then $G_{r_1,r_2,\ldots,r_k}$ denotes the subgraph obtained by deleting vertices $v_{r_1}, v_{r_2}, \ldots, v_{r_k}$ from G. We use $G[h_r]$ to denote the graph obtained from G by attaching to the vertex v_r a loop of weight h_r. Similarly, $G[h_r, h_s]$ stands for the graph obtained by attaching to both v_r and v_s loops of weight h_r and h_s, respectively. Finally, $G[h_1, h_2, \ldots, h_n]$ is the graph obtained by attaching a loop of weight h_r to vertex v_r for each $r = 1, 2, \ldots, n$. The adjacency matrix $A(G[h_{r_1}, h_{r_2}, \ldots, h_{r_s}])$ of $G[h_{r_1}, h_{r_2}, \ldots, h_{r_s}]$ is defined as the $n \times n$ matrix (a_{ij}) with

$$a_{ij} = \begin{cases} h_r, & \text{if } i = j = r \text{ and } r \in \{r_1, r_2, \ldots, r_s\}, \\ 1, & \text{if } i \neq j \text{ and } v_i v_j \in E(G), \\ 0, & \text{otherwise.} \end{cases}$$

By Lemmas 1 and 2, expanding along the rth column, we can obtain the recursion relation

$$\pi(A(G[h_r]), x) = \pi(A(G), x) - h_r \pi(A(G_r), x). \tag{1}$$

For example, expanding along the first column of $\pi(A(G[h_1]), x)$, we have

$$\begin{aligned} \pi(A(G[h_1]), x) &= \mathrm{per}(xI_n - A(G[h_1])) \\ &= \mathrm{per}\begin{bmatrix} x - h_1 & \boldsymbol{u} \\ \boldsymbol{v} & xI_{n-1} - A(G_1) \end{bmatrix} \\ &= \mathrm{per}\begin{bmatrix} x & \boldsymbol{u} \\ \boldsymbol{v} & xI_{n-1} - A(G_1) \end{bmatrix} + \mathrm{per}\begin{bmatrix} -h_1 & \boldsymbol{u} \\ \boldsymbol{0} & xI_{n-1} - A(G_1) \end{bmatrix} \\ &= \pi(A(G), x) - h_1 \mathrm{per}(xI_{n-1} - A(G_1)) \\ &= \pi(A(G), x) - h_1 \pi(A(G_1), x). \end{aligned}$$

By repeated application of (1) for $G[h_r, h_s]$, we have

$$\begin{aligned} &\pi(A(G[h_r, h_s]), x) \\ &= \pi(A(G[h_r]), x) - h_s \pi(A(G_s[h_r]), x) \\ &= \pi(A(G), x) - h_r \pi(A(G_r), x) - h_s(\pi(A(G_s), x) - h_r \pi(A(G_{r,s}), x)) \\ &= \pi(A(G), x) - h_r \pi(A(G_r), x) - h_s \pi(A(G_s), x) + h_r h_s \pi(A(G_{r,s}), x). \end{aligned}$$

Additional iterations can be made to take into account loops on additional vertices. For loops on all n vertices, the expression becomes

$$\pi(A(G[h_1,h_2,\ldots,h_n]),x) = \pi(A(G),x) + \sum_{k=1}^{n}(-1)^k \sum_{1\le r_1<\cdots<r_k\le n} h_{r_1}\cdots h_{r_k}\pi(A(G_{r_1,\ldots,r_k}),x). \quad (2)$$

Let $A_\mu(G) := A(G) - \mu D(G)$. We see that the generalized permanental polynomial $P_G(x,\mu)$ of G is the permanental polynomial of $A_\mu(G)$; that is, $\pi(A_\mu(G),x)$. If the degree sequence of G is $(d_1,d_2,\ldots,d_n)$, then $A_\mu(G)$ is precisely the adjacency matrix of $G[-\mu d_1,-\mu d_2,\ldots,-\mu d_n]$. Hence, we obtain a relation between the generalized and ordinary permanental polynomials as an immediate consequence of (2).

Theorem 1. *Let G be a graph on n vertices. Then,*

$$P_G(x,\mu) = \pi(A_\mu(G),x) = \pi(A(G),x) + \sum_{k=1}^{n}\mu^k \sum_{1\le r_1<\cdots<r_k\le n} d_{r_1}\cdots d_{r_k}\pi(A(G_{r_1,\ldots,r_k}),x).$$

Theorem 1 was inspired by Gutman's method [22] for obtaining a general relation between the Laplacian and the ordinary characteristic polynomials of graphs. From Theorem 1, one can easily give a coefficient formula between the generalized and the ordinary permanental polynomials.

Theorem 2. *Suppose that $\pi(A(G),x) = \sum_{i=0}^{n} a_i(G)x^{n-i}$ and $P_G(x,\mu) = \sum_{i=0}^{n} c_{\mu i}(G)x^{n-i}$. Then,*

$$c_{\mu i}(G) = a_i(G) + \sum_{k=1}^{n}\mu^k \sum_{1\le r_1<\cdots<r_k\le n} d_{r_1}\cdots d_{r_k}a_{i-k}(G_{r_1,\ldots,r_k}), \quad 1\le i\le n.$$

2.2. The First Five Coefficients of $P_G(x,\mu)$

In what follows, we use t_G and q_G to denote respectively the number of triangles (i.e., cycles of length 3) and quadrangles (i.e., cycles of length 4) of G, and $t_G(v)$ denotes the number of triangles containing the vertex v of G.

Liu and Zhang [15] obtained combinatorial expressions for the first five coefficients of the permanental polynomial of a graph.

Lemma 3 ([15])**.** *Let G be a graph with n vertices and m edges, and let $(d_1,d_2,\ldots,d_n)$ be the degree sequence of G. Suppose that $\pi(A(G),x) = \sum_{i=0}^{n} a_i(G)x^{n-i}$. Then,*

$$a_0(G)=1,\ a_1(G)=0,\ a_2(G)=m,\ a_3(G)=-2t_G,\ a_4(G)=\binom{m}{2}-\sum_{i=1}^{n}\binom{d_i}{2}+2q_G.$$

Theorem 3. *Let G be a graph with n vertices and m edges, and let $(d_1,d_2,\ldots,d_n)$ be the degree sequence of G. Suppose that $P_G(x,\mu) = \sum_{i=0}^{n} c_{\mu i}(G)x^{n-i}$. Then*

$$c_{\mu 0}(G)=1,\quad c_{\mu 1}(G)=2\mu m,\quad c_{\mu 2}(G)=2\mu^2m^2+m-\frac{1}{2}\mu^2\sum_{i=1}^{n}d_i^2,$$

$$c_{\mu 3}(G)=\frac{1}{3}\mu^3\sum_{i=1}^{n}d_i^3-(\mu^3m+\mu)\sum_{i=1}^{n}d_i^2+\frac{4}{3}\mu^3m^3+2\mu m^2-2t_G,$$

$$c_{\mu 4}(G) = -\frac{1}{4}\mu^4 \sum_{i=1}^{n} d_i^4 + \left(\frac{2}{3}\mu^4 m + \mu^2\right) \sum_{i=1}^{n} d_i^3 - \frac{1}{2}(2\mu^4 m^2 + 5\mu^2 m + 1) \sum_{i=1}^{n} d_i^2$$

$$+ \frac{1}{8}\mu^4 \left(\sum_{i=1}^{n} d_i^2\right)^2 + \mu^2 \sum_{v_i v_j \in E(G)} d_i d_j + 2\mu \sum_{i=1}^{n} d_i t_G(v_i) + 2q_G - 4\mu m\, t_G$$

$$+ \frac{2}{3}\mu^4 m^4 + 2\mu^2 m^3 + \frac{1}{2}m^2 + \frac{1}{2}m.$$

Proof. It is obvious that $c_{\mu 0}(G) = 1$. By Theorem 2 and Lemma 3, we have

$$\begin{aligned}
c_{\mu 1}(G) &= a_1(G) + \mu \textstyle\sum_i d_i a_0(G_i) = 0 + \mu \sum_i d_i = 2\mu m, \\
c_{\mu 2}(G) &= a_2(G) + \mu \textstyle\sum_i d_i a_1(G_i) + \mu^2 \sum_{i<j} d_i d_j a_0(G_{i,j}) = m + 0 + \mu^2 \sum_{i<j} d_i d_j \\
&= m + \tfrac{1}{2}\mu^2 \left((\textstyle\sum_i d_i)^2 - \sum_i d_i^2 \right) = 2\mu^2 m^2 + m - \tfrac{1}{2}\mu^2 \sum_i d_i^2, \\
c_{\mu 3}(G) &= a_3(G) + \mu \textstyle\sum_i d_i a_2(G_i) + \mu^2 \sum_{i<j} d_i d_j a_1(G_{i,j}) + \mu^3 \sum_{i<j<k} d_i d_j d_k a_0(G_{i,j,k}) \\
&= -2t_G + \mu \textstyle\sum_i d_i (m - d_i) + 0 + \mu^3 \sum_{i<j<k} d_i d_j d_k \\
&= -2t_G + \mu m \textstyle\sum_i d_i - \mu \sum_i d_i^2 + \frac{1}{6}\mu^3 \left((\sum_i d_i)^3 - 3 \sum_i \sum_{j \neq i} d_i^2 d_j - \sum_i d_i^3 \right) \\
&= -2t_G + 2\mu m^2 - \mu \textstyle\sum_i d_i^2 + \frac{4}{3}\mu^3 m^3 - \frac{1}{2}\mu^3 \left((\sum_i d_i^2)(\sum_j d_j) - \sum_i d_i^3 \right) - \frac{1}{6}\mu^3 \sum_i d_i^3 \\
&= \tfrac{1}{3}\mu^3 \textstyle\sum_i d_i^3 - (\mu^3 m + \mu) \sum_i d_i^2 + \frac{4}{3}\mu^3 m^3 + 2\mu m^2 - 2t_G, \\
c_{\mu 4}(G) &= a_4(G) + \mu \textstyle\sum_i d_i a_3(G_i) + \mu^2 \sum_{i<j} d_i d_j a_2(G_{i,j}) + \mu^3 \sum_{i<j<k} d_i d_j d_k a_1(G_{i,j,k}) \\
&\quad + \mu^4 \textstyle\sum_{i<j<k<l} d_i d_j d_k d_l a_0(G_{i,j,k,l}) \\
&= \tbinom{m}{2} - \textstyle\sum_i \binom{d_i}{2} + 2q_G - 2\mu \sum_i d_i (t_G - t_G(v_i)) + \mu^2 \sum_{i<j} d_i d_j |E(G_{i,j})| + 0 \\
&\quad + \mu^4 \textstyle\sum_{i<j<k<l} d_i d_j d_k d_l.
\end{aligned} \tag{3}$$

By a straightforward calculation, we have

$$\begin{aligned}
\textstyle\sum_{i<j} d_i d_j |E(G_{i,j})| &= \sum_{\substack{i<j \\ v_i v_j \in E(G)}} d_i d_j |E(G_{i,j})| + \sum_{\substack{i<j \\ v_i v_j \notin E(G)}} d_i d_j |E(G_{i,j})| \\
&= \sum_{\substack{i<j \\ v_i v_j \in E(G)}} d_i d_j (m - d_i - d_j + 1) + \sum_{\substack{i<j \\ v_i v_j \notin E(G)}} d_i d_j (m - d_i - d_j) \\
&= \textstyle\sum_{i<j} d_i d_j (m - d_i - d_j) + \sum_{v_i v_j \in E(G)} d_i d_j \\
&= m \textstyle\sum_{i<j} d_i d_j - \sum_i \sum_{j \neq i} d_i^2 d_j + \sum_{v_i v_j \in E(G)} d_i d_j \\
&= \tfrac{m}{2} \left(4m^2 - \textstyle\sum_i d_i^2\right) - \left(2m \sum_i d_i^2 - \sum_i d_i^3\right) + \sum_{v_i v_j \in E(G)} d_i d_j \\
&= \textstyle\sum_i d_i^3 - \frac{5}{2} m \sum_i d_i^2 + \sum_{v_i v_j \in E(G)} d_i d_j + 2m^3,
\end{aligned} \tag{4}$$

and

$$
\begin{aligned}
&\textstyle\sum_{i<j<k<l} d_i d_j d_k d_l \\
&= \tfrac{1}{24}\left(\left(\sum_i d_i\right)^4 - 12\sum_i \sum_{\substack{j \\ j\neq i}} \sum_{\substack{k \\ k\neq i, k\neq j}} d_i^2 d_j d_k - 4\sum_i \sum_{\substack{j \\ j\neq i}} d_i^3 d_j - 6\sum_{i<j} d_i^2 d_j^2 - \sum_i d_i^4 \right) \\
&= \tfrac{2}{3}m^4 - \tfrac{1}{2}\times\tfrac{1}{2}\left(\left(\sum_i d_i^2\right)\left(\sum_i d_i\right)^2 - \sum_i d_i^4 - 2\sum_{i<j} d_i^2 d_j^2 - 2\sum_i \sum_{\substack{j \\ j\neq i}} d_i^3 d_j \right) \\
&\quad -\tfrac{1}{6}\sum_i \sum_{\substack{j \\ j\neq i}} d_i^3 d_j - \tfrac{1}{4}\sum_{i<j} d_i^2 d_j^2 - \tfrac{1}{24}\sum_i d_i^4 \qquad (5) \\
&= \tfrac{2}{3}m^4 - m^2\sum_i d_i^2 + \tfrac{5}{24}\sum_i d_i^4 + \tfrac{1}{4}\sum_{i<j} d_i^2 d_j^2 + \tfrac{1}{3}\sum_i \sum_{\substack{j \\ j\neq i}} d_i^3 d_j \\
&= \tfrac{2}{3}m^4 - m^2\sum_i d_i^2 + \tfrac{5}{24}\sum_i d_i^4 + \tfrac{1}{4}\times\tfrac{1}{2}\left(\left(\sum_i d_i^2\right)^2 - \sum_i d_i^4 \right) + \tfrac{1}{3}\left(\left(\sum_i d_i^3\right)\left(\sum_i d_i\right) - \sum_i d_i^4 \right) \\
&= -\tfrac{1}{4}\sum_i d_i^4 + \tfrac{2}{3}m\sum_i d_i^3 - m^2\sum_i d_i^2 + \tfrac{1}{8}\left(\sum_i d_i^2\right)^2 + \tfrac{2}{3}m^4.
\end{aligned}
$$

Substituting (4) and (5) into (3), we obtain

$$
\begin{aligned}
c_{\mu 4}(G) = & -\frac{1}{4}\mu^4 \sum_{i=1}^{n} d_i^4 + \left(\frac{2}{3}\mu^4 m + \mu^2\right)\sum_{i=1}^{n} d_i^3 - \frac{1}{2}(2\mu^4 m^2 + 5\mu^2 m + 1)\sum_{i=1}^{n} d_i^2 \\
& + \frac{1}{8}\mu^4\left(\sum_{i=1}^{n} d_i^2\right)^2 + \mu^2 \sum_{v_i v_j \in E(G)} d_i d_j + 2\mu \sum_{i=1}^{n} d_i t_G(v_i) + 2q_G - 4\mu m\, t_G \\
& + \frac{2}{3}\mu^4 m^4 + 2\mu^2 m^3 + \frac{1}{2}m^2 + \frac{1}{2}m.
\end{aligned}
$$

This completes the proof. □

Since $\pi(L(G), x) = (-1)^{|V(G)|} P_G(-x, 1)$ and $\pi(Q(G), x) = P_G(x, -1)$, we immediately obtain the combinatorial expressions for the first five coefficients of $\pi(L(G), x)$ and $\pi(Q(G), x)$ by Theorem 3.

Corollary 1. *Let G be a graph with n vertices and m edges, and let $(d_1, d_2, \ldots, d_n)$ be the degree sequence of G. Suppose that $\pi(L(G), x) = \sum_{i=0}^{n} p_i(G) x^{n-i}$, then*

$$
\begin{aligned}
& p_0(G) = 1, \quad p_1(G) = -2m, \quad p_2(G) = 2m^2 + m - \frac{1}{2}\sum_{i=1}^{n} d_i^2, \\
& p_3(G) = -\frac{1}{3}\sum_{i=1}^{n} d_i^3 + (m+1)\sum_{i=1}^{n} d_i^2 - \frac{4}{3}m^3 - 2m^2 + 2t_G,
\end{aligned}
$$

$$
\begin{aligned}
p_4(G) = & -\frac{1}{4}\sum_{i=1}^{n} d_i^4 + \left(\frac{2}{3}m + 1\right)\sum_{i=1}^{n} d_i^3 - \frac{1}{2}(2m^2 + 5m + 1)\sum_{i=1}^{n} d_i^2 + \frac{1}{8}\left(\sum_{i=1}^{n} d_i^2\right)^2 \\
& + \sum_{v_i v_j \in E(G)} d_i d_j + 2\sum_{i=1}^{n} d_i t_G(v_i) + 2q_G - 4m\, t_G + \frac{2}{3}m^4 + 2m^3 + \frac{1}{2}m^2 + \frac{1}{2}m.
\end{aligned}
$$

Corollary 2. *Let G be a graph with n vertices and m edges, and let $(d_1, d_2, \ldots, d_n)$ be the degree sequence of G. Suppose that $\pi(Q(G), x) = \sum_{i=0}^{n} q_i(G) x^{n-i}$. Then,*

$$q_0(G) = 1, \qquad q_1(G) = -2m, \qquad q_2(G) = 2m^2 + m - \frac{1}{2}\sum_{i=1}^{n} d_i^2,$$
$$q_3(G) = -\frac{1}{3}\sum_{i=1}^{n} d_i^3 + (m+1)\sum_{i=1}^{n} d_i^2 - \frac{4}{3}m^3 - 2m^2 - 2t_G,$$
$$q_4(G) = -\frac{1}{4}\sum_{i=1}^{n} d_i^4 + \left(\frac{2}{3}m+1\right)\sum_{i=1}^{n} d_i^3 - \frac{1}{2}(2m^2+5m+1)\sum_{i=1}^{n} d_i^2 + \frac{1}{8}\left(\sum_{i=1}^{n} d_i^2\right)^2$$
$$+ \sum_{v_iv_j \in E(G)} d_i d_j - 2\sum_{i=1}^{n} d_i t_G(v_i) + 2q_G + 4m\, t_G + \frac{2}{3}m^4 + 2m^3 + \frac{1}{2}m^2 + \frac{1}{2}m.$$

2.3. Examples

In this subsection, by applying Theorem 3, we obtain the first five coefficients of the generalized permanental polynomials of some specific graphs: Paths, cycles, complete graphs, complete bipartite graphs, star graphs, and wheel graphs.

Example 1. *Let P_n ($n \geq 3$) be the path on n vertices. We see at once that $t_{P_n} = q_{P_n} = 0$, and $t_{P_n}(v) = 0$ for each vertex v of P_n. By Theorem 3, we have*

$$c_{\mu 0}(P_n) = 1, \quad c_{\mu 1}(P_n) = 2(n-1)\mu, \quad c_{\mu 2}(P_n) = (2n^2 - 6n + 5)\mu^2 + n - 1,$$
$$c_{\mu 3}(P_n) = \frac{2}{3}(2n^2 - 8n + 9)(n-2)\mu^3 + 2(n-2)^2\mu,$$
$$c_{\mu 4}(P_n) = \frac{2}{3}(n^2 - 5n + 7)(n-3)(n-2)\mu^4 + (2n^2 - 10n + 13)(n-3)\mu^2 + \frac{1}{2}(n-3)(n-2).$$

Example 2. *Let C_n ($n \geq 5$) be the cycle on n vertices. We see at once that $t_{C_n} = q_{C_n} = 0$, and $t_{C_n}(v) = 0$ for each vertex v of C_n. By Theorem 3, we have*

$$c_{\mu 0}(C_n) = 1, \quad c_{\mu 1}(C_n) = 2n\mu, \quad c_{\mu 2}(C_n) = 2n(n-1)\mu^2 + n,$$
$$c_{\mu 3}(C_n) = \frac{4}{3}n(n-1)(n-2)\mu^3 + 2n(n-2)\mu,$$
$$c_{\mu 4}(C_n) = \frac{2}{3}n(n-1)(n-2)(n-3)\mu^4 + 2n(n-2)(n-3)\mu^2 + \frac{1}{2}n(n-3).$$

Example 3. *Let K_n ($n \geq 4$) be the complete graph on n vertices. It is easy to check that $t_{K_n} = \binom{n}{3} = n(n-1)(n-2)/6$, $q_{K_n} = 3\binom{n}{4} = n(n-1)(n-2)(n-3)/8$, and $t_{K_n}(v) = \binom{n-1}{2} = (n-1)(n-2)/2$ for each vertex v of K_n. By Theorem 3, we have*

$$c_{\mu 0}(K_n) = 1, \quad c_{\mu 1}(K_n) = n(n-1)\mu, \quad c_{\mu 2}(K_n) = \frac{1}{2}n(n-1)^3\mu^2 + \frac{1}{2}n(n-1),$$
$$c_{\mu 3}(K_n) = \frac{1}{6}n(n-2)(n-1)^4\mu^3 + \frac{1}{2}n(n-2)(n-1)^2\mu - \frac{1}{3}n(n-1)(n-2),$$
$$c_{\mu 4}(K_n) = \frac{1}{24}n(n-2)(n-3)(n-1)^5\mu^4 + \frac{1}{4}n(n-2)(n-3)(n-1)^3\mu^2 -$$
$$\frac{1}{3}n(n-2)(n-3)(n-1)^2\mu + \frac{3}{8}n(n-1)(n-2)(n-3).$$

Example 4. *Let $K_{a,b}$ ($a \geq b \geq 2$) be the complete bipartite graph with partition sets of sizes a and b. We see at once that $t_{K_{a,b}} = 0$, $q_{K_{a,b}} = \binom{a}{2}\binom{b}{2} = ab(a-1)(b-1)/4$, and $t_{K_{a,b}}(v) = 0$ for each vertex v of $K_{a,b}$. By Theorem 3, we have*

$$c_{\mu 0}(K_{a,b}) = 1, \quad c_{\mu 1}(K_{a,b}) = 2ab\mu, \quad c_{\mu 2}(K_{a,b}) = \frac{1}{2}ab(4ab - a - b)\mu^2 + ab,$$
$$c_{\mu 3}(K_{a,b}) = \frac{1}{3}ab(4a^2b^2 - 3a^2b - 3ab^2 + a^2 + b^2)\mu^3 + ab(2ab - a - b)\mu,$$
$$c_{\mu 4}(K_{a,b}) = \frac{1}{24}ab(16a^3b^3 - 24a^3b^2 - 24a^2b^3 + 19a^3b + 6a^2b^2 + 19ab^3 - 6a^3 - 6b^3)\mu^4 +$$
$$\frac{1}{2}ab(4a^2b^2 - 5a^2b - 5ab^2 + 2a^2 + 2ab + 2b^2)\mu^2 + ab(a-1)(b-1).$$

Example 5. *Let S_n ($n \geq 3$) be the star graph with $n+1$ vertices and n edges. We see at once that $t_{S_n} = q_{S_n} = 0$, and $t_{S_n}(v) = 0$ for each vertex v of S_n. By Theorem 3, we have*

$$c_{\mu 0}(S_n) = 1, \quad c_{\mu 1}(S_n) = 2n\mu, \quad c_{\mu 2}(S_n) = \frac{1}{2}n(3n-1)\mu^2 + n,$$
$$c_{\mu 3}(S_n) = \frac{1}{3}n(2n-1)(n-1)\mu^3 + n(n-1)\mu,$$
$$c_{\mu 4}(S_n) = \frac{1}{24}n(n-1)(n-2)(5n-3)\mu^4 + \frac{1}{2}n(n-1)(n-2)\mu^2.$$

Example 6. *Let W_n ($n \geq 5$) be the wheel graph with $n+1$ vertices and $2n$ edges. It is obvious that $t_{W_n} = q_{W_n} = n$. Let v_0 be the hub (i.e., the vertex of degree n) of W_n. We see that $t_{W_n}(v_0) = n$ and $t_{W_n}(v) = 2$ for other vertices v of W_n. By Theorem 3, we have*

$$c_{\mu 0}(W_n) = 1, \quad c_{\mu 1}(W_n) = 4n\mu, \quad c_{\mu 2}(W_n) = \frac{3}{2}n(5n-3)\mu^2 + 2n,$$
$$c_{\mu 3}(W_n) = 9n(n-1)^2\mu^3 + n(7n-9)\mu - 2n,$$
$$c_{\mu 4}(W_n) = \frac{9}{8}n(n-1)(n-2)(7n-9)\mu^4 + 6n(2n-3)(n-2)\mu^2 - 6n(n-2)\mu + \frac{3}{2}n(n-1).$$

3. Numerical Results

In this section, by computer we enumerate the generalized permanental polynomials for all graphs on at most 10 vertices, and we count the numbers of such graphs for which there is another graph with the same generalized permanental polynomial.

Two graphs G and H are said to be generalized co-permanental if they have the same generalized permanental polynomial. If a graph H is generalized co-permanental but non-isomorphic to G, then H is called a generalized co-permanental mate of G.

In order to compute the generalized permanental polynomials of graphs, we, first of all, have to generate the graphs by computer. We use nauty and Traces [23] to generate all graphs on at most 10 vertices. Next, the generalized permanental polynomials of these graphs are calculated by a Maple procedure. Finally, we count the numbers of generalized co-permanental graphs.

The results are summarized in Table 1. Table 1 lists, for $n \leq 10$, the total number of graphs on n vertices, the total number of distinct generalized permanental polynomials of such graphs, the number of such graphs with a generalized co-permanental mate, the fraction of such graphs with a generalized co-permanental mate, and the size of the largest family of generalized co-permanental graphs.

In Table 1, we see that the smallest generalized co-permanental graphs, with respect to the order, contain 10 vertices. Even more striking is that out of 12,005,168 graphs with 10 vertices, only 106 graphs could not be discriminated by the generalized permanental polynomial.

From Table 1 in [9], we see that the smallest graphs that cannot be distinguished by the bivariate permanent polynomial, introduced by Balasubramanian and Parthasarathy, contain 8 vertices. By comparing the present data of Table 1 with that of Table 1 in [9], we find that the generalized permanental polynomial is more efficient than the bivariate permanent polynomial when we use them to distinguish graphs. From Tables 2 and 3 in [5], it is seen that the generalized permanental polynomial is more efficient than the graph invariants based on the zeros of permanental polynomials of graphs. Comparing the present data of Table 1 with that of Table 1 in [24], we see that the generalized permanental polynomial is also superior to the the generalized characteristic polynomial when distinguishing graphs. So, the generalized permanental polynomial is quite efficient in distinguishing graphs.

Table 1. Graphs on at most 10 vertices.

n	# Graphs	# Generalized Perm. Pols	# with Mate	Frac. with Mate	Max. Family
1	1	1	0	0	1
2	2	2	0	0	1
3	4	4	0	0	1
4	11	11	0	0	1
5	34	34	0	0	1
6	156	156	0	0	1
7	1044	1044	0	0	1
8	12,346	12,346	0	0	1
9	274,668	274,668	0	0	1
10	12,005,168	12,005,115	106	8.83×10^{-6}	2

We enumerate all graphs on 10 vertices with a generalized co-permanental mate for each possible number of edges in Appendix A. We see that the generalized co-permanental graphs G_1 and H_1 with 10 edges are disconnected (see Figure 1), the generalized co-permanental graphs G_2 and H_2 with 11 edges, and G_3 and H_3 with 12 edges are all bipartite (see Figures 2 and 3), and two pairs (G_4, H_4) and (G_5, H_5) of generalized co-permanental graphs with 14 edges are all non-bipartite (see Figure 4). The common generalized permanental polynomial of the smallest generalized co-permanental graphs G_1 and H_1 is

$$\begin{aligned}
P_{G_1}(x,\mu) &= P_{H_1}(x,\mu) \\
&= x^{10} + 20\mu x^9 + (178\mu^2 + 10)x^8 + (928\mu^3 + 156\mu)x^7 + (3137\mu^4 + 1050\mu^2 + 37)x^6 \\
&+ (7180\mu^5 + 3980\mu^3 + 416\mu)x^5 + (11260\mu^6 + 9284\mu^4 + 1912\mu^2 + 60)x^4 \\
&+ (11936\mu^7 + 13632\mu^5 + 4592\mu^3 + 416\mu)x^3 + (8176\mu^8 + 12288\mu^6 + 6068\mu^4 + 1048\mu^2 + 36)x^2 \\
&+ (3264\mu^9 + 6208\mu^7 + 4176\mu^5 + 1136\mu^3 + 96\mu)x + 576\mu^{10} + 1344\mu^8 + 1168\mu^6 + 448\mu^4 + 64\mu^2.
\end{aligned}$$

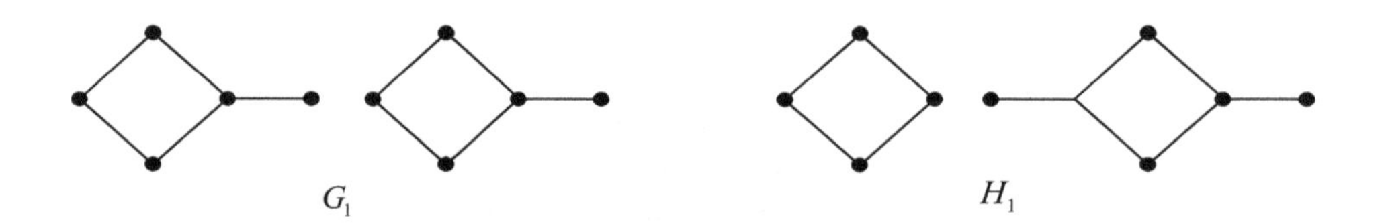

Figure 1. Two generalized co-permanental graphs with 10 vertices and 10 edges.

Figure 2. Two generalized co-permanental graphs with 10 vertices and 11 edges.

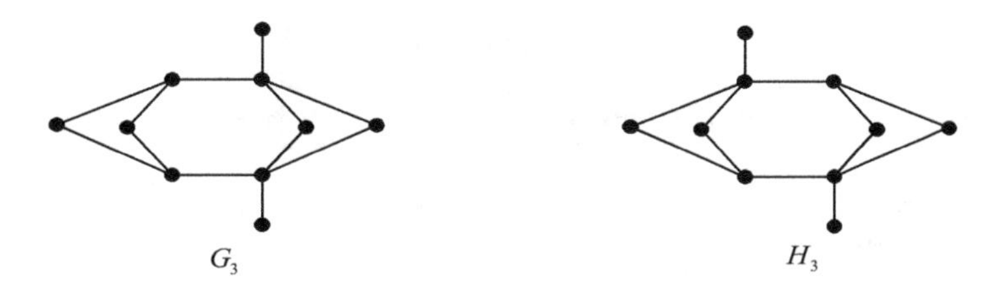

Figure 3. Two generalized co-permanental graphs with 10 vertices and 12 edges.

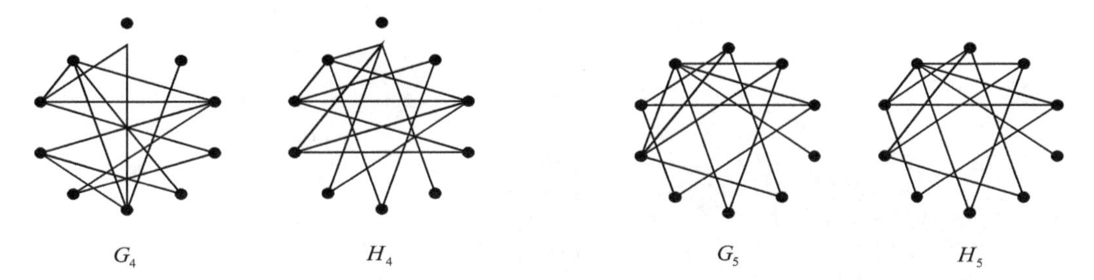

Figure 4. Two pairs of generalized co-permanental graphs with 10 vertices and 14 edges.

4. Conclusions

This paper is a continuance of the research relating to the search of almost-complete graph invariants. In order to find an almost-complete graph invariant, we introduce the generalized permanental polynomials of graphs. As can be seen, the generalized permanental polynomial is quite efficient in distinguishing graphs (networks). It may serve as a powerful tool for dealing with graph isomorphisms. We also obtain the combinatorial expressions for the first five coefficients of the generalized permanental polynomials of graphs.

Appendix A

In the Appendix, we enumerate all graphs on 10 vertices with a generalized co-permanental mate for each possible number m of edges. Since the coefficient of x^{n-1} in $P_G(x,\mu)$ is $2\mu m$, two graphs with a distinct number of edges must have distinct generalized permanental polynomials. So, the enumeration can be implemented for each possible number of edges. We list the numbers of graphs with 10 vertices for all numbers m of edges, the numbers of distinct generalized permanental polynomials of such graphs, the numbers of such graphs with a generalized co-permanental mate, and the maximum size of a family of generalized co-permanental graphs (see Table A1).

Table A1. Graphs on 10 vertices.

m	# Graphs	# Generalized Perm. Pols	# with Mate	Max. Family
0	1	1	0	1
1	1	1	0	1
2	2	2	0	1
3	5	5	0	1
4	11	11	0	1
5	26	26	0	1
6	66	66	0	1
7	165	165	0	1
8	428	428	0	1

Table A1. *Cont.*

m	# Graphs	# Generalized Perm. Pols	# with Mate	Max. Family
9	1103	1103	0	1
10	2769	2768	2	2
11	6759	6758	2	2
12	15,772	15,771	2	2
13	34,663	34,663	0	1
14	71,318	71,316	4	2
15	136,433	136,429	8	2
16	241,577	241,575	4	2
17	395,166	395,162	8	2
18	596,191	596,183	16	2
19	828,728	828,723	10	2
20	1,061,159	1,061,154	10	2
21	1,251,389	1,251,381	16	2
22	1,358,852	1,358,848	8	2
23	1,358,852	1,358,850	4	2
24	1,251,389	1,251,385	8	2
25	1,061,159	1,061,157	4	2
26	828,728	828,728	0	1
27	596,191	596,191	0	1
28	395,166	395,166	0	1
29	241,577	241,577	0	1
30	136,433	136,433	0	1
31	71,318	71,318	0	1
32	34,663	34,663	0	1
33	15,772	15,772	0	1
34	6759	6759	0	1
35	2769	2769	0	1
36	1103	1103	0	1
37	428	428	0	1
38	165	165	0	1
39	66	66	0	1
40	26	26	0	1
41	11	11	0	1
42	5	5	0	1
43	2	2	0	1
44	1	1	0	1
45	1	1	0	1

References

1. Dehmer, M.; Emmert-Streib, F.; Grabner, M. A computational approach to construct a multivariate complete graph invariant. *Inf. Sci.* **2014**, *260*, 200–208. [CrossRef]
2. Shi, Y.; Dehmer, M.; Li, X.; Gutman, I. *Graph Polynomials*; CRC Press: Boca Raton, FL, USA, 2017.
3. Noy, M. Graphs determined by polynomial invariants. *Theor. Comput. Sci.* **2003**, *307*, 365–384. [CrossRef]
4. Dehmer, M.; Moosbrugger, M.; Shi, Y. Encoding structural information uniquely with polynomial-based descriptors by employing the Randić matrix. *Appl. Math. Comput.* **2015**, *268*, 164–168. [CrossRef]
5. Dehmer, M.; Emmert-Streib, F.; Hu, B.; Shi, Y.; Stefu, M.; Tripathi, S. Highly unique network descriptors based on the roots of the permanental polynomial. *Inf. Sci.* **2017**, *408*, 176–181. [CrossRef]
6. Dehmer, M.; Chen, Z.; Emmert-Streib, F.; Shi, Y.; Tripathi, S. Graph measures with high discrimination power revisited: A random polynomial approach. *Inf. Sci.* **2018**, *467*, 407–414. [CrossRef]
7. Balasubramanian, K.; Parthasarathy, K.R. In search of a complete invariant for graphs. In *Lecture Notes in Mathematics*; Springer: Berlin/Heidelberg, Germany, 1981; Volume 885, pp. 42–59.
8. Parthasarathy, K.R. Graph characterising polynomials. *Discret. Math.* **1999**, *206*, 171–178. [CrossRef]
9. Liu, S. On the bivariate permanent polynomials of graphs. *Linear Algebra Appl.* **2017**, *529*, 148–163. [CrossRef]
10. Valiant, L.G. The complexity of computing the permanent. *Theor. Comput. Sci.* **1979**, *8*, 189–201. [CrossRef]

11. Merris, R.; Rebman, K.R.; Watkins, W. Permanental polynomials of graphs. *Linear Algebra Appl.* **1981**, *38*, 273–288. [CrossRef]
12. Kasum, D.; Trinajstić, N.; Gutman, I. Chemical graph theory. III. On permanental polynomial. *Croat. Chem. Acta* **1981**, *54*, 321–328.
13. Cash, G.G. Permanental polynomials of smaller fullerenes. *J. Chem. Inf. Comput. Sci.* **2000**, *40*, 1207–1209. [CrossRef]
14. Tong, H.; Liang, H.; Bai, F. Permanental polynomials of the larger fullerenes. *MATCH Commun. Math. Comput. Chem.* **2006**, *56*, 141–152.
15. Liu, S.; Zhang, H. On the characterizing properties of the permanental polynomials of graphs. *Linear Algebra Appl.* **2013**, *438*, 157–172. [CrossRef]
16. Liu, S.; Zhang, H. Characterizing properties of permanental polynomials of lollipop graphs. *Linear Multilinear Algebra* **2014**, *62*, 419–444. [CrossRef]
17. Wu, T.; Zhang, H. Per-spectral characterization of graphs with extremal per-nullity. *Linear Algebra Appl.* **2015**, *484*, 13–26. [CrossRef]
18. Wu, T.; Zhang, H. Per-spectral and adjacency spectral characterizations of a complete graph removing six edges. *Discret. Appl. Math.* **2016**, *203*, 158–170. [CrossRef]
19. Zhang, H.; Wu, T.; Lai, H. Per-spectral characterizations of some edge-deleted subgraphs of a complete graph. *Linear Multilinear Algebra* **2015**, *63*, 397–410. [CrossRef]
20. Faria, I. Permanental roots and the star degree of a graph. *Linear Algebra Appl.* **1985**, *64*, 255–265. [CrossRef]
21. Li, W.; Liu, S.; Wu, T.; Zhang, H. On the permanental polynomials of graphs. In *Graph Polynomials*; Shi, Y., Dehmer, M., Li, X., Gutman, I., Eds.; CRC Press: Boca Raton, FL, USA, 2017; pp. 101–122.
22. Gutman, I. Relation between the Laplacian and the ordinary characteristic polynomial. *MATCH Commun. Math. Comput. Chem.* **2003**, *47*, 133–140.
23. McKay, B.D.; Piperno, A. Practical graph isomorphism, II. *J. Symb. Comput.* **2014**, *60*, 94–112. [CrossRef]
24. Van Dam, E.R.; Haemers, W.H.; Koolen, J.H. Cospectral graphs and the generalized adjacency matrix. *Linear Algebra Appl.* **2007**, *423*, 33–41. [CrossRef]

4

Edge-Version Atom-Bond Connectivity and Geometric Arithmetic Indices of Generalized Bridge Molecular Graphs

Xiujun Zhang [1], Xinling Wu [2], Shehnaz Akhter [3], Muhammad Kamran Jamil [4], Jia-Bao Liu [5,*] and Mohammad Reza Farahani [6]

1 Key Laboratory of Pattern Recognition and Intelligent Information Processing, Institutions of Higher Education of Sichuan Province, Chengdu University, Chengdu 610106, China; woodszhang@cdu.edu.cn
2 South China Business College, Guang Dong University of Foreign Studies, Guangzhou 510545, China; xinlingwu.guangzhou@gmail.com
3 Department of Mathematics, School of Natural Sciences (SNS), National University of Sciences and Technology (NUST), Sector H-12, Islamabad 44000, Pakistan; shehnazakhter36@yahoo.com
4 Department of Mathematics, Riphah Institute of Computing and Applied Sciences, Riphah International University Lahore, Lahore 54660, Pakistan; m.kamran.sms@gmail.com
5 School of Mathematics and Physics, Anhui Jianzhu University, Hefei 230601, China
6 Department of Applied Mathematics, Iran University of Science and Technology, Narmak, Tehran 16844, Iran; mrfarahani88@gmail.com
* Correspondence: liujiabao@ahjzu.edu.cn

Abstract: Topological indices are graph invariants computed by the distance or degree of vertices of the molecular graph. In chemical graph theory, topological indices have been successfully used in describing the structures and predicting certain physicochemical properties of chemical compounds. In this paper, we propose a definition of generalized bridge molecular graphs that can model more kinds of long chain polymerization products than the bridge molecular graphs, and provide some results of the edge versions of atom-bond connectivity (ABC_e) and geometric arithmetic (GA_e) indices for some generalized bridge molecular graphs, which have regular, periodic and symmetrical structures. The results of this paper offer promising prospects in the applications for chemical and material engineering, especially in chemical industry research.

Keywords: atom-bond connectivity index; geometric arithmetic index; line graph; generalized bridge molecular graph

---

1. Introduction

Let G be an undirected simple graph without loops or multiple edges. We denote by $V(G)$ the vertex set of G and we denote by $E(G)$ the edge set of G. We denote by $e = uv$ the edge connect vertices u and v or vertices u and v adjacent. We denote by P_n, C_n, and S_n the path, cycle, and star of n vertices, respectively. We denote by $N(v)$ the open neighborhood of vertex v, i.e., $N(v) = \{u|uv \in E(G)\}$. We denote by $d(v)$ or $d_G(v)$ the degree of a vertex v of a graph G, i.e., $d(v) = |\{u \in N(v)\}|$. Let $L(G)$ or G^L be a line graph of G, so each vertex of $L(G)$ corresponds an edge of G. Two vertices of $L(G)$ are adjacent if and only if a common endpoint is shared by their corresponding edges in G [1]. The degree of edge e in G is denoted by $d_{L(G)}(e)$, which is the number of edges that share common endpoint with edge e in G; it is also the degree of vertex e in $L(G)$. We give simple a illustration to explain the relationship of original graph and corresponding line graph in Figure 1. We can see u, v, w denote corresponding vertexes, and e, f, g, h, i, j denote corresponding edges in original graph G and denote corresponding vertices in line graph $L(G)$. We get $d(u) = d(v) = 3, d(w) = 2$. $d_{L(G)}(e)$, which is the degree of vertex e in $L(G)$, is also the degree of edge e in G, thus $d_{L(G)}(e) = 4$ in Figure 1.

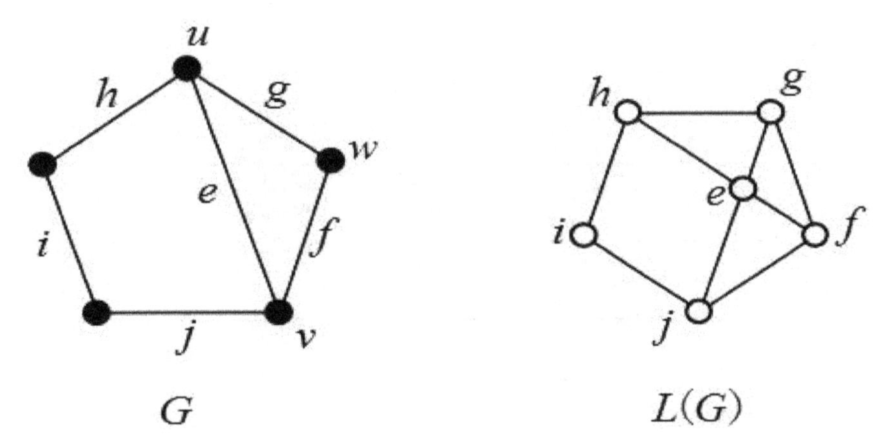

Figure 1. The original graph G and corresponding line graph $L(G)$.

Topological indices are graph invariants, which are obtained by performing some numerical operations on the distance or degree of vertices of the molecular graph. In chemical graph theory, topological indices are the molecular descriptors. They have been successfully used in describing the structures and predicting certain physicochemical properties of chemical compounds. To study the relationship between molecular structure and physical properties of saturated hydrocarbons, Wiener index was first published in 1947 [2], and the edge version of Wiener index, which can be considered as the Wiener index of line graph of G, was proposed by Iranmanesh et al. in 2009 [3]. As the important role of topological indices in chemical research has been confirmed, more topological indices appeared, which include atom-bond connectivity index and geometric arithmetic index.

In chemical graph theory, hydrogen atoms are usually ignored when the topological indices are calculated, which is very similar to how organic chemists usually simply write a benzene ring as a hexagon [4]. Now, three types of graphs of $C_{24}H_{28}$ are illustrated in Figure 2.

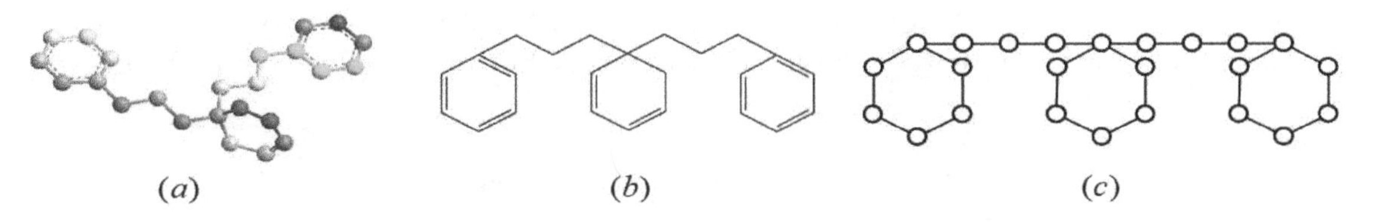

Figure 2. (**a**) $C_{24}H_{28}$ ball and stick model graph in 3D; (**b**) $C_{24}H_{28}$ chemical structure graph; and (**c**) $C_{24}H_{28}$ model graph in chemical graph theory.

To explore the properties of simple short chain compound products, Gao et al. [5] defined some join graphs such as $P_n + C_m$, $P_n + S_m$, $C_m + P_n + C_m$, $S_m + P_n + S_m$, and $C_m + P_n + S_r$, created by P_n, C_n and S_n and obtained the ABC_e and GA_e indices of these graphs. In another paper, Gao et al. [6] defined the bridge molecular structures, which can be used to research some long chain polymerization products, and the forgotten indices ($F(G)$) formulae of some simple bridge molecular structures constructed by P_2, C_6 or K_3 are presented. The forgotten index is defined as $F(G) = \sum_{v \in V(G)} (d(v)^3)$ [7]. In this paper, we define generalized bridge molecular graphs that could cover more kinds of long chain polymerization products, and the edge-version atom-bond connectivity and geometric arithmetic indices of generalized bridge molecular graphs are calculated.

To facilitate the reader, the topological indices discussed in this thesis are all given in Table 1.

Table 1. The definition of topological indices.

Index Name	Definition	Proposed	Recent Studied
atom-bond connection index	$ABC(G) = \sum_{uv \in E(G)} \sqrt{\frac{d(u)+d(v)-2}{d(u)d(v)}}$	[8]	[9–11]
edge version of ABC index	$ABC_e(G) = \sum_{e_1e_2 \in E(L(G))} \sqrt{\frac{d_{L(G)}(e_1)+d_{L(G)}(e_2)-2}{d_{L(G)}(e_1) \times d_{L(G)}(e_2)}}$	[12]	[5,13,14]
geometric arithmetic index	$GA(G) = \sum_{uv \in E(G)} \frac{2\sqrt{d_G(u)d_G(v)}}{d_G(u)+d_G(v)}$	[15]	[16–18]
edge version of GA index	$GA_e(G) = \sum_{e_1e_2 \in E(L(G))} \frac{2\sqrt{d_{L(G)}(e_1)d_{L(G)}(e_2)}}{d_{L(G)}(e_1)+d_{L(G)}(e_2)}$	[19]	[5,12,19–21]

In Table 1, $d_G(u)$ and $d_G(v)$ are the degrees of the vertices u and v in G, and $d_{L(G)}(e_1)$ and $d_{L(G)}(e_2)$ are the degrees of the edges e_1 and e_2 in G.

2. Main Results and Proofs

2.1. Definition of the Generalized Bridge Molecular Graph

Before we start a discussion, we give the definition of the generalized bridge molecular graph as follows. For a positive integer d, d pairwise disjoint molecular graphs $\{G^{(1)}, G^{(2)}, \cdots, G^{(d)}\}$ with $v^{(i)} \in V(G^{(i)})$ for each $i = 1, 2, \cdots, d$, and $d-1$ pairwise disjoint path molecular graphs $P^{(1)}, P^{(2)}, \cdots, P^{(d-1)}$ (called bridges), the generalized bridge molecular graph $GBG(G^{(1)}, v^{(1)}, G^{(2)}, v^{(2)}, \cdots, G^{(d)}, v^{(d)}; P^{(1)}, P^{(2)}, \cdots, P^{(d-1)})$ is the graph obtained by connecting the vertices $v^{(i)}$ and $v^{(i+1)}$ by a path $P^{(i)}$ for which two end vertices are identified with $v^{(i)}$ and $v^{(i+1)}$ for $i = 1, 2, ..., d-1$ (See Figure 3). When $G := G^{(i)}$, $P := P^{(i)}$, $v := v^{(i)}$ for each i, we simplify $GBG(G^{(1)}, v^{(1)}, G^{(2)}, v^{(2)}, \cdots, G^{(d)}, v^{(d)}; P^{(1)}, P^{(2)}, \cdots, P^{(d-1)})$ to be $GBG(G, v; P; d)$. In this paper, if G is a star, then v is the central vertex and if G is a cycle, v is considered as any vertex. In such cases, we further simplify $GBG(G, v; P; d)$ to be $GBG(G, P; d)$. The bridge molecular graph's bridge is strictly P_2 in [6], which limits the scope of modeling objects. The generalized bridge molecular graphs can model more kinds of long chain polymerization products than the bridge molecular graphs, because the bridge can be either P_2 or P_n and $n \geq 3$.

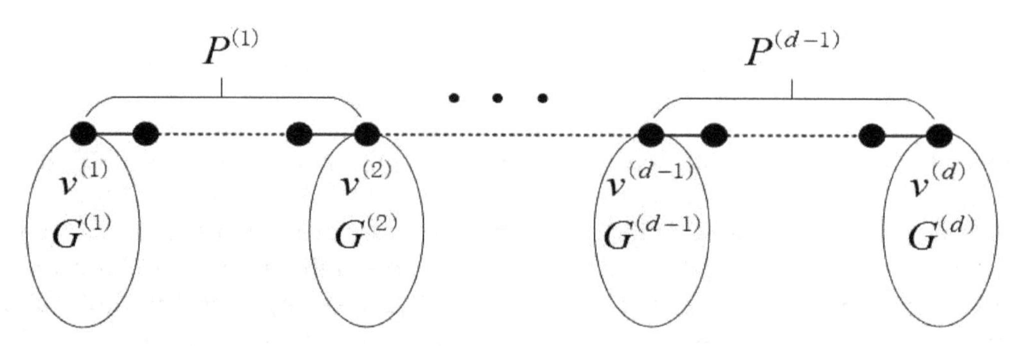

Figure 3. The generalized bridge molecular graph $GBG(G^{(1)}, v^{(1)}, G^{(2)}, v^{(2)}, ..., G^{(d)}, v^{(d)}; P^{(1)}, P^{(2)}, ..., P^{(d-1)})$.

2.2. Results and Discussion

In the following, we discuss the edge-version atom-bond connectivity and geometric arithmetic indices of some generalized bridge molecular graph. The line graph $GBG^L(S_m, P_n; d)$ of $GBG(S_m, P_n; d)$ is illustrated in Figure 4.

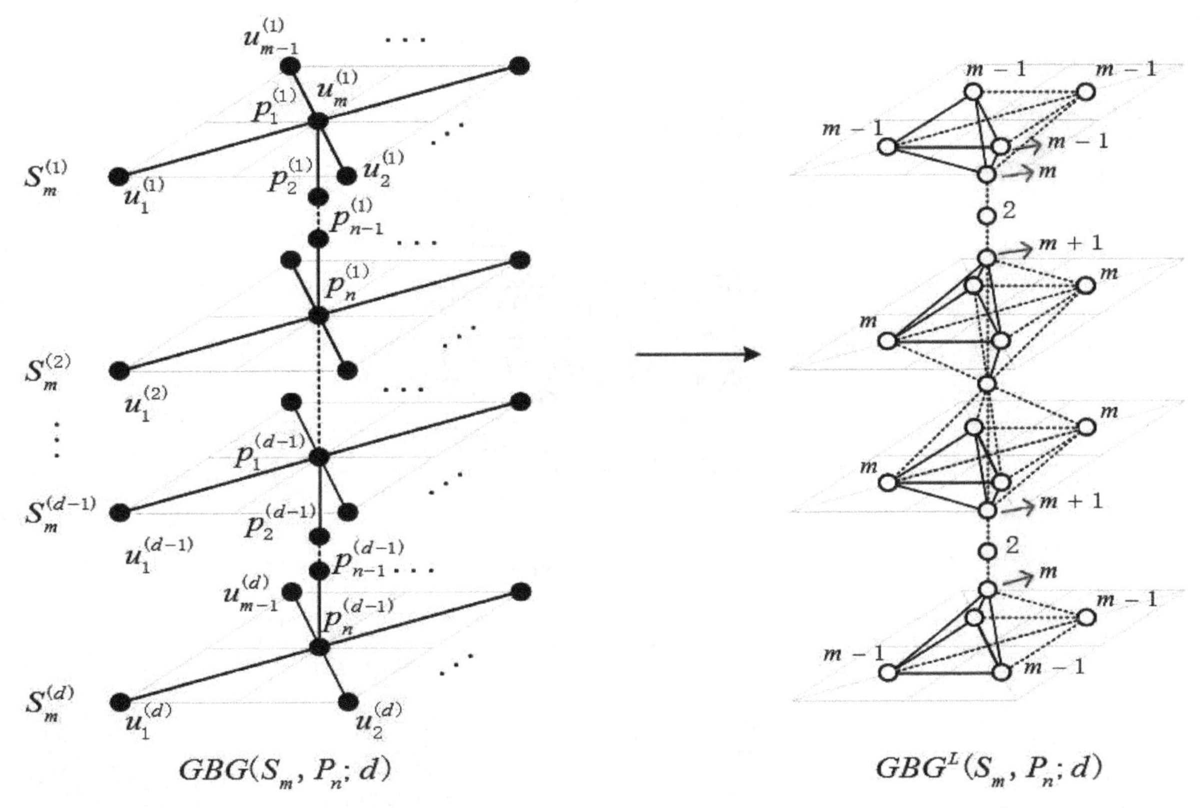

Figure 4. The generalized bridge molecular graph of $GBG(S_m, P_n; d)$ and $GBG^L(S_m, P_n; d)$.

Theorem 1. *Let $GBG(S_m, P_n; d)$ be the generalized bridge molecular graph for $n \geq 4$, $d \geq 2$ and $m \geq 2$ (see Figure 4), then the ABC_e and GA_e of $GBG(S_m, P_n; d)$ are*

$$\begin{aligned} ABC_e(GBG(S_m, P_n; d)) &= \frac{\sqrt{2}}{2}(d-1)(n-2) + \frac{d-2}{m+1}\sqrt{2m} \\ &+2(m-1)\sqrt{\frac{2m-3}{m^2-m}} + 2(d-2)(m-1)\sqrt{\frac{2m-1}{m^2+m}} \\ &+(m-2)\sqrt{2m-4} + \frac{(d-2)(m-1)(m-2)\sqrt{2m-2}}{2m}, \end{aligned}$$

$$\begin{aligned} GA_e(GBG(S_m, P_n; d)) &= \frac{4\sqrt{2m}}{m+2} + \frac{4(d-2)\sqrt{2m+2}}{m+3} + (d-1)(n-4) + (d-2) \\ &+4(m-1)\frac{\sqrt{m^2-m}}{2m-1} + 4(d-2)(m-1)\frac{\sqrt{m^2+m}}{2m+1} \\ &+\frac{d}{2}(m-1)(m-2). \end{aligned}$$

Proof. This line graph has $2 - 2m - n + \frac{d}{2}(m^2 + m + 2n - 4)$ edges. If $d_{L(G)}(e_1)$ and $d_{L(G)}(e_2)$ are the degree of edge of e_1 and e_2, then there are 2 edges of type $d_{L(G)}(e_1) = m, d_{L(G)}(e_2) = 2, 2(d-2)$ edges of type $d_{L(G)}(e_1) = m+1, d_{L(G)}(e_2) = 2$, $(d-1)(n-4)$ edges of type $d_{L(G)}(e_1) = d_{L(G)}(e_2) = 2, d-2$ edges of type $d_{L(G)}(e_1) = d_{L(G)}(e_2) = m+1, 2(m-1)$ edges of type $d_{L(G)}(e_1) = m, d_{L(G)}(e_2) = m-1$, $2(d-2)(m-1)$ edges of type $d_{L(G)}(e_1) = m$, $d_{L(G)}(e_2) = m+1$, $(m-1)(m-2)$ edges of type $d_{L(G)}(e_1) = d_{L(G)}(e_2) = m-1$, and $\frac{d-2}{2}(m-1)(m-2)$ edges of type $d_{L(G)}(e_1) = d_{L(G)}(e_2) = m$. Hence, we get

$$
\begin{aligned}
ABC_e(GBG(S_m, P_n; d)) &= 2\left(\sqrt{\frac{m+2-2}{m\times 2}}\right) + 2(d-2)\left(\sqrt{\frac{m+1+2-2}{(m+1)\times 2}}\right) \\
&\quad +(d-1)(n-4)\left(\sqrt{\frac{2+2-2}{2\times 2}}\right) \\
&\quad +(d-2)\left(\sqrt{\frac{m+1+m+1-2}{(m+1)\times(m+1)}}\right) \\
&\quad +2(m-1)\left(\sqrt{\frac{m+m-1-2}{m\times(m-1)}}\right) \\
&\quad +2(d-2)(m-1)\left(\sqrt{\frac{m+m+1-2}{m\times(m+1)}}\right) \\
&\quad +(m-1)(m-2)\left(\sqrt{\frac{m-1+m-1-2}{(m-1)\times(m-1)}}\right) \\
&\quad +\frac{d-2}{2}(m-1)(m-2)\left(\sqrt{\frac{m+m-2}{m\times m}}\right) \\
&= \frac{\sqrt{2}}{2}(d-1)(n-2) + \frac{d-2}{m+1}\sqrt{2m} \\
&\quad +2(m-1)\sqrt{\frac{2m-3}{m^2-m}} + 2(d-2)(m-1)\sqrt{\frac{2m-1}{m^2+m}} \\
&\quad +(m-2)\sqrt{2m-4} + \frac{(d-2)(m-1)(m-2)\sqrt{2m-2}}{2m},
\end{aligned}
$$

$$
\begin{aligned}
GA_e(GBG(S_m, P_n; d)) &= 2\left(\frac{2\sqrt{m\times 2}}{m+2}\right) + 2(d-2)\left(\frac{2\sqrt{(m+1)\times 2}}{m+1+2}\right) \\
&\quad +(d-1)(n-4)\left(\frac{2\sqrt{2\times 2}}{2+2}\right) \\
&\quad +(d-2)\left(\frac{2\sqrt{(m+1)\times(m+1)}}{m+1+m+1}\right) \\
&\quad +2(m-1)\left(\frac{2\sqrt{m\times(m-1)}}{m+m-1}\right) \\
&\quad +2(d-2)(m-1)\left(\frac{2\sqrt{m\times(m+1)}}{m+m+1}\right) \\
&\quad +(m-1)(m-2)\left(\frac{2\sqrt{(m-1)\times(m-1)}}{m-1+m-1}\right) \\
&\quad +\frac{d-2}{2}(m-1)(m-2)\left(\frac{2\sqrt{m\times m}}{m+m}\right) \\
&= \frac{4\sqrt{2m}}{m+2} + \frac{4(d-2)\sqrt{2m+2}}{m+3} + (d-1)(n-4) + (d-2) \\
&\quad +4(m-1)\frac{\sqrt{m^2-m}}{2m-1} + 4(d-2)(m-1)\frac{\sqrt{m^2+m}}{2m+1} \\
&\quad +\frac{d}{2}(m-1)(m-2).
\end{aligned}
$$

The proof is complete. □

For Example 1, in Figure 5, 2,7,7,12 − *tetramethyltridecane* can be modeled by $GBG(S_3, P_6; 3)$, so $ABC_e(GBG(S_3, P_6; 3)) \approx 13.76052$ and $GA_e(GBG(S_3, P_6; 3)) \approx 19.72337$.

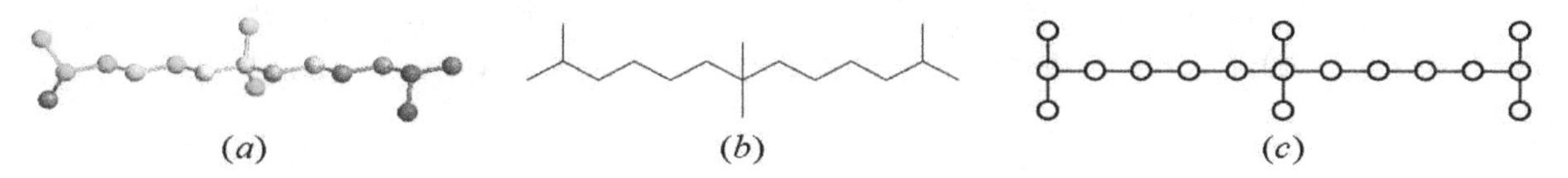

Figure 5. (**a**) 2,7,7,12-*tetramethyltridecane* ball and stick model graph in 3D; (**b**) 2,7,7,12-*tetramethyltridecane* chemical structure graph; and (**c**) 2,7,7,12-*tetramethyltridecane* model graph in chemical graph theory.

Theorem 2. *Let $GBG(S_m, P_3; d)$ be the generalized bridge molecular graph for $n = 3$, $d \geq 3$ and $m \geq 2$ (see Figure 6), then the ABC_e and GA_e of $GBG(S_m, P_3; d)$ are*

$$\begin{aligned} ABC_e(GBG(S_m, P_3; d)) &= 2\sqrt{\frac{2m-1}{m^2+m}} + \frac{2d-5}{m+1}\sqrt{2m} + 2(m-1)\sqrt{\frac{2m-3}{m^2-m}} \\ &+2(d-2)(m-1)\sqrt{\frac{2m-1}{m^2+m}} + (m-2)\sqrt{2m-4} \\ &+\frac{(d-2)(m-1)(m-2)}{2m}\sqrt{2m-2}, \end{aligned}$$

$$\begin{aligned} GA_e(GBG(S_m, P_3; d)) &= \frac{4}{2m+1}\sqrt{m^2+m} + (2d-5) \\ &+\frac{4}{2m-1}(m-1)\sqrt{m^2-m} \\ &+\frac{4}{2m+1}(d-2)(m-1)\sqrt{m^2+m} \\ &+\frac{d}{2}(m-1)(m-2). \end{aligned}$$

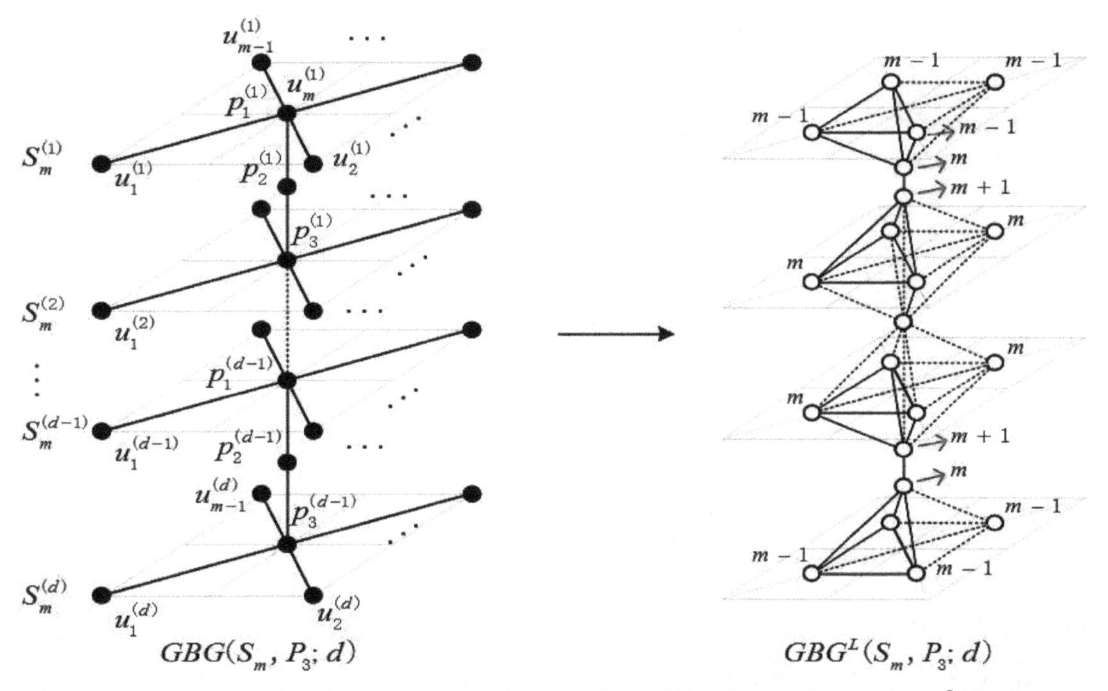

Figure 6. The generalized bridge molecular graph of $GBG(S_m, P_3; d)$ and $GBG^L(S_m, P_3; d)$.

Proof. This line graph has $\frac{d}{2}(m^2+m+2)-2m-1$ edges. If $d_{L(G)}(e_1)$ and $d_{L(G)}(e_2)$ are the degree of edge of e_1 and e_2, then there are 2 edges of type $d_{L(G)}(e_1)=m$, $d_{L(G)}(e_2)=m+1$, $2d-5$ edges of type $d_{L(G)}(e_1)=m+1$, $d_{L(G)}(e_2)=m+1$, $2(m-1)$ edges of type $d_{L(G)}(e_1)=m$, $d_{L(G)}(e_2)=m-1$, $2(d-2)(m-1)$ edges of type $d_{L(G)}(e_1)=m$,$d_{L(G)}(e_2)=m+1$, $(m-1)(m-2)$ edges of type $d_{L(G)}(e_1)=d_{L(G)}(e_2)=m-1$, and $\frac{d-2}{2}(m-1)(m-2)$ edges of type $d_{L(G)}(e_1)=d_{L(G)}(e_2)=m$. Hence, we get

$$\begin{aligned}
ABC_e(GBG(S_m,P_3;d)) &= 2\left(\sqrt{\frac{m+m+1-2}{m\times(m+1)}}\right)+(2d-5)\left(\sqrt{\frac{m+1+m+1-2}{(m+1)\times(m+1)}}\right)\\
&+2(m-1)\left(\sqrt{\frac{m+m-1-2}{m\times(m-1)}}\right)\\
&+2(d-2)(m-1)\left(\sqrt{\frac{m+m+1-2}{m\times(m+1)}}\right)\\
&+(m-1)(m-2)\left(\sqrt{\frac{m-1+m-1-2}{(m-1)\times(m-1)}}\right)\\
&+\frac{d-2}{2}(m-1)(m-2)\left(\sqrt{\frac{m+m-2}{m\times m}}\right)\\
&= 2\sqrt{\frac{2m-1}{m^2+m}}+\frac{2d-5}{m+1}\sqrt{2m}+2(m-1)\sqrt{\frac{2m-3}{m^2-m}}\\
&+2(d-2)(m-1)\sqrt{\frac{2m-1}{m^2+m}}+(m-2)\sqrt{2m-4}\\
&+\frac{(d-2)(m-1)(m-2)}{2m}\sqrt{2m-2},
\end{aligned}$$

$$\begin{aligned}
GA_e(GBG(S_m,P_3;d)) &= 2\left(\frac{2\sqrt{m\times(m+1)}}{m+m+1}\right)+(2d-5)\left(\frac{2\sqrt{(m+1)\times(m+1)}}{m+1+m+1}\right)\\
&+2(m-1)\left(\frac{2\sqrt{m\times(m-1)}}{m+m-1}\right)\\
&+2(d-2)(m-1)\left(\frac{2\sqrt{m\times(m+1)}}{m+m+1}\right)\\
&+(m-1)(m-2)\left(\frac{2\sqrt{(m-1)\times(m-1)}}{m-1+m-1}\right)\\
&+\frac{d-2}{2}(m-1)(m-2)\left(\frac{2\sqrt{m\times m}}{m+m}\right)\\
&= \frac{4}{2m+1}\sqrt{m^2+m}+(2d-5)\\
&+\frac{4}{2m-1}(m-1)\sqrt{m^2-m}\\
&+\frac{4}{2m+1}(d-2)(m-1)\sqrt{m^2+m}\\
&+\frac{d}{2}(m-1)(m-2).
\end{aligned}$$

The proof is complete. □

For Example 2, in Figure 7, 2,4,4,6 − *tetramethylheptane* can be modeled by $GBG(S_3,P_3;3)$, so $ABC_e(GBG(S_3,P_3;3))\approx 9.394663$ and $GA_e(GBG(S_3,P_3;3))\approx 13.85764$.

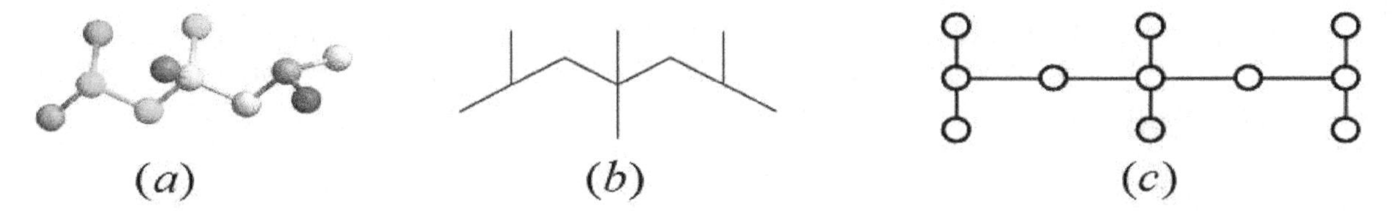

Figure 7. (**a**) 2, 4, 4, 6-*tetramethylheptane* ball and stick model graph in 3*D*; (**b**) 2, 4, 4, 6-*tetramethylheptane* chemical structure graph; and (**c**) 2, 4, 4, 6-*tetramethylheptane* model graph in chemical graph theory.

Theorem 3. *Let $GBG(S_m, P_2; d)$ be the generalized bridge molecular graph for $n = 2$, $d \geq 4$ and $m \geq 2$ (see Figure 8), then the ABC_e and GA_e of $GBG(S_m, P_2; d)$ are*

$$\begin{aligned} ABC_e(GBG(S_m, P_2; d)) &= 2(m-1)\sqrt{\frac{3m-4}{2m^2-3m+1}} + 2(m-1)\sqrt{\frac{3m-3}{2m^2-m}} \\ &+2\frac{(d-3)(m-1)}{m}\sqrt{\frac{3m-2}{2}} + (m-2)\sqrt{2m-4} \\ &+\frac{(d-2)(m-1)(m-2)}{2m}\sqrt{2m-2} \\ &+2\sqrt{\frac{4m-3}{4m^2-2m}} + \frac{d-4}{2m}\sqrt{4m-2}, \end{aligned}$$

$$\begin{aligned} GA_e(GBG(S_m, P_2; d)) &= \frac{4(m-1)}{3m-2}\sqrt{2m^2-3m+1} + \frac{4(m-1)}{3m-1}\sqrt{2m^2-m} \\ &+\frac{4\sqrt{2}}{3}(d-3)(m-1) + \frac{d}{2}(m-1)(m-2) \\ &+\frac{4(\sqrt{4m^2-2m})}{4m-1} + (d-4). \end{aligned}$$

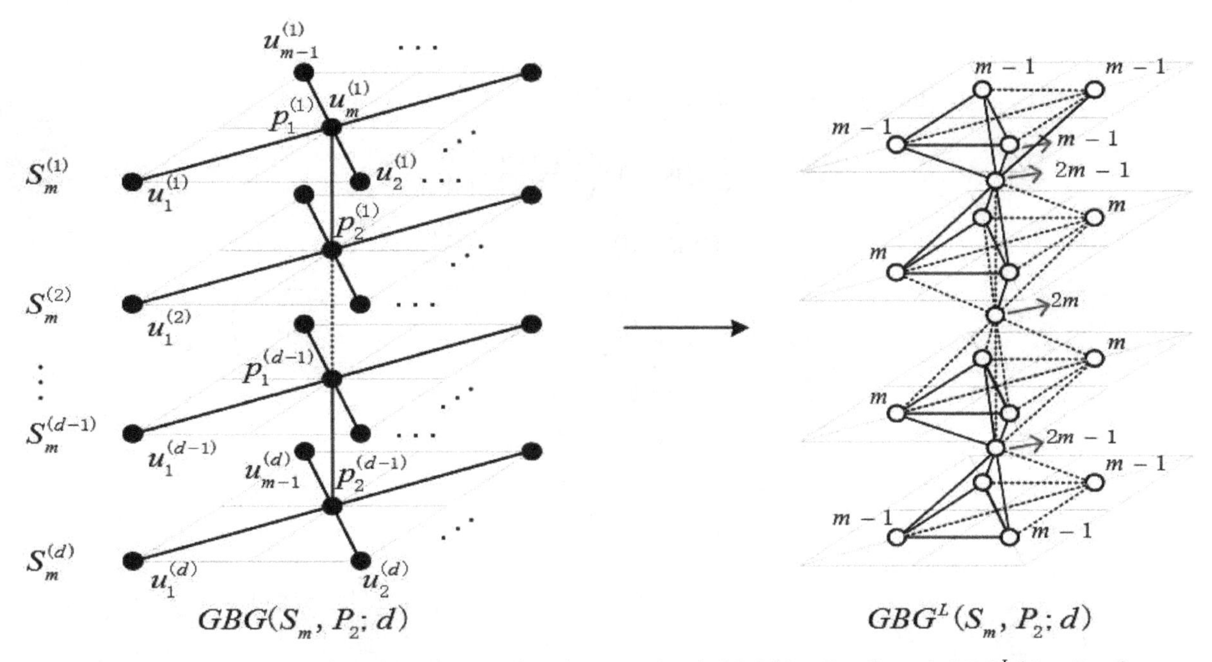

Figure 8. The generalized bridge molecular graph of $GBG(S_m, P_2; d)$ and $GBG^L(S_m, P_2; d)$.

Proof. This line graph has $\frac{1}{2}m(dm + d - 4)$ edges. If $d_{L(G)}(e_1)$ and $d_{L(G)}(e_2)$ are the degree of edge of e_1 and e_2, then there are $2(m-1)$ edges of type $d_{L(G)}(e_1) = 2m-1, d_{L(G)}(e_2) = m-1$, $2(m-1)$ edges of type $d_{L(G)}(e_1) = 2m-1, d_{L(G)}(e_2) = m$, $2(d-3)(m-1)$ edges of type $d_{L(G)}(e_1) = 2m, d_{L(G)}(e_2) = m$, $(m-1)(m-2)$ edges of type $d_{L(G)}(e_1) = d_{L(G)}(e_2) = m-1$, $\frac{d-2}{2}(m-1)(m-2)$ edges of type $d_{L(G)}(e_1) = d_{L(G)}(e_2) = m$, 2 edges of type $d_{L(G)}(e_1) = 2m-1, d_{L(G)}(e_2) = 2m$, and $d-4$ edges of type $d_{L(G)}(e_1) = d_{L(G)}(e_2) = 2m$. Hence, we get

$$\begin{aligned}
ABC_e(GBG(S_m,P_2;d)) &= 2(m-1)\left(\sqrt{\frac{2m-1+m-1-2}{(2m-1)\times(m-1)}}\right) \\
&\quad +2(m-1)\left(\sqrt{\frac{2m-1+m-2}{(2m-1)\times m}}\right) \\
&\quad +2(d-3)(m-1)\left(\sqrt{\frac{2m+m-2}{2m\times m}}\right) \\
&\quad +(m-1)(m-2)\left(\sqrt{\frac{m-1+m-1-2}{(m-1)\times(m-1)}}\right) \\
&\quad +\frac{d-2}{2}(m-1)(m-2)\left(\sqrt{\frac{m+m-2}{m\times m}}\right) \\
&\quad +2\left(\sqrt{\frac{2m-1+2m-2}{(2m-1)\times 2m}}\right) \\
&\quad +(d-4)\left(\sqrt{\frac{2m+2m-2}{2m\times 2m}}\right) \\
&= 2(m-1)\sqrt{\frac{3m-4}{2m^2-3m+1}}+2(m-1)\sqrt{\frac{3m-3}{2m^2-m}} \\
&\quad +2\frac{(d-3)(m-1)}{m}\sqrt{\frac{3m-2}{2}}+(m-2)\sqrt{2m-4} \\
&\quad +\frac{(d-2)(m-1)(m-2)}{2m}\sqrt{2m-2} \\
&\quad +2\sqrt{\frac{4m-3}{4m^2-2m}}+\frac{d-4}{2m}\sqrt{4m-2},
\end{aligned}$$

$$\begin{aligned}
GA_e(GBG(S_m,P_2;d)) &= 2(m-1)\left(\frac{2\sqrt{(2m-1)\times(m-1)}}{2m-1+m-1}\right) \\
&\quad +2(m-1)\left(\frac{2\sqrt{(2m-1)\times m}}{2m-1+m}\right) \\
&\quad +2(d-3)(m-1)\left(\frac{2\sqrt{2m\times m}}{2m+m}\right) \\
&\quad +(m-1)(m-2)\left(\frac{2\sqrt{(m-1)\times(m-1)}}{m-1+m-1}\right) \\
&\quad +\frac{d-2}{2}(m-1)(m-2)\left(\frac{2\sqrt{m\times m}}{m+m}\right) \\
&\quad +2\left(\frac{2\sqrt{(2m-1)\times 2m}}{2m-1+2m}\right) \\
&\quad +(d-4)\left(\frac{2\sqrt{2m\times 2m}}{2m+2m}\right) \\
&= \frac{4(m-1)}{3m-2}\sqrt{2m^2-3m+1}+\frac{4(m-1)}{3m-1}\sqrt{2m^2-m} \\
&\quad +\frac{4\sqrt{2}}{3}(d-3)(m-1)+\frac{d}{2}(m-1)(m-2) \\
&\quad +\frac{4(\sqrt{4m^2-2m})}{4m-1}+(d-4).
\end{aligned}$$

The proof is complete. □

For Example 3, in Figure 9, 2,3,3,4-*tetramethylpentane* can be modeled by $GBG(S_3, P_2; 4)$, so $ABC_e(GBG(S_3, P_2; 4)) \approx 11.69568$ and $GA_e(GBG(S_3, P_2; 4)) \approx 17.24996952$.

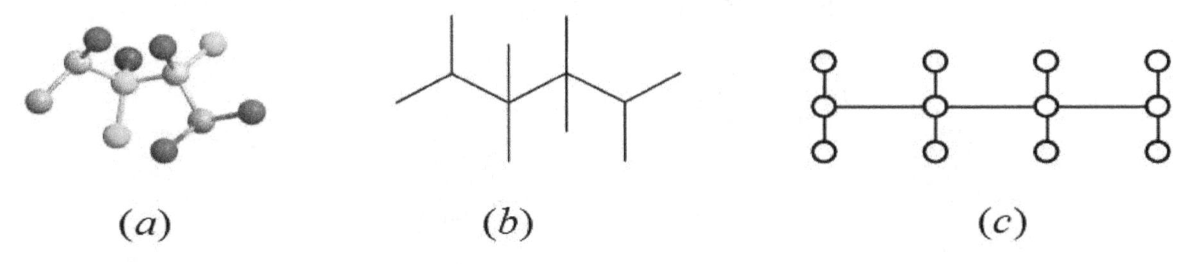

Figure 9. (**a**) 2,3,3,4-*tetramethylpentane* ball and stick model graph in 3D; (**b**) 2,3,3,4-*tetramethylpentane* chemical structure graph; and (**c**) 2,3,3,4-*tetramethylpentane* model graph in chemical graph theory.

Theorem 4. *Let $GBG(C_m, P_n; d)$ be the generalized bridge molecular graph for $n \geq 4$, $d \geq 2$ and $m \geq 3$ (see Figure 10), then the ABC_e and GA_e of $GBG(C_m, P_n; d)$ are*

$$\begin{aligned} ABC_e(GBG(C_m, P_n; d)) &= \frac{\sqrt{2}}{2}(d(m-3) + (d-1)(n-4)) + (2\sqrt{2} + \frac{3\sqrt{6}}{2})d \\ &\quad -\sqrt{2} - 3\sqrt{6} + 4, \end{aligned}$$

$$\begin{aligned} GA_e(GBG(C_m, P_n; d)) &= d(m-3) + (d-1)(n-4) \\ &\quad +(\frac{8\sqrt{2}}{3} + 6)(d-2) + \frac{12\sqrt{6}}{5} + 5. \end{aligned}$$

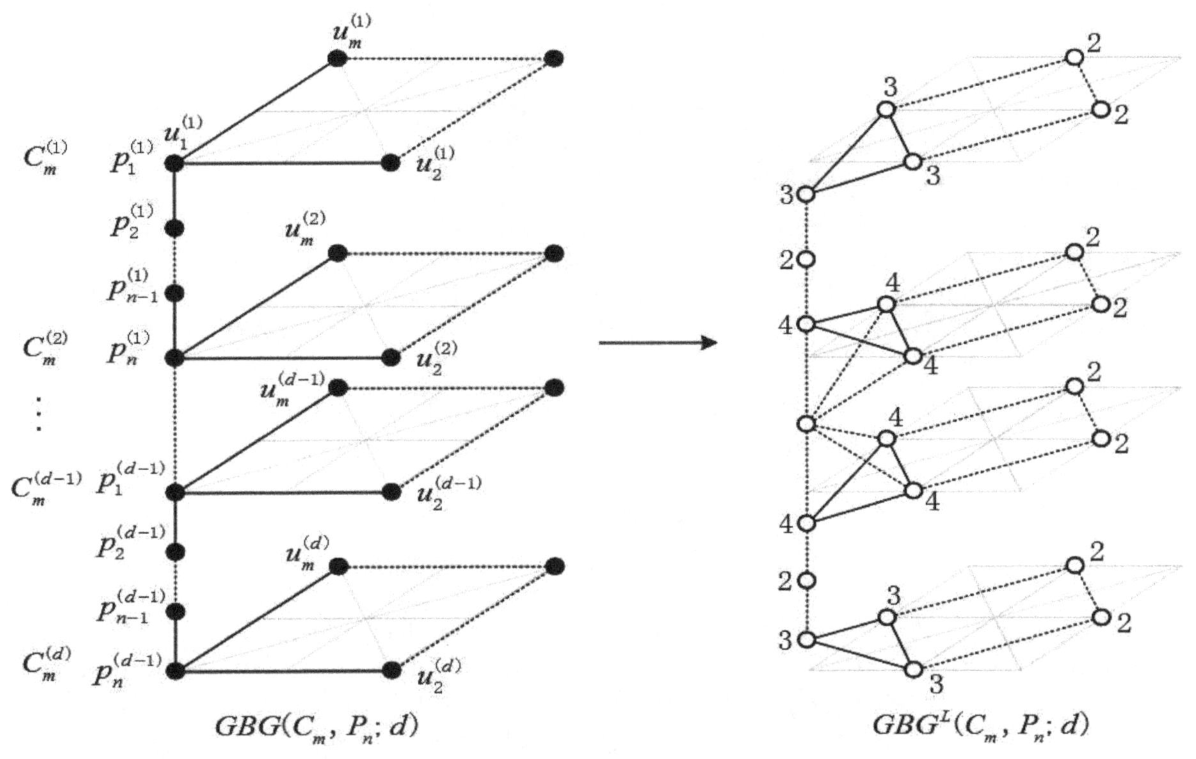

Figure 10. The generalized bridge molecular graph of $GBG(C_m, P_n; d)$ and $GBG^L(C_m, P_n; d)$.

Proof. In Figure 10, the degrees of vertices in line graph $G^L(G_d(C_m + P_n))$ are displayed near by the corresponding vertices. This line graph has $d(m+n+3) - n - 4$ edges. In addition, there are $d(m-3) + (d-1)(n-4)$ edges of type $d_{L(G)}(e_1) = d_{L(G)}(e_2) = 2$, 6 edges of type $d_{L(G)}(e_1) = 2$ and $d_{L(G)}(e_2) = 3$, 6 edges of type $d_{L(G)}(e_1) = d_{L(G)}(e_2) = 3$, $4(d-2)$ edges of type $d_{L(G)}(e_1) = 2$ and $d_{L(G)}(e_2) = 4$, and $6(d-2)$ edges of type $d_{L(G)}(e_1) = d_{L(G)}(e_2) = 4$. Hence, we have

$$\begin{aligned}
ABC_e(GBG(C_m, P_n; d)) &= \Big(d(m-3) + (d-1)(n-4)\Big)\left(\sqrt{\frac{2+2-2}{2\times 2}}\right) \\
&+6\left(\sqrt{\frac{2+3-2}{2\times 3}}\right) + 6\left(\sqrt{\frac{3+3-2}{3\times 3}}\right) \\
&+4(d-2)\left(\sqrt{\frac{2+4-2}{2\times 4}}\right) + 6(d-2)\left(\sqrt{\frac{4+4-2}{4\times 4}}\right) \\
&= \frac{\sqrt{2}}{2}(d(m-3) + (d-1)(n-4)) + (2\sqrt{2} + \frac{3\sqrt{6}}{2})d \\
&-\sqrt{2} - 3\sqrt{6} + 4,
\end{aligned}$$

$$\begin{aligned}
GA_e(GBG(C_m, P_n; d)) &= \Big(d(m-3) + (d-1)(n-4)\Big)\left(\frac{2\sqrt{2\times 2}}{2+2}\right) \\
&+6\left(\frac{2\sqrt{2\times 3}}{2+3}\right) + 6\left(\frac{2\sqrt{3\times 3}}{3+3}\right) \\
&+4(d-2)\left(\frac{2\sqrt{2\times 4}}{2+4}\right) + 6(d-2)\left(\frac{2\sqrt{4\times 4}}{4+4}\right) \\
&= d(m-3) + (d-1)(n-4) \\
&+(\frac{8\sqrt{2}}{3} + 6)(d-2) + \frac{12\sqrt{6}}{5} + 5.
\end{aligned}$$

The proof is complete. □

For Example 4, in Figure 2, $C_{24}H_{28}$ is (*cyclohexa-2,4-diene-1,1-diylbis(propane-3,1-diyl))dibenzene*, which can be modeled by $GBG(C_6, P_5; 3)$, so $ABC_e(GBG(C_6, P_5; 3)) \approx 22.52347702$ and $GA_e(GBG(C_6, P_5; 3)) \approx 31.65001155$.

Theorem 5. *Let $GBG(C_m, P_3; d)$ be the generalized bridge molecular graph for $n = 3$, $d \geq 3$, and $m \geq 3$ (see Figure 11), then the ABC_e and GA_e of $GBG(C_m, P_3; d)$ are*

$$ABC_e(GBG(C_m, P_3; d)) = \tfrac{\sqrt{2}}{2}d(m-3) + (\sqrt{2} + 7\tfrac{\sqrt{6}}{4})d + 4 + \tfrac{\sqrt{15}}{3} - \tfrac{15\sqrt{6}}{4},$$
$$GA_e(GBG(C_m, P_3; d)) = d(m-3) + (\tfrac{4\sqrt{2}}{3} + 7)d + \tfrac{8\sqrt{6}}{5} + 6 + \tfrac{8\sqrt{3}}{7} - \tfrac{8\sqrt{2}}{3} - 15.$$

Proof. In Figure 11, the degrees of vertices in line graph $G^L(GBG(C_m, P_3; d))$ are displayed near by the corresponding vertices. This line graph has $d(m+6) - 7$ edges. In addition, there are $d(m-3)$ edges of type $d_{L(G)}(e_1) = d_{L(G)}(e_2) = 2$, 4 edges of type $d_{L(G)}(e_1) = 2$ and $d_{L(G)}(e_2) = 3$, $2(d-2)$ edges of type $d_{L(G)}(e_1) = 2$ and $d_{L(G)}(e_2) = 4$, 6 edges of type $d_{L(G)}(e_1) = d_{L(G)}(e_2) = 3$, 2 edges of type $d_{L(G)}(e_1) = 3$ and $d_{L(G)}(e_2) = 4$, and $7d - 15$ edges of type $d_{L(G)}(e_1) = d_{L(G)}(e_2) = 4$. Hence, we have

$$\begin{aligned}
ABC_e(GBG(C_m, P_3; d)) &= d(m-3)\left(\sqrt{\frac{2+2-2}{2\times 2}}\right)+4\left(\sqrt{\frac{2+3-2}{2\times 3}}\right)\\
&\quad +2(d-2)\left(\sqrt{\frac{2+4-2}{2\times 4}}\right)+6\left(\sqrt{\frac{3+3-2}{3\times 3}}\right)\\
&\quad +2\left(\sqrt{\frac{3+4-2}{3\times 4}}\right)+(7d-15)\left(\sqrt{\frac{4+4-2}{4\times 4}}\right)\\
&= \frac{\sqrt{2}}{2}d(m-3)+(\sqrt{2}+7\frac{\sqrt{6}}{4})d+4\\
&\quad +\frac{\sqrt{15}}{3}-\frac{15\sqrt{6}}{4},
\end{aligned}$$

$$\begin{aligned}
GA_e(GBG(C_m, P_3; d)) &= d(m-3)\left(\frac{2\sqrt{2\times 2}}{2+2}\right)+4\left(\frac{2\sqrt{2\times 3}}{2+3}\right)\\
&\quad +2(d-2)\left(\frac{2\sqrt{2\times 4}}{2+4}\right)+6\left(\frac{2\sqrt{3\times 3}}{3+3}\right)\\
&\quad +2\left(\frac{2\sqrt{3\times 4}}{3+4}\right)+(7d-15)\left(\frac{2\sqrt{4\times 4}}{4+4}\right)\\
&= d(m-3)+(\frac{4\sqrt{2}}{3}+7)d+\frac{8\sqrt{6}}{5}+6\\
&\quad +\frac{8\sqrt{3}}{7}-\frac{8\sqrt{2}}{3}-15.
\end{aligned}$$

The proof is complete. □

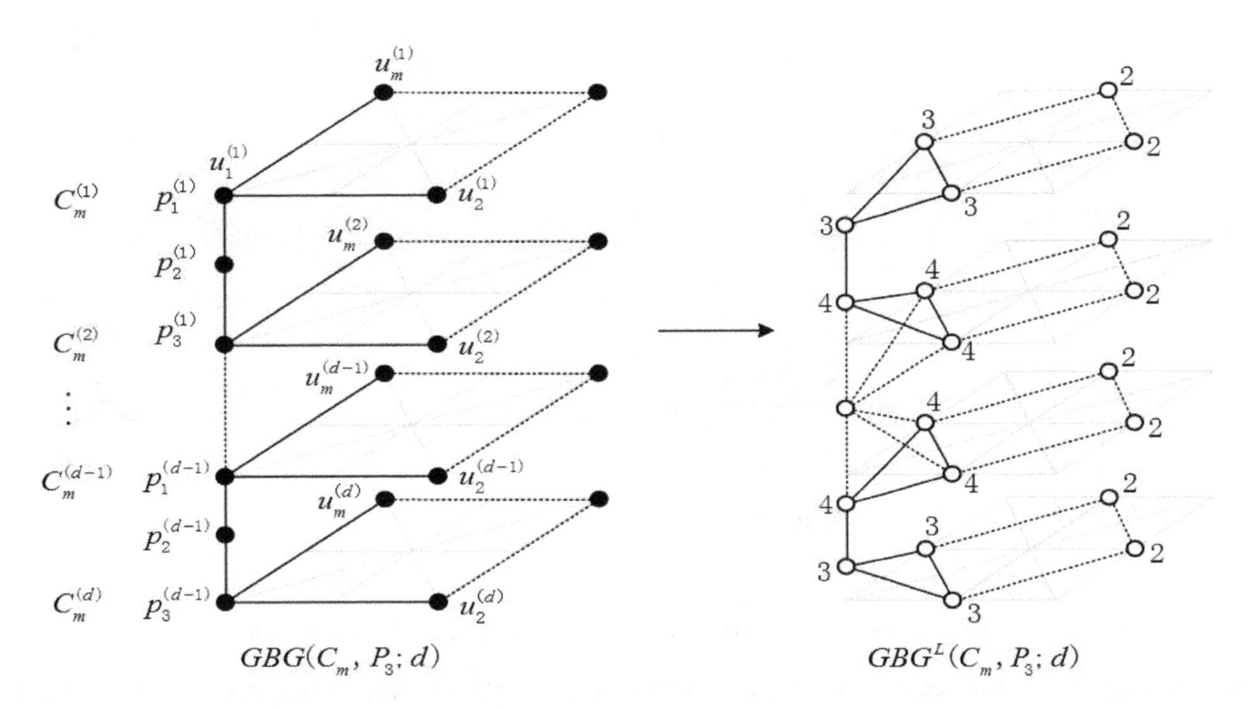

Figure 11. The generalized bridge molecular graph of $GBG(C_m, P_3; d)$ and $GBG^L(C_m, P_3; d)$.

For Example 5, in Figure 12, *(cyclohexane-1,1-diylbis(methylene))dicyclohexane* can be modeled by $GBG(C_6, P_3; 3)$, so $ABC_e(GBG(C_6, P_3; 3)) \approx 19.57183078$ and $GA_e(GBG(C_6, P_3; 3)) \approx 28.78428831$.

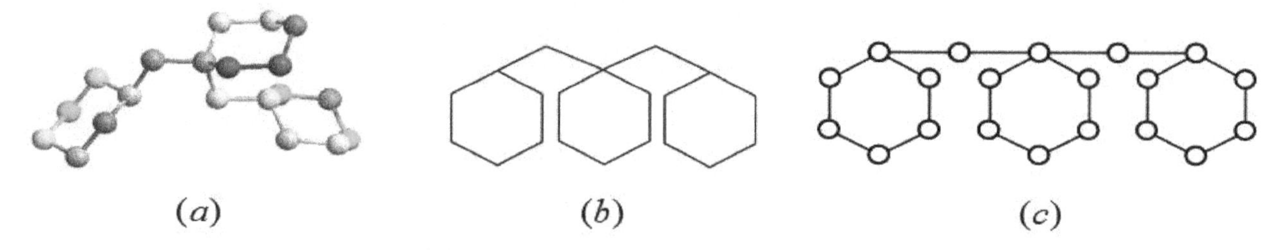

Figure 12. **(a)** *(cyclohexane-1, 1-diylbis(methylene))dicyclohexane* ball and stick model graph in *3D*; **(b)** *(cyclohexane-1, 1-diylbis(methylene))dicyclohexane* chemical structure graph; and **(c)** *(cyclohexane-1, 1-diylbis(methylene))dicyclohexane* model graph in chemical graph theory.

Theorem 6. *Let $GBG(C_m, P_2; d)$ be the generalized bridge molecular graph for $n = 2$, $d \geq 4$, and $m \geq 3$ (see Figure 13), then the ABC_e and GA_e of $GBG(C_m, P_2; d)$ are*

$$\begin{aligned} ABC_e(GBG(C_m, P_2; d)) &= \frac{\sqrt{2}}{2}dm + (\sqrt{2} + \frac{\sqrt{6}}{4} + \frac{4\sqrt{3}}{3} + \frac{\sqrt{10}}{6} - \frac{3\sqrt{2}}{2})d \\ &\quad + \frac{4}{3} + 4\sqrt{\frac{2}{5}} + 2\sqrt{\frac{7}{5}} + 2\sqrt{\frac{3}{10}} - \frac{\sqrt{6}}{2} - 4\sqrt{3} - \frac{2\sqrt{10}}{3}, \end{aligned}$$

$$\begin{aligned} GA_e(GBG(C_m, P_2; d)) &= dm + (\frac{4\sqrt{2}}{3} + \frac{8\sqrt{6}}{5} - 1)d + \sqrt{15} + \frac{16\sqrt{5}}{9} + \frac{4\sqrt{30}}{11} \\ &\quad - \frac{16\sqrt{6}}{5} - \frac{8\sqrt{2}}{3} - 4. \end{aligned}$$

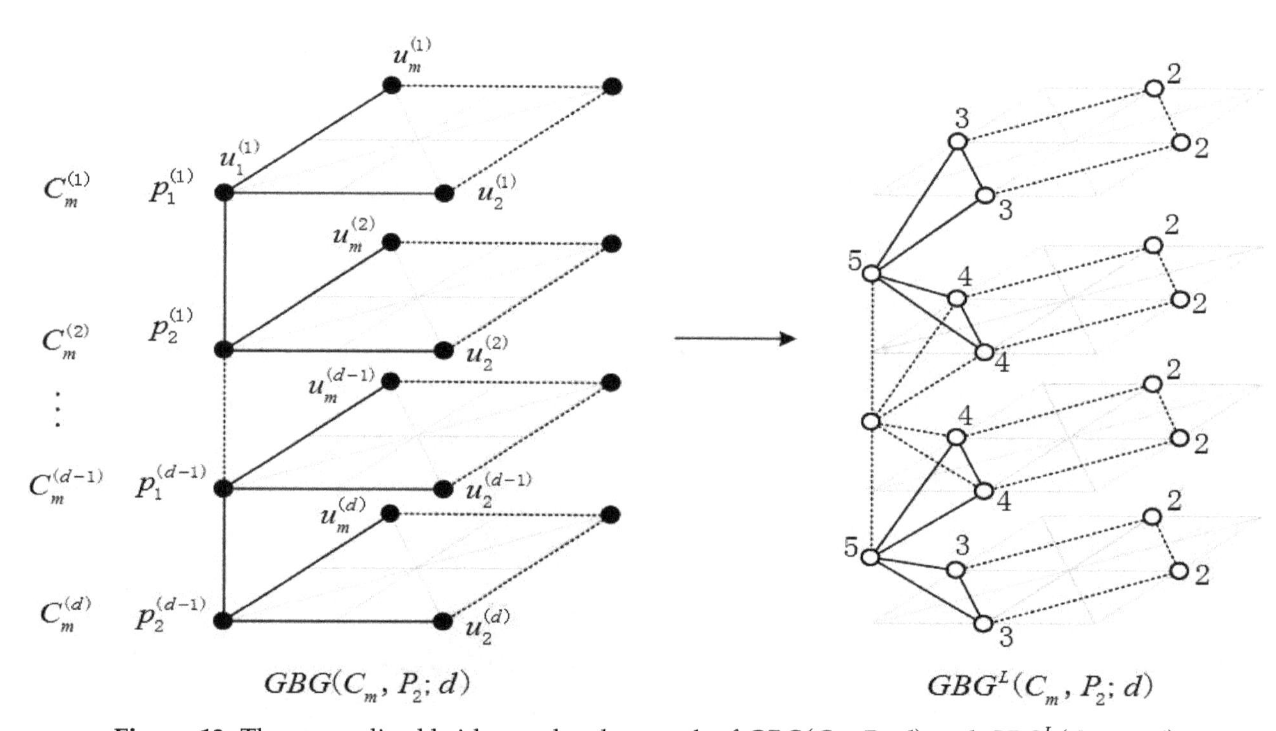

Figure 13. The generalized bridge molecular graph of $GBG(C_m, P_2; d)$ and $GBG^L(C_m, P_2; d)$.

Proof. In Figure 13, the degrees of vertices in line graph $G^L(GBG(C_m, P_2; d))$ are displayed near by the corresponding vertices. This line graph has $d(m-5) - 6$ edges. In addition, there are $d(m-3)$ edges of type $d_{L(G)}(e_1) = d_{L(G)}(e_2) = 2$, 4 edges of type $d_{L(G)}(e_1) = 2$, $d_{L(G)}(e_2) = 3$, $2(d-2)$ edges of type $d_{L(G)}(e_1) = 2$,$d_{L(G)}(e_2) = 4$, 2 edges of type $d_{L(G)}(e_1) = d_{L(G)}(e_2) = 3$, 4 edges of type $d_{L(G)}(e_1) = 3$, $d_{L(G)}(e_2) = 5$, $d-2$ edges of type $d_{L(G)}(e_1) = d_{L(G)}(e_2) = 4$, 4 edges of type

$d_{L(G)}(e_1) = 4$, $d_{L(G)}(e_2) = 5$, $4(d-3)$ edges of type $d_{L(G)}(e_1) = 4$, $d_{L(G)}(e_2) = 6$, 2 edges of type $d_{L(G)}(e_1) = 5$, $d_{L(G)}(e_2) = 6$, and $d-4$ edges of type $d_{L(G)}(e_1) = d_{L(G)}(e_2) = 6$. Hence, we have

$$\begin{aligned} ABC_e(GBG(C_m, P_2; d)) &= d(m-3)\Big(\sqrt{\frac{2+2-2}{2\times 2}}\Big) + 4\Big(\sqrt{\frac{2+3-2}{2\times 3}}\Big) \\ &+2(d-2)\Big(\sqrt{\frac{2+4-2}{2\times 4}}\Big) + 2\Big(\sqrt{\frac{3+3-2}{3\times 3}}\Big) \\ &+4\Big(\sqrt{\frac{3+5-2}{3\times 5}}\Big) + (d-2)\Big(\sqrt{\frac{4+4-2}{4\times 4}}\Big) \\ &+4\Big(\sqrt{\frac{4+5-2}{4\times 5}}\Big) + 4(d-3)\Big(\sqrt{\frac{4+6-2}{4\times 6}}\Big) \\ &+2\Big(\sqrt{\frac{5+6-2}{5\times 6}}\Big) + (d-4)\Big(\sqrt{\frac{6+6-2}{6\times 6}}\Big) \\ &= \frac{\sqrt{2}}{2}dm + (\sqrt{2} + \frac{\sqrt{6}}{4} + \frac{4\sqrt{3}}{3} + \frac{\sqrt{10}}{6} - \frac{3\sqrt{2}}{2})d \\ &+\frac{4}{3} + 4\sqrt{\frac{2}{5}} + 2\sqrt{\frac{7}{5}} + 2\sqrt{\frac{3}{10}} - \frac{\sqrt{6}}{2} - 4\sqrt{3} - \frac{2\sqrt{10}}{3}, \end{aligned}$$

$$\begin{aligned} GA_e(GBG(C_m, P_2; d)) &= d(m-3)\Big(\frac{2\sqrt{2\times 2}}{2+2}\Big) + 4\Big(\frac{2\sqrt{2\times 3}}{2+3}\Big) \\ &+2(d-2)\Big(\frac{2\sqrt{2\times 4}}{2+4}\Big) + 2\Big(\frac{2\sqrt{3\times 3}}{3+3}\Big) \\ &+4\Big(\frac{2\sqrt{3\times 5}}{3+5}\Big) + (d-2)\Big(\frac{2\sqrt{4\times 4}}{4+4}\Big) \\ &+4\Big(\frac{2\sqrt{4\times 5}}{4+5}\Big) + 4(d-3)\Big(\frac{2\sqrt{4\times 6}}{4+6}\Big) \\ &+2\Big(\frac{2\sqrt{5\times 6}}{5+6}\Big) + (d-4)\Big(\frac{2\sqrt{6\times 6}}{6+6}\Big) \\ &= dm + (\frac{4\sqrt{2}}{3} + \frac{8\sqrt{6}}{5} - 1)d + \sqrt{15} + \frac{16\sqrt{5}}{9} + \frac{4\sqrt{30}}{11} \\ &- \frac{16\sqrt{6}}{5} - \frac{8\sqrt{2}}{3} - 4. \end{aligned}$$

The proof is complete.$\gamma_i(C_6 \Box C_n)$ □

For Example 6, in Figure 14, $2'H,2''H$-$1,1' : 1',1'' : 1'',1'''$-*quaterphenyl* can be modeled by $GBG(C_6, P_2; 4)$, so $ABC_e(GBG(C_6, P_2; 4)) \approx 25.00131406$ and $GA_e(GBG(C_6, P_2; 4)) \approx 37.44953704$.

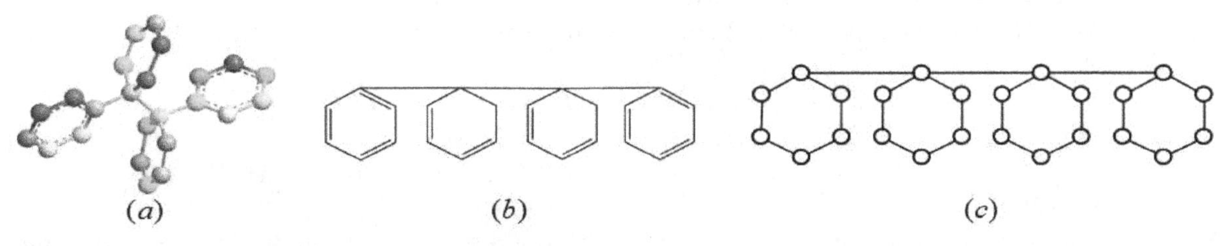

Figure 14. (**a**) $2'H,2''H$-$1,1' : 1',1'' : 1'',1'''$-*quaterphenyl* ball and stick model graph in $3D$; (**b**) $2'H,2''H$-$1,1' : 1',1'' : 1'',1'''$-*quaterphenyl* chemical structure graph; and (**c**) $2'H,2''H$-$1,1' : 1',1'' : 1'',1'''$-*quaterphenyl* model graph in chemical graph theory.

3. Conclusions

Topological indices are proven to be very helpful to test the chemical properties of new chemical or physical materials. To describe more kinds of long chain polymerization products than the bridge molecular graphs, we propose the generalized bridge molecular graph structures. In this paper, we focus on some generalized bridge molecular graphs such as $GBG(S_m, P_n; d)$ and $GBG(C_m, P_n; d)$ and give the formulas of the edge version ABC and GA indices of these generalized bridge molecular graphs. By demonstrating the calculation of real molecules, we find that some long chain molecular graphs can be quickly modeled and their topological indices can be calculated using generalized bridge molecular graphs. The results of this paper also offer promising prospects in the applications for chemical and material engineering, especially in chemical industry research.

Author Contributions: X.Z. contribute for conceptualization, designing the experiments, wrote original draft preparation and funding; J.-B.L. contribute for supervision, methodology, project administration and formal analysing; X.W. contribute for performed experiments, validation. M.K.J. for resources, software, some computations; S.A. reviewed and edited initial draft and wrote the final draft; M.R.F. analyzed the data curation; All authors read and approved the final version of the paper.

Acknowledgments: The authors are grateful to the anonymous reviewers and the editor for the valuable comments and suggestions.

References

1. Harary, F.; Norman, R.Z. Some properties of line digraphs. *Rendiconti del Circolo Matematico di Palermo* **1960**, *9*, 161–169. [CrossRef]
2. Wiener, H. Structural determination of paraffin boiling points. *J. Am. Chem. Soc.* **1947**, *69*, 7–20. [CrossRef]
3. Iranmanesh, A.; Gutman, I.; Khormali, O.; Mahmiani, A. The edge versions of the wiener index. *MATCH Commun. Math. Comput. Chem.* **2009**, *61*, 663–672.
4. Devillers, J.; Balaban, A.T. *Topological Indices and Related Descriptors in QSAR and QSPR*; CRC Press: Boca Raton, FL, USA, 1999.
5. Gao, W.; Farahani, M.R.; Wang, S.H.; Husin, M.N. On the edge-version atom-bond connectivity and geometric arithmetic indices of certain graph operations. *Appl. Math. Comput.* **2017**, *308*, 11–17. [CrossRef]
6. Gao, W.; Siddiqui, M.K.; Imran, M.; Jamil, M.K.; Farahani, M. R. Forgotten topological index of chemical structure in drugs. *Saudi Pharm. J.* **2016**, *24*, 258. [CrossRef] [PubMed]
7. Nikolić, S.; Kovačević, G.; Miličević, A.; Trinajstić, N. The Zagreb indices 30 years after. *Croat. Chem. Acta* **2003**, *76*, 113–124.
8. Estrada, E.; Torres, L.; Rodriguez, L.; Gutman, I. An atom-bond connectivity index: Modelling the enthalpy of formation of Alkanes. *Indian J. Chem.* **1998**, *37A*, 849–855.
9. Dimitrov, D. Efficient computation of trees with minimal atom–bond connectivity index. *Appl. Math. Comput.* **2013**, *224*, 663–670. [CrossRef]
10. Shao, Z.; Wu, P.; Gao, Y.; Gutman, I.; Zhang, X.J. On the maximum ABC index of graphs without pendent vertices. *Appl. Math. Comput.* **2017**, *315*, 298–312. [CrossRef]
11. Shao, Z.; Wu, P.; Zhang, X.; Dimitrov, D.; Liu, J.B. On the maximum ABC index of graphs with prescribed size and without pendent vertices. *IEEE Access* **2018**, *6*, 27604–27616. [CrossRef]
12. Farahani, M.R. The edge version of atom bond connectivity index of connected graph. *Acta Univ. Apul.* **2012**, *36*, 277–284.
13. Gao, W.; Husin, M.N.; Farahani, M.R.; Imran, M. On the Edges Version of Atom-Bond Connectivity Index of Nanotubes. *J. Comput. Theor. Nanosci.* **2017**, *13*, 6733–6740. [CrossRef]
14. Farahani, M.R. The Edge Version of Geometric-Arithmetic Index of Benzenoid Graph. *Proc. Romanian Acad. Ser. B* **2013**, 15, 95–98.
15. Vukicevic, D.; Furtula, B. Topological index based on the ratios of geometric and arithmetical means of end-vertex degrees of edges. *J. Math. Chem.* **2009**, *46*, 1369–1376. [CrossRef]
16. Martińez-Peŕez, A.; Rodriǵuez, J.M,; Sigarreta, J.M. CMMSE: A new approximation to the geometric–arithmetic index. *J. Math. Chem.* **2017**, *4*, 1–19.

17. Baig, A.Q.; Imran, M.; Khalid, W.; Naeem, M. Molecular description of carbon graphite and crystal cubic carbon structures. *Can. J. Chem.* **2017**, *95*, 674–686. [CrossRef]
18. Gutman, I.; Furtula, B.; Das, K.C. Extended energy and its dependence on molecular structure. *Can. J. Chem.* **2017**, *95*, 526–529. [CrossRef]
19. Mahmiani, A.; Khormali, O.; Iranmanesh, A. On the edge version of geometric-arithmetic index. *Dig. J. Nanomater. Biostruct.* **2012**, *7*, 411–414.
20. Zafar, S.; Nadeem, M.F.; Zahid, Z. On the edge version of geometric-arithmetic index of nanocones. *Stud. Univ. Babes-Bolyai Chem.* **2016**, *61*, 273–282.
21. Gao, W.; Husin, M.N.; Farahani,M.R.; Imran, M. On the Edges Version of Atom-Bond Connectivity and Geometric Arithmetic Indices of Nanocones CNCk. *J. Comput. Theor. Nanosci.* **2016**, *13*, 6741–6746. [CrossRef]

5

Novel Three-Way Decisions Models with Multi-Granulation Rough Intuitionistic Fuzzy Sets

Zhan-Ao Xue [1,2,*], Dan-Jie Han [1,2,*], Min-Jie Lv [1,2] and Min Zhang [1,2]

1 College of Computer and Information Engineering, Henan Normal University, Xinxiang 453007, China; lmj2921419592@163.com (M.-J.L.); zhang_min95@163.com (M.Z.)
2 Engineering Lab of Henan Province for Intelligence Business & Internet of Things, Henan Normal University, Xinxiang 453007, China
* Correspondence: 121017@htu.edu.cn (Z.-A.X.); handanjie2017@163.com (D.-J.H.)

Abstract: The existing construction methods of granularity importance degree only consider the direct influence of single granularity on decision-making; however, they ignore the joint impact from other granularities when carrying out granularity selection. In this regard, we have the following improvements. First of all, we define a more reasonable granularity importance degree calculating method among multiple granularities to deal with the above problem and give a granularity reduction algorithm based on this method. Besides, this paper combines the reduction sets of optimistic and pessimistic multi-granulation rough sets with intuitionistic fuzzy sets, respectively, and their related properties are shown synchronously. Based on this, to further reduce the redundant objects in each granularity of reduction sets, four novel kinds of three-way decisions models with multi-granulation rough intuitionistic fuzzy sets are developed. Moreover, a series of concrete examples can demonstrate that these joint models not only can remove the redundant objects inside each granularity of the reduction sets, but also can generate much suitable granularity selection results using the designed comprehensive score function and comprehensive accuracy function of granularities.

Keywords: three-way decisions; intuitionistic fuzzy sets; multi-granulation rough intuitionistic fuzzy sets; granularity importance degree

---

1. Introduction

Pawlak [1,2] proposed rough sets theory in 1982 as a method of dealing with inaccuracy and uncertainty, and it has been developed into a variety of theories [3–6]. For example, the multi-granulation rough sets (MRS) model is one of the important developments [7,8]. The MRS can also be regarded as a mathematical framework to handle granular computing, which is proposed by Qian et al. [9]. Thereinto, the problem of granularity reduction is a vital research aspect of MRS. Considering the test cost problem of granularity structure selection in data mining and machine learning, Yang et al. constructed two reduction algorithms of cost-sensitive multi-granulation decision-making system based on the definition of approximate quality [10]. Through introducing the concept of distribution reduction [11] and taking the quality of approximate distribution as the measure in the multi-granulation decision rough sets model, Sang et al. proposed an α-lower approximate distribution reduction algorithm based on multi-granulation decision rough sets, however, the interactions among multiple granularities were not considered [12]. In order to overcome the problem of updating reduction, when the large-scale data vary dynamically, Jing et al. developed an incremental attribute reduction approach based on knowledge granularity with a multi-granulation view [13]. Then other multi-granulation reduction methods have been put forward one after another [14–17].

The notion of intuitionistic fuzzy sets (IFS), proposed by Atanassov [18,19], was initially developed in the framework of fuzzy sets [20,21]. Within the previous literature, how to get reasonable membership and non-membership functions is a key issue. In the interest of dealing with fuzzy information better, many experts and scholars have expanded the IFS model. Huang et al. combined IFS with MRS to obtain intuitionistic fuzzy MRS [22]. On the basis of fuzzy rough sets, Liu et al. constructed covering-based multi-granulation fuzzy rough sets [23]. Moreover, multi-granulation rough intuitionistic fuzzy cut sets model was structured by Xue et al. [24]. In order to reduce the classification errors and the limitation of ordering by single theory, they further combined IFS with graded rough sets theory based on dominance relation and extended them to a multi-granulation perspective. [25]. Under the optimistic multi-granulation intuitionistic fuzzy rough sets, Wang et al. proposed a novel method to solve multiple criteria group decision-making problems [26]. However, the above studies rarely deal with the optimal granularity selection problem in intuitionistic fuzzy environments. The measure of similarity between intuitionistic fuzzy sets is also one of the hot areas of research for experts, and some similarity measures about IFS are summarized in references [27–29], whereas these metric formulas cannot measure the importance degree of multiple granularities in the same IFS.

For further explaining the semantics of decision-theoretic rough sets (DTRS), Yao proposed a three-way decisions theory [30,31], which vastly pushed the development of rough sets. As a risk decision-making method, the key strategy of three-way decisions is to divide the domain into acceptance, rejection, and non-commitment. Up to now, researchers have accumulated a vast literature on its theory and application. For instance, in order to narrow the applications limits of three-way decisions model in uncertainty environment, Zhai et al. extended the three-way decisions models to tolerance rough fuzzy sets and rough fuzzy sets, respectively, the target concepts are relatively extended to tolerance rough fuzzy sets and rough fuzzy sets [32,33]. To accommodate the situation where the objects or attributes in a multi-scale decision table are sequentially updated, Hao et al. used sequential three-way decisions to investigate the optimal scale selection problem [34]. Subsequently, Luo et al. applied three-way decisions theory to incomplete multi-scale information systems [35]. With respect to multiple attribute decision-making, Zhang et al. study the inclusion relations of neutrosophic sets in their case in reference [36]. For improving the classification correct rate of three-way decisions, Zhang et al. proposed a novel three-way decisions model with DTRS by considering the new risk measurement functions through the utility theory [37]. Yang et al. combined three-way decisions theory with IFS to obtain novel three-way decision rules [38]. At the same time, Liu et al. explored the intuitionistic fuzzy three-way decision theory based on intuitionistic fuzzy decision systems [39]. Nevertheless, Yang et al. [38] and Liu et al. [39] only considered the case of a single granularity, and did not analyze the decision-making situation of multiple granularities in an intuitionistic fuzzy environment. The DTRS and three-way decisions theory are both used to deal with decision-making problems, so it is also enlightening for us to study three-way decisions theory through DTRS. An extension version that can be used to multi-periods scenarios has been introduced by Liang et al. using intuitionistic fuzzy decision- theoretic rough sets [40]. Furthermore, they introduced the intuitionistic fuzzy point operator into DTRS [41]. The three-way decisions are also applied in multiple attribute group decision making [42], supplier selection problem [43], clustering analysis [44], cognitive computer [45], and so on. However, they have not applied the three-way decisions theory to the optimal granularity selection problem. To solve this problem, we have expanded the three-way decisions models.

The main contributions of this paper include four points:

(1) The new granularity importance degree calculating methods among multiple granularities (i.e., $sig\prime_{in}^{\Delta}(A_i, A\prime, D)$ and $sig\prime_{out}^{\Delta}(A_i, A\prime, D)$) are given respectively, which can generate more discriminative granularities.

(2) Optimistic optimistic multi-granulation rough intuitionistic fuzzy sets (OOMRIFS) model, optimistic pessimistic multi-granulation rough intuitionistic fuzzy sets (OIMRIFS) model,

pessimistic optimistic multi-granulation rough intuitionistic fuzzy sets (IOMRIFS) model and pessimistic pessimistic multi-granulation rough intuitionistic fuzzy sets (IIMRIFS) model are constructed by combining intuitionistic fuzzy sets with the reduction of the optimistic and pessimistic multi-granulation rough sets. These four models can reduce the subjective errors caused by a single intuitionistic fuzzy set.

(3) We put forward four kinds of three-way decisions models based on the proposed four multi-granulation rough intuitionistic fuzzy sets (MRIFS), which can further reduce the redundant objects in each granularity of reduction sets.

(4) Comprehensive score function and comprehensive accuracy function based on MRIFS are constructed. Based on this, we can obtain the optimal granularity selection results.

The rest of this paper is organized as follows. In Section 2, some basic concepts of MRS, IFS, and three-way decisions are briefly reviewed. In Section 3, we propose two new granularity importance degree calculating methods and a granularity reduction Algorithm 1. At the same time, a comparative example is given. Four novel MRIFS models are constructed in Section 4, and the properties of the four models are verified by Example 2. Section 5 proposes some novel three-way decisions models based on above four new MRIFS, and the comprehensive score function and comprehensive accuracy function based on MRIFS are built. At the same time, through Algorithm 2, we make the optimal granularity selection. In Section 6, we use Example 3 to study and illustrate the three-way decisions models based on new MRIFS. Section 7 concludes this paper.

2. Preliminaries

The basic notions of MRS, IFS, and three-way decisions theory are briefly reviewed in this section. Throughout the paper, we denote U as a nonempty object set, i.e., the universe of discourse and $A = \{A_1, A_2, \cdots, A_m\}$ is an attribute set.

Definition 1 ([9]). *Suppose $IS = < U, A, V, f >$ is a consistent information system, $A = \{A_1, A_2, \cdots, A_m\}$ is an attribute set. And R_{A_i} is an equivalence relation generated by A. $[x]_{A_i}$ is the equivalence class of R_{A_i}, $\forall X \subseteq U$, the lower and upper approximations of optimistic multi-granulation rough sets (OMRS) of X are defined by the following two formulas:*

$$\underline{\sum_{i=1}^{m} A_i}^{O}(X) = \{x \in U | [x]_{A_1} \subseteq X \vee [x]_{A_2} \subseteq X \vee [x]_{A_3} \subseteq X \ldots \vee [x]_{A_m} \subseteq X\};$$
$$\overline{\sum_{i=1}^{m} A_i}^{O}(X) = \sim (\underline{\sum_{i=1}^{m} A_i}^{O}(\sim X)).$$

where $\vee$ is a disjunction operation, $\sim X$ is a complement of X, if $\underline{\sum_{i=1}^{m} A_i}^{O}(X) \neq \overline{\sum_{i=1}^{m} A_i}^{O}(X)$, the pair $(\underline{\sum_{i=1}^{m} A_i}^{O}(X), \overline{\sum_{i=1}^{m} A_i}^{O}(X))$ is referred to as an optimistic multi-granulation rough set of X.

Definition 2 ([9]). *Let $IS = < U, A, V, f >$ be an information system, where $A = \{A_1, A_2, \cdots, A_m\}$ is an attribute set, and R_{A_i} is an equivalence relation generated by A. $[x]_{A_i}$ is the equivalence class of R_{A_i}, $\forall X \subseteq U$, the pessimistic multi-granulation rough sets (IMRS) of X with respect to A are defined as follows:*

$$\underline{\sum_{i=1}^{m} A_i}^{I}(X) = \{x \in U | [x]_{A_1} \subseteq X \wedge [x]_{A_2} \subseteq X \wedge [x]_{A_3} \subseteq X \wedge \ldots \wedge [x]_{A_m} \subseteq X\};$$
$$\overline{\sum_{i=1}^{m} A_i}^{I}(X) = \sim (\underline{\sum_{i=1}^{m} A_i}^{I}(\sim X)).$$

where $[x]_{A_i}(1 \leq i \leq m)$ *is equivalence class of x for* A_i, $\wedge$ *is a conjunction operation, if* $\underline{\sum_{i=1}^{m} A_i}^{I}(X) \neq \overline{\sum_{i=1}^{m} A_i}^{I}(X)$, *the pair* $(\underline{\sum_{i=1}^{m} A_i}^{I}(X), \overline{\sum_{i=1}^{m} A_i}^{I}(X))$ *is referred to as a pessimistic multi-granulation rough set of X.*

Definition 3 ([18,19]). *Let U be a finite non-empty universe set, then the IFS E in U are denoted by:*

$$E = \{< x, \mu_E(x), \nu_E(x) > | x \in U\},$$

where $\mu_E(x) : U \to [0,1]$ *and* $\nu_E(x) : U \to [0,1]$. $\mu_E(x)$ *and* $\nu_E(x)$ *are called membership and non-membership functions of the element x in E with* $0 \leq \mu_E(x) + \nu_E(x) \leq 1$. *For* $\forall\, x \in U$, *the hesitancy degree function is defined as* $\pi_E(x) = 1 - \mu_E(x) - \nu_E(x)$, *obviously,* $\pi_E(x) : U \to [0,1]$. *Suppose* $\forall\, E_1,\ E_2 \in IFS(U)$, *the basic operations of* E_1 *and* E_2 *are given as follows:*

(1) $E_1 \subseteq E_2 \Leftrightarrow \mu_{E_1}(x) \leq \mu_{E_2}(x),\ \nu_{E_1}(x) \geq \nu_{E_2}(x),\ \forall x \in U;$
(2) $A = B \Leftrightarrow \mu_A(x) = \mu_B(x),\ \nu_A(x) = \nu_B(x),\ \forall x \in U;$
(3) $E_1 \cup E_2 = \{< x, \max\{\mu_{E_1}(x), \mu_{E_2}(x)\}, \min\{\nu_{E_1}(x), \nu_{E_2}(x)\} > | x \in U\};$
(4) (4) $E_1 \cap E_2 = \{< x, \min\{\mu_{E_1}(x), \mu_{E_2}(x)\}, \max\{\nu_{E_1}(x), \nu_{E_2}(x)\} > | x \in U\};$
(5) (5) $\sim E_1 = \{< x, \nu_{E_1}(x), \mu_{E_1}(x) > | x \in U\}.$

Definition 4 ([30,31]). *Let* $U = \{x_1, x_2, \cdots, x_n\}$ *be a universe of discourse,* $\xi = \{\omega_P, \omega_N, \omega_B\}$ *represents the decisions of dividing an object x into receptive* $POS(X)$, *rejective* $NEG(X)$, *and boundary regions* $BND(X)$, *respectively. The cost functions* λ_{PP}, λ_{NP} *and* λ_{BP} *are used to represent the three decision- making costs of* $\forall x \in U$, *and the cost functions* λ_{PN}, λ_{NN} *and* λ_{BN} *are used to represent the three decision-making costs of* $\forall x \notin U$, *as shown in Table 1.*

Table 1. Cost matrix of decision actions.

Decision Actions	**Decision Functions**	
	X	**~X**
ω_P	λ_{PP}	λ_{PN}
ω_B	λ_{BP}	λ_{BN}
ω_N	λ_{NP}	λ_{NN}

According to the minimum-risk principle of Bayesian decision procedure, three-way decisions rules can be obtained as follows:

(P): If $P(X|[x]) \geq \alpha$, then $x \in POS(X)$;
(N): If $P(X|[x]) \leq \beta$, then $x \in NEG(X)$;
(B): If $\beta < P(X|[x]) < \alpha$, then $x \in BND(X)$.
Here α, β and γ represent respectively:

$$\alpha = \frac{\lambda_{PN} - \lambda_{BN}}{(\lambda_{PN} - \lambda_{BN}) + (\lambda_{BP} - \lambda_{PP})};$$

$$\beta = \frac{\lambda_{BN} - \lambda_{NN}}{(\lambda_{BN} - \lambda_{NN}) + (\lambda_{NP} - \lambda_{BP})};$$

$$\gamma = \frac{\lambda_{PN} - \lambda_{NN}}{(\lambda_{PN} - \lambda_{NN}) + (\lambda_{NP} - \lambda_{PP})}.$$

3. Granularity Reduction Algorithm Derives from Granularity Importance Degree

Definition 5 ([10,12]). *Let $DIS = (U, C \cup D, V, f)$ be a decision information system, $A = \{A_1, A_2, \cdots, A_m\}$ are m sub-attributes of condition attributes C. $U/D = \{X_1, X_2, \cdots, X_s\}$ is the partition induced by the decision attributes D, then approximation quality of U/D about granularity set A is defined as:*

$$\gamma(A, D) = \frac{\left| \cup \left\{ \underline{\sum_{i=1}^{m} A_i^{\Delta}}(X_t) | 1 \leq t \leq s \right\} \right|}{|U|}.$$

where $|X|$ denotes the cardinal number of set X. $\Delta \in \{O, I\}$ represents two cases of optimistic and pessimistic multi-granulation rough sets, the same as the following.

Definition 6 ([12]). *Let $DIS = (U, C \cup D, V, f)$ be a decision information system, $A = \{A_1, A_2, \cdots, A_m\}$ are m sub-attributes of C, $A' \subseteq A$, $X \in U/D$,*

(1) *If $\underline{\sum_{i=1, A_i \in A}^{m} A_i^{\Delta}}(X) \neq \underline{\sum_{i=1, A_i \in A - A'}^{m} A_i^{\Delta}}(X)$, then A' is important in A for X;*

(2) *If $\underline{\sum_{i=1, A_i \in A}^{m} A_i^{\Delta}}(X) = \underline{\sum_{i=1, A_i \in A - A'}^{m} A_i^{\Delta}}(X)$, then A' is not important in A for X.*

Definition 7 ([10,12]). *Suppose $DIS = (U, C \cup D, V, f)$ is a decision information system, $A = \{A_1, A_2, \cdots, A_m\}$ are m sub-attributes of C, $A' \subseteq A$. $\forall A_i \in A'$, on the granularity sets A', the internal importance degree of A_i for D can be defined as follows:*

$$sig_{in}^{\Delta}(A_i, A', D) = |\gamma(A', D) - \gamma(A' - \{A_i\}, D)|.$$

Definition 8 ([10,12]). *Let $DIS = (U, C \cup D, V, f)$ be a decision information system, $A = \{A_1, A_2, \cdots, A_m\}$ are m sub-attributes of C, $A' \subseteq A$. $\forall A_i \in A - A'$, on the granularity sets A', the external importance degree of A_i for D can be defined as follows:*

$$sig_{out}^{\Delta}(A_i, A', D) = |\gamma(A_i \cup A', D) - \gamma(A', D)|.$$

Theorem 1. *Let $DIS = (U, C \cup D, V, f)$ be a decision information system, $A = \{A_1, A_2, \cdots, A_m\}$ are m sub-attributes of C, $A' \subseteq A$.*

(1) *For $\forall A_i \in A'$, on the basis of attribute subset family A', the granularity importance degree of A_i in A' with respect to D is expressed as follows:*

$$sig_{in}^{\Delta}(A_i, A', D) = \frac{1}{m-1} \sum |sig_{in}^{\Delta}(\{A_k, A_i\}, A', D) - sig_{in}^{\Delta}(A_k, A' - \{A_i\}, D)|.$$

where $1 \leq k \leq m$, $k \neq i$, the same as the following.

(2) *For $\forall A_i \in A - A'$, on the basis of attribute subset family A', the granularity importance degree of A_i in $A - A'$ with respect to D, we have:*

$$sig_{out}^{\Delta}(A_i, A', D) = \frac{1}{m-1} \sum |sig_{out}^{\Delta}(\{A_k, A_i\}, \{A_i\} \cup A', D) - sig_{out}^{\Delta}(A_k, A', D)|.$$

Proof. (1) According to Definition 7, then

$$
\begin{aligned}
sig_{in}^{\Delta}(A_i, A\prime, D) &= |\gamma(A\prime, D) - \gamma(A\prime - \{A_i\}, D)| \\
&= \tfrac{m-1}{m-1}|\gamma(A\prime, D) - \gamma(A\prime - \{A_i\}, D)| + \sum|\gamma(A\prime - \{A_k, A_i\}, D) - \gamma(A\prime - \{A_k, A_i\}, D)| \\
&= \tfrac{1}{m-1}\sum(|\gamma(A\prime, D) - \gamma(A\prime - \{A_k, A_i\}, D) - (\gamma(A\prime - \{A_i\}, D) - \gamma(A\prime - \{A_k, A_i\}, D)|) \\
&= \tfrac{1}{m-1}\sum|sig_{in}^{\Delta}(\{A_k, A_i\}, A\prime, D) - sig_{in}^{\Delta}(A_k, A\prime - \{A_i\}, D)|.
\end{aligned}
$$

(2) According to Definition 8, we can get:

$$
\begin{aligned}
sig_{out}^{\Delta}(A_i, A\prime, D) &= |\gamma(\{A_i\} \cup A\prime, D) - \gamma(A\prime, D)| \\
&= \tfrac{m-1}{m-1}|\gamma(\{A_i\} \cup A\prime, D) - \gamma(A\prime, D)| - \sum|\gamma(A\prime - \{A_k\}, D) - \gamma(A\prime - \{A_k\}, D)| \\
&= \tfrac{1}{m-1}\sum(|\gamma(\{A_i\} \cup A\prime, D) - \gamma(A\prime - \{A_k\}, D)| - |(\gamma(A\prime - \{A_k\}, D) - \gamma(A\prime, D)|) \\
&= \tfrac{1}{m-1}\sum|sig_{out}^{\Delta}(\{A_k, A_i\}, \{A_i\} \cup A\prime, D) - sig_{out}^{\Delta}(A_k, A\prime, D)|.
\end{aligned}
$$

□

In Definitions 7 and 8, only the direct effect of a single granularity on the whole granularity sets is given, without considering the indirect effect of the remaining granularities on decision-making. The following Definitions 9 and 10 synthetically analyze the interdependence between multiple granularities and present two new methods for calculating granularity importance degree.

Definition 9. *Let $DIS = (U, C \cup D, V, f)$ be a decision information system, $A = \{A_1, A_2, \cdots, A_m\}$ are m sub-attributes of C, $A\prime \subseteq A$. $\forall A_i, A_k \in A\prime$, on the attribute subset family, A, the new internal importance degree of A_i relative to D is defined as follows:*

$$sig\prime_{in}^{\Delta}(A_i, A\prime, D) = sig_{in}^{\Delta}(A_i, A\prime, D) + \frac{1}{m-1}\sum|sig_{in}^{\Delta}(A_k, A\prime - \{A_i\}, D) - sig_{in}^{\Delta}(A_k, A\prime, D)|.$$

$sig_{in}^{\Delta}(A_i, A\prime, D)$ and $\frac{1}{m-1}\sum|sig_{in}^{\Delta}(A_k, A\prime - \{A_i\}, D) - sig_{in}^{\Delta}(A_k, A\prime, D)|$ respectively indicate the direct and indirect effects of granularity A_i on decision-making. When $|sig_{in}^{\Delta}(A_k, A\prime - \{A_i\}, D) - sig_{in}^{\Delta}(A_k, A\prime, D)|$ > 0 is satisfied, it is shown that the granularity importance degree of A_k is increased by the addition of A_i in attribute subset $A\prime - \{A_i\}$, so the granularity importance degree of A_k should be added to A_i. Therefore, when there are m sub-attributes, we should add $\frac{1}{m-1}\sum|sig_{in}^{\Delta}(A_k, A\prime - \{A_i\}, D) - sig_{in}^{\Delta}(A_k, A\prime, D)|$ to the granularity importance degree of A_i.

If $|sig_{in}^{\Delta}(A_k, A\prime - \{A_i\}, D) - sig_{in}^{\Delta}(A_k, A\prime, D)| = 0$ and $k \neq i$, then it shows that there is no interaction between granularity A_i and other granularities, which means $sig\prime_{in}^{\Delta}(A_i, A\prime, D) = sig_{in}^{\Delta}(A_i, A\prime, D)$.

Definition 10. *Let $DIS = (U, C \cup D, V, f)$ be a decision information system, $A = \{A_1, A_2, \cdots, A_m\}$ be m sub-attributes of C, $A\prime \subseteq A$. $\forall A_i \in A - A\prime$, the new external importance degree of A_i relative to D is defined as follows:*

$$sig\prime_{out}^{\Delta}(A_i, A\prime, D) = sig_{out}^{\Delta}(A_i, A\prime, D) + \frac{1}{m-1}\sum|sig_{out}^{\Delta}(A_k, A\prime, D) - sig_{out}^{\Delta}(A_k, \{A_i\} \cup A\prime, D)|.$$

Similarly, the new external importance degree calculation formula has a similar effect.

Theorem 2. *Let $DIS = (U, C \cup D, V, f)$ be a decision information system, $A = \{A_1, A_2, \cdots, A_m\}$ be m sub-attributes of C, $A\prime \subseteq A$, $\forall A_i \in A\prime$. The improved internal importance can be rewritten as:*

$$sig\prime_{in}^{\Delta}(A_i, A\prime, D) = \frac{1}{m-1}\sum sig_{in}^{\Delta}(A_i, A\prime - \{A_k\}, D).$$

Proof.

$$sig'^{\Delta}_{in}(A_i, A\prime, D) = sig^{\Delta}_{in}(A_i, A\prime, D) + \tfrac{1}{m-1}\sum |sig^{\Delta}_{in}(A_k, A\prime - \{A_i\}, D) - sig^{\Delta}_{in}(A_k, A\prime, D)|$$

$$= \tfrac{m-1}{m-1}|\gamma(A\prime, D) - \gamma(A\prime - \{A_i\}, D)| + \tfrac{1}{m-1}\sum ||\gamma(A\prime - \{A_i\}, D) -$$

$$\gamma(A\prime - \{A_k, A_i\}, D)| - |\gamma(A\prime, D) - \gamma(A\prime - \{A_k\}, D)||$$

$$= \tfrac{1}{m-1}\sum |\gamma(A\prime - \{A_k\}, D) - \gamma(A\prime - \{A_k, A_i\}, D)|$$

$$= \tfrac{1}{m-1}\sum sig^{\Delta}_{in}(A_i, A\prime - \{A_k\}, D).$$

□

Theorem 3. *Let $DIS = (U, C \cup D, V, f)$ be a decision information system, $A = \{A_1, A_2, \cdots, A_m\}$ are m sub-attributes of C, $A\prime \subseteq A$. The improved external importance can be expressed as follows:*

$$sig'^{\Delta}_{out}(A_i, A\prime, D) = \frac{1}{m-1}\sum sig^{\Delta}_{out}(A_i, \{A_k\} \cup A\prime, D).$$

Proof.

$$sig'^{\Delta}_{out}(A_i, A\prime, D) = sig^{\Delta}_{out}(A_i, A\prime, D) + \tfrac{1}{m-1}\sum |(sig^{\Delta}_{out}(A_k, A\prime, D) - sig^{\Delta}_{out}(A_k, \{A_i\} \cup A\prime, D))|$$

$$= \tfrac{m-1}{m-1}|\gamma(\{A_i\} \cup A\prime, D) - \gamma(A\prime, D)| + \tfrac{1}{m-1}\sum ||\gamma(A\prime, D) - \gamma(\{A_k\} \cup A\prime, D)| -$$

$$|\gamma(\{A_i\} \cup A\prime, D)||$$

$$= \tfrac{1}{m-1}\sum |\gamma(\{A_i, A_k\} \cup A\prime, D) - \gamma(\{A_i\} \cup A\prime, D)|$$

$$= \tfrac{1}{m-1}\sum sig^{\Delta}_{out}(A_i, \{A_k\} \cup A\prime, D).$$

□

Theorems 2 and 3 show that when $sig^{\Delta}_{in}(A_i, A\prime - \{A_k\}, D) = 0$ ($sig^{\Delta}_{out}(A_i, \{A_k\} \cup A\prime, D) = 0$) is satisfied, having $sig'^{\Delta}_{in}(A_i, A\prime, D) = 0$ ($sig'^{\Delta}_{out}(A_i, A\prime, D) = 0$). And each granularity importance degree is calculated on the basis of removing A_k from $A\prime$, which makes it more convenient for us to choose the required granularity.

According to [10,12], we can get optimistic and pessimistic multi-granulation lower approximations L^O and L^I. The granularity reduction algorithm based on improved granularity importance degree is derived from Theorems 2 and 3, as shown in Algorithm 1.

Algorithm 1. Granularity reduction algorithm derives from granularity importance degree

Input: $DIS = (U, C \cup D, V, f)$, $A = \{A_1, A_2, \cdots, A_m\}$ *are m sub-attributes of* C, $A\prime \subseteq A, \forall A_i \in A\prime$, $U/D = \{X_1, X_2, \cdots, X_s\}$;
Output: A granularity reduction set A_i^{Δ} of this information system.
1: set up $A_i^{\Delta} \leftarrow \phi, 1 \leq h \leq m$;
2: compute U/D, optimistic and pessimistic multi-granulation lower approximations L^{Δ};
3: **for** $\forall A_i \in A$
4: compute $sig\prime_{in}^{\Delta}(A_i, A\prime, D)$ via Definition 9;
5: **if** $(sig\prime_{in}^{\Delta}(A_i, A\prime, D) > 0)$ **then** $A_i^{\Delta} = A_i^{\Delta} \cup A_i$;
6: **end**
7: **for** $\forall A_i \in A - A_i^{\Delta}$
8: **if** $\gamma(A_i^{\Delta}, D) = \gamma(A, D)$ **then** compute $sig_{out}^{\prime,\Delta}(A_i, A\prime, D)$ via Definition 10;
9: **end**
10: **if** $sig\prime_{out}^{\Delta}(A_h, A\prime, D) = \max\{sig\prime_{out}^{\Delta}(A_h, A\prime, D)\}$ **then** $A_i^{\Delta} = A_i^{\Delta} \cup A_h$;
11: **end**
12: **end**
13: **for** $\forall A_i \in A_i^{\Delta}$,
14: **if** $\gamma(A_i^{\Delta} - A_i, D) = \gamma(A, D)$ **then** $A_i^{\Delta} = A_i^{\Delta} - A_i$;
15: **end**
16: **end**
17: **return** granularity reduction set A_i^{Δ};
18: **end**

Therefore, we can obtain two reductions by utilizing Algorithm 1.

Example 1. *This paper calculates the granularity importance of 10 on-line investment schemes given in Reference [12]. After comparing and analyzing the obtained granularity importance degree, we can obtain the reduction results of 5 evaluation sites through Algorithm 1, and the detailed calculation steps are as follows.*

According to [12], we can get $A = \{A_1, A_2, A_3, A_4, A_5\}$, $A\prime \subseteq A$, $U/D = \{\{x_1, x_2, x_4, x_6, x_8\}, \{x_3, x_5, x_7, x_9, x_{10}\}\}$.

(1) Reduction set of OMRS

First of all, we can calculate the internal importance degree of OMRS by Theorem 2 as shown in Table 2.

Table 2. Internal importance degree of optimistic multi-granulation rough sets (OMRS).

	A_1	A_2	A_3	A_4	A_5
$sig_{in}^{O}(A_i, A\prime, D)$	0	0.15	0.05	0	0.05
$sig\prime_{in}^{O}(A_i, A\prime, D)$	0.025	0.375	0.225	0	0

Then, according to Algorithm 1, we can deduce the initial granularity set is $\{A_1, A_2, A_3\}$. Inspired by Definition 5, we obtain $r^O(\{A_2, A_3\}, D) = r^O(A, D) = 1$. So, the reduction set of the OMRS is $A_i^O = \{A_2, A_3\}$.

As shown in Table 2, when using the new method to calculate internal importance degree, more discriminative granularities can be generated, which are more convenient for screening out the required granularities. In literature [12], the approximate quality of granularity A_2 in the reduction set is different from that of the whole granularity set, so it is necessary to calculate the external importance degree again. When calculating the internal and external importance degree, References [10,12] only considered the direct influence of the single granularity on the granularity A_2, so the influence of the granularity A_2 on the overall decision-making can't be fully reflected.

(2) Reduction set of IMRS

Similarly, by using Theorem 2, we can get the internal importance degree of each site under IMRS, as shown in Table 3.

Table 3. Internal importance degree of pessimistic multi-granulation rough sets (IMRS).

	A_1	A_2	A_3	A_4	A_5
$sig_{in}^{I}(A_i, A\prime, D)$	0	0.05	0	0	0
$sig\prime_{in}^{I}(A_i, A\prime, D)$	0	0.025	0	0.025	0.025

According to Algorithm 1, the sites 2, 4, and 5 with internal importance degrees greater than 0, which are added to the granularity reduction set as the initial granularity set, and then the approximate quality of it can be calculated as follows:

$$r^I(\{A_2, A_4\}, D) = r^I(\{A_4, A_5\}, D) = r^I(A, D) = 0.2.$$

Namely, the reduction set of IMRS is $A_i^I = \{A_2, A_4\}$ or $A_i^I = \{A_4, A_5\}$ without calculating the external importance degree.

In this paper, when calculating the internal and external importance degree of each granularity, the influence of removing other granularities on decision-making is also considered. According to Theorem 2, after calculating the internal importance degree of OMRS and IMRS, if the approximate quality of each granularity in the reduction sets are the same as the overall granularities, it is not necessary to calculate the external importance degree again, which can reduce the amount of computation.

4. Novel Multi-Granulation Rough Intuitionistic Fuzzy Sets Models

In Example 1, two reduction sets are obtained under IMRS, so we need a novel method to obtain more accurate granularity reduction results by calculating granularity reduction.

In order to obtain the optimal determined site selection result, we combine the optimistic and pessimistic multi-granulation reduction sets based on Algorithm 1 with IFS, respectively, and construct the following four new MRIFS models.

Definition 11 ([22,25]). *Suppose $IS = (U, A, V, f)$ is an information system, $A = \{A_1, A_2, \cdots, A_m\}$. $\forall E \subseteq U$, E are IFS. Then the lower and upper approximations of optimistic MRIFS of A_i are respectively defined by:*

$$\underline{\sum_{i=1}^{m} R_{A_i}}^{O}(E) = \{< x, \mu_{\underline{\sum_{i=1}^{m} R_{A_i}}^{O}(E)}(x), \nu_{\underline{\sum_{i=1}^{m} R_{A_i}}^{O}(E)}(x) > | x \in U\};$$

$$\overline{\sum_{i=1}^{m} R_{A_i}}^{O}(E) = \{< x, \mu_{\overline{\sum_{i=1}^{m} R_{A_i}}^{O}(E)}(x), \nu_{\overline{\sum_{i=1}^{m} R_{A_i}}^{O}(E)}(x) > | x \in U\}.$$

where

$$\mu_{\underline{\sum_{i=1}^{m} R_{A_i}}^{O}(E)}(x) = \bigvee_{i=1}^{m} \inf_{y \in [x]_{A_i}} \mu_E(y),\ \nu_{\underline{\sum_{i=1}^{m} R_{A_i}}^{O}(E)}(x) = \bigwedge_{i=1}^{m} \sup_{y \in [x]_{A_i}} \nu_E(y);$$

$$\mu_{\overline{\sum_{i=1}^{m} R_{A_i}}^{O}(E)}(x) = \bigwedge_{i=1}^{m} \sup_{y \in [x]_{A_i}} \mu_E(y),\ \nu_{\overline{\sum_{i=1}^{m} R_{A_i}}^{O}(E)}(x) = \bigvee_{i=1}^{m} \inf_{y \in [x]_{A_i}} \nu_E(y).$$

where R_{A_i} is an equivalence relation of x in A, $[x]_{A_i}$ is the equivalence class of R_{A_i}, and $\vee$ is a disjunction operation.

Definition 12 ([22,25]). *Suppose $IS =< U, A, V, f >$ is an information system, $A = \{A_1, A_2, \cdots, A_m\}$. $\forall E \subseteq U$, E are IFS. Then the lower and upper approximations of pessimistic MRIFS of A_i can be described as follows:*

$$\underline{\sum_{i=1}^{m} R_{A_i}}^{I}(E) = \{< x, \mu_{\underline{\sum_{i=1}^{m} R_{A_i}}^{I}(E)}(x), \nu_{\underline{\sum_{i=1}^{m} R_{A_i}}^{I}(E))}(x) > | x \in U\};$$

$$\overline{\sum_{i=1}^{m} R_{A_i}}^{I}(E) = \{< x, \mu_{\overline{\sum_{i=1}^{m} R_{A_i}}^{I}(E)}(x), \nu_{\overline{\sum_{i=1}^{m} R_{A_i}}^{I}(E)}(x) > | x \in U\}.$$

where

$$\mu_{\underline{\sum_{i=1}^{m} R_{A_i}}^{I}(E)}(x) = \bigwedge_{i=1}^{m} \inf_{y \in [x]_{A_i}} \mu_E(y),\ \nu_{\underline{\sum_{i=1}^{m} R_{A_i}}^{I}(E)}(x) = \bigvee_{i=1}^{m} \sup_{y \in [x]_{A_i}} \nu_E(y);$$

$$\mu_{\overline{\sum_{i=1}^{m} R_{A_i}}^{I}(E)}(x) = \bigvee_{i=1}^{m} \sup_{y \in [x]_{A_i}} \mu_E(y),\ \nu_{\overline{\sum_{i=1}^{m} R_{A_i}}^{I}(E)}(x) = \bigwedge_{i=1}^{m} \inf_{y \in [x]_{A_i}} \nu_E(y).$$

where $[x]_{A_i}$ is the equivalence class of x about the equivalence relation R_{A_i}, and $\wedge$ is a conjunction operation.

Definition 13. *Suppose $IS =< U, A, V, f >$ is an information system, $A_i^O = \{A_1, A_2, \cdots, A_r\} \subseteq A$, $A = \{A_1, A_2, \cdots, A_m\}$. And $R_{A_i^O}$ is an equivalence relation of x with respect to the attribute reduction set A_i^O under OMRS, $[x]_{A_i^O}$ is the equivalence class of $R_{A_i^O}$. Let E be IFS of U and they can be characterized by a pair of lower and upper approximations:*

$$\underline{\sum_{i=1}^{r} R_{A_i^O}}^{O}(E) = \{< x, \mu_{\underline{\sum_{i=1}^{r} R_{A_i^O}}^{O}(E)}(x), \nu_{\underline{\sum_{i=1}^{r} R_{A_i^O}}^{O}(E)}(x) > | x \in U\};$$

$$\overline{\sum_{i=1}^{r} R_{A_i^O}}^{O}(E) = \{< x, \mu_{\overline{\sum_{i=1}^{r} R_{A_i^O}}^{O}(E)}(x), \nu_{\overline{\sum_{i=1}^{r} R_{A_i^O}}^{O}(E)}(x) > | x \in U\}.$$

where

$$\mu_{\underline{\sum_{i=1}^{r} R_{A_i^O}}^{O}(E)}(x) = \bigvee_{i=1}^{r} \inf_{y \in [x]_{A_i^O}} \mu_E(y),\ \nu_{\underline{\sum_{i=1}^{r} R_{A_i^O}}^{O}(E)}(x) = \bigwedge_{i=1}^{r} \sup_{y \in [x]_{A_i^O}} \nu_E(y);$$

$$\mu_{\overline{\sum_{i=1}^{r} R_{A_i^O}}^{O}(E)}(x) = \bigwedge_{i=1}^{r} \sup_{y \in [x]_{A_i^O}} \mu_E(y),\ \nu_{\overline{\sum_{i=1}^{r} R_{A_i^O}}^{O}(E)}(x) = \bigvee_{i=1}^{r} \inf_{y \in [x]_{A_i^O}} \nu_E(y).$$

If $\underline{\sum_{i=1}^{r} R_{A_i^O}}^{O}(E) \neq \overline{\sum_{i=1}^{r} R_{A_i^O}}^{O}(E)$, then E can be called OOMRIFS.

Definition 14. *Suppose $IS =< U, A, V, f >$ is an information system, $\forall E \subseteq U$, E are IFS. $A_i^O = \{A_1, A_2, \cdots, A_r\} \subseteq A$, $A = \{A_1, A_2, \cdots, A_m\}$. where A_i^O is an optimistic multi-granulation attribute reduction set. Then the lower and upper approximations of pessimistic MRIFS under optimistic multi-granulation environment can be defined as follows:*

$$\underline{\sum_{i=1}^{r} R_{A_i^O}}^{I}(E) = \{< x, \mu_{\underline{\sum_{i=1}^{r} R_{A_i^O}}^{I}(E)}(x), \nu_{\underline{\sum_{i=1}^{r} R_{A_i^O}}^{I}(E)}(x) > | x \in U\};$$

$$\overline{\sum_{i=1}^{r} R_{A_i^O}}^{I}(E) = \{< x, \mu_{\overline{\sum_{i=1}^{r} R_{A_i^O}}^{I}(E)}(x), \nu_{\overline{\sum_{i=1}^{r} R_{A_i^O}}^{I}(E)}(x) > | x \in U\}.$$

where

$$\mu_{\underline{\sum_{i=1}^{r} R_{A_i^O}}^{I}(E)}(x) = \bigwedge_{i=1}^{r} \inf_{y \in [x]_{A_i^O}} \mu_E(y), \nu_{\underline{\sum_{i=1}^{r} R_{A_i^O}}^{I}(E)}(x) = \bigvee_{i=1}^{r} \sup_{y \in [x]_{A_i^O}} \nu_E(y);$$

$$\mu_{\overline{\sum_{i=1}^{r} R_{A_i^O}}^{I}(E)}(x) = \bigvee_{i=1}^{r} \sup_{y \in [x]_{A_i^O}} \mu_E(y), \nu_{\overline{\sum_{i=1}^{r} R_{A_i^O}}^{I}(E)}(x) = \bigwedge_{i=1}^{r} \inf_{y \in [x]_{A_i^O}} \nu_E(y).$$

The pair $(\underline{\sum_{i=1}^{r} R_{A_i^O}}^{I}(E), \overline{\sum_{i=1}^{r} R_{A_i^O}}^{I}(E))$ *are called OIMRIFS, if* $\underline{\sum_{i=1}^{r} R_{A_i^O}}^{I}(E) \neq \overline{\sum_{i=1}^{r} R_{A_i^O}}^{I}(E)$.

According to Definitions 13 and 14, the following theorem can be obtained.

Theorem 4. *Let* $IS = < U, A, V, f >$ *be an information system,* $A_i^O = \{A_1, A_2, \cdots, A_r\} \subseteq A$, $A = \{A_1, A_2, \cdots, A_m\}$, *and* E_1, E_2 *be IFS on U. Comparing with Definitions 13 and 14, the following proposition is obtained.*

(1) $\underline{\sum_{i=1}^{r} R_{A_i^O}}^{O}(E_1) = \bigcup_{i=1}^{r} \underline{R_{A_i^O}}^{O}(E_1)$;

(2) $\overline{\sum_{i=1}^{r} R_{A_i^O}}^{O}(E_1) = \bigcap_{i=1}^{r} \overline{R_{A_i^O}}^{O}(E_1)$;

(3) $\underline{\sum_{i=1}^{r} R_{A_i^O}}^{I}(E_1) = \bigcap_{i=1}^{r} \underline{R_{A_i^O}}^{I}(E_1)$;

(4) $\underline{\sum_{i=1}^{r} R_{A_i^O}}^{I}(E_1) = \bigcup_{i=1}^{r} \underline{R_{A_i^O}}^{I}(E_1)$;

(5) $\underline{\sum_{i=1}^{r} R_{A_i^O}}^{I}(E_1) \subseteq \underline{\sum_{i=1}^{r} R_{A_i^O}}^{O}(E_1)$;

(6) $\overline{\sum_{i=1}^{r} R_{A_i^O}}^{O}(E_1) \subseteq \overline{\sum_{i=1}^{r} R_{A_i^O}}^{I}(E_1)$;

(7) $\underline{\sum_{i=1}^{r} R_{A_i^O}}^{O}(E_1 \cap E_2) = \underline{\sum_{i=1}^{r} R_{A_i^O}}^{O}(E_1) \cap \underline{\sum_{i=1}^{r} R_{A_i^O}}^{O}(E_2), \underline{\sum_{i=1}^{r} R_{A_i^O}}^{I}(E_1 \cap E_2) = \underline{\sum_{i=1}^{r} R_{A_i^O}}^{I}(E_1) \cap \underline{\sum_{i=1}^{r} R_{A_i^O}}^{I}(E_2)$;

(8) $\overline{\sum_{i=1}^{r} R_{A_i^O}}^{O}(E_1 \cup E_2) = \overline{\sum_{i=1}^{r} R_{A_i^O}}^{O}(E_1) \cup \overline{\sum_{i=1}^{r} R_{A_i^O}}^{O}(E_2), \overline{\sum_{i=1}^{r} R_{A_i^O}}^{I}(E_1 \cup E_2) = \overline{\sum_{i=1}^{r} R_{A_i^O}}^{I}(E_1) \cup \overline{\sum_{i=1}^{r} R_{A_i^O}}^{I}(E_2)$;

(9) $\underline{\sum_{i=1}^{r} R_{A_i^O}}^{O}(E_1 \cup E_2) \supseteq \underline{\sum_{i=1}^{r} R_{A_i^O}}^{O}(E_1) \cup \underline{\sum_{i=1}^{r} R_{A_i^O}}^{O}(E_2), \underline{\sum_{i=1}^{r} R_{A_i^O}}^{I}(E_1 \cup E_2) \supseteq \underline{\sum_{i=1}^{r} R_{A_i^O}}^{I}(E_1) \cup \underline{\sum_{i=1}^{r} R_{A_i^O}}^{I}(E_2)$;

(10) $\overline{\sum_{i=1}^{r} R_{A_i^O}}^{O}(E_1 \cap E_2) \subseteq \overline{\sum_{i=1}^{r} R_{A_i^O}}^{O}(E_1) \cap \overline{\sum_{i=1}^{r} R_{A_i^O}}^{O}(E_2), \overline{\sum_{i=1}^{r} R_{A_i^O}}^{I}(E_1 \cap E_2) \subseteq \overline{\sum_{i=1}^{r} R_{A_i^O}}^{I}(E_1) \cap \overline{\sum_{i=1}^{r} R_{A_i^O}}^{I}(E_2)$.

Proof. It is easy to prove by the Definitions 13 and 14. □

Definition 15. *Let* $IS = < U, A, V, f >$ *be an information system, and E be IFS on U.* $A_i^I = \{A_1, A_2, \cdots, A_r\} \subseteq A$, $A = \{A_1, A_2, \cdots, A_m\}$, *where* A_i^I *is a pessimistic multi-granulation attribute reduction set. Then, the pessimistic optimistic lower and upper approximations of E with respect to equivalence relation* $R_{A_i^I}$ *are defined by the following formulas:*

$$\underline{\sum_{i=1}^{r} R_{A_i^I}}^{O}(E) = \{< x, \mu_{\underline{\sum_{i=1}^{r} R_{A_i^I}}^{O}(E)}(x), \nu_{\underline{\sum_{i=1}^{r} R_{A_i^I}}^{O}(E)}(x) > | x \in U\};$$

$$\overline{\sum_{i=1}^{r} R_{A_i^I}}^{O}(E) = \{< x, \mu_{\overline{\sum_{i=1}^{r} R_{A_i^I}}^{O}(E)}(x), \nu_{\overline{\sum_{i=1}^{r} R_{A_i^I}}^{O}(E)}(x) > | x \in U\}.$$

where

$$\mu_{\underline{\sum_{i=1}^{r} R_{A_i^I}}^{O}(E)}(x) = \bigvee_{i=1}^{r} \inf_{y\in[x]_{A_i^I}} \mu_E(y), \nu_{\underline{\sum_{i=1}^{r} R_{A_i^I}}^{O}(E)}(x) = \bigwedge_{i=1}^{r} \sup_{y\in[x]_{A_i^I}} \nu_E(y);$$

$$\mu_{\overline{\sum_{i=1}^{r} R_{A_i^I}}^{O}(E)}(x) = \bigwedge_{i=1}^{r} \sup_{y\in[x]_{A_i^I}} \mu_E(y), \nu_{\overline{\sum_{i=1}^{r} R_{A_i^I}}^{O}(E)}(x) = \bigvee_{i=1}^{r} \inf_{y\in[x]_{A_i^I}} \nu_E(y).$$

If $\underline{\sum_{i=1}^{r} R_{A_i^I}}^{O}(E) \neq \overline{\sum_{i=1}^{r} R_{A_i^I}}^{O}(E)$, *then E can be called IOMRIFS.*

Definition 16. *Let* $IS =< U, A, V, f >$ *be an information system, and E be IFS on U.* $A_i^I = \{A_1, A_2, \cdots, A_r\} \subseteq A$, $A = \{A_1, A_2, \cdots, A_m\}$, *where* A_i^I *is a pessimistic multi-granulation attribute reduction set. Then, the pessimistic lower and upper approximations of E under IMRS are defined by the following formulas:*

$$\underline{\sum_{i=1}^{r} R_{A_i^I}}^{I}(E) = \{< x, \mu_{\underline{\sum_{i=1}^{r} R_{A_i^I}}^{I}(E)}(x), \nu_{\underline{\sum_{i=1}^{r} R_{A_i^I}}^{I}(E)}(x) > | x \in U\};$$

$$\overline{\sum_{i=1}^{r} R_{A_i^I}}^{I}(E) = \{< x, \mu_{\overline{\sum_{i=1}^{r} R_{A_i^I}}^{I}(E)}(x), \nu_{\overline{\sum_{i=1}^{r} R_{A_i^I}}^{I}(E)}(x) > | x \in U\}.$$

where

$$\mu_{\underline{\sum_{i=1}^{r} R_{A_i^I}}^{I}(E)}(x) = \bigwedge_{i=1}^{r} \inf_{y\in[x]_{A_i^I}} \mu_E(y), \nu_{\underline{\sum_{i=1}^{r} R_{A_i^I}}^{I}(E)}(x) = \bigvee_{i=1}^{r} \sup_{y\in[x]_{A_i^I}} \nu_E(y);$$

$$\mu_{\overline{\sum_{i=1}^{r} R_{A_i^I}}^{I}(E)}(x) = \bigvee_{i=1}^{r} \sup_{y\in[x]_{A_i^I}} \mu_E(y), \nu_{\overline{\sum_{i=1}^{r} R_{A_i^I}}^{I}(E)}(x) = \bigwedge_{i=1}^{r} \inf_{y\in[x]_{A_i^I}} \nu_E(y).$$

where $R_{A_i^I}$ *is an equivalence relation of x about the attribute reduction set* A_i^I *under IMRS,* $[x]_{A_i^O}$ *is the equivalence class of* $R_{A_i^I}$.

If $\underline{\sum_{i=1}^{r} R_{A_i^I}}^{I}(E) \neq \overline{\sum_{i=1}^{r} R_{A_i^I}}^{I}(E)$, *then the pair* $(\underline{\sum_{i=1}^{r} R_{A_i^I}}^{I}(E), \overline{\sum_{i=1}^{r} R_{A_i^I}}^{I}(E))$ *is said to be IIMRIFS.*

According to Definitions 15 and 16, the following theorem can be captured.

Theorem 5. *Let* $IS =< U, A, V, f >$ *be an information system,* $A_i^I = \{A_1, A_2, \cdots, A_r\} \subseteq A$, $A = \{A_1, A_2, \cdots, A_m\}$, *and* E_1, E_2 *be IFS on U. Then IOMRIFS and IIOMRIFS models have the following properties:*

(1) $\underline{\sum_{i=1}^{r} R_{A_i^I}}^{O}(E_1) = \bigcup_{i=1}^{r} \underline{R_{A_i^I}}^{O}(E_1)$;

(2) $\overline{\sum_{i=1}^{r} R_{A_i^I}}^{O}(E_1) = \bigcap_{i=1}^{r} \overline{R_{A_i^I}}^{O}(E_1)$;

(3) $\underline{\sum_{i=1}^{r} R_{A_i^I}}^{I}(E_1) = \bigcup_{i=1}^{r} \underline{R_{A_i^I}}^{I}(E_1)$;

(4) $\underline{\sum_{i=1}^{r} R_{A_i^I}}^{I}(E_1) = \bigcup_{i=1}^{r} \underline{R_{A_i^I}}^{I}(E_1)$;

(5) $\underline{\sum_{i=1}^{r} R_{A_i^I}}^{I}(E_1) \subseteq \underline{\sum_{i=1}^{r} R_{A_i^I}}^{O}(E_1)$;

(6) $\overline{\sum_{i=1}^{r} R_{A_i^I}}^{O}(E_1) \subseteq \overline{\sum_{i=1}^{r} R_{A_i^I}}^{I}(E_1)$.

(7) $\underline{\sum_{i=1}^{r} R_{A_i^I}}^{O}(E_1 \cap E_2) = \underline{\sum_{i=1}^{r} R_{A_i^I}}^{O}(E_1) \cap \underline{\sum_{i=1}^{r} R_{A_i^I}}^{O}(E_2), \underline{\sum_{i=1}^{r} R_{A_i^I}}^{I}(E_1 \cap E_2) = \underline{\sum_{i=1}^{r} R_{A_i^I}}^{I}(E_1) \cap \underline{\sum_{i=1}^{r} R_{A_i^I}}^{I}(E_2)$;

(8) $\overline{\sum_{i=1}^{r} R_{A_i^I}}^{O}(E_1 \cup E_2) = \overline{\sum_{i=1}^{r} R_{A_i^I}}^{O}(E_1) \cup \overline{\sum_{i=1}^{r} R_{A_i^I}}^{O}(E_2), \overline{\sum_{i=1}^{r} R_{A_i^I}}^{I}(E_1 \cup E_2) = \overline{\sum_{i=1}^{r} R_{A_i^I}}^{I}(E_1) \cup \overline{\sum_{i=1}^{r} R_{A_i^I}}^{I}(E_2)$;

(9) $\underline{\sum_{i=1}^{r} R_{A_i^!}}^{O}(E_1 \cup E_2) \supseteq \underline{\sum_{i=1}^{r} R_{A_i^!}}^{O}(E_1) \cup \underline{\sum_{i=1}^{r} R_{A_i^!}}^{O}(E_2), \underline{\sum_{i=1}^{r} R_{A_i^!}}^{I}(E_1 \cup E_2) \supseteq \underline{\sum_{i=1}^{r} R_{A_i^!}}^{I}(E_1) \cup \underline{\sum_{i=1}^{r} R_{A_i^!}}^{I}(E_2)$;

(10) $\overline{\sum_{i=1}^{r} R_{A_i^!}}^{O}(E_1 \cap E_2) \subseteq \overline{\sum_{i=1}^{r} R_{A_i^!}}^{O}(E_1) \cap \overline{\sum_{i=1}^{r} R_{A_i^!}}^{O}(E_2), \overline{\sum_{i=1}^{r} R_{A_i^!}}^{I}(E_1 \cap E_2) \subseteq \overline{\sum_{i=1}^{r} R_{A_i^!}}^{I}(E_1) \cap \overline{\sum_{i=1}^{r} R_{A_i^!}}^{I}(E_2)$.

Proof. It can be derived directly from Definitions 15 and 16. □

The characteristics of the proposed four models are further verified by Example 2 below.

Example 2. (Continued with Example 1). *From Example 1, we know that these 5 sites are evaluated by 10 investment schemes respectively. Suppose they have the following IFS with respect to 10 investment schemes*

$$E = \left\{ \frac{[0.25,0.43]}{x_1}, \frac{[0.51,0.28]}{x_2}, \frac{[0.54,0.38]}{x_3}, \frac{[0.37,0.59]}{x_4}, \frac{[0.49,0.35]}{x_5}, \frac{[0.92,0.04]}{x_6}, \frac{[0.09,0.86]}{x_7}, \frac{[0.15,0.46]}{x_8}, \frac{[0.72,0.12]}{x_9}, \frac{[0.67,0.23]}{x_{10}} \right\}.$$

(1) *In OOMRIFS, the lower and upper approximations of OOMRIFS can be calculated as follows:*

$$\underline{\sum_{i=1}^{r} R_{A_i^O}}^{O}(E) = \left\{ \frac{[0.25,0.59]}{x_1}, \frac{[0.49,0.38]}{x_2}, \frac{[0.49,0.38]}{x_3}, \frac{[0.25,0.59]}{x_4}, \frac{[0.49,0.38]}{x_5}, \frac{[0.25,0.46]}{x_6}, \frac{[0.09,0.86]}{x_7}, \frac{[0.15,0.46]}{x_8}, \frac{[0.15,0.46]}{x_9}, \frac{[0.67,0.23]}{x_{10}} \right\},$$

$$\overline{\sum_{i=1}^{r} R_{A_i^O}}^{O}(E) = \left\{ \frac{[0.51,0.28]}{x_1}, \frac{[0.51,0.28]}{x_2}, \frac{[0.54,0.35]}{x_3}, \frac{[0.51,0.28]}{x_4}, \frac{[0.54,0.35]}{x_5}, \frac{[0.92,0.04]}{x_6}, \frac{[0.54,0.35]}{x_7}, \frac{[0.15,0.46]}{x_8}, \frac{[0.72,0.12]}{x_9}, \frac{[0.67,0.23]}{x_{10}} \right\}.$$

(2) *Similarly, in OIMRIFS, we have:*

$$\underline{\sum_{i=1}^{r} R_{A_i^O}}^{I}(E) = \left\{ \frac{[0.25,0.59]}{x_1}, \frac{[0.25,0.59]}{x_2}, \frac{[0.09,0.86]}{x_3}, \frac{[0.25,0.59]}{x_4}, \frac{[0.09,0.86]}{x_5}, \frac{[0.15,0.59]}{x_6}, \frac{[0.09,0.86]}{x_7}, \frac{[0.15,0.46]}{x_8}, \frac{[0.09,0.86]}{x_9}, \frac{[0.09,0.86]}{x_{10}} \right\},$$

$$\overline{\sum_{i=1}^{r} R_{A_i^O}}^{I}(E) = \left\{ \frac{[0.92,0.04]}{x_1}, \frac{[0.54,0.28]}{x_2}, \frac{[0.54,0.28]}{x_3}, \frac{[0.92,0.04]}{x_4}, \frac{[0.54,0.28]}{x_5}, \frac{[0.92,0.04]}{x_6}, \frac{[0.72,0.12]}{x_7}, \frac{[0.92,0.04]}{x_8}, \frac{[0.92,0.04]}{x_9}, \frac{[0.72,0.12]}{x_{10}} \right\}.$$

From the above results, Figure 1 can be drawn as follows:

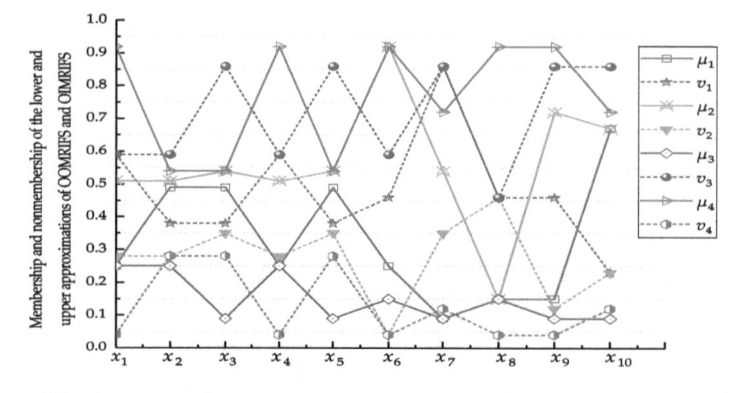

Figure 1. The lower and upper approximations of OOMRIFS and OIMRIFS.

Note that
$\mu_1 = \underline{\mu^{OO}}(x_j)$ *and* $\nu_1 = \underline{\nu^{OO}}(x_j)$ *represent the lower approximation of OOMRIFS;*
$\mu_2 = \overline{\mu^{OO}}(x_j)$ *and* $\nu_2 = \overline{\nu^{OO}}(x_j)$ *represent the upper approximation of OOMRIFS;*

$\mu_3 = \underline{\mu^{OI}}(x_j)$ *and* $\nu_3 = \underline{\nu^{OI}}(x_j)$ *represent the lower approximation of OIMRIFS;*
$\mu_4 = \overline{\mu^{OI}}(x_j)$ *and* $\nu_4 = \overline{\nu^{OI}}(x_j)$ *represent the upper approximation of OIMRIFS.*
Regarding Figure 1, we can get,

$$\overline{\mu^{OI}}(x_j) \geq \overline{\mu^{OO}}(x_j) \geq \underline{\mu^{OO}}(x_j) \geq \underline{\mu^{OI}}(x_j);\ \underline{\nu^{OI}}(x_j) \geq \underline{\nu^{OO}}(x_j) \geq \overline{\nu^{OO}}(x_j) \geq \overline{\nu^{OI}}(x_j).$$

As shown in Figure 1, the rules of Theorem 4 are satisfied. By constructing the OOMRIFS and OIMRIFS models, we can reduce the subjective scoring errors of experts under intuitionistic fuzzy conditions.

(3) *Similar to* (1), *in IOMRIFS, we have:*

$$\underline{\sum_{i=1}^{r} R_{A_i^I}}^{O}(E) = \left\{\frac{[0.25,0.43]}{x_1}, \frac{[0.25,0.43]}{x_2}, \frac{[0.25,0.43]}{x_3}, \frac{[0.37,0.59]}{x_4}, \frac{[0.25,0.43]}{x_5}, \frac{[0.25,0.46]}{x_6}, \frac{[0.09,0.86]}{x_7}, \frac{[0.15,0.46]}{x_8}, \frac{[0.67,0.23]}{x_9}, \frac{[0.67,0.23]}{x_{10}}\right\},$$

$$\overline{\sum_{i=1}^{r} R_{A_i^I}}^{O}(E) = \left\{\frac{[0.51,0.28]}{x_1}, \frac{[0.51,0.28]}{x_2}, \frac{[0.54,0.35]}{x_3}, \frac{[0.37,0.59]}{x_4}, \frac{[0.49,0.35]}{x_5}, \frac{[0.92,0.04]}{x_6}, \frac{[0.51,0.35]}{x_7}, \frac{[0.49,0.35]}{x_8}, \frac{[0.72,0.12]}{x_9}, \frac{[0.67,0.23]}{x_{10}}\right\}.$$

(4) *The same as* (1), *in IIMRIFS, we can get:*

$$\underline{\sum_{i=1}^{r} R_{A_i^I}}^{I}(E) = \left\{\frac{[0.25,0.59]}{x_1}, \frac{[0.09,0.86]}{x_2}, \frac{[0.09,0.86]}{x_3}, \frac{[0.25,0.59]}{x_4}, \frac{[0.09,0.86]}{x_5}, \frac{[0.09,0.86]}{x_6}, \frac{[0.09,0.86]}{x_7}, \frac{[0.09,0.86]}{x_8}, \frac{[0.15,0.46]}{x_9}, \frac{[0.67,0.23]}{x_{10}}\right\},$$

$$\overline{\sum_{i=1}^{r} R_{A_i^I}}^{I}(E) = \left\{\frac{[0.92,0.04]}{x_1}, \frac{[0.54,0.28]}{x_2}, \frac{[0.92,0.04]}{x_3}, \frac{[0.92,0.04]}{x_4}, \frac{[0.54,0.28]}{x_5}, \frac{[0.92,0.04]}{x_6}, \frac{[0.92,0.04]}{x_7}, \frac{[0.92,0.04]}{x_8}, \frac{[0.92,0.04]}{x_9}, \frac{[0.72,0.12]}{x_{10}}\right\}.$$

From (3) *and* (4), *we can obtain Figure 2 as shown:*

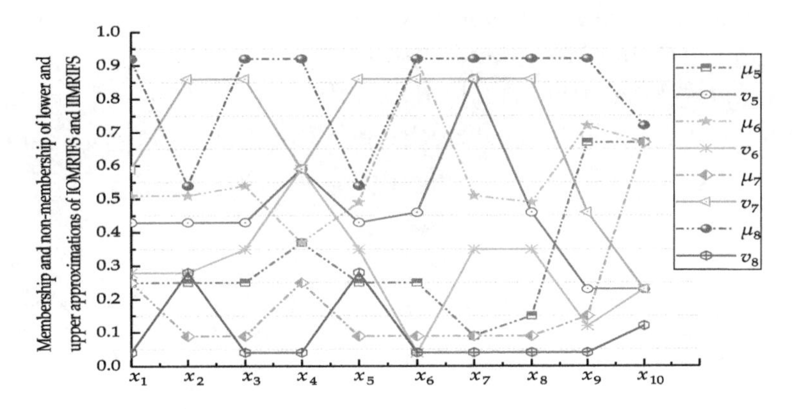

Figure 2. The lower and upper approximations of IOMRIFS and IIMRIFS.

Note that
$\mu_5 = \underline{\mu^{IO}}(x_j)$ *and* $\nu_5 = \underline{\nu^{IO}}(x_j)$ *represent the lower approximation of IOMRIFS;*
$\mu_6 = \overline{\mu^{IO}}(x_j)$ *and* $\nu_6 = \overline{\nu^{IO}}(x_j)$ *represent the upper approximation of IOMRIFS;*
$\mu_7 = \underline{\mu^{II}}(x_j)$ *and* $\nu_7 = \underline{\nu^{II}}(x_j)$ *represent the lower approximation of IIMRIFS;*
$\mu_8 = \overline{\mu^{II}}(x_j)$ *and* $\nu_8 = \overline{\nu^{II}}(x_j)$ *represent the upper approximation of IIMRIFS.*
For Figure 2, we can get,

$$\overline{\mu^{II}}(x_j) \geq \overline{\mu^{IO}}(x_j) \geq \underline{\mu^{IO}}(x_j) \geq \underline{\mu^{II}}(x_j);\ \underline{\nu^{II}}(x_j) \geq \underline{\nu^{IO}}(x_j) \geq \overline{\nu^{IO}}(x_j) \geq \overline{\nu^{II}}(x_j).$$

As shown in Figure 2, the rules of Theorem 5 are satisfied.

Through the Example 2, we can obtain four relatively more objective MRIFS models, which are beneficial to reduce subjective errors.

5. Three-Way Decisions Models Based on MRIFS and Optimal Granularity Selection

In order to obtain the optimal granularity selection results in the case of optimistic and pessimistic multi-granulation sets, it is necessary to further distinguish the importance degree of each granularity in the reduction sets. We respectively combine the four MRIFS models mentioned above with three-way decisions theory to get four new three-way decisions models. By extracting the rules, the redundant objects in the reduction sets are removed, and the decision error is further reduced. Then the optimal granularity selection results in two cases are obtained respectively by constructing the comprehensive score function and comprehensive accuracy function measurement formulas of each granularity of the reduction sets.

5.1. Three-Way Decisions Model Based on OOMRIFS

Suppose A_i^O is the reduction set under OMRS. According to reference [46], the expected loss function $R^{OO}(\omega_*|[x]_{A_i^O})(* = P, B, N)$ of object x can be obtained:

$$R^{OO}(\omega_P|[x]_{A_i^O}) = \lambda_{PP} \cdot \mu^{OO}(x) + \lambda_{PN} \cdot \nu^{OO}(x) + \lambda_{PB} \cdot \pi^{OO}(x);$$
$$R^{OO}(\omega_N|[x]_{A_i^O}) = \lambda_{NP} \cdot \mu^{OO}(x) + \lambda_{NN} \cdot \nu^{OO}(x) + \lambda_{NB} \cdot \pi^{OO}(x);$$
$$R^{OO}(\omega_B|[x]_{A_i^O}) = \lambda_{BP} \cdot \mu^{OO}(x) + \lambda_{BN} \cdot \nu^{OO}(x) + \lambda_{BB} \cdot \pi^{OO}(x).$$

where

$$\mu^{OO}(x) = \mu_{\underline{\sum_{i=1}^{r} R_{A_i^O}}^{O}(E)}(x) = \bigvee_{i=1}^{r} \inf_{y \in [x]_{A_i^O}} \mu_E(y),\ \nu^{OO}(x) = \nu_{\underline{\sum_{i=1}^{r} R_{A_i^O}}^{O}(E)}(x) = \bigwedge_{i=1}^{r} \sup_{y \in [x]_{A_i^O}} \nu_E(y),\ \pi^{OO}(x) = 1 - \mu_{\underline{\sum_{i=1}^{r} R_{A_i^O}}^{O}(E)}(x) - \nu_{\underline{\sum_{i=1}^{r} R_{A_i^O}}^{O}(E)}(x);$$

or

$$\mu^{OO}(x) = \mu_{\overline{\sum_{i=1}^{r} R_{A_i^O}}^{O}(E)}(x) = \bigwedge_{i=1}^{r} \sup_{y \in [x]_{A_i^O}} \mu_E(y),\ \nu^{OO}(x) = \nu_{\overline{\sum_{i=1}^{r} R_{A_i^O}}^{O}(E)}(x) = \bigvee_{i=1}^{r} \inf_{y \in [x]_{A_i^O}} \nu_E(y),\ \pi^{OO}(x) = 1 - \mu_{\overline{\sum_{i=1}^{r} R_{A_i^O}}^{O}(E)}(x) - \nu_{\overline{\sum_{i=1}^{r} R_{A_i^O}}^{O}(E)}(x).$$

The minimum-risk decision rules derived from the Bayesian decision process are as follows:
(P′): If $R'(\omega_P|[x]_{A_i^O}) \leq R'(\omega_B|[x]_{A_i^O})$ and $R'(\omega_P|[x]_{A_i^O}) \leq R'(\omega_N|[x]_{A_i^O})$, then $x \in POS(X)$;
(N′): If $R'(\omega_N|[x]_{A_i^O}) \leq R'(\omega_P|[x]_{A_i^O})$ and $R'(\omega_N|[x]_{A_i^O}) \leq R'(\omega_B|[x]_{A_i^O})$, then $x \in NEG(X)$;
(B′): If $R'(\omega_B|[x]_{A_i^O}) \leq R'(\omega_N|[x]_{A_i^O})$ and $R'(\omega_B|[x]_{A_i^O}) \leq R'(\omega_P|[x]_{A_i^O})$, then $x \in BND(X)$.
Thus, the decision rules (P′)-(B′) can be re-expressed concisely as:
(P′) rule satisfies:

$$(\mu^{OO}(x) \leq (1 - \pi^{OO}(x)) \cdot \frac{\lambda_{NN} - \lambda_{PN}}{(\lambda_{PP} - \lambda_{NP}) + (\lambda_{PN} - \lambda_{NN})}) \wedge (\mu^{OO}(x) \leq (1 - \pi^{OO}(x)) \cdot \frac{\lambda_{BN} - \lambda_{PN}}{(\lambda_{PP} - \lambda_{BP}) + (\lambda_{PN} - \lambda_{BN})});$$

(N′) rule satisfies:

$$(\mu^{OO}(x) < (1 - \pi^{OO}(x)) \cdot \frac{\lambda_{PN} - \lambda_{NN}}{(\lambda_{NP} - \lambda_{PP}) + (\lambda_{PN} - \lambda_{NN})}) \wedge (\mu^{OO}(x) < (1 - \pi^{OO}(x)) \cdot \frac{\lambda_{BN} - \lambda_{NN}}{(\lambda_{NP} - \lambda_{BP}) + (\lambda_{BN} - \lambda_{NN})});$$

(B′) rule satisfies:

$$(\mu^{OO}(x) > (1 - \pi^{OO}(x)) \cdot \frac{\lambda_{BN} - \lambda_{PN}}{(\lambda_{PN} - \lambda_{BN}) + (\lambda_{BP} - \lambda_{PP})}) \wedge (\mu^{OO}(x) \geq (1 - \pi^{OO}(x)) \cdot \frac{\lambda_{BN} - \lambda_{NN}}{(\lambda_{BN} - \lambda_{NN}) + (\lambda_{NP} - \lambda_{BP})}).$$

Therefore, the three-way decisions rules based on OOMRIFS are as follows:
(P1): If $\mu^{OO}(x) \geq (1 - \pi^{OO}(x)) \cdot \alpha$, then $x \in POS(X)$;
(N1): If $\mu^{OO}(x) \leq (1 - \pi^{OO}(x)) \cdot \beta$, then $x \in NEG(X)$;
(B1): If $(1 - \pi^{OO}(x)) \cdot \beta \leq \mu^{OO}(x)$ and $\mu^{OO}(x) \leq (1 - \pi^{OO}(x)) \cdot \alpha$, then $x \in BND(X)$.

5.2. Three-Way Decisions Model Based on OIMRIFS

Suppose A_i^O is the reduction set under OMRS. According to reference [46], the expected loss functions $R^{OO}(\omega_*|[x]_{A_i^O})(* = P, B, N)$ of an object x are presented as follows:

$$R^{OI}(\omega_P|[x]_{A_i^O}) = \lambda_{PP} \cdot \mu^{OI}(x) + \lambda_{PN} \cdot \nu^{OI}(x) + \lambda_{PB} \cdot \pi^{OI}(x);$$
$$R^{OI}(\omega_N|[x]_{A_i^O}) = \lambda_{NP} \cdot \mu^{OI}(x) + \lambda_{NN} \cdot \nu^{OI}(x) + \lambda_{NB} \cdot \pi^{OI}(x);$$
$$R^{OI}(\omega_B|[x]_{A_i^O}) = \lambda_{BP} \cdot \mu^{OI}(x) + \lambda_{BN} \cdot \nu^{OI}(x) + \lambda_{BB} \cdot \pi^{OI}(x).$$

where

$$\mu^{OI}(x) = \mu_{\underline{\sum_{i=1}^{r} R_{A_i^O}}^{I}(E)}(x) = \bigwedge_{i=1}^{r} \inf_{y \in [x]_{A_i O}} \mu_E(y),\ \nu^{OI}(x) = v_{\underline{\sum_{i=1}^{r} R_{A_i^O}}^{I}(E)}(x) = \bigvee_{i=1}^{r} \sup_{y \in [x]_{A_i O}} v_E(y),\ \pi^{OI}(x) = 1 - \mu_{\underline{\sum_{i=1}^{r} R_{A_i^O}}^{I}(E)}(x) - v_{\underline{\sum_{i=1}^{r} R_{A_i^O}}^{I}(E)}(x);$$

or

$$\mu^{OI}(x) = \mu_{\overline{\sum_{i=1}^{r} R_{A_i^O}}^{I}(E)}(x) = \bigvee_{i=1}^{r} \sup_{y \in [x]_{A_i O}} \mu_E(y),\ \nu^{OI}(x) = \nu_{\overline{\sum_{i=1}^{r} R_{A_i^O}}^{I}(E)}(x) = \bigwedge_{i=1}^{r} \inf_{y \in [x]_{A_i O}} \nu_E(y),\ \pi^{OI}(x) = 1 - \mu_{\overline{\sum_{i=1}^{r} R_{A_i^O}}^{I}(E)}(x) - \nu_{\overline{\sum_{i=1}^{r} R_{A_i^O}}^{I}(E)}(x).$$

Therefore, the three-way decisions rules based on OIMRIFS are as follows:
(P2): If $\mu^{OI}(x) \geq (1 - \pi^{OI}(x)) \cdot \alpha$, then $x \in POS(X)$;
(N2): If $\mu^{OI}(x) \leq (1 - \pi^{OI}(x)) \cdot \beta$, then $x \in NEG(X)$;
(B2): If $(1 - \pi^{OI}(x)) \cdot \beta \leq \mu^{OI}(x)$ and $\mu^{OI}(x) \leq (1 - \pi^{OI}(x)) \cdot \alpha$, then $x \in BND(X)$.

5.3. Three-Way Decisions Model Based on IOMRIFS

Suppose $\overset{I}{A_i}$ is the reduction set under IMRS. According to reference [46], the expected loss functions $R^{IO}(\omega_*|[x]_{A_i^I})(* = P, B, N)$ of an object x are as follows:

$$R^{IO}(\omega_P|[x]_{A_i^I}) = \lambda_{PP} \cdot \mu^{IO}(x) + \lambda_{PN} \cdot \nu^{IO}(x) + \lambda_{PB} \cdot \pi^{IO}(x);$$
$$R^{IO}(\omega_N|[x]_{A_i^I}) = \lambda_{NP} \cdot \mu^{IO}(x) + \lambda_{NN} \cdot \nu^{IO}(x) + \lambda_{NB} \cdot \pi^{IO}(x);$$
$$R^{IO}(\omega_B|[x]_{A_i^I}) = \lambda_{BP} \cdot \mu^{IO}(x) + \lambda_{BN} \cdot \nu^{IO}(x) + \lambda_{BB} \cdot \pi^{IO}(x).$$

where

$$\mu^{IO}(x) = \mu_{\underline{\sum_{i=1}^{r} R_{A_i^I}}^{O}(E)}(x) = \bigvee_{i=1}^{r} \inf_{y \in [x]_{A_i I}} \mu_E(y),\ \nu^{IO}(x) = \nu_{\underline{\sum_{i=1}^{r} R_{A_i^I}}^{O}(E)}(x) = \bigwedge_{i=1}^{r} \sup_{y \in [x]_{A_i I}} \nu_E(y),\ \pi^{IO}(x) = 1 - \mu_{\underline{\sum_{i=1}^{r} R_{A_i^I}}^{O}(E)}(x) - \nu_{\underline{\sum_{i=1}^{r} R_{A_i^I}}^{O}(E)}(x);$$

or

$$\mu^{IO}(x) = \mu_{\overline{\sum_{i=1}^{r} R_{A_i^I}}^{O}(E)}(x) = \bigwedge_{i=1}^{r} \sup_{y \in [x]_{A_i I}} \mu_E(y),\ \nu^{IO}(x) = \nu_{\overline{\sum_{i=1}^{r} R_{A_i^I}}^{O}(E)}(x) = \bigvee_{i=1}^{r} \inf_{y \in [x]_{A_i I}} \nu_E(y),\ \pi^{IO}(x) = 1 - \mu_{\overline{\sum_{i=1}^{r} R_{A_i^I}}^{O}(E)}(x) - \nu_{\overline{\sum_{i=1}^{r} R_{A_i^I}}^{O}(E)}(x).$$

Therefore, the three-way decisions rules based on IOMRIFS are as follows:
(P3): If $\mu^{IO}(x) \geq (1 - \pi^{IO}(x)) \cdot \alpha$, then $x \in POS(X)$;
(N3): If $\mu^{IO}(x) \leq (1 - \pi^{IO}(x)) \cdot \beta$, then $x \in NEG(X)$;
(B3): If $(1 - \pi^{IO}(x)) \cdot \beta \leq \mu^{IO}(x)$ and $\mu^{IO}(x) \leq (1 - \pi^{IO}(x)) \cdot \alpha$, then $x \in BND(X)$.

5.4. Three-Way Decisions Model Based on IIMRIFS

Suppose A_i^I is the reduction set under IMRS. Like Section 5.1, the expected loss functions $R^{II}(\omega_*|[x]_{A_i^I})(* = P, B, N)$ of an object x are as follows:

$$R^{II}(\omega_P|[x]_{A_i^I}) = \lambda_{PP} \cdot \mu^{II}(x) + \lambda_{PN} \cdot \nu^{II}(x) + \lambda_{PB} \cdot \pi^{II}(x);$$

$$R^{II}(\omega_N|[x]_{A_i^I}) = \lambda_{NP} \cdot \mu^{II}(x) + \lambda_{NN} \cdot \nu^{II}(x) + \lambda_{NB} \cdot \pi^{II}(x);$$

$$R^{II}(\omega_B|[x]_{A_i^I}) = \lambda_{BP} \cdot \mu^{II}(x) + \lambda_{BN} \cdot \nu^{II}(x) + \lambda_{BB} \cdot \pi^{II}(x).$$

where

$$\mu^{II}(x) = \mu_{\underline{\sum_{i=1}^{r} R_{A_i^I}}^{I}(E)}(x) = \bigwedge_{i=1}^{r} \inf_{y \in [x]_{A_i^I}} \mu_E(y),\ \nu^{II}(x) = \nu_{\underline{\sum_{i=1}^{r} R_{A_i^I}}^{I}(E)}(x) = \bigvee_{i=1}^{r} \sup_{y \in [x]_{A_i^I}} \nu_E(y),\ \pi^{II}(x) = 1 - \mu_{\underline{\sum_{i=1}^{r} R_{A_i^I}}^{I}(E)}(x) - \nu_{\underline{\sum_{i=1}^{r} R_{A_i^I}}^{I}(E)}(x);$$

or

$$\mu^{II}(x) = \mu_{\overline{\sum_{i=1}^{r} R_{A_i^I}}^{I}(E)}(x) = \bigvee_{i=1}^{r} \sup_{y \in [x]_{A_i^I}} \mu_E(y),\ \nu^{II}(x) = \nu_{\overline{\sum_{i=1}^{r} R_{A_i^I}}^{I}(E)}(x) = \bigwedge_{i=1}^{r} \inf_{y \in [x]_{A_i^I}} \nu_E(y),\ \pi^{II}(x) = 1 - \mu_{\overline{\sum_{i=1}^{r} R_{A_i^I}}^{I}(E)}(x) - \nu_{\overline{\sum_{i=1}^{r} R_{A_i^I}}^{I}(E)}(x).$$

Therefore, the three-way decisions rules based on IIMRIFS are captured as follows:
(P4): If $\mu^{II}(x) \geq (1 - \pi^{II}(x)) \cdot \alpha$, then $x \in POS(X)$;
(N4): If $\mu^{II}(x) \leq (1 - \pi^{II}(x)) \cdot \beta$, then $x \in NEG(X)$;
(B4): If $(1 - \pi^{II}(x)) \cdot \beta \leq \mu^{II}(x)$ and $\mu^{II}(x) \leq (1 - \pi^{II}(x)) \cdot \alpha$, then $x \in BND(X)$.
By constructing the above three decision models, the redundant objects in the reduction sets can be removed, which is beneficial to the optimal granular selection.

5.5. Comprehensive Measuring Methods of Granularity

Definition 17 ([40]). *Let an intuitionistic fuzzy number $\widetilde{E}(f_1) = (\mu_{\widetilde{E}}(f_1), \nu_{\widetilde{E}}(f_1))$, $f_1 \in U$, then the score function of $\widetilde{E}(f_1)$ is calculated as:*

$$S(\widetilde{E}(f_1)) = \mu_{\widetilde{E}}(f_1) - \nu_{\widetilde{E}}(f_1).$$

The accuracy function of $\widetilde{E}(f_1)$ is defined as:

$$H(\widetilde{E}(f_1)) = \mu_{\widetilde{E}}(f_1) + \nu_{\widetilde{E}}(f_1).$$

where $-1 \leq S(\widetilde{E}(f_1)) \leq 1$ and $0 \leq H(\widetilde{E}(f_1)) \leq 1$.

Definition 18. *Let $DIS = (U, C \cup D)$ be a decision information system, $A = \{A_1, A_2, \cdots, A_m\}$ are m sub-attributes of C. Suppose E are IFS on the universe $U = \{x_1, x_2, \cdots, x_n\}$, defined by $\mu_{A_i}(x_j)$ and $\nu_{A_i}(x_j)$, where $\mu_{A_i}(x_j)$ and $\nu_{A_i}(x_j)$ are their membership and non-membership functions respectively. $|[x_j]_{A_i}|$ is the number of equivalence classes of x_j on granularity A_i, $U/D = \{X_1, X_2, \cdots, X_s\}$ is the partition induced by the decision attributes D. Then, the comprehensive score function of granularity A_i is captured as:*

$$CSF_{A_i}(E) = \frac{1}{s} \times \sum_{j=1, n \in [x_j]_{A_i}}^{n} \frac{|\mu_{A_i}(x_j) - \nu_{A_i}(x_j)|}{|[x_j]_{A_i}|}.$$

The comprehensive accuracy function of granularity A_i is captured as:

$$CAF_{A_i}(E) = \frac{1}{s} \times \sum_{j=1, n \in [x_j]_{A_i}}^{n} \frac{|\mu_{A_i}(x_j) + \nu_{A_i}(x_j)|}{|[x_j]_{A_i}|}.$$

where $-1 \leq CSF_{A_i}(E) \leq 1$ and $0 \leq CAF_{A_i}(E) \leq 1$.

With respect to Definition 19, according to references [27,39], we can deduce the following rules.

Definition 19. *Let two granularities A_1, A_2, then we have:*

(1) *If $CSF_{A_1}(E) > CSF_{A_2}(E)$, then A_2 is smaller than A_1, expressed as $A_1 > A_2$;*
(2) *If $CSF_{A_1}(E) < CSF_{A_2}(E)$, then A_1 is smaller than A_2, expressed as $A_1 < A_2$;*
(3) *If $CSF_{A_1}(E) = CSF_{A_2}(E)$, then*

   (i) *If $CSF_{A_1}(E) = CSF_{A_2}(E)$, then A_2 is equal to A_1, expressed as $A_1 = A_2$;*
   (ii) *If $CSF_{A_1}(E) > CSF_{A_2}(E)$, then A_2 is smaller than A_1, expressed as $A_1 > A_2$;*
   (iii) *If $CSF_{A_1}(E) < CSF_{A_2}(E)$, then A_1 is smaller than A_2, expressed as $A_1 < A_2$.*

5.6. Optimal Granularity Selection Algorithm to Derive Three-Way Decisions from MRIFS

Suppose the reduction sets of optimistic and IMRS are A_i^O and A_i^I respectively. In this section, we take the reduction set under OMRS as an example to make the result $A_i^{O\prime}$ of optimal granularity selection.

Algorithm 2. Optimal granularity selection algorithm to derive three-way decisions from MRIFS

Input: $DIS = (U, C \cup D, V, f)$, $A = \{A_1, A_2, \cdots, A_m\}$ *be m sub-attributes of condition attributes C*, $\forall A_i \in A'$, $U/D = \{X_1, X_2, \cdots, X_s\}$, IFS E;
Output: Optimal granularity selection result $A_i^{O\prime}$.
1: compute via Algorithm 1;
2: **if** $|A_i^O| > 1$
3: **for** $\forall A_i \in A_i^O$
4: compute $\mu_{\underline{\sum_{i=1}^{r} R_{A_i^O}}^{\Delta}(E)}(x_j)$, $\nu_{\underline{\sum_{i=1}^{r} R_{A_i^O}}^{\Delta}(E)}(x_j)$, $\mu_{\overline{\sum_{i=1}^{r} R_{A_i^O}}^{\Delta}(E)}(x_j)$ and $\nu_{\overline{\sum_{i=1}^{r} R_{A_i^O}}^{\Delta}(E)}(x_j)$;
5: according (P1)-(B1) and (P2)-(B2), compute $POS(\underline{X^{O\Delta}})$, $NEG(\underline{X^{O\Delta}})$, $BND(\underline{X^{O\Delta}})$, $POS(\overline{X^{O\Delta}})$, $NEG(\overline{X^{O\Delta}})$, $BND(\overline{X^{O\Delta}})$;
6: **if** $NEG(X^{O\Delta}) \neq U$ or $NEG(\overline{X^{O\Delta}}) \neq U$
7: compute $U/\underline{A_i^{O\Delta}}$, $CSF_{\underline{A_i^{O\Delta}}}(E)$, $CAF_{\underline{A_i^{O\Delta}}}(E)$ or $(U/\overline{A_i^{O\Delta}})$, $(CSF_{\overline{A_i^{O\Delta}}}(E)$, $CAF_{\overline{A_i^{O\Delta}}}(E)$;
8: according to Definition 19 to get $A_i^{O\prime}$;
9: **return** $A_i^{O\prime} = A_i$;
10: **end**
11: **else**
12: **return** NULL;
13: **end**
14: **end**
15: **end**
16: **else**
17: **return** $A_i^{O\prime} = A_i^O$;
18: **end**

---

6. Example Analysis 3 (Continued with Example 2)

In Example 1, only site 1 can be ignored under optimistic and pessimistic multi-granulation conditions, so it can be determined that site 1 does not need to be evaluated, while sites 2 and 3 need to be further investigated under the environment of optimistic multi-granulation. At the same time, with respect to the environment of pessimistic multi-granulation, comprehensive considera- tion site 3 can ignore the assessment and sites 2, 4 and 5 need to be further investigated.

According to Example 1, we can get that the reduction set of OMRS is $\{A_2, A_3\}$, but in the case of IMRS, there are two reduction sets, which are contradictory. Therefore, two reduction sets should be reconsidered simultaneously, so the joint reduction set under IMRS is $\{A_2, A_4, A_5\}$.

Where the corresponding granularity structures of sites 2, 3, 4 and 5 are divided as follows:

$$U/A_2 = \{\{x_1, x_2, x_4\}, \{x_3, x_5, x_7\}, \{x_6, x_8, x_9\}, \{x_{10}\}\},$$
$$U/A_3 = \{\{x_1, x_4, x_6\}, \{x_2, x_3, x_5\}, \{x_8\}, \{x_7, x_9, x_{10}\}\},$$
$$U/A_4 = \{\{x_1, x_2, x_3, x_5\}, \{x_4\}, \{x_6, x_7, x_8\}, \{x_9, x_{10}\}\},$$
$$U/A_5 = \{\{x_1, x_3, x_4, x_6\}, \{x_2, x_7\}, \{x_5, x_8\}, \{x_9, x_{10}\}\}.$$

According to reference [11], we can get:
$\alpha = \frac{8-2}{(8-2)+(2-0)} = 0.75$; $\beta = \frac{2-0}{(2-0)+(6-2)} = 0.33$.
The optimal site selection process under optimistic and IMRS is as follows:

(1) Optimal site selection based on OOMRIFS

According to the Example 2, we can get the values of evaluation functions $\underline{\mu^{OO}}(x_j)$, $(1-\underline{\pi^{OO}}(x_j))\cdot\alpha$, $(1-\underline{\pi^{OO}}(x_j))\cdot\beta$, $\overline{\mu^{OO}}(x_j)$, $(1-\overline{\pi^{OO}}(x_j))\cdot\alpha$ and $(1-\overline{\pi^{OO}}(x_j))\cdot\beta$ of OOMRIFS, as shown in Table 4.

Table 4. The values of evaluation functions for OOMRIFS.

	$\underline{\mu^{OO}}(x_j)$	$(1-\underline{\pi^{OO}}(x_j))\cdot\alpha$	$(1-\underline{\pi^{OO}}(x_j))\cdot\beta$	$\overline{\mu^{OO}}(x_j)$	$(1-\overline{\pi^{OO}}(x_j))\cdot\alpha$	$(1-\overline{\pi^{OO}}(x_j))\cdot\beta$
x_1	0.25	0.63	0.2772	0.51	0.5925	0.2607
x_2	0.49	0.6525	0.2871	0.51	0.5925	0.2607
x_3	0.49	0.6525	0.2871	0.54	0.6675	0.2937
x_4	0.25	0.63	0.2772	0.51	0.5925	0.2607
x_5	0.49	0.6525	0.2871	0.54	0.6675	0.2937
x_6	0.25	0.5325	0.2343	0.92	0.72	0.3168
x_7	0.09	0.7125	0.3135	0.54	0.6675	0.2937
x_8	0.15	0.4575	0.2013	0.15	0.4575	0.2013
x_9	0.15	0.4575	0.2013	0.72	0.63	0.2772
x_{10}	0.67	0.675	0.297	0.67	0.675	0.297

We can get decision results of the lower and upper approximations of OOMRIFS by three-way decisions of the Section 5.1, as follows:

$POS(\underline{X^{OO}}) = \phi$,
$NEG(\underline{X^{OO}}) = \{x_1, x_4, x_7, x_8, x_9\}$,
$BND(\underline{X^{OO}}) = \{x_2, x_3, x_5, x_6, x_{10}\}$;
$POS(\overline{X^{OO}}) = \{x_6, x_9\}$,
$NEG(\overline{X^{OO}}) = \{x_8\}$,
$BND(\overline{X^{OO}}) = \{x_2, x_3, x_5\}$.

In the light of three-way decisions rules based on OOMRIFS, after getting rid of the objects in the rejection domain, we choose to fuse the objects in the delay domain with those in the acceptance domain for the optimal granularity selection. Therefore, the new granularities A_2, A_3 are as follows:

$U/\underline{A_2^{OI}} = \{\{x_2\}, \{x_3, x_5\}, \{x_6\}, \{x_{10}\}\}$,
$U/\underline{A_3^{OI}} = \{\{x_2, x_3, x_5\}, \{x_6\}, \{x_{10}\}\}$;
$U/\overline{A_2^{OI}} = \{\{x_1, x_2, x_4\}, \{x_3, x_5, x_7\}, \{x_6, x_9\}, \{x_{10}\}\}$,
$U/\overline{A_3^{OI}} = \{\{x_1, x_4, x_6\}, \{x_2, x_3, x_5\}, \{x_7, x_9, x_{10}\}\}$.

Then, according to Definition 18, we can get:

$$\begin{aligned} CSF_{\underline{A_2^{OO}}}(E) &= \tfrac{1}{s} \times \sum_{j=1, n\in[x_j]_{A_i}}^{n} \frac{|\mu_{A_i}(x_j)-\nu_{A_i}(x_j)|}{|[x_j]_{A_i}|} \\ &= \tfrac{1}{4} \times \sum_{j=1, n\in[x_j]_{\underline{A_2^{OO}}}}^{10} \frac{|\mu_{\underline{A_2^{OO}}}(x_j)-\nu_{\underline{A_2^{OO}}}(x_j)|}{|[x_j]_{\underline{A_2^{OO}}}|} \\ &= \tfrac{1}{4} \times ((0.49-0.38) + \tfrac{(0.49-0.38)+(0.49-0.38)}{2} + (0.25-0.46) + (0.67-0.23)) \\ &= 0.1125, \end{aligned}$$

$$\begin{aligned} CSF_{\underline{A_3^{OO}}}(E) &= \tfrac{1}{s} \times \sum_{j=1, n\in[x_j]_{A_i}}^{n} \frac{|\mu_{A_i}(x_j)-\nu_{A_i}(x_j)|}{|[x_j]_{A_i}|} \\ &= \tfrac{1}{3} \times \sum_{j=1, n\in[x_j]_{\underline{A_3^{OO}}}}^{10} \frac{|\mu_{\underline{A_3^{OO}}}(x_j)-\nu_{\underline{A_3^{OO}}}(x_j)|}{|[x_j]_{\underline{A_3^{OO}}}|} \\ &= \tfrac{1}{3} \times ((0.25-0.46) + \tfrac{(0.49-0.38)+(0.49-0.38)+(0.49-0.38)}{3} + (0.81-0.14)) \\ &= 0.1133; \end{aligned}$$

Similarly, we have:

$CSF_{\overline{A_2^{OO}}}(E) = 0.4$, $CSF_{\overline{A_3^{OO}}}(E) = 0.3533$.

From the above results, in OOMRIFS, we can see that we can't get the selection result of sites 2 and 3 only according to the comprehensive score function of granularities A_2 and A_3. Therefore, we need to further calculate the comprehensive accuracies to get the results as follows:

$$\begin{aligned} CAF_{\underline{A_2^{OO}}}(E) &= \frac{1}{s} \times \sum_{j=1, n\in[x_j]_{A_i}}^{n} \frac{|\mu_{A_i}(x_j)+\nu_{A_i}(x_j)|}{|[x_j]_{A_i}|} \\ &= \frac{1}{4} \times \sum_{j=1, n\in[x_j]_{\underline{A_2^{OO}}}}^{10} \frac{|\mu_{\underline{A_2^{OO}}}(x_j)+\nu_{\underline{A_2^{OO}}}(x_j)|}{|[x_j]_{\underline{A_2^{OO}}}|} \\ &= \frac{1}{4} \times ((0.49+0.38) + \frac{(0.49+0.38)+(0.49+0.38)}{2} + (0.25+0.46) + (0.67+0.23)) \\ &= 0.8375, \end{aligned}$$

$$\begin{aligned} CAF_{\underline{A_3^{OO}}}(E) &= \frac{1}{s} \times \sum_{j=1, n\in[x_j]_{A_i}}^{n} \frac{|\mu_{A_i}(x_j)+\nu_{A_i}(x_j)|}{|[x_j]_{A_i}|} \\ &= \frac{1}{3} \times \sum_{j=1, n\in[x_j]_{\underline{A_3^{OO}}}}^{10} \frac{|\mu_{\underline{A_3^{OO}}}(x_j)+\nu_{\underline{A_3^{OO}}}(x_j)|}{|[x_j]_{\underline{A_3^{OO}}}|} \\ &= \frac{1}{3} \times ((0.25+0.46) + \frac{(0.49+0.38)+(0.49+0.38)+(0.49+0.38)}{3} + (0.81+0.14)) \\ &= 0.8267; \end{aligned}$$

Analogously, we have:
$CAF_{\overline{A_2^{OO}}}(E) = 0.87$, $CAF_{\overline{A_3^{OO}}}(E) = 0.86$.
Through calculation above, we know that the comprehensive accuracy of the granularity A_3 is higher, so the site 3 is selected as the selection result.

(2) Optimal site selection based on OIMRIFS

The same as (1), we can get the values of evaluation functions $\underline{\mu^{OI}}(x_j)$, $(1-\underline{\pi^{OI}}(x_j))\cdot\alpha$, $(1-\underline{\pi^{OI}}(x_j))\cdot\beta$, $\overline{\mu^{OI}}(x_j)$, $(1-\overline{\pi^{OI}}(x_j))\cdot\alpha$ and $(1-\overline{\pi^{OI}}(x_j))\cdot\beta$ of OIMRIFS listed in Table 5.

Table 5. The values of evaluation functions for OIMRIFS.

	$\underline{\mu^{OI}}(x_j)$	$(1-\underline{\pi^{OI}}(x_j))\cdot\alpha$	$(1-\underline{\pi^{OI}}(x_j))\cdot\beta$	$\overline{\mu^{OI}}(x_j)$	$(1-\overline{\pi^{OI}}(x_j))\cdot\alpha$	$(1-\overline{\pi^{OI}}(x_j))\cdot\beta$
x_1	0.25	0.63	0.2772	0.92	0.72	0.3168
x_2	0.25	0.63	0.2772	0.54	0.615	0.2706
x_3	0.09	0.7125	0.3135	0.54	0.615	0.2706
x_4	0.25	0.63	0.2772	0.92	0.72	0.3168
x_5	0.09	0.7125	0.3135	0.54	0.615	0.2706
x_6	0.15	0.555	0.2442	0.92	0.72	0.3168
x_7	0.09	0.7125	0.3135	0.72	0.63	0.2772
x_8	0.15	0.4575	0.2013	0.92	0.72	0.3168
x_9	0.09	0.7125	0.3135	0.92	0.72	0.3168
x_{10}	0.09	0.7125	0.3135	0.72	0.63	0.2772

We can get decision results of the lower and upper approximations of OIMRIFS by three-way decisions in the Section 5.2, as follows:
$POS(\underline{X^{OI}}) = \phi$,
$NEG(\underline{X^{OI}}) = U$,
$BND(\underline{X^{OI}}) = \phi$;
$POS(\overline{X^{OI}}) = \{x_1, x_4, x_6, x_7, x_8, x_9, x_{10}\}$,
$NEG(\overline{X^{OI}}) = \phi$,
$BND(\overline{X^{OI}}) = \{x_2, x_3, x_5\}$.
Hence, in the upper approximations of OIMRIFS, the new granularities A_2, A_3 are as follows:

$U/\overline{A_2^{OI}} = \{\{x_1, x_2, x_4\}, \{x_3, x_5, x_7\}, \{x_6, x_8, x_9\}, \{x_{10}\}\},$
$U/\overline{A_3^{OI}} = \{\{x_1, x_4, x_6\}, \{x_2, x_3, x_5\}, \{x_8\}, \{x_7, x_9, x_{10}\}\}.$
According to Definition 18, we can calculate that
$CSF_{\underline{A_2^{OI}}}(E) = CSF_{\underline{A_3^{OI}}}(E) = 0;$
$CAF_{\underline{A_2^{OI}}}(E) = CAF_{\underline{A_3^{OI}}}(E) = 0;$
$CSF_{\overline{A_2^{OI}}}(E) = 0.6317, CSF_{\overline{A_3^{OI}}}(E) = 0.6783;$
$CAF_{\overline{A_2^{OI}}}(E) = 0.885, CAF_{\overline{A_3^{OI}}}(E) = 0.905.$

In OIMRIFS, the comprehensive score and comprehensive accuracy of the granularity A_3 are both higher than the granularity A_2. So, we choose site 3 as the evaluation site.

In reality, we are more inclined to select the optimal granularity in the case of more stringent requirements. According to (1) and (2), we can find that the granularity A_3 is a better choice when the requirements are stricter in four cases of OMRS. Therefore, we choose site 3 as the optimal evaluation site.

(3) Optimal site selection based on IOMRIFS

Similar to (1), we can obtain the values of evaluation functions $\underline{\mu^{IO}}(x_j)$, $(1 - \underline{\pi^{IO}}(x_j)) \cdot \alpha$, $(1 - \underline{\pi^{IO}}(x_j)) \cdot \beta$, $\overline{\mu^{IO}}(x_j)$, $(1 - \overline{\pi^{IO}}(x_j)) \cdot \alpha$ and $(1 - \overline{\pi^{IO}}(x_j)) \cdot \beta$ of IOMRIFS, as described in Table 6.

Table 6. The values of evaluation functions for IOMRIFS.

	$\underline{\mu^{IO}}(x_j)$	$(1-\underline{\pi^{IO}}(x_j))\cdot\alpha$	$(1-\underline{\pi^{IO}}(x_j))\cdot\beta$	$\overline{\mu^{IO}}(x_j)$	$(1-\overline{\pi^{IO}}(x_j))\cdot\alpha$	$(1-\overline{\pi^{IO}}(x_j))\cdot\beta$
x_1	0.25	0.51	0.2244	0.51	0.5925	0.2607
x_2	0.25	0.51	0.2244	0.51	0.5925	0.2607
x_3	0.25	0.51	0.2244	0.54	0.6675	0.2937
x_4	0.37	0.72	0.3168	0.37	0.72	0.3168
x_5	0.25	0.51	0.2244	0.49	0.63	0.2772
x_6	0.25	0.5325	0.2343	0.92	0.72	0.3168
x_7	0.09	0.7125	0.3135	0.51	0.645	0.2838
x_8	0.15	0.4575	0.2013	0.49	0.63	0.2772
x_9	0.67	0.675	0.297	0.72	0.63	0.2772
x_{10}	0.67	0.675	0.297	0.67	0.675	0.297

We can get decision results of the lower and upper approximations of IOMRIFS by three-way decisions in the Section 5.3, as follows:
$POS(\underline{X^{IO}}) = \phi,$
$NEG(\underline{X^{IO}}) = \{x_7, x_8\},$
$BND(\underline{X^{IO}}) = \{x_1, x_2, x_3, x_4, x_5, x_6, x_9, x_{10}\};$
$POS(\overline{X^{IO}}) = \{x_6, x_9\},$
$NEG(\overline{X^{IO}}) = \phi,$
$BND(\overline{X^{IO}}) = \{x_1, x_2, x_3, x_4, x_5, x_7, x_8, x_{10}\}.$
Therefore, the granularities A_2, A_4, A_5 can be rewritten as follows:
$U/\underline{A_2^{IO}} = \{\{x_1, x_2, x_4\}, \{x_3, x_5\}, \{x_6, x_9\}, \{x_{10}\}\},$
$U/\underline{A_4^{IO}} = \{\{x_1, x_2, x_3, x_5\}, \{x_4\}, \{x_6\}, \{x_9, x_{10}\}\},$
$U/\underline{A_5^{IO}} = \{\{x_1, x_3, x_4, x_6\}, \{x_2\}, \{x_5\}, \{x_9, x_{10}\}\};$
$U/\overline{A_2^{IO}} = \{\{x_1, x_2, x_4\}, \{x_3, x_5, x_7\}, \{x_6, x_8, x_9\}, \{x_{10}\}\},$
$U/\overline{A_4^{IO}} = \{\{x_1, x_2, x_3, x_5\}, \{x_4\}, \{x_6, x_7, x_8\}, \{x_9, x_{10}\}\},$
$U/\overline{A_5^{IO}} = \{\{x_1, x_3, x_4, x_6\}, \{x_2, x_7\}, \{x_5, x_8\}, \{x_9, x_{10}\}\}.$
According to Definition 18, one can see that the results are captured as follows:
$CSF_{\underline{A_2^{IO}}}(E) = 0.0454, CSF_{\underline{A_4^{IO}}}(E) = -0.0567, CSF_{\underline{A_5^{IO}}}(E) = -0.0294;$
$CSF_{\overline{A_2^{IO}}}(E) = 0.3058, CSF_{\overline{A_4^{IO}}}(E) = 0.2227, CSF_{\overline{A_5^{IO}}}(E) = 0.2813.$

In summary, the comprehensive score function of the granularity A_2 is higher than the granularity A_3 in IOMRIFS, so we choose site 2 as the result of granularity selection.

(4) Optimal site selection based on IIMRIFS

In the same way as (1), we can get the values of evaluation functions $\underline{\mu^{II}}(x_j)$, $(1-\underline{\pi^{II}}(x_j))\cdot\alpha$, $(1-\underline{\pi^{II}}(x_j))\cdot\beta$, $\overline{\mu^{II}}(x_j)$, $(1-\overline{\pi^{II}}(x_j))\cdot\alpha$ and $(1-\overline{\pi^{II}}(x_j))\cdot\beta$ of IIMRIFS, as shown in Table 7.

Table 7. The values of evaluation functions for IIMRIFS.

	$\underline{\mu^{II}}(x_j)$	$(1-\underline{\pi^{II}}(x_j))\cdot\alpha$	$(1-\underline{\pi^{II}}(x_j))\cdot\beta$	$\overline{\mu^{II}}(x_j)$	$(1-\overline{\pi^{II}}(x_j))\cdot\alpha$	$(1-\overline{\pi^{II}}(x_j))\cdot\beta$
x_1	0.25	0.63	0.2772	0.92	0.72	0.3168
x_2	0.09	0.7125	0.3135	0.54	0.615	0.2706
x_3	0.09	0.7125	0.3135	0.92	0.72	0.3168
x_4	0.25	0.63	0.2772	0.92	0.72	0.3168
x_5	0.09	0.7125	0.3135	0.54	0.615	0.2706
x_6	0.09	0.7125	0.3135	0.92	0.72	0.3168
x_7	0.09	0.7125	0.3135	0.92	0.72	0.3168
x_8	0.09	0.7125	0.3135	0.92	0.72	0.3168
x_9	0.15	0.4575	0.2013	0.92	0.72	0.3168
x_{10}	0.67	0.675	0.297	0.72	0.63	0.2772

We can get decision results of the lower and upper approximations of IIMRIFS by three-way decisions in the Section 5.4, as follows:

$POS(\underline{X^{II}}) = \phi$,
$NEG(\underline{X^{II}}) = \{x_1, x_2, x_3, x_4, x_5, x_6, x_7, x_8, x_9\}$,
$BND(\underline{X^{II}}) = \{x_{10}\}$;
$POS(\overline{X^{II}}) = \{x_1, x_3, x_4, x_6, x_7, x_8, x_9, x_{10}\}$,
$NEG(\overline{X^{II}}) = \phi$,
$BND(\overline{X^{II}}) = \{x_2, x_5\}$.

Therefore, the granularity structures of A_2, A_4, A_5 can be rewritten as follows:

$U/\underline{A_2^{II}} = U/\underline{A_4^{II}} = U/\underline{A_5^{II}} = \{x_{10}\}$;
$U/\overline{A_2^{II}} = \{\{x_1, x_2, x_4\}, \{x_3, x_5, x_7\}, \{x_6, x_8, x_9\}, \{x_{10}\}\}$,
$U/\overline{A_4^{II}} = \{\{x_1, x_2, x_3, x_5\}, \{x_4\}, \{x_6, x_7, x_8\}, \{x_9, x_{10}\}\}$,
$U/\overline{A_5^{II}} = \{\{x_1, x_3, x_4, x_6\}, \{x_2, x_7\}, \{x_5, x_8\}, \{x_9, x_{10}\}\}$.

According to Definition 18, one can see that the results are captured as follows:

$CSF_{\underline{A_2^{II}}}(E) = CSF_{\underline{A_4^{II}}}(E) = CSF_{\underline{A_5^{II}}}(E) = 0.44$;
$CAF_{\underline{A_2^{II}}}(E) = CAF_{\underline{A_4^{II}}}(E) = CAF_{\underline{A_5^{II}}}(E) = 0.9$;
$CSF_{\overline{A_2^{II}}}(E) = 0.7067, CSF_{\overline{A_4^{II}}}(E) = 0.7675, CSF_{\overline{A_5^{II}}}(E) = 0.69$;
$CAF_{\overline{A_2^{II}}}(E) = 0.9067,\ CAF_{\overline{A_4^{II}}}(E) = 0.9275, CAF_{\overline{A_5^{II}}}(E) = 0.91$.

In IIMRIFS, the values of the comprehensive score and comprehensive accuracy of granularity A_4 are higher than A_2 and A_5, so site 4 is chosen as the evaluation site.

Considering (3) and (4) synthetically, we find that the results of granularity selection in IOMRIFS and IIMRIFS are inconsistent, so we need to further compute the comprehensive accuracies of IIMRIFS.

$CAF_{\underline{A_2^{IO}}}(E) = 0.7896, CAF_{\underline{A_4^{IO}}}(E) = 0.8125, CAF_{\underline{A_5^{IO}}}(E) = 0.7544$;
$CAF_{\overline{A_2^{IO}}}(E) = 0.8725, CAF_{\overline{A_4^{IO}}}(E) = 0.886,\ CAF_{\overline{A_5^{IO}}}(E) = 0.8588$.

Through the above calculation results, we can see that the comprehensive score and comprehensive accuracy of granularity A_4 are higher than A_2 and A_5 in the case of pessimistic multi- granulation when the requirements are stricter. Therefore, the site 4 is eventually chosen as the optimal evaluation site.

7. Conclusions

In this paper, we propose two new granularity importance degree calculating methods among multiple granularities, and a granularity reduction algorithm is further developed. Subsequently, we design four novel MRIFS models based on reduction sets under optimistic and IMRS, i.e., OOMRIFS, OIMRIFS, IOMRIFS, and IIMRIFS, and further demonstrate their relevant properties. In addition, four three-way decisions models with novel MRIFS for the issue of internal redundant objects in reduction sets are constructed. Finally, we designe the comprehensive score function and the comprehensive precision function for the optimal granularity selection results. Meanwhile, the validity of the proposed models is verified by algorithms and examples. The works of this paper expand the application scopes of MRIFS and three-way decisions theory, which can solve issues such as spam e-mail filtering, risk decision, investment decisions, and so on. A question worth considering is how to extend the methods of this article to fit the big data environment. Moreover, how to combine the fuzzy methods based on triangular or trapezoidal fuzzy numbers with the methods proposed in this paper is also a research problem. These issues will be investigated in our future work.

Author Contributions: Z.-A.X. and D.-J.H. initiated the research and wrote the paper, M.-J.L. participated in some of these search work, and M.Z. supervised the research work and provided helpful suggestions.

Acknowledgments: This work is supported by the National Natural Science Foundation of China under Grant Nos. 61772176, 61402153, and the Scientific And Technological Project of Henan Province of China under Grant Nos. 182102210078, 182102210362, and the Plan for Scientific Innovation of Henan Province of China under Grant No. 18410051003, and the Key Scientific And Technological Project of Xinxiang City of China under Grant No. CXGG17002.

References

1. Pawlak, Z. Rough sets. *Int. J. Comput. Inf. Sci.* **1982**, *11*, 341–356. [CrossRef]
2. Pawlak, Z.; Skowron, A. Rough sets: some extensions. *Inf. Sci.* **2007**, *177*, 28–40. [CrossRef]
3. Yao, Y.Y. Probabilistic rough set approximations. *Int. J. Approx. Reason.* **2008**, *49*, 255–271. [CrossRef]
4. Slezak, D.; Ziarko, W. The investigation of the Bayesian rough set model. *Int. J. Approx. Reason.* **2005**, *40*, 81–91. [CrossRef]
5. Ziarko, W. Variable precision rough set model. *J. Comput. Syst. Sci.* **1993**, *46*, 39–59. [CrossRef]
6. Zhu, W. Relationship among basic concepts in covering-based rough sets. *Inf. Sci.* **2009**, *179*, 2478–2486. [CrossRef]
7. Ju, H.R.; Li, H.X.; Yang, X.B.; Zhou, X.Z. Cost-sensitive rough set: A multi-granulation approach. *Knowl.-Based Syst.* **2017**, *123*, 137–153. [CrossRef]
8. Qian, Y.H.; Liang, J.Y.; Dang, C.Y. Incomplete multi-granulation rough set. *IEEE Trans. Syet. Man Cybern. A* **2010**, *40*, 420–431. [CrossRef]
9. Qian, Y.H.; Liang, J.Y.; Yao, Y.Y.; Dang, C.Y. MGRS: A multi-granulation rough set. *Inf. Sci.* **2010**, *180*, 949–970. [CrossRef]
10. Yang, X.B.; Qi, Y.S.; Song, X.N.; Yang, J.Y. Test cost sensitive multigranulation rough set: model and mini-mal cost selection. *Inf. Sci.* **2013**, *250*, 184–199. [CrossRef]
11. Zhang, W.X.; Mi, J.S.; Wu, W.Z. Knowledge reductions in inconsistent information systems. *Chinese J. Comput.* **2003**, *26*, 12–18. (In Chinese)
12. Sang, Y.L.; Qian, Y.H. Granular structure reduction approach to multigranulation decision-theoretic rough sets. *Comput. Sci.* **2017**, *44*, 199–205. (In Chinese)
13. Jing, Y.G.; Li, T.R.; Fujita, H.; Yu, Z.; Wang, B. An incremental attribute reduction approach based on knowledge granularity with a multi-granulation view. *Inf. Sci.* **2017**, *411*, 23–38. [CrossRef]
14. Feng, T.; Fan, H.T.; Mi, J.S. Uncertainty and reduction of variable precision multigranulation fuzzy rough sets based on three-way decisions. *Int. J. Approx. Reason.* **2017**, *85*, 36–58. [CrossRef]
15. Tan, A.H.; Wu, W.Z.; Tao, Y.Z. On the belief structures and reductions of multigranulation spaces with decisions. *Int. J. Approx. Reason.* **2017**, *88*, 39–52. [CrossRef]

16. Kang, Y.; Wu, S.X.; Li, Y.W.; Liu, J.H.; Chen, B.H. A variable precision grey-based multi-granulation rough set model and attribute reduction. *Knowl.-Based Syst.* **2018**, *148*, 131–145. [CrossRef]
17. Xu, W.H.; Li, W.T.; Zhang, X.T. Generalized multigranulation rough sets and optimal granularity selection. *Granul. Comput.* **2017**, *2*, 271–288. [CrossRef]
18. Atanassov, K.T. More on intuitionistic fuzzy sets. *Fuzzy Set Syst.* **1989**, *33*, 37–45. [CrossRef]
19. Atanassov, K.T.; Rangasamy, P. Intuitionistic fuzzy sets. *Fuzzy Sets Syst.* **1986**, *20*, 87–96. [CrossRef]
20. Zadeh, L.A. Fuzzy sets. *Inf. Control* **1965**, *8*, 338–353. [CrossRef]
21. Zhang, X.H. Fuzzy anti-grouped filters and fuzzy normal filters in pseudo-BCI algebras. *J. Intell. Fuzzy Syst.* **2017**, *33*, 1767–1774. [CrossRef]
22. Huang, B.; Guo, C.X.; Zhang, Y.L.; Li, H.X.; Zhou, X.Z. Intuitionistic fuzzy multi-granulation rough sets. *Inf. Sci.* **2014**, *277*, 299–320. [CrossRef]
23. Liu, C.H.; Pedrycz, W. Covering-based multi-granulation fuzzy rough sets. *J. Intell. Fuzzy Syst.* **2016**, *30*, 303–318. [CrossRef]
24. Xue, Z.A.; Wang, N.; Si, X.M.; Zhu, T.L. Research on multi-granularity rough intuitionistic fuzzy cut sets. *J. Henan Normal Univ. (Nat. Sci. Ed.)* **2016**, *44*, 131–139. (In Chinese)
25. Xue, Z.A.; Lv, M.J.; Han, D.J.; Xin, X.W. Multi-granulation graded rough intuitionistic fuzzy sets models based on dominance relation. *Symmetry* **2018**, *10*, 446. [CrossRef]
26. Wang, J.Q.; Zhang, X.H. Two types of intuitionistic fuzzy covering rough sets and an application to multiple criteria group decision making. *Symmetry* **2018**, *10*, 462. [CrossRef]
27. Boran, F.E.; Akay, D. A biparametric similarity measure on intuitionistic fuzzy sets with applications to pattern recognition. *Inf. Sci.* **2014**, *255*, 45–57. [CrossRef]
28. Intarapaiboon, P. A hierarchy-based similarity measure for intuitionistic fuzzy sets. *Soft Comput.* **2016**, *20*, 1–11. [CrossRef]
29. Ngan, R.T.; Le, H.S.; Cuong, B.C.; Mumtaz, A. H-max distance measure of intuitionistic fuzzy sets in decision making. *Appl. Soft Comput.* **2018**, *69*, 393–425. [CrossRef]
30. Yao, Y.Y. The Superiority of Three-way decisions in probabilistic rough set models. *Inf. Sci.* **2011**, *181*, 1080–1096. [CrossRef]
31. Yao, Y.Y. Three-way decisions with probabilistic rough sets. *Inf. Sci.* **2010**, *180*, 341–353. [CrossRef]
32. Zhai, J.H.; Zhang, Y.; Zhu, H.Y. Three-way decisions model based on tolerance rough fuzzy set. *Int. J. Mach. Learn. Cybern.* **2016**, *8*, 1–9. [CrossRef]
33. Zhai, J.H.; Zhang, S.F. Three-way decisions model based on rough fuzzy set. *J. Intell. Fuzzy Syst.* **2018**, *34*, 2051–2059. [CrossRef]
34. Hao, C.; Li, J.H.; Fan, M.; Liu, W.Q.; Tsang, E.C.C. Optimal scale selection in dynamic multi-scale decision tables based on sequential three-way decisions. *Inf. Sci.* **2017**, *415*, 213–232. [CrossRef]
35. Luo, C.; Li, T.R.; Huang, Y.Y.; Fujita, H. Updating three-way decisions in incomplete multi-scale information systems. *Inf. Sci.* **2018**. [CrossRef]
36. Zhang, X.H.; Bo, C.X.; Smarandache, F.; Dai, J.H. New inclusion relation of neutrosophic sets with applications and related lattice structure. *Int. J. Mach. Learn. Cybern.* **2018**, *9*, 1753–1763. [CrossRef]
37. Zhang, Q.H.; Xie, Q.; Wang, G.Y. A novel three-way decision model with decision-theoretic rough sets using utility theory. *Knowl.-Based Syst.* **2018**. [CrossRef]
38. Yang, X.P.; Tan, A.H. Three-way decisions based on intuitionistic fuzzy sets. In Proceedings of the International Joint Conference on Rough Sets, Olsztyn, Poland, 3–7 July 2017.
39. Liu, J.B.; Zhou, X.Z.; Huang, B.; Li, H.X. A three-way decision model based on intuitionistic fuzzy decision systems. In Proceedings of the International Joint Conference on Rough Sets, Olsztyn, Poland, 3–7 July 2017.
40. Liang, D.C.; Liu, D. Deriving three-way decisions from intuitionistic fuzzy decision-theoretic rough sets. *Inf. Sci.* **2015**, *300*, 28–48. [CrossRef]
41. Liang, D.C.; Xu, Z.S.; Liu, D. Three-way decisions with intuitionistic fuzzy decision-theoretic rough sets based on point operators. *Inf. Sci.* **2017**, *375*, 18–201. [CrossRef]
42. Sun, B.Z.; Ma, W.M.; Li, B.J.; Li, X.N. Three-way decisions approach to multiple attribute group decision making with linguistic information-based decision-theoretic rough fuzzy set. *Int. J. Approx. Reason.* **2018**, *93*, 424–442. [CrossRef]
43. Abdel-Basset, M.; Gunasekaran, M.; Mai, M.; Chilamkurti, N. Three-way decisions based on neutrosophic sets and AHP-QFD framework for supplier selection problem. *Future Gener. Comput. Syst.* **2018**, *89*. [CrossRef]

44. Yu, H.; Zhang, C.; Wang, G.Y. A tree-based incremental overlapping clustering method using the three- way decision theory. *Knowl.-Based Syst.* **2016**, *91*, 189–203. [CrossRef]
45. Li, J.H.; Huang, C.C.; Qi, J.J.; Qian, Y.H.; Liu, W.Q. Three-way cognitive concept learning via multi-granulation. *Inf. Sci.* **2017**, *378*, 244–263. [CrossRef]
46. Xue, Z.A.; Zhu, T.L.; Xue, T.Y.; Liu, J. Methodology of attribute weights acquisition based on three-way decision theory. *Comput. Sci.* **2015**, *42*, 265–268. (In Chinese)

6

On the Crossing Numbers of the Joining of a Specific Graph on Six Vertices with the Discrete Graph

Michal Staš

Faculty of Electrical Engineering and Informatics, Technical University of Košice, 042 00 Košice, Slovakia; michal.stas@tuke.sk

Abstract: In the paper, we extend known results concerning crossing numbers of join products of small graphs of order six with discrete graphs. The crossing number of the join product $G^* + D_n$ for the graph G^* on six vertices consists of one vertex which is adjacent with three non-consecutive vertices of the 5-cycle. The proofs were based on the idea of establishing minimum values of crossings between two different subgraphs that cross the edges of the graph G^* exactly once. These minimum symmetrical values are described in the individual symmetric tables.

Keywords: graph; good drawing; crossing number; join product; cyclic permutation

---

1. Introduction

An investigation on the crossing number of graphs is a classical and very difficult problem. Garey and Johnson [1] proved that this problem is NP-complete. Recall that the exact values of the crossing numbers are known for only a few families of graphs. The purpose of this article is to extend the known results concerning this topic. In this article, we use the definitions and notation of the crossing numbers of graphs presented by Klešč in [2]. Kulli and Muddebihal [3] described the characterization for all pairs of graphs which join product of a planar graph. In the paper, some parts of proofs are also based on Kleitman's result [4] on the crossing numbers for some complete bipartite graphs. More precisely, he showed that

$$\mathrm{cr}(K_{m,n}) = \left\lfloor \frac{m}{2} \right\rfloor \left\lfloor \frac{m-1}{2} \right\rfloor \left\lfloor \frac{n}{2} \right\rfloor \left\lfloor \frac{n-1}{2} \right\rfloor, \quad \text{for} \quad m \leq 6.$$

Again, by Kleitman's result [4], the crossing numbers for the join of two different paths, the join of two different cycles, and also for the join of path and cycle, were established in [2]. Further, the exact values for crossing numbers of $G + D_n$ and of $G + P_n$ for all graphs G on less than five vertices were determined in [5]. At present, the crossing numbers of the graphs $G + D_n$ are known only for few graphs G of order six in [6–9]. In all these cases, the graph G is usually connected and includes at least one cycle.

The methods in the paper mostly use the combinatorial properties of cyclic permutations. For the first time, the idea of configurations is converted from the family of subgraphs which do not cross the edges of the graph G^* of order six onto the family of subgraphs whose edges cross the edges of G^* just once. According to this algebraic topological approach, we can extend known results for the crossing numbers of new graphs. Some of the ideas and methods were used for the first time in [10]. In [6,8,9], some parts of proofs were done with the help of software which is described in detail in [11].

It is important to recall that the methods presented in [5,7,12] do not suffice to determine the crossing number of the graph $G^* + D_n$. Also in this article, some parts of proofs can be simplified by utilizing the work of the software that generates all cyclic permutations in [11]. Its C++ version is located also on the website http://web.tuke.sk/fei-km/coga/, and the list with all short names of 120 cyclic permutations of six elements have already been collected in Table 1 of [8].

2. Cyclic Permutations and Corresponding Configurations of Subgraphs

Let G^* be the connected graph on six vertices consisting of one vertex which is adjacent with three non-consecutive vertices of the 5-cycle. We consider the join product of the graph G^* with the discrete graph D_n on n vertices. It is not difficult to see that the graph $G^* + D_n$ consists of just one copy of the graph G^* and of n vertices $t_1, \ldots, t_n$, where any vertex t_j, $j = 1, \ldots, n$, is adjacent to every vertex of the graph G^*. Let T^j, $j = 1, \ldots, n$, denote the subgraph which is uniquely induced by the six edges incident with the fixed vertex t_j. This means that the graph $T^1 \cup \cdots \cup T^n$ is isomorphic with $K_{6,n}$ and

$$G^* + D_n = G^* \cup K_{6,n} = G^* \cup \left(\bigcup_{j=1}^{n} T^j \right). \tag{1}$$

In the paper, the definitions and notation of the cyclic permutations and of the corresponding configurations of subgraphs for a good drawing D of the graph $G^* + D_n$ presented in [8] are used. The *rotation* $\mathrm{rot}_D(t_j)$ of a vertex t_j in the drawing D is the cyclic permutation that records the (cyclic) counter-clockwise order in which the edges leave t_j, see [10]. We use the notation (123456) if the counterclockwise order of the edges incident with the vertex t_j is t_jv_1, t_jv_2, t_jv_3, t_jv_4, t_jv_5, and t_jv_6. Recall that a rotation is a cyclic permutation. Moreover, as we have already mentioned, we separate all subgraphs T^j, $j = 1, \ldots, n$, of the graph $G^* + D_n$ into three mutually-disjoint families depending on how many times the edges of G^* are crossed by the edges of the considered subgraph T^j in D. This means, for $j = 1, \ldots, n$, let $R_D = \{T^j : \mathrm{cr}_D(G^*, T^j) = 0\}$ and $S_D = \{T^j : \mathrm{cr}_D(G^*, T^j) = 1\}$. The edges of G^* are crossed by each other subgraph T^j at least twice in D. For $T^j \in R_D \cup S_D$, let F^j denote the subgraph $G^* \cup T^j$, $j \in \{1, 2, \ldots, n\}$, of $G^* + D_n$, and let $D(F^j)$ be its subdrawing induced by D.

If we would like to obtain an optimal drawing D of $G^* + D_n$, then the set $R_D \cup S_D$ must be nonempty provided by the arguments in Theorem 1. Thus, we only consider drawings of the graph G^* for which there is a possibility of obtaining a subgraph $T^j \in R_D \cup S_D$. Since the graph G^* contains the 6-cycle as a subgraph (for brevity, we can write $C_6(G^*)$), we have to assume only crossings between possible subdrawings of the subgraph $C_6(G^*)$ and two remaining edges of G^*. Of course, the edges of the cycle $C_6(G^*)$ can cross themselves in the considered subdrawings. The vertex notation of G^* will be substantiated later in all drawings in Figure 1.

First, assume a good drawing D of $G^* + D_n$ in which the edges of G^* do not cross each other. In this case, without loss of generality, we can consider the drawing of G^* with the vertex notation like that in Figure 1a. Clearly, the set R_D is empty. Our aim is to list all possible rotations $\mathrm{rot}_D(t_j)$ which can appear in D if the edges of G^* are crossed by the edges of T^j just once. There is only one possible subdrawing of $F^j \setminus \{v_4\}$ represented by the rotation (16532), which yields that there are exactly five ways of obtaining the subdrawing of $G \cup T^j$ depending on which edge of the graph G^* can be crossed by the edge t_jv_4. We denote these five possibilities by $\mathcal{A}_k$, for $k = 1, \ldots, 5$. For our considerations over the number of crossings of $G^* + D_n$, it does not play a role in which of the regions is unbounded. So we can assume the drawings shown in Figure 2. Thus, the configurations $\mathcal{A}_1$, $\mathcal{A}_2$, $\mathcal{A}_3$, $\mathcal{A}_4$, and $\mathcal{A}_5$ are represented by the cyclic permutations (165324), (165432), (146532), (165342), and (164532), respectively. Of course, in a fixed drawing of the graph $G^* + D_n$, some configurations from $\mathcal{M} = \{\mathcal{A}_1, \mathcal{A}_2, \mathcal{A}_3, \mathcal{A}_4, \mathcal{A}_5\}$ need not appear. We denote by $\mathcal{M}_D$ the set of all configurations that exist in the drawing D belonging to the set $\mathcal{M}$.

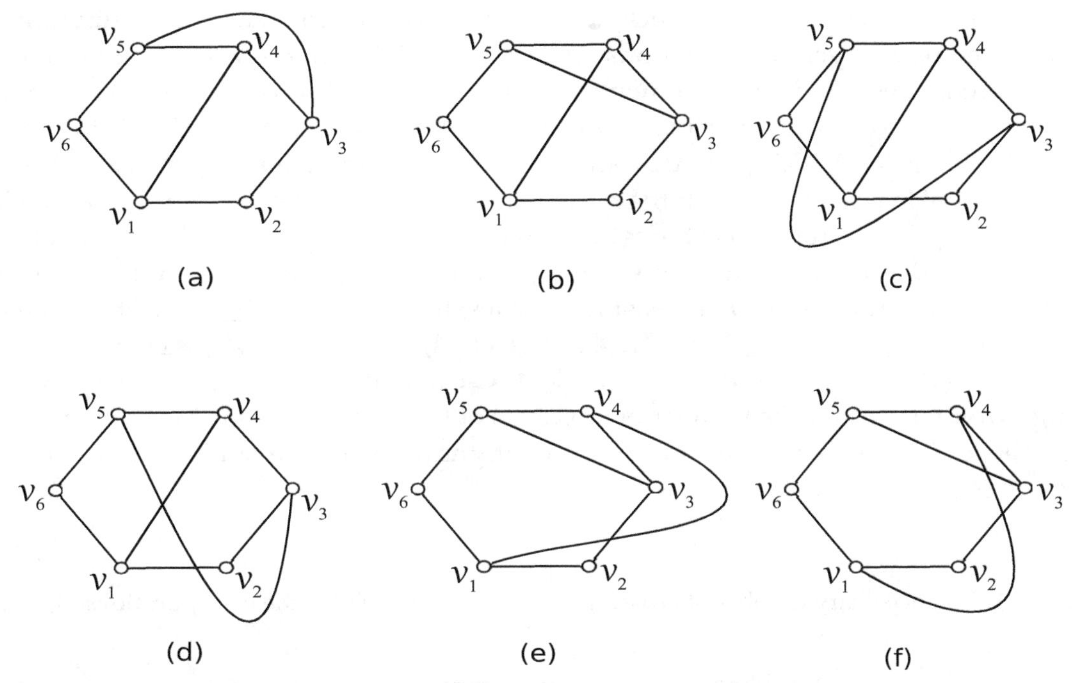

Figure 1. Six possible drawings of G^* with no crossing among edges of $C_6(G^*)$. **(a)**: the planar drawing of G^*; **(b)**: the drawing of G^* with $\mathrm{cr}_D(G^*) = 1$ and without crossing on edges of $C_6(G^*)$; **(c)**: the drawing of G^* only with two crossings on edges of $C_6(G^*)$; **(d)**: the drawing of G^* with $\mathrm{cr}_D(G^*) = 2$ and with one crossing on edges of $C_6(G^*)$; **(e)**: the drawing of G^* only with one crossing on edges of $C_6(G^*)$; **(f)**: the drawing of G^* with $\mathrm{cr}_D(G^*) = 2$ and with one crossing on edges of $C_6(G^*)$.

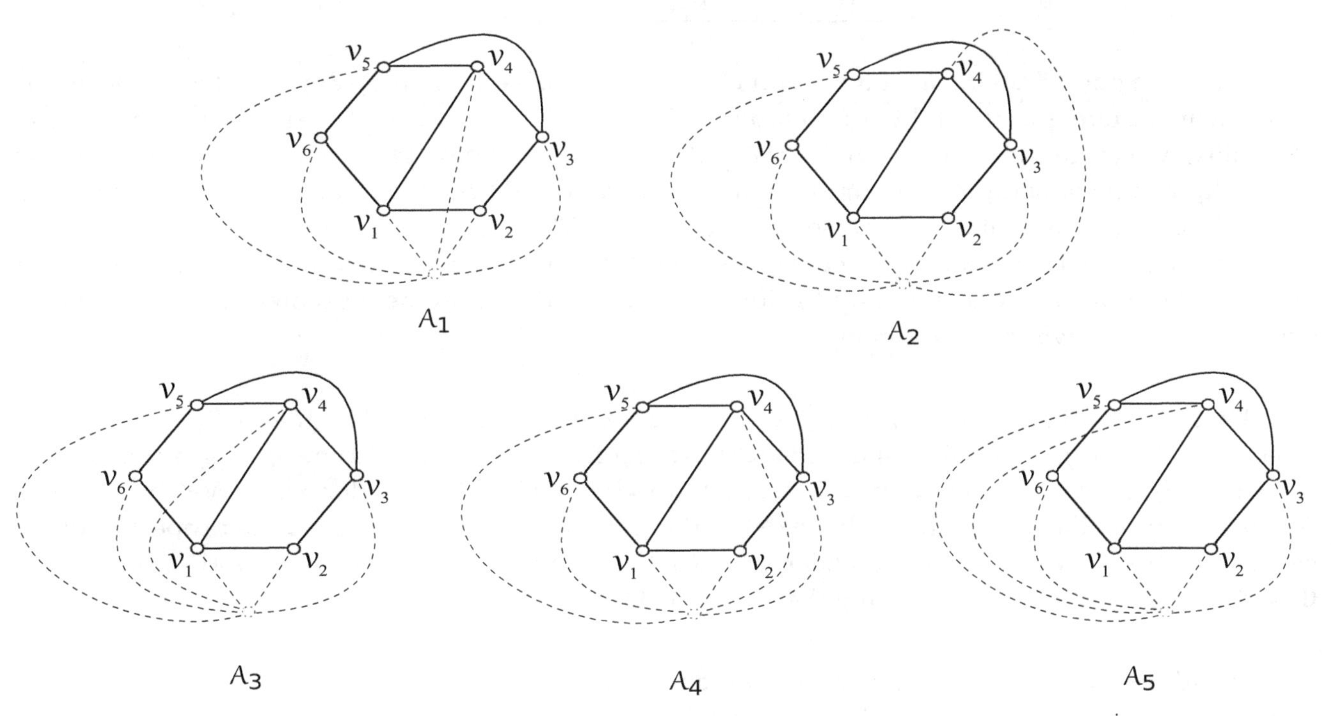

Figure 2. Drawings of five possible configurations from $\mathcal{M}$ of the subgraph F^j.

Recall that we are able to extend the idea of establishing minimum values of crossings between two different subgraphs onto the family of subgraphs which cross the edges of G^* exactly once. Let $\mathcal{X}$ and $\mathcal{Y}$ be the configurations from $\mathcal{M}_D$. We denote by $\mathrm{cr}_D(\mathcal{X}$ and $\mathcal{Y})$ the number of crossings in D between T^i and T^j for different $T^i, T^j \in S_D$ such that F^i and F^j have configurations $\mathcal{X}$ and $\mathcal{Y}$, respectively. Finally, let $\mathrm{cr}(\mathcal{X}, \mathcal{Y}) = \min\{\mathrm{cr}_D(\mathcal{X}, \mathcal{Y})\}$ over all possible good drawings of $G^* + D_n$ with $\mathcal{X}, \mathcal{Y} \in \mathcal{M}_D$.

Our aim is to determine $\mathrm{cr}(\mathcal{X},\mathcal{Y})$ for all such pairs $\mathcal{X},\mathcal{Y}\in\mathcal{M}$. In particular, the configurations $\mathcal{A}_1$ and $\mathcal{A}_2$ are represented by the cyclic permutations (165324) and (165432), respectively. Since the minimum number of interchanges of adjacent elements of (165324) required to produce cyclic permutation (165432) is two, we need at least four interchanges of adjacent elements of (165432) to produce cyclic permutation $\overline{(165324)} = (142356)$. (Let T^x and T^y be two different subgraphs represented by their $\mathrm{rot}(t_x)$ and $\mathrm{rot}(t_y)$ of length m, $m \geq 3$. If the minimum number of interchanges of adjacent elements of $\mathrm{rot}(t_x)$ required to produce $\mathrm{rot}(t_y)$ is at most z, then $\mathrm{cr}_D(T^x,T^y) \geq \lfloor \frac{m}{2} \rfloor \lfloor \frac{m-1}{2} \rfloor - z$. Details have been worked out by Woodall [13].) So any subgraph T^j with the configuration $\mathcal{A}_2$ of F^j crosses the edges of T^i with the configuration $\mathcal{A}_1$ of F^i at least four times; that is, $\mathrm{cr}(\mathcal{A}_1,\mathcal{A}_2)\geq 4$. The same reasoning gives $\mathrm{cr}(\mathcal{A}_1,\mathcal{A}_3)\geq 5$, $\mathrm{cr}(\mathcal{A}_1,\mathcal{A}_4)\geq 5$, $\mathrm{cr}(\mathcal{A}_1,\mathcal{A}_5)\geq 4$, $\mathrm{cr}(\mathcal{A}_2,\mathcal{A}_3)\geq 4$, $\mathrm{cr}(\mathcal{A}_2,\mathcal{A}_4)\geq 5$, $\mathrm{cr}(\mathcal{A}_2,\mathcal{A}_5)\geq 5$, $\mathrm{cr}(\mathcal{A}_3,\mathcal{A}_4)\geq 4$, $\mathrm{cr}(\mathcal{A}_3,\mathcal{A}_5)\geq 5$, and $\mathrm{cr}(\mathcal{A}_4,\mathcal{A}_5)\geq 4$. Clearly, also $\mathrm{cr}(\mathcal{A}_i,\mathcal{A}_i)\geq 6$ for any $i=1,\dots,5$. All resulting lower bounds for the number of crossings of two configurations from $\mathcal{M}$ are summarized in the symmetric Table 1 (here, $\mathcal{A}_k$ and $\mathcal{A}_l$ are configurations of the subgraphs F^i and F^j, where $k,l\in\{1,2,3,4,5\}$).

Table 1. The necessary number of crossings between T^i and T^j for the configurations $\mathcal{A}_k$, $\mathcal{A}_l$.

-	$\mathcal{A}_1$	$\mathcal{A}_2$	$\mathcal{A}_3$	$\mathcal{A}_4$	$\mathcal{A}_5$
$\mathcal{A}_1$	6	4	5	5	4
$\mathcal{A}_2$	4	6	4	5	5
$\mathcal{A}_3$	5	4	6	4	5
$\mathcal{A}_4$	5	5	4	6	4
$\mathcal{A}_5$	4	5	5	4	6

Assume a good drawing D of the graph G^*+D_n with just one crossing among edges of the graph G^* (in which there is a possibility of obtaining of subgraph $T^j\in R_D\cup S_D$). At first, without loss of generality, we can consider the drawing of G^* with the vertex notation like that in Figure 1b. Of course, the set R_D can be nonempty, but our aim will be also to list all possible rotations $\mathrm{rot}_D(t_j)$ which can appear in D if the edges of G^* are crossed by the edges of T^j just once. Since the edges v_1v_2, v_2v_3, v_1v_6, and v_5v_6 of G^* can be crossed by the edges t_jv_3, t_jv_1, t_jv_5, and t_jv_1, respectively, these four ways under our consideration can be denoted by $\mathcal{B}_k$, for $k=1,2,3,4$. Based on the aforementioned arguments, we assume the drawings shown in Figure 3.

Thus, the configurations $\mathcal{B}_1$, $\mathcal{B}_2$, $\mathcal{B}_3$, and $\mathcal{B}_4$ are uniquely represented by the cyclic permutations (165423), (126543), (156432), and (154326), respectively. Because some configurations from $\mathcal{N}=\{\mathcal{B}_1,\mathcal{B}_2,\mathcal{B}_3,\mathcal{B}_4\}$ may not appear in a fixed drawing of G^*+D_n, we denote by $\mathcal{N}_D$ the subset of $\mathcal{N}$ consisting of all configurations that exist in the drawing D. Further, due to the properties of the cyclic rotations, we can easily verify that $\mathrm{cr}(\mathcal{B}_i,\mathcal{B}_j)\geq 4$ for any $i,j\in\{1,2,3,4\}$, $i\neq j$. (Let us note that this idea was used for an establishing the values in Table 1)

In addition, without loss of generality, we can consider the drawing of G^* with the vertex notation like that in Figure 1e. In this case, the set R_D is also empty. Hence, our aim is to list again all possible rotations $\mathrm{rot}_D(t_j)$ which can appear in D if $T^j\in S_D$. Since there is only one subdrawing of $F^j\setminus\{v_3\}$ represented by the rotation (16542), there are four ways to obtain the subdrawing of F^j depending on which edge of G^* is crossed by the edge t_jv_3. These four possibilities under our consideration are denoted by $\mathcal{E}_k$, for $k=1,2,3,4$. Again, based on the aforementioned arguments, we assume the drawings shown in Figure 4.

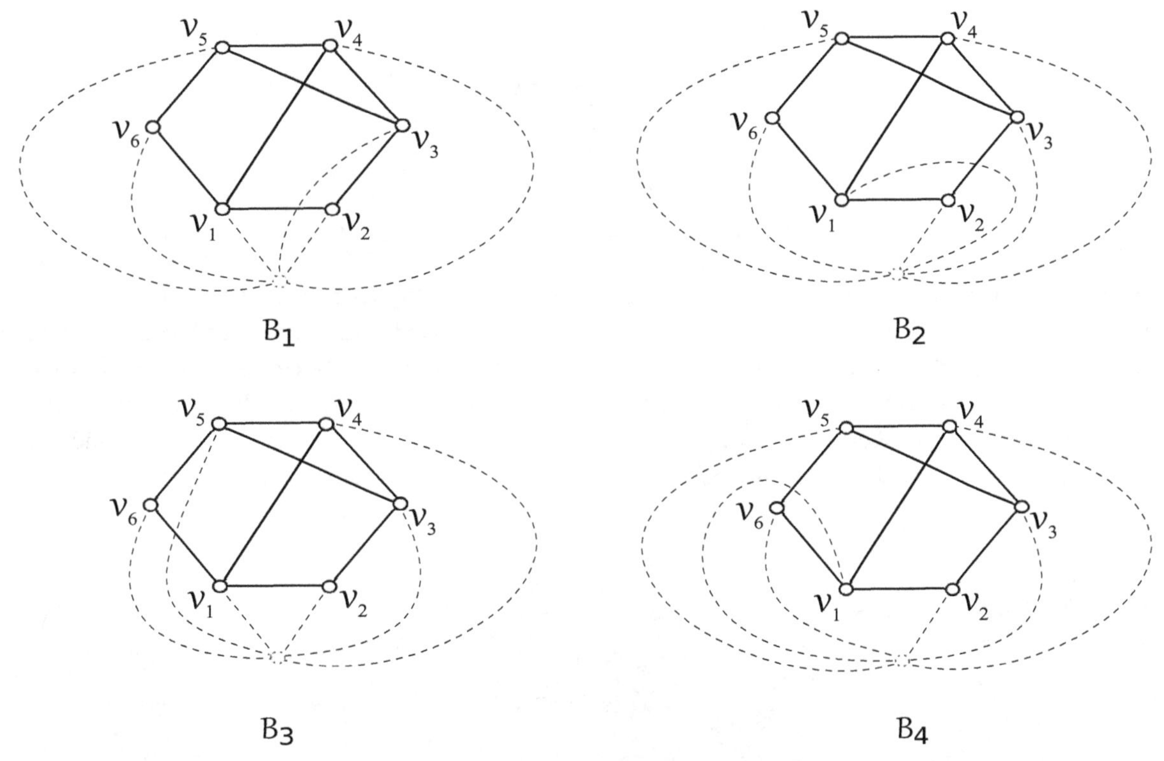

Figure 3. Drawings of four possible configurations from $\mathcal{N}$ of the subgraph F^j.

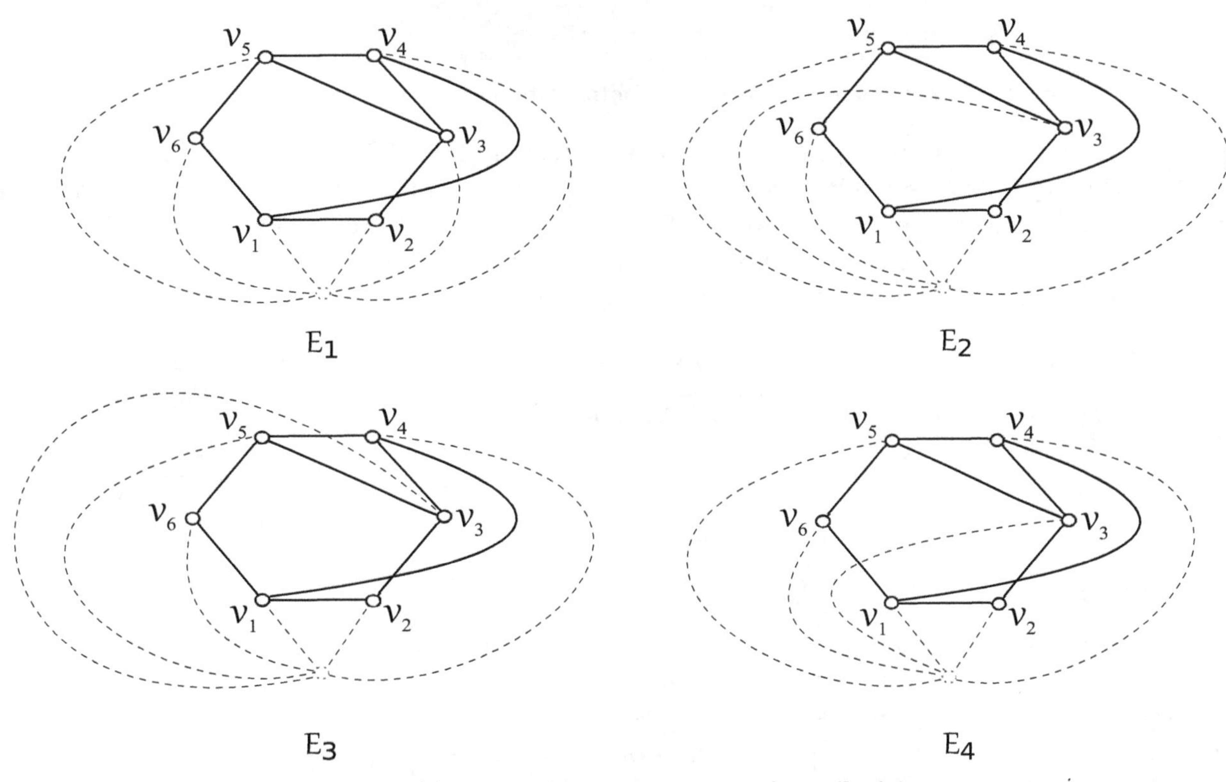

Figure 4. Drawings of four possible configurations from $\mathcal{O}$ of the subgraph F^j.

Thus, the configurations $\mathcal{E}_1$, $\mathcal{E}_2$, $\mathcal{E}_3$, and $\mathcal{E}_4$ are represented by the cyclic permutations (165432), (163542), (165342), and (136542), respectively. Again, we denote by $\mathcal{O}_D$ the subset of $\mathcal{O} = \{\mathcal{E}_1, \mathcal{E}_2, \mathcal{E}_3, \mathcal{E}_4\}$ consisting of all configurations that exist in the drawing D. Further, due to the properties of the cyclic rotations, all lower-bounds of number of crossings of two configurations from $\mathcal{O}$ can be summarized in the symmetric Table 2 (here, $\mathcal{E}_k$ and $\mathcal{E}_l$ are configurations of the subgraphs F^i and F^j, where $k, l \in \{1, 2, 3, 4\}$).

Table 2. The necessary number of crossings between T^i and T^j for the configurations $\mathcal{E}_k$, $\mathcal{E}_l$.

-	$\mathcal{E}_1$	$\mathcal{E}_2$	$\mathcal{E}_3$	$\mathcal{E}_4$
$\mathcal{E}_1$	6	4	5	4
$\mathcal{E}_2$	4	6	5	5
$\mathcal{E}_3$	5	5	6	4
$\mathcal{E}_4$	4	5	4	6

Finally, without loss of generality, we can consider the drawing of G^* with the vertex notation like that in Figure 1f. In this case, the set R_D is also empty. So our aim will be to list again all possible rotations $\mathrm{rot}_D(t_j)$ which can appear in D if $T^j \in S_D$. Since there is only one subdrawing of $F^j \setminus \{v_2\}$ represented by the rotation (16543), there are three ways to obtain the subdrawing of F^j depending on which edge of G^* is crossed by the edge t_jv_2. These three possibilities under our consideration are denoted by $\mathcal{F}_k$, for $k = 1, 2, 3$. Again, based on the aforementioned arguments, we assume the drawings shown in Figure 5.

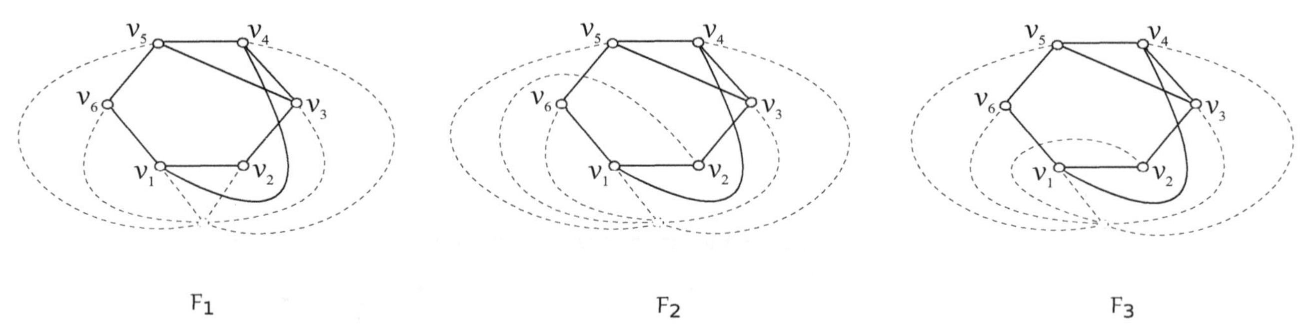

Figure 5. Drawings of three possible configurations from $\mathcal{P}$ of the subgraph F^j.

Thus, the configurations $\mathcal{F}_1$, $\mathcal{F}_2$, and $\mathcal{F}_3$ are represented by the cyclic permutations (165432), (162543), and (126543), respectively. Again, we denote by $\mathcal{P}_D$ the subset of $\mathcal{P} = \{\mathcal{F}_1, \mathcal{F}_2, \mathcal{F}_3\}$ consisting of all configurations that exist in the drawing D. Further, due to the properties of the cyclic rotations, all lower-bounds of number of crossings of two configurations from $\mathcal{P}$ can be summarized in the symmetric Table 3 (here, $\mathcal{F}_k$ and $\mathcal{F}_l$ are configurations of the subgraphs F^i and F^j, where $k, l \in \{1, 2, 3\}$).

Table 3. The necessary number of crossings between T^i and T^j for the configurations $\mathcal{F}_k$ and $\mathcal{F}_l$.

-	$\mathcal{F}_1$	$\mathcal{F}_2$	$\mathcal{F}_3$
$\mathcal{F}_1$	6	4	5
$\mathcal{F}_2$	4	6	5
$\mathcal{F}_3$	5	5	6

3. The Crossing Number of $G^* + D_n$

Recall that two vertices t_i and t_j of $G^* + D_n$ are *antipodal* in a drawing D of $G^* + D_n$ if the subgraphs T^i and T^j do not cross. A drawing is *antipodal-free* if it has no antipodal vertices. For easier and more accurate labeling in the proofs of assertions, let us define notation of regions in some subdrawings of $G^* + D_n$. The unique drawing of G^* as shown in Figure 1a contains four different regions. Let us denote these four regions by $\omega_{1,2,3,4}$, $\omega_{1,4,5,6}$, $\omega_{3,4,5}$, and $\omega_{1,2,3,5,6}$ depending on which of vertices are located on the boundary of the corresponding region.

Lemma 1. *Let D be a good and antipodal-free drawing of $G^* + D_n$, for $n > 3$, with the drawing of G^* with the vertex notation like that in Figure 1a. If $T^u, T^v, T^t \in S_D$ are three different subgraphs such that F^u, F^v, and F^t have three different configurations from the set $\{\mathcal{A}_i, \mathcal{A}_j, \mathcal{A}_k\} \subseteq \mathcal{M}_D$ with $i + 2 \equiv j + 1 \equiv k \pmod 5$, then*

$$\mathrm{cr}_D(G^* \cup T^u \cup T^v \cup T^t, T^m) \geq 6 \qquad \text{for any } T^m \notin S_D.$$

Proof of Lemma 1. Let us assume the configurations $\mathcal{A}_1$ of F^u, $\mathcal{A}_2$ of F^v, and $\mathcal{A}_3$ of F^t. It is obvious that $\mathrm{cr}_D(T^u \cup T^v \cup T^t, T^m) \geq 3$ holds for any subgraph T^m, $m \neq u, v, t$. Further, if $\mathrm{cr}_D(G^*, T^m) > 2$, then we obtain the desired result $\mathrm{cr}_D(G^* \cup T^u \cup T^v \cup T^t, T^m) \geq 3 + 3 = 6$. To finish the proof, let us suppose that there is a subgraph $T^m \notin S_D$ such that T^m crosses exactly once the edges of each subgraph T^u, T^v, and T^t, and let also consider $\mathrm{cr}_D(G^*, T^m) = 2$. As $\mathrm{cr}_D(T^u, T^m) = 1$, the vertex t_m must be placed in the quadrangular region with four vertices of G^* on its boundary; that is, $t_m \in \omega_{1,4,5,6}$. Similarly, the assumption $\mathrm{cr}_D(T^t, T^m) = 1$ enforces that $t_m \in \omega_{1,2,3,4}$. Since the vertex t_m cannot be placed simultaneously in both regions, we obtain a contradiction. The proof proceeds in the similar way also for the remaining possible cases of the configurations of subgraphs F^u, F^v, and F^t, and the proof is done. □

Now we are able to prove the main result of the article. We can calculate the exact values of crossing numbers for small graphs using an algorithm located on a website http://crossings.uos.de/. It uses an ILP formulation based on Kuratowski subgraphs. The system also generates verifiable formal proofs like those described in [14]. Unfortunately, the capacity of this system is limited.

Lemma 2. $\mathrm{cr}(G^* + D_1) = 1$ *and* $\mathrm{cr}(G^* + D_2) = 3$.

Theorem 1. $\mathrm{cr}(G^* + D_n) = 6\left\lfloor \frac{n}{2} \right\rfloor \left\lfloor \frac{n-1}{2} \right\rfloor + n + \left\lfloor \frac{n}{2} \right\rfloor$ *for* $n \geq 1$.

Proof of Theorem 1. Figure 6 offers the drawing of $G^* + D_n$ with exactly $6\lfloor \frac{n}{2} \rfloor \lfloor \frac{n-1}{2} \rfloor + n + \lfloor \frac{n}{2} \rfloor$ crossings. Thus, $\mathrm{cr}(G^* + D_n) \leq 6\lfloor \frac{n}{2} \rfloor \lfloor \frac{n-1}{2} \rfloor + n + \lfloor \frac{n}{2} \rfloor$. We prove the reverse inequality by induction on n. By Lemma 2, the result is true for $n = 1$ and $n = 2$. Now suppose that, for some $n \geq 3$, there is a drawing D with

$$\mathrm{cr}_D(G^* + D_n) < 6\left\lfloor \frac{n}{2} \right\rfloor \left\lfloor \frac{n-1}{2} \right\rfloor + n + \left\lfloor \frac{n}{2} \right\rfloor \tag{2}$$

and that

$$\mathrm{cr}(G^* + D_m) \geq 6\left\lfloor \frac{m}{2} \right\rfloor \left\lfloor \frac{m-1}{2} \right\rfloor + m + \left\lfloor \frac{m}{2} \right\rfloor \qquad \text{for any integer } m < n. \tag{3}$$

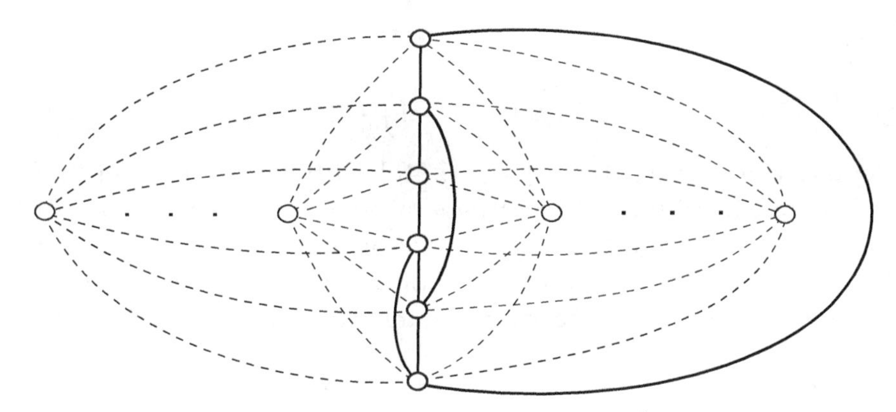

Figure 6. The good drawing of $G^* + D_n$ with $6\lfloor \frac{n}{2} \rfloor \lfloor \frac{n-1}{2} \rfloor + n + \lfloor \frac{n}{2} \rfloor$ crossings.

Let us first show that the considered drawing D must be antipodal-free. For a contradiction, suppose, without loss of generality, that $\mathrm{cr}_D(T^{n-1}, T^n) = 0$. If at least one of T^{n-1} and T^n, say T^n, does not cross G^*, it is not difficult to verify in Figure 1 that T^{n-1} must cross $G^* \cup T^n$ at least trice; that is, $\mathrm{cr}_D(G^*, T^{n-1} \cup T^n) \geq 3$. From [4], we already know that $\mathrm{cr}(K_{6,3}) = 6$, which yields that the edges of the subgraph $T^{n-1} \cup T^n$ are crossed by any T^k, $k = 1, 2, \ldots, n-2$, at least six times. So, for the number of crossings in D we have:

$$\mathrm{cr}_D(G^* + D_n) = \mathrm{cr}_D\left(G^* + D_{n-2}\right) + \mathrm{cr}_D(T^{n-1} \cup T^n) + \mathrm{cr}_D(K_{6,n-2}, T^{n-1} \cup T^n) + \mathrm{cr}_D(G^*, T^{n-1} \cup T^n)$$

$$\geq 6\left\lfloor \frac{n-2}{2} \right\rfloor \left\lfloor \frac{n-3}{2} \right\rfloor + n - 2 + \left\lfloor \frac{n-2}{2} \right\rfloor + 6(n-2) + 3 = 6\left\lfloor \frac{n}{2} \right\rfloor \left\lfloor \frac{n-1}{2} \right\rfloor + n + \left\lfloor \frac{n}{2} \right\rfloor.$$

This contradiction with the assumption (2) confirms that D is antipodal-free. Moreover, if $r = |R_D|$ and $s = |S_D|$, the assumption (3) together with $\text{cr}(K_{6,n}) = 6\lfloor \frac{n}{2} \rfloor \lfloor \frac{n-1}{2} \rfloor$ imply that, in D, if $r = 0$, then there are at least $\lceil \frac{n}{2} \rceil + 1$ subgraphs T^j for which the edges of G^* are crossed just once by them. More precisely:

$$\text{cr}_D(G^*) + \text{cr}_D(G^*, K_{6,n}) \leq \text{cr}_D(G^*) + 0r + 1s + 2(n - r - s) < n + \left\lfloor \frac{n}{2} \right\rfloor;$$

that is,

$$s + 2(n - r - s) < n + \left\lfloor \frac{n}{2} \right\rfloor. \tag{4}$$

This enforces that $2r + s \geq n - \lfloor \frac{n}{2} \rfloor + 1$, and if $r = 0$, then $s \geq n - \lfloor \frac{n}{2} \rfloor + 1 = \lceil \frac{n}{2} \rceil + 1$. Now, for $T^j \in R_D \cup S_D$, we discuss the existence of possible configurations of subgraphs $F^j = G^* \cup T^j$ in D.

Case 1: $\text{cr}_D(G^*) = 0$. Without loss of generality, we can consider the drawing of G^* with the vertex notation like that in Figure 1a. It is obvious that the set R_D is empty; that is, $r = 0$. Thus, we deal with only the configurations belonging to the nonempty set $\mathcal{M}_D$ and we discuss over all cardinalities of the set $\mathcal{M}_D$ in the following subcases:

i. $|\mathcal{M}_D| \geq 3$. We consider two subcases. Let us first assume that $\{\mathcal{A}_i, \mathcal{A}_j, \mathcal{A}_k\} \subseteq \mathcal{M}_D$ with $i + 2 \equiv j + 1 \equiv k \pmod 5$. Without lost of generality, let us consider three different subgraphs T^{n-2}, T^{n-1}, $T^n \in S_D$ such that F^{n-2}, F^{n-1} and F^n have configurations $\mathcal{A}_i$, $\mathcal{A}_j$, and $\mathcal{A}_k$, respectively. Then, $\text{cr}_D(T^{n-2} \cup T^{n-1} \cup T^n, T^m) \geq 14$ holds for any $T^m \in S_D$ with $m \neq n-2, n-1, n$ by summing the values in all columns in the considered three rows of Table 1. Moreover, $\text{cr}_D(G^* \cup T^{n-2} \cup T^{n-1} \cup T^n, T^m) \geq 6$ is fulfilling for any subgraph $T^m \notin S_D$ by Lemma 1. $\text{cr}_D(T^{n-2} \cup T^{n-1} \cup T^n) \geq 13$ holds by summing of three corresponding values of Table 1 between the considered configurations $\mathcal{A}_i$, $\mathcal{A}_j$, and $\mathcal{A}_k$, by fixing the subgraph $G^* \cup T^{n-2} \cup T^{n-1} \cup T^n$,

$$\text{cr}_D(G^* + D_n) = \text{cr}_D(K_{6,n-3}) + \text{cr}_D(K_{6,n-3}, G^* \cup T^{n-2} \cup T^{n-1} \cup T^n) + \text{cr}_D(G^* \cup T^{n-2} \cup T^{n-1} \cup T^n)$$

$$\geq 6\left\lfloor \frac{n-3}{2} \right\rfloor \left\lfloor \frac{n-4}{2} \right\rfloor + 15(s-3) + 6(n-s) + 13 + 3 = 6\left\lfloor \frac{n-3}{2} \right\rfloor \left\lfloor \frac{n-4}{2} \right\rfloor + 6n + 9s - 29$$

$$\geq 6\left\lfloor \frac{n-3}{2} \right\rfloor \left\lfloor \frac{n-4}{2} \right\rfloor + 6n + 9\left(\left\lceil \frac{n}{2} \right\rceil + 1\right) - 29 \geq 6\left\lfloor \frac{n}{2} \right\rfloor \left\lfloor \frac{n-1}{2} \right\rfloor + n + \left\lfloor \frac{n}{2} \right\rfloor.$$

In addition, let us assume that $\mathcal{M}_D = \{\mathcal{A}_i, \mathcal{A}_j, \mathcal{A}_k\}$ with $i + 1 \equiv j \pmod 5$, $j + 1 \not\equiv k \pmod 5$, and $k + 1 \not\equiv i \pmod 5$. Without lost of generality, let us consider two different subgraphs T^{n-1}, $T^n \in S_D$ such that F^{n-1} and F^n have mentioned configurations $\mathcal{A}_i$ and $\mathcal{A}_j$, respectively. Then, $\text{cr}_D(G^* \cup T^{n-1} \cup T^n, T^m) \geq 1 + 10 = 11$ holds for any $T^m \in S_D$ with $m \neq n-1, n$ also, by summing the values in Table 1. Hence, by fixing the subgraph $G^* \cup T^{n-1} \cup T^n$,

$$\text{cr}_D(G^* + D_n) = \text{cr}_D(K_{6,n-2}) + \text{cr}_D(K_{6,n-2}, G^* \cup T^{n-1} \cup T^n) + \text{cr}_D(G^* \cup T^{n-1} \cup T^n)$$

$$\geq 6\left\lfloor \frac{n-2}{2} \right\rfloor \left\lfloor \frac{n-3}{2} \right\rfloor + 11(s-2) + 4(n-s) + 4 + 2 = 6\left\lfloor \frac{n-2}{2} \right\rfloor \left\lfloor \frac{n-3}{2} \right\rfloor + 4n + 7s - 16$$

$$\geq 6\left\lfloor \frac{n-2}{2} \right\rfloor \left\lfloor \frac{n-3}{2} \right\rfloor + 4n + 7\left(\left\lceil \frac{n}{2} \right\rceil + 1\right) - 16 \geq 6\left\lfloor \frac{n}{2} \right\rfloor \left\lfloor \frac{n-1}{2} \right\rfloor + n + \left\lfloor \frac{n}{2} \right\rfloor.$$

ii. $|\mathcal{M}_D| = 2$; that is, $\mathcal{M}_D = \{\mathcal{A}_i, \mathcal{A}_j\}$ for some $i, j \in \{1, \ldots, 5\}$ with $i \neq j$. Without lost of generality, let us consider two different subgraphs T^{n-1}, $T^n \in S_D$ such that F^{n-1} and F^n have mentioned configurations $\mathcal{A}_i$ and $\mathcal{A}_j$, respectively. Then, $\text{cr}_D(G^* \cup T^{n-1} \cup T^n, T^m) \geq 1 + 10 = 11$ holds for any $T^m \in S_D$ with $m \neq n-1, n$ also by Table 1. Thus, by fixing the subgraph $G^* \cup T^{n-1} \cup T^n$, we are able to use the same inequalities as in the previous subcase.

iii. $|\mathcal{M}_D| = 1$; that is, $\mathcal{M}_D = \{\mathcal{A}_j\}$ for only one $j \in \{1,\ldots,5\}$. Without lost of generality, let us assume that $T^n \in S_D$ with the configuration $\mathcal{A}_j \in \mathcal{M}_D$ of the subgraph F^n. As $\mathcal{M}_D = \{\mathcal{A}_j\}$, we have $\mathrm{cr}_D(G^* \cup T^n, T^k) \geq 1+6 = 7$ for any $T^k \in S_D$, $k \neq n$ provided that $\mathrm{rot}_D(t_n) = \mathrm{rot}_D(t_k)$, for more see [13]. Hence, by fixing the subgraph $G^* \cup T^n$,

$$\mathrm{cr}_D(G^* + D_n) = \mathrm{cr}_D(K_{6,n-1}) + \mathrm{cr}_D(K_{6,n-1}, G^* \cup T^n) + \mathrm{cr}_D(G^* \cup T^n)$$

$$\geq 6\left\lfloor\frac{n-1}{2}\right\rfloor\left\lfloor\frac{n-2}{2}\right\rfloor + 7(s-1) + 3(n-s) + 1 = 6\left\lfloor\frac{n-1}{2}\right\rfloor\left\lfloor\frac{n-2}{2}\right\rfloor + 3n + 4s - 6$$

$$\geq 6\left\lfloor\frac{n-1}{2}\right\rfloor\left\lfloor\frac{n-2}{2}\right\rfloor + 3n + 4\left(\left\lceil\frac{n}{2}\right\rceil + 1\right) - 6 \geq 6\left\lfloor\frac{n}{2}\right\rfloor\left\lfloor\frac{n-1}{2}\right\rfloor + n + \left\lfloor\frac{n}{2}\right\rfloor.$$

Case 2: $\mathrm{cr}_D(G^*) = 1$ with $\mathrm{cr}_D(C_6(G^*)) = 0$. At first, without loss of generality, we can consider the drawing of G^* with the vertex notation like that in Figure 1b. Since the set R_D can be nonempty, two possible subcases may occur:

i. Let R_D be the nonempty set; that is, there is a subgraph $T^i \in R_D$. Now, for a $T^i \in R_D$, the reader can easily see that the subgraph $F^i = G^* \cup T^i$ is uniquely represented by $\mathrm{rot}_D(t_i) = (165432)$, and $\mathrm{cr}_D(T^i, T^j) \geq 6$ for any $T^j \in R_D$ with $j \neq i$ provided that $\mathrm{rot}_D(t_i) = \mathrm{rot}_D(t_j)$; for more see [13]. Moreover, it is not difficult to verify by a discussion over all possible drawings D that $\mathrm{cr}_D(G^* \cup T^i, T^k) \geq 5$ holds for any subgraph $T^k \in S_D$, and $\mathrm{cr}_D(G^* \cup T^i, T^k) \geq 4$ is also fulfilling for any subgraph $T^k \notin R_D \cup S_D$. Thus, by fixing the subgraph $G^* \cup T^i$,

$$\mathrm{cr}_D(G^* + D_n) \geq 6\left\lfloor\frac{n-1}{2}\right\rfloor\left\lfloor\frac{n-2}{2}\right\rfloor + 6(r-1) + 5s + 4(n-r-s) + 1 = 6\left\lfloor\frac{n-1}{2}\right\rfloor\left\lfloor\frac{n-2}{2}\right\rfloor$$

$$+4n + (2r+s) - 5 \geq 6\left\lfloor\frac{n-1}{2}\right\rfloor\left\lfloor\frac{n-2}{2}\right\rfloor + 4n + \left(n - \left\lfloor\frac{n}{2}\right\rfloor + 1\right) - 5 \geq 6\left\lfloor\frac{n}{2}\right\rfloor\left\lfloor\frac{n-1}{2}\right\rfloor + n + \left\lfloor\frac{n}{2}\right\rfloor.$$

ii. Let R_D be the empty set; that is, each subgraph T^j crosses the edges of G^* at least once in D. Thus, we deal with the configurations belonging to the nonempty set $\mathcal{N}_D$. Let us consider a subgraph $T^j \in S_D$ with the configuration $\mathcal{B}_i \in \mathcal{N}_D$ of F^j, where $i \in \{1,2,3,4\}$. Then, the lower-bounds of number of crossings of two configurations from $\mathcal{N}$ confirm that $\mathrm{cr}_D(G^* \cup T^j, T^k) \geq 1+4 = 5$ holds for any $T^k \in S_D$, $k \neq j$. Moreover, one can also easily verify over all possible drawings D that $\mathrm{cr}_D(G^* \cup T^j, T^k) \geq 4$ is true for any subgraph $T^k \notin S_D$. Hence, by fixing the subgraph $G^* \cup T^j$,

$$\mathrm{cr}_D(G^* + D_n) \geq 6\left\lfloor\frac{n-1}{2}\right\rfloor\left\lfloor\frac{n-2}{2}\right\rfloor + 5(s-1) + 4(n-s) + 1 + 1 = 6\left\lfloor\frac{n-1}{2}\right\rfloor\left\lfloor\frac{n-2}{2}\right\rfloor$$

$$+4n + s - 3 \geq 6\left\lfloor\frac{n-1}{2}\right\rfloor\left\lfloor\frac{n-2}{2}\right\rfloor + 4n + \left(\left\lceil\frac{n}{2}\right\rceil + 1\right) - 3 \geq 6\left\lfloor\frac{n}{2}\right\rfloor\left\lfloor\frac{n-1}{2}\right\rfloor + n + \left\lfloor\frac{n}{2}\right\rfloor.$$

In addition, without loss of generality, we can consider the drawing of G^* with the vertex notation like that in Figure 1e. It is obvious that the set R_D is empty; that is, the set S_D cannot be empty. Thus, we deal with the configurations belonging to the nonempty set $\mathcal{O}_D$. Note that the lower-bounds of number of crossings of two configurations from $\mathcal{O}$ were already established in Table 2. Since there is a possibility to find a subdrawing of $G^* \cup T^j \cup T^k$, in which $\mathrm{cr}_D(G^* \cup T^j, T^k) = 3$ with $T^j \in S_D$ and $T^k \notin S_D$, we discuss four following subcases:

i. $\mathcal{E}_4 \in \mathcal{O}_D$. Without lost of generality, let us assume that $T^n \in S_D$ with the configuration $\mathcal{E}_4 \in \mathcal{O}_D$ of F^n. Only for this subcase, one can easily verify over all possible drawings D for which $\mathrm{cr}_D(G^* \cup T^n, T^k) \geq 4$ is true for any subgraph $T^k \notin S_D$. Thus, by fixing the subgraph $G^* \cup T^n$,

$$\mathrm{cr}_D(G^* + D_n) \geq 6\left\lfloor\frac{n-1}{2}\right\rfloor\left\lfloor\frac{n-2}{2}\right\rfloor + 5(s-1) + 4(n-s) + 1 + 1 = 6\left\lfloor\frac{n-1}{2}\right\rfloor\left\lfloor\frac{n-2}{2}\right\rfloor$$

$$+4n+s-3 \geq 6\left\lfloor\frac{n-1}{2}\right\rfloor\left\lfloor\frac{n-2}{2}\right\rfloor+4n+\left(\left\lceil\frac{n}{2}\right\rceil+1\right)-3 \geq 6\left\lfloor\frac{n}{2}\right\rfloor\left\lfloor\frac{n-1}{2}\right\rfloor+n+\left\lfloor\frac{n}{2}\right\rfloor.$$

ii. $\mathcal{E}_4 \notin \mathcal{O}_D$ and $\mathcal{E}_3 \in \mathcal{O}_D$. Without lost of generality, let us assume that $T^n \in S_D$ with the configuration $\mathcal{E}_3 \in \mathcal{O}_D$ of F^n. In this subcase, $\mathrm{cr}_D(G^* \cup T^n, T^k) \geq 1+5=6$ holds for any subgraph $T^k \in S_D$, $k \neq n$ by the remaining values in the third row of Table 2. Hence, by fixing the subgraph $G^* \cup T^n$,

$$\mathrm{cr}_D(G^*+D_n) \geq 6\left\lfloor\frac{n-1}{2}\right\rfloor\left\lfloor\frac{n-2}{2}\right\rfloor+6(s-1)+3(n-s)+1+1=6\left\lfloor\frac{n-1}{2}\right\rfloor\left\lfloor\frac{n-2}{2}\right\rfloor$$

$$+3n+3s-4 \geq 6\left\lfloor\frac{n-1}{2}\right\rfloor\left\lfloor\frac{n-2}{2}\right\rfloor+3n+3\left(\left\lceil\frac{n}{2}\right\rceil+1\right)-4 \geq 6\left\lfloor\frac{n}{2}\right\rfloor\left\lfloor\frac{n-1}{2}\right\rfloor+n+\left\lfloor\frac{n}{2}\right\rfloor.$$

iii. $\mathcal{O}_D=\{\mathcal{E}_1,\mathcal{E}_2\}$. Without lost of generality, let us consider two different subgraphs T^{n-1}, $T^n \in S_D$ such that F^{n-1} and F^n have mentioned configurations $\mathcal{E}_1$ and $\mathcal{E}_2$, respectively. Then, $\mathrm{cr}_D(G^* \cup T^{n-1} \cup T^n, T^k) \geq 1+10=11$ holds for any $T^k \in S_D$ with $k \neq n-1, n$ also by Table 2. Thus, by fixing the subgraph $G^* \cup T^{n-1} \cup T^n$,

$$\mathrm{cr}_D(G^*+D_n) \geq 6\left\lfloor\frac{n-2}{2}\right\rfloor\left\lfloor\frac{n-3}{2}\right\rfloor+11(s-2)+4(n-s)+4+2=6\left\lfloor\frac{n-2}{2}\right\rfloor\left\lfloor\frac{n-3}{2}\right\rfloor$$

$$+4n+7s-16 \geq 6\left\lfloor\frac{n-2}{2}\right\rfloor\left\lfloor\frac{n-3}{2}\right\rfloor+4n+7\left(\left\lceil\frac{n}{2}\right\rceil+1\right)-16 \geq 6\left\lfloor\frac{n}{2}\right\rfloor\left\lfloor\frac{n-1}{2}\right\rfloor+n+\left\lfloor\frac{n}{2}\right\rfloor.$$

iv. $\mathcal{O}_D=\{\mathcal{E}_i\}$ for only one $i \in \{1,2\}$. Without lost of generality, let us assume that $T^n \in S_D$ with the configuration $\mathcal{E}_1$ of F^n. In this subcase, $\mathrm{cr}_D(G^* \cup T^n, T^k) \geq 1+6=7$ holds for any $T^k \in S_D$, $k \neq n$ provided that $\mathrm{rot}_D(t_n)=\mathrm{rot}_D(t_k)$. Hence, by fixing the subgraph $G^* \cup T^n$,

$$\mathrm{cr}_D(G^*+D_n) \geq 6\left\lfloor\frac{n-1}{2}\right\rfloor\left\lfloor\frac{n-2}{2}\right\rfloor+7(s-1)+3(n-s)+1=6\left\lfloor\frac{n-1}{2}\right\rfloor\left\lfloor\frac{n-2}{2}\right\rfloor$$

$$+3n+4s-6 \geq 6\left\lfloor\frac{n-1}{2}\right\rfloor\left\lfloor\frac{n-2}{2}\right\rfloor+3n+4\left(\left\lceil\frac{n}{2}\right\rceil+1\right)-6 \geq 6\left\lfloor\frac{n}{2}\right\rfloor\left\lfloor\frac{n-1}{2}\right\rfloor+n+\left\lfloor\frac{n}{2}\right\rfloor.$$

Case 3: $\mathrm{cr}_D(G^*)=2$ with $\mathrm{cr}_D(C_6(G^*))=0$. At first, without loss of generality, we can consider the drawing of G^* with the vertex notation like that in Figure 1c. It is obvious that the set R_D is empty, that is, the set S_D cannot be empty. Our aim is to list again all possible rotations $\mathrm{rot}_D(t_j)$ which can appear in D if a subgraph $T^j \in S_D$. Since there is only one subdrawing of $F^j \setminus \{v_1\}$ represented by the rotation (26543), there are three ways to obtain the subdrawing of F^j depending on which edge of G^* is crossed by the edge t_jv_1. These three possible ways under our consideration can be denoted by $\mathcal{C}_k$, for $k=1,2,3$. Based on the aforementioned arguments, we assume the drawings shown in Figure 7.

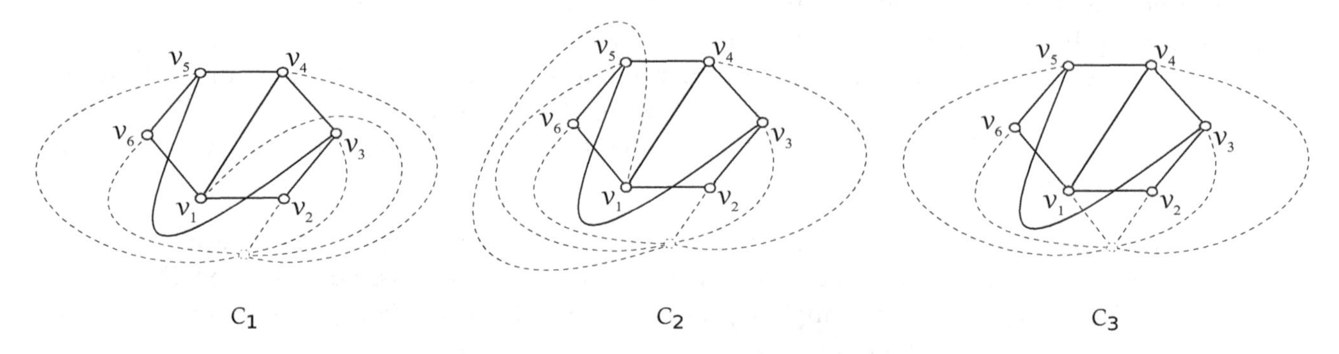

Figure 7. Drawings of three possible configurations of the subgraph F^j.

Thus the configurations $\mathcal{C}_1$, $\mathcal{C}_2$, and $\mathcal{C}_3$ are represented by the cyclic permutations (132654), (143265), and (165432), respectively. Further, due to the properties of the cyclic rotations we can easily verify that $\mathrm{cr}(\mathcal{C}_i, \mathcal{C}_j) \geq 4$ for any $i, j \in \{1,2,3\}$. Moreover, one can also easily verify over all possible drawings D that $\mathrm{cr}_D(G^* \cup T^j, T^k) \geq 4$ holds for any subgraph $T^k \notin S_D$, where $T^j \in S_D$ with some configuration $\mathcal{C}_i$ of F^j. As there is a $T^j \in S_D$, by fixing the subgraph $G^* \cup T^j$,

$$\mathrm{cr}_D(G^* + D_n) \geq 6\left\lfloor\frac{n-1}{2}\right\rfloor\left\lfloor\frac{n-2}{2}\right\rfloor + 5(s-1) + 4(n-s) + 2 + 1 = 6\left\lfloor\frac{n-1}{2}\right\rfloor\left\lfloor\frac{n-2}{2}\right\rfloor$$

$$+4n + s - 2 \geq 6\left\lfloor\frac{n-1}{2}\right\rfloor\left\lfloor\frac{n-2}{2}\right\rfloor + 4n + \left(\left\lceil\frac{n}{2}\right\rceil + 1\right) - 2 \geq 6\left\lfloor\frac{n}{2}\right\rfloor\left\lfloor\frac{n-1}{2}\right\rfloor + n + \left\lfloor\frac{n}{2}\right\rfloor.$$

In addition, without loss of generality, we can consider the drawing of G^* with the vertex notation like that in Figure 1d. In this case, by applying the same process, we obtain two possible forms of rotation $\mathrm{rot}_D(t_j)$ for $T^j \in S_D$. Namely, the rotations (165423) and (165432) if the edge $t_j v_2$ crosses either the edge $v_3 v_4$ or the edge $v_3 v_5$ of G^*, respectively. Further, they satisfy also the same properties like in the previous subcase, i.e., the same lower bounds of numbers of crossings on the edges of the subgraph $G^* \cup T^j$ by any T^k, $k \neq j$. Hence, we are able to use the same fixing of the subgraph $G^* \cup T^j$ for obtaining a contradiction with the number of crossings in D.

Finally, without loss of generality, we can consider the drawing of G^* with the vertex notation like that in Figure 1f. In this case, the set R_D is empty; that is, the set S_D cannot be empty. Thus, we can deal with the configurations belonging to the nonempty set $\mathcal{P}_D$. Recall that the lower-bounds of number of crossings of two configurations from $\mathcal{P}$ were already established in Table 3. Further, we can apply the same idea and also the same arguments as for the configurations $\mathcal{E}_i \in \mathcal{O}_D$, with $i = 1,2,3$, in the subcases ii.–iv. of Case 2.

Case 4: $\mathrm{cr}_D(G^*) \geq 1$ with $\mathrm{cr}_D(C_6(G^*)) \geq 1$. For all possible subdrawings of the graph G^* with at least one crossing among edges of $C_6(G^*)$, and also with the possibility of obtaining a subgraph T^j that crosses the edges of G^* at most once, one of the ideas of the previous subcases can be applied.

We have shown, in all cases, that there is no good drawing D of the graph $G^* + D_n$ with fewer than $6\lfloor\frac{n}{2}\rfloor\lfloor\frac{n-1}{2}\rfloor + n + \lfloor\frac{n}{2}\rfloor$ crossings. This completes the proof of the main theorem. □

4. Conclusions

Determining the crossing number of a graph $G + D_n$ is an essential step in establishing the so far unknown values of the numbers of crossings of graphs $G + P_n$ and $G + C_n$, where P_n and C_n are the path and the cycle on n vertices, respectively. Using the result in Theorem 1 and the optimal drawing of $G^* + D_n$ in Figure 6, we are able to postulate that $\mathrm{cr}(G^* + P_n)$ and $\mathrm{cr}(G^* + C_n)$ are at least one more than $\mathrm{cr}(G^* + D_n) = 6\lfloor\frac{n}{2}\rfloor\lfloor\frac{n-1}{2}\rfloor + n + \lfloor\frac{n}{2}\rfloor$.

Acknowledgments: This work was supported by the internal faculty research project number FEI-2017-39.

References

1. Garey, M.R.; Johnson, D.S. Crossing number is NP-complete. *SIAM J. Algebraic. Discrete Methods* **1983**, *4*, 312–316.
2. Klešč, M. The join of graphs and crossing numbers. *Electron. Notes Discret. Math.* **2007**, *28*, 349–355.
3. Kulli, V.R.; Muddebihal, M.H. Characterization of join graphs with crossing number zero. *Far East J. Appl. Math.* **2001**, *5*, 87–97.
4. Kleitman, D.J. The crossing number of $K_{5,n}$. *J. Comb. Theory* **1970**, *9*, 315–323.
5. Klešč, M.; Schrötter, Š. The crossing numbers of join products of paths with graphs of order four. *Discuss. Math. Graph Theory* **2011**, *31*, 312–331.

6. Berežný, Š.; Staš, M. Cyclic permutations and crossing numbers of join products of symmetric graph of order six. *Carpathian J. Math.* **2018**, *34*, 143–155.
7. Klešč, M. The crossing numbers of join of the special graph on six vertices with path and cycle. *Discret. Math.* **2010**, *310*, 1475–1481.
8. Staš, M. Cyclic permutations: Crossing numbers of the join products of graphs. In Proceedings of the Aplimat 2018: 17th Conference on Applied Mathematics, Bratislava, Slovakia, 6–8 February 2018; pp. 979–987.
9. Staš, M. Determining crossing numbers of graphs of order six using cyclic permutations. *Bull. Aust. Math. Soc.* **2018**, *98*, 353–362.
10. Hernández-Vélez, C.; Medina, C.; Salazar G. The optimal drawing of $K_{5,n}$. *Electron. J. Comb.* **2014**, *21*, 29.
11. Berežný, Š.; Buša J., Jr.; Staš, M. Software solution of the algorithm of the cyclic-order graph. *Acta Electrotech. Inform.* **2018**, *18*, 3–10.
12. Klešč, M.; Schrötter, Š. The crossing numbers of join of paths and cycles with two graphs of order five. In *Lecture Notes in Computer Science: Mathematical Modeling and Computational Science*; Springer: Berlin/Heidelberg, Germany, 2012; Volume 7125, pp. 160–167.
13. Woodall, D.R. Cyclic-order graphs and Zarankiewicz's crossing number conjecture. *J. Graph Theory* **1993**, *17*, 657–671.
14. Chimani, M.; Wiedera, T. An ILP-based proof system for the crossing number problem. In Proceedings of the 24th Annual European Symposium on Algorithms (ESA 2016), Aarhus, Denmark, 22–24 August 2016; Volume 29, pp. 1–13.

7

Invariant Graph Partition Comparison Measures

Fabian Ball * and Andreas Geyer-Schulz

Karlsruhe Institute of Technology, Institute of Information Systems and Marketing, Kaiserstr. 12, 76131 Karlsruhe, Germany; andreas.geyer-schulz@kit.edu

* Correspondence: fabian.ball@kit.edu.

Abstract: Symmetric graphs have non-trivial automorphism groups. This article starts with the proof that all partition comparison measures we have found in the literature fail on symmetric graphs, because they are not invariant with regard to the graph automorphisms. By the construction of a pseudometric space of equivalence classes of permutations and with Hausdorff's and von Neumann's methods of constructing invariant measures on the space of equivalence classes, we design three different families of invariant measures, and we present two types of invariance proofs. Last, but not least, we provide algorithms for computing invariant partition comparison measures as pseudometrics on the partition space. When combining an invariant partition comparison measure with its classical counterpart, the decomposition of the measure into a structural difference and a difference contributed by the group automorphism is derived.

Keywords: graph partitioning; graph clustering; invariant measures; partition comparison; finite automorphism groups; graph automorphisms

---

1. Introduction

Partition comparison measures are routinely used in a variety of tasks in cluster analysis: finding the proper number of clusters, assessing the stability and robustness of solutions of cluster algorithms, comparing different solutions of randomized cluster algorithms or comparing optimal solutions of different cluster algorithms in benchmarks [1], or in competitions like the 10th DIMACS graph-clustering challenge [2]. Their development has been for more than a century an active area of research in statistics, data analysis and machine learning. One of the oldest and still very well-known measure is the one of Jaccard [3]; more recent approaches were by Horta and Campello [4] and Romano et al. [5]. For an overview of many of these measures, see Appendix B. Besides the need to compare clustering partitions, there is an ongoing discussion of what actually are the best clusters [6,7]. Another problem often addressed is how to measure cluster validity [8,9].

However, the comparison of graph partitions leads to new challenges because of the need to handle graph automorphisms properly. The following small example shows that standard partition comparison measures have unexpected results when applied to graph partitions: in Figure 1, we show two different ways of partitioning the cycle graph C_4 (Figure 1a,d). Partitioning means grouping the nodes into non-overlapping clusters. The nodes are arbitrarily labeled with 1 to 4 (Figure 1b,e), and then, there are four possibilities of relabeling the nodes so that the edges stay the same. One possibility is relabeling 1 by 2, 2 by 3, 3 by 4 and 4 by 1, and the images resulting from this relabeling are shown in Figure 1c,f. The relabeling corresponds to a counterclockwise rotation of the graph by 90°, and formal details are given in Section 2. The effects of this relabeling on the partitions $\mathcal{P}_1$ and $\mathcal{Q}_1$ are different:

1. Partition $\mathcal{P}_1 = \{\{1,2\},\{3,4\}\}$ is mapped to the structurally equivalent partition $\mathcal{P}_2 = \{\{1,4\},\{2,3\}\}$.
2. Partition $\mathcal{Q}_1 = \{\{1,3\},\{2,4\}\}$ is mapped to the identical partition $\mathcal{Q}_2$.

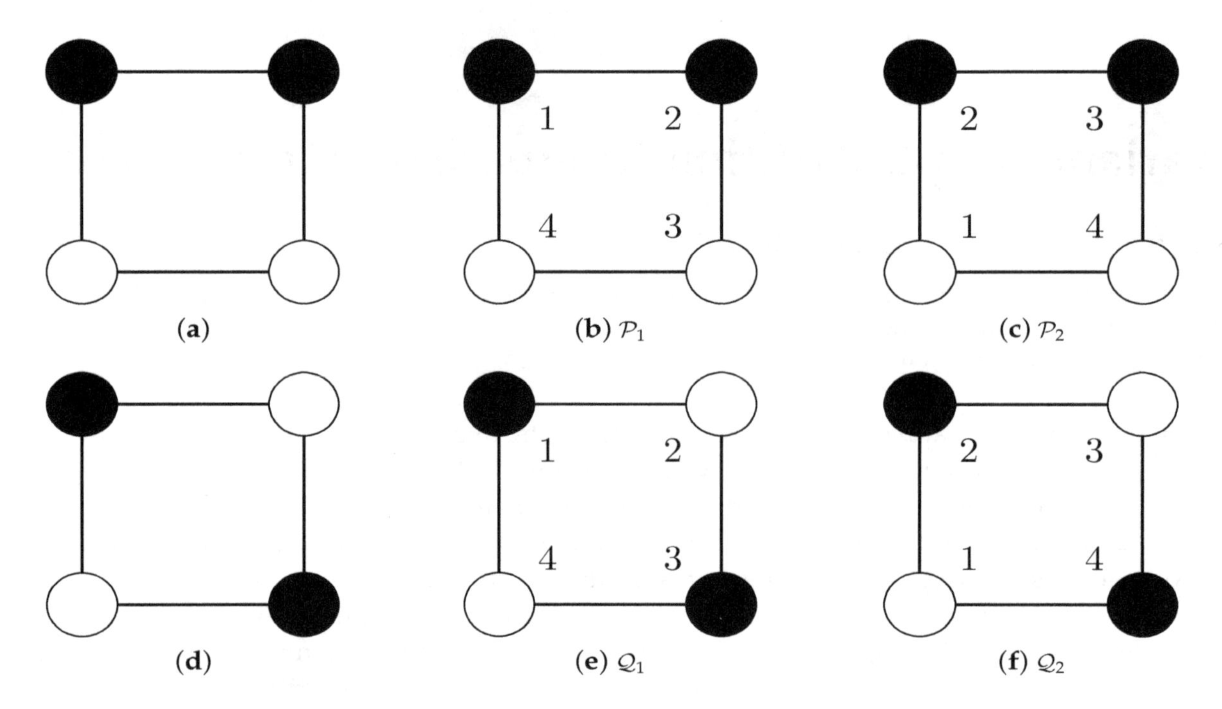

Figure 1. Two structurally different partitions of the cycle graph C_4: grouping pairs of neighbors (**a**) and grouping pairs of diagonals (**d**). Equally-colored nodes represent graph clusters, and the choice of colors is arbitrary. Adding, again arbitrary, but fixed, node labels impacts the node partitions and results in the failure to recognize the structural difference when comparing these partitions with partition comparison measures (see Table 1). The different images (**b**,**c**) ($\mathcal{P}_1 = \{\{1,2\},\{3,4\}\}$, $\mathcal{P}_2 = \{\{1,4\},\{2,3\}\}$) and (**e**,**f**) ($\mathcal{Q}_1 = \mathcal{Q}_2 = \{\{1,3\},\{2,4\}\}$) emerge from the graph's symmetry.

Table 1 illustrates the failure of partition comparison measures (here, the Rand Index (RI)) to recognize structural differences:

1. Because $\mathcal{P}_1$ and $\mathcal{P}_2$ are structurally equivalent, the RI should be one (as for Cases 1, 2 and 3) instead of 1/3.
2. Comparisons of structurally different different partitions (Cases 4 and 5) and comparisons of structurally equivalent partitions (Case 6) should not result in the same value.

Table 1. The Rand index is $RI = \frac{N_{11}+N_{00}}{N_{11}+N_{10}+N_{01}+N_{00}}$. N_{11} indicates the number of nodes that are in both partitions together in a cluster; N_{10} and N_{01} are the number of nodes that are together in a cluster in one partition, but not in the other; and N_{00} are the number of nodes that are in both partitions in different clusters. See Appendix B for the formal definitions. Partitions $\mathcal{P}_1$ and $\mathcal{P}_2$ are equivalent (yet not equal, denoted "$\sim$"), and partitions $\mathcal{Q}_1$ and $\mathcal{Q}_2$ are identical (thus, also equivalent, denoted "$=$"). However, the comparison of the structurally different partitions (denoted "$\neq$") $\mathcal{P}_i$ and $\mathcal{Q}_j$ yields the same result as the comparison between the equivalent partitions $\mathcal{P}_1$ and $\mathcal{P}_2$. This makes the recognition of structural differences impossible.

Case	Compared Partitions	Relation	N_{11}	N_{10}	N_{01}	N_{00}	RI
1	$\mathcal{P}_1, \mathcal{P}_1$	$=$	2	0	0	4	1
2	$\mathcal{P}_2, \mathcal{P}_2$	$=$	2	0	0	4	1
3	$\mathcal{Q}_1, \mathcal{Q}_1$ or $\mathcal{Q}_1, \mathcal{Q}_2$ or $\mathcal{Q}_2, \mathcal{Q}_2$	$=$	2	0	0	4	1
4	$\mathcal{P}_1, \mathcal{Q}_1$ or $\mathcal{P}_1, \mathcal{Q}_2$	$\neq$	0	2	2	2	$\frac{1}{3}$
5	$\mathcal{P}_2, \mathcal{Q}_1$ or $\mathcal{P}_2, \mathcal{Q}_2$	$\neq$	0	2	2	2	$\frac{1}{3}$
6	$\mathcal{P}_1, \mathcal{P}_2$	$\sim$	0	2	2	2	$\frac{1}{3}$

One may argue that graphs in real applications contain symmetries only rarely. However, recent investigations of graph symmetries in real graph datasets show that a non-negligible proportion of these graphs contain symmetries. MacArthur et al. [10] state that "a certain degree of symmetry is also ubiquitous in complex systems" [10] (p. 3525). Their study includes a small number of biological, technological and social networks. In addition, Darga et al. [11] studied automorphism groups in very large sparse graphs (circuits, road networks and the Internet router network), with up to five million nodes with eight million links with execution times below 10 s. Katebi et al. [12] reported symmetries in 268 of 432 benchmark graphs. A recent large-scale study conducted by the authors of this article for approximately 1700 real-world graphs revealed that about three quarters of these graphs contain symmetries [13].

The rather frequent occurrence of symmetries in graphs and the obvious deficiencies of classic partition comparison measures demonstrated above have motivated our analysis of the effects of graph automorphisms on partition comparison measures.

Our contribution has the following structure: Permutation groups and graph automorphisms are introduced in Section 2. The full automorphism group of the butterfly graph serves as a motivating example for the formal definition of stable partitions, stable with regard to the actions of the automorphism group of a graph. In Section 3, we first provide a definition that captures the property that a measure is invariant with regard to the transformations in an automorphism group. Based on this definition, we first give a simple proof by counterexample for each partition comparison measure in Appendix B, that these measures based on the comparison of two partitions are not invariant to the effects of automorphisms on partitions. The non-existence of partition comparison measures for which the identity and the invariance axioms hold simultaneously is proven subsequently. In Section 4, we construct three families of invariant partition comparison measures by a two-step process: First, we define a pseudometric space by defining equivalence classes of partitions as the orbit of a partition under the automorphism group $Aut(G)$. Second, the definitions of the invariant counterpart of a partition comparison measure are given: we define them as the computation of the maximum, the minimum and the average of the direct product of the two equivalence classes. The section also contains a proof of the equivalence of several variants of the computation of the invariant measures, which—by exploiting the group properties of $Aut(G)$—differ in the complexity of the computation. In Section 5, we introduce the decomposition of the measures into a structurally stable and unstable part, as well as upper bounds for instability. In Section 6, we present an application of the decomposition of measures for analyzing partitions of the Karate graph. The article ends with a short discussion, conclusion and outlook in Section 7.

2. Graphs, Permutation Groups and Graph Automorphisms

We consider connected, undirected, unweighted and loop-free graphs. Let $G = (V, E)$ denote a graph where V is a finite set of nodes and E is a set of edges. An edge is represented as $\{u, v\} \in \{\{x, y\} \mid (x, y) \in V \times V \wedge x \neq y\}$. Nodes adjacent to $u \in V$ (there exists an edge between u and those nodes) are called neighbors. A partition $\mathcal{P}$ of a graph G is a set of subsets $C_i, i = 1, \ldots, k$ of V with the usual properties: (i) $C_i \cap C_j = \emptyset\ (i \neq j)$, (ii) $\bigcup_i C_i = V$ and (iii) $C_i \neq \emptyset$. Each subset is called a cluster, and it is identified by its labeled nodes.

As a partition quality criterion, we use the well-known modularity measure Q of Newman and Girvan [14] (see Appendix A). It is a popular optimization criterion for unsupervised graph clustering algorithms, which try to partition the nodes of the graph in a way that the connectivity within the clusters is maximized and the number of edges connecting the clusters is minimized. For a fast and efficient randomized state-of-the-art algorithm, see Ovelgönne and Geyer-Schulz [15].

Partitions are compared by comparison measures, which are functions of the form $m : P(V) \times P(V) \to \mathbb{R}$ where $P(V)$ denotes the set of all possible partitions of the set V. A survey of many of these measures is given in Appendix B.

A permutation on V is a bijection $g : V \rightarrow V$. We denote permutations by the symbols f, g and h. Each permutation can be written in cycle form: for a permutation with a single cycle of length r, we write $c = (v_1\, v_2\, \ldots\, v_r)$. c maps v_i to v_{i+1} $(i = 1, \ldots, r-1)$, v_r to v_1 and leave all other nodes fixed. Permutations with more than one cycle are written as a product of disjoint cycles (i.e., no two cycles have a common element). (v_k) means that the element v_k remains fixed, and for brevity, these elements are omitted.

Permutations are applied from the right: The image of u under the permutation g is ug. The composition of g and h is $h \circ g$, with $\circ$ being the permutation composition symbol. For brevity, $h \circ g$ is written as gh, so that $u(gh) = (ug)h$ holds. Computer scientists call this a postfix notation; in prefix notation, we have $h(g(u))$. Often, we also find u^g, which we will use in the following. For k compositions $g \circ g \circ g \circ \ldots$, we write g^k and $g^0 = id$.

A set of permutation functions forms a permutation group H, if the usual group axioms hold [16]:

1. Closure: $\forall g, h \in H : g \circ h \in H$
2. Unit element: The identity function $id \in H$ acts as the neutral element: $\forall g \in H : id \circ g = g \circ id = g$
3. Inverse element: For any g in H, the inverse permutation function $g^{-1} \in H$ is the inverse of g: $\forall g \in H : g \circ g^{-1} = g^{-1} \circ g = id$
4. Associativity: The associative law holds: $\forall f, g, h \in H : f \circ (g \circ h) = (f \circ g) \circ h$

If H_1 is a subset of H and if H_1 is a group, H_1 is a subgroup of H (written $H_1 \leq H$). The set of all permutations of V is denoted by $Sym(V)$. $Sym(V)$ is a group, and it is called the symmetric group (see [17]). $Sym(V) \sim Sym(V')$ iff $|V| = |V'|$ with $\sim$ denoting isomorphism. A generator of a finite permutation group H is a subset of the permutations of H from which all permutations in H can be generated by application of the group axioms [18].

An action of H on V (H acts on V) is called the group action of a set [19] (p. 5):

1. $u^{id} = u$, $\forall u \in V$
2. $(u^g)^h = u^{gh}$, $\forall u \in V, \forall g, h \in H$

Groups acting on a set V also act on combinatorial structures defined on V [20] (p. 149), for example the power set 2^V, the set of all partitions $P(V)$ or the set of graphs $G(V)$. We denote combinatorial structures as capital calligraphic letters; in the following, only partitions ($\mathcal{P}$) are of interest because they are the results of graph cluster algorithms. The action of a permutation g on a combinatorial structure is performed by pointwise application of g. For instance, for $\mathcal{P}$, the image of g is $\mathcal{P}^g = \{\{u^g \mid u \in C\} \mid C \in \mathcal{P}\}$.

Let H be a permutation group. When H acts on V, a node u is mapped by the elements of H onto other nodes. The set of these images is called the orbit of u under H:

$$u^H = \left\{u^h \mid h \in H\right\}.$$

The group of permutations H_u that fixes u is called the stabilizer of u under H:

$$H_u = \{h \in H \mid u^h = u\}.$$

The orbit stabilizer theorem is given without proof [16]. It links the order of a permutation group with the cardinality of an orbit and the order of the stabilizer:

Theorem 1. *The relation:*

$$|H| = |u^H| \cdot |H_u|$$

holds.

The action of H on V induces an equivalence relation on the set: for $u_1, u_2 \in V$, let $u_1 \sim u_2$ iff there exists $h \in H$ so that $u_1 = u_2^h$. All elements of an orbit are equivalent, and the orbits of a group partition the set V. An orbit of length one (in terms of set cardinality) is called trivial. Analogously, for a partition $\mathcal{P}$, the definition is:

Definition 1. *The image of the action of H on a partition $\mathcal{P}$ (or the orbit of $\mathcal{P}$ under H) is the set of all equivalent partitions of partition $\mathcal{P}$ under H*

$$\mathcal{P}^H = \left\{\mathcal{P}^h \mid h \in H\right\}.$$

A graph automorphism f is a permutation that preserves edges, i.e., $\{u^f, v^f\} \in E \Leftrightarrow \{u,v\} \in E$, $\forall u, v \in V$.

The automorphism group of a graph contains all permutations of vertices that map edges to edges and non-edges to non-edges. The automorphism group of G is defined as:

$$Aut(G) = \left\{f \in Sym(V) \mid E^f = E\right\}$$

where $E^f = \left\{\{u^f, v^f\} \mid \{u,v\} \in E\right\}$. Of course, $Aut(G) \leq Sym(V)$.

Example 1. *Let G_{bf} be the butterfly graph (Figure 2, e.g., Erdős et al. [21], Burr et al. [22]) whose full automorphism group is given in Table 2 (first column). The permutation $(2\,5)$ is not an automorphism, because it does not preserve the edges from 1 to 2 and from 5 to 4. The butterfly graph has the two orbits $\{1,2,4,5\}$ and $\{3\}$. The group $H = \{id, g_1, g_2, g_3\}$ is a subgroup of $Aut(G_{bf})$.*

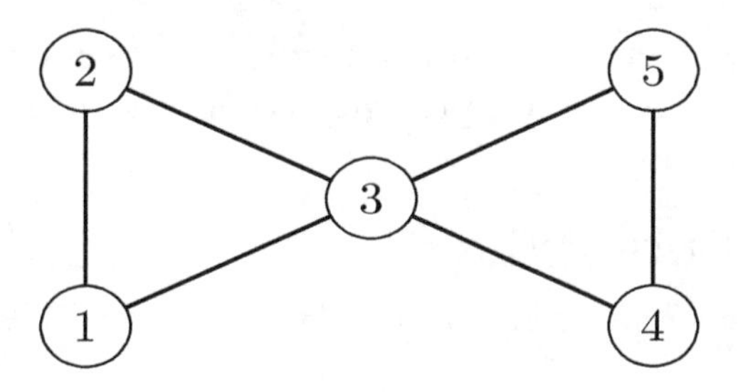

Figure 2. The butterfly graph (five nodes, with two node pairs connected by the bridging node 3).

Table 2. The full automorphism group $Aut(G_{bf}) = \{id, g_1, \ldots, g_7\}$ of the butterfly graph in Figure 2 and its effect on three partitions. Bold partitions are distinct. A possible generator is $\{g_1, g_4\}$.

Permutation	$\mathcal{P}_1$, $Q = 0$	$\mathcal{P}_2$, $Q = \frac{1}{9}$	$\mathcal{P}_3$, $Q = -\frac{1}{18}$
$id = (1)(2)(3)(4)(5)$	**{1,2},{3},{4,5}**	**{1,2,3},{4,5}**	**{1,2,3,4},{5}**
$g_1 = (1\,2)$	{2,1},{3},{4,5}	{2,1,3},{4,5}	{2,1,3,4},{5}
$g_2 = (4\,5)$	{1,2},{3},{5,4}	{1,2,3},{5,4}	**{1,2,3,5},{4}**
$g_3 = (1\,2)(4\,5)$	{2,1},{3},{5,4}	{2,1,3},{5,4}	{2,1,3,5},{4}
$g_4 = (1\,4)(2\,5)$	{4,5},{3},{1,2}	**{4,5,3},{1,2}**	**{4,5,3,1},{2}**
$g_5 = (1\,5)(2\,4)$	{5,4},{3},{2,1}	{5,4,3},{2,1}	{5,4,3,2},{1}
$g_6 = (1\,4\,2\,5)$	{4,5},{3},{2,1}	{4,5,3},{2,1}	**{4,5,3,2},{1}**
$g_7 = (1\,5\,2\,4)$	{5,4},{3},{1,2}	{5,4,3},{1,2}	{5,4,3,1},{2}

Definition 2. *Let $G = (V, E)$ be a graph. A partition $\mathcal{P}$ is called stable, if $|\mathcal{P}^{Aut(G)}| = 1$, otherwise it is called unstable.*

Stability here means that the automorphism group of the graph does not affect the given partition by tearing apart clusters.

Example 2. *Only $\mathcal{P}_1$ in Table 2 is stable because its orbit is trivial. The two modularity optimal partitions (e.g., $\mathcal{P}_2^{id}$ and $\mathcal{P}_2^{g_4}$) are not stable because $|\mathcal{P}_2^{Aut(G_{bf})}| = 2$. Furthermore, $|\mathcal{P}_3^{Aut(G_{bf})}| = 4$.*

For the evaluation of graph clustering solutions, the effects of graph automorphisms on graph partitions are of considerable importance:

1. Automorphisms may lead to multiple equivalent optimal solutions as the butterfly graph shows ($\mathcal{P}_2^{id}$ and $\mathcal{P}_2^{g_4}$ in Table 2).
2. Partition comparison measures are not invariant with regard to automorphisms, as we show in Section 3.

3. Graph Partition Comparison Measures Are Not Invariant

When comparing graph partitions, a natural requirement is that the partition comparison measure is invariant under automorphism.

Definition 3. *A partition comparison measure $m : P(V) \times P(V) \to \mathbb{R}$ is invariant under automorphism, if:*

$$m(\mathcal{P}, \mathcal{Q}) = m(\tilde{\mathcal{P}}, \tilde{\mathcal{Q}})$$

for all $\mathcal{P}, \mathcal{Q} \in P(V)$ and $\tilde{\mathcal{P}} \in \mathcal{P}^{Aut(G)}$, $\tilde{\mathcal{Q}} \in \mathcal{Q}^{Aut(G)}$.

Observe that if $\mathcal{Q} \in \mathcal{P}^{Aut(G)}$, then such a measure m cannot distinguish between $\mathcal{P}$ and $\mathcal{Q}$, since $m(\mathcal{P}, \mathcal{Q}) = m(\mathcal{P}, \mathcal{P})$ by definition.

However, unfortunately, as we show in the rest of this section, such a partition comparison measure does not exist. In the following, we present two proofs of this fact, which differ both in their level of generality and sophistication.

3.1. Variant 1: Construction of a Counterexample

Theorem 2. *The measures for comparing partitions defined in Appendix B do not fulfill Definition 3 in general.*

Proof. We choose the cycle graph C_{36} and compute all modularity maximal partitions with $Q = 2/3$. Each of these six partitions has six clusters, and each of these clusters consists of a chain of six nodes (see Figure 3).

Clearly, since all partitions are equivalent, an invariant partition comparison measure should identify them as equivalent:

$$m(\mathcal{P}_0, \mathcal{P}_0^{g^0}) = \ldots = m(\mathcal{P}_0, \mathcal{P}_0^{g^5}) \tag{1}$$

Computing $m(\mathcal{P}_0, \mathcal{P}_0^{g^k})$ for $k = 0, \ldots, 5$ produces Table 3. Because the values in each row differ (in contrast to the requirements defined by Equation (1)), each row of Table 3 contains the counterexample for the measure used. □

3.2. Variant 2: Inconsistency of the Identity and the Invariance Axiom

Theorem 3. *Let $G = (V, E)$ be a graph with $|V| > 2$ and nontrivial $Aut(G)$. For partition comparison measures $m : P(V) \times P(V) \to \mathbb{R}$, it is impossible to fulfill jointly the identity axiom $m(\mathcal{P}, \mathcal{Q}) = c$, if and only if $\mathcal{P} = \mathcal{Q}$ (e.g., for a distance measure $c = 0$, for a similarity measure $c = 1$, etc.) for all $\mathcal{P}, \mathcal{Q} \in P(V)$ and the axiom of invariance (from Definition 3) $m(\mathcal{P}, \mathcal{Q}) = c, \forall \mathcal{Q} \in \mathcal{P}^{Aut(G)}$.*

Proof.

1. Since $Aut(G)$ is nontrivial, a nontrivial orbit with at least two different partitions, namely $\mathcal{P}$ and $\mathcal{Q}$, exists because $|\mathcal{P}^{Aut(G)}| > 1$. It follows from the invariance axiom that $m(\mathcal{P}, \mathcal{Q}) = c$.
2. The identity axiom implies that it follows from $m(\mathcal{P}, \mathcal{Q}) = c$ that $\mathcal{P} = \mathcal{Q}$.

3. This contradicts the assumption that $\mathcal{P}$ and $\mathcal{Q}$ are different.

□

Table 3. Comparing the modularity maximizing partitions of the cycle graph C_{36} with modularity $Q = \frac{2}{3}$. The six optimal partitions consist of six clusters (see Figure 3). The number of pairs in the same cluster in both partitions is denoted by N_{11}, in different clusters by N_{00} and in the same cluster in one partition, but not in the other, by N_{01} or N_{10}. For the definitions of all partition comparison measures, see Appendix B. To compute this table, the R package `partitionComparison` has been used [23].

Measure	$m(\mathcal{P}_0, \mathcal{P}_0^{g^k})$ with $g = (1\,2\,3\,\ldots\,35\,36)$ for k:					
	0	1	2	3	4	5
Pair counting measures ($f(N_{11}, N_{00}, N_{01}, N_{10})$; see Tables A1 and A2)						
RI	1.0	0.90476	0.84762	0.82857	0.84762	0.90476
ARI	1.0	0.61111	0.37778	0.3	0.37778	0.61111
H	1.0	0.80952	0.69524	0.65714	0.69524	0.80952
CZ	1.0	0.66667	0.46667	0.4	0.46667	0.66667
K	1.0	0.66667	0.46667	0.4	0.46667	0.66667
MC	1.0	0.33333	−0.06667	−0.2	−0.06667	0.33333
P	1.0	0.61111	0.37778	0.3	0.37778	0.61111
W_I	1.0	0.66667	0.46667	0.4	0.46667	0.66667
W_{II}	1.0	0.66667	0.46667	0.4	0.46667	0.66667
FM	1.0	0.66667	0.46667	0.4	0.46667	0.66667
Γ	1.0	0.61111	0.37778	0.3	0.37778	0.61111
SS1	1.0	0.80556	0.68889	0.65	0.68889	0.80556
B1	1.0	0.91383	0.87084	0.85796	0.87084	0.91383
GL	1.0	0.95	0.91753	0.90625	0.91753	0.95
SS2	1.0	0.33333	0.17949	0.14286	0.17949	0.33333
SS3	1.0	0.62963	0.42519	0.36	0.42519	0.62963
RT	1.0	0.82609	0.73554	0.70732	0.73554	0.82609
GK	1.0	0.94286	0.79937	0.71429	0.79937	0.94286
J	1.0	0.5	0.30435	0.25	0.30435	0.5
RV	1.0	0.61039	0.37662	0.29870	0.37662	0.61039
RR	0.14286	0.09524	0.06667	0.05714	0.06667	0.09524
M	0.0	120.0	192.0	216.0	192.0	120.0
Mi	0.0	0.81650	1.03280	1.09545	1.03280	0.81650
Pe	0.00002	0.00001	0.00001	0.00001	0.00001	0.00001
B2	0.12245	0.07483	0.04626	0.03673	0.04626	0.07483
LI	24.37212	14.89407	9.20724	7.31163	9.20724	14.89407
NLI	1.0	0.61111	0.37778	0.3	0.37778	0.61111
FMG	0.94730	0.61396	0.41396	0.34730	0.41396	0.61396
Set-based comparison measures (see Table A3)						
LA	1.0	0.83333	0.66667	0.5	0.66667	0.83333
d_{CE}	0.0	0.16667	0.33333	0.5	0.33333	0.16667
D	0.0	12.0	24.0	36.0	24.0	12.0
Information theory-based measures (see Table A4)						
MI	1.79176	1.34120	1.15525	1.09861	1.15525	1.34120
NMI (max)	1.0	0.74854	0.64475	0.61315	0.64475	0.74854
NMI (min)	1.0	0.74854	0.64475	0.61315	0.64475	0.74854
NMI (Σ)	1.0	0.74854	0.64475	0.61315	0.64475	0.74854
VI	0.0	0.90112	1.27303	1.38629	1.27303	0.90112

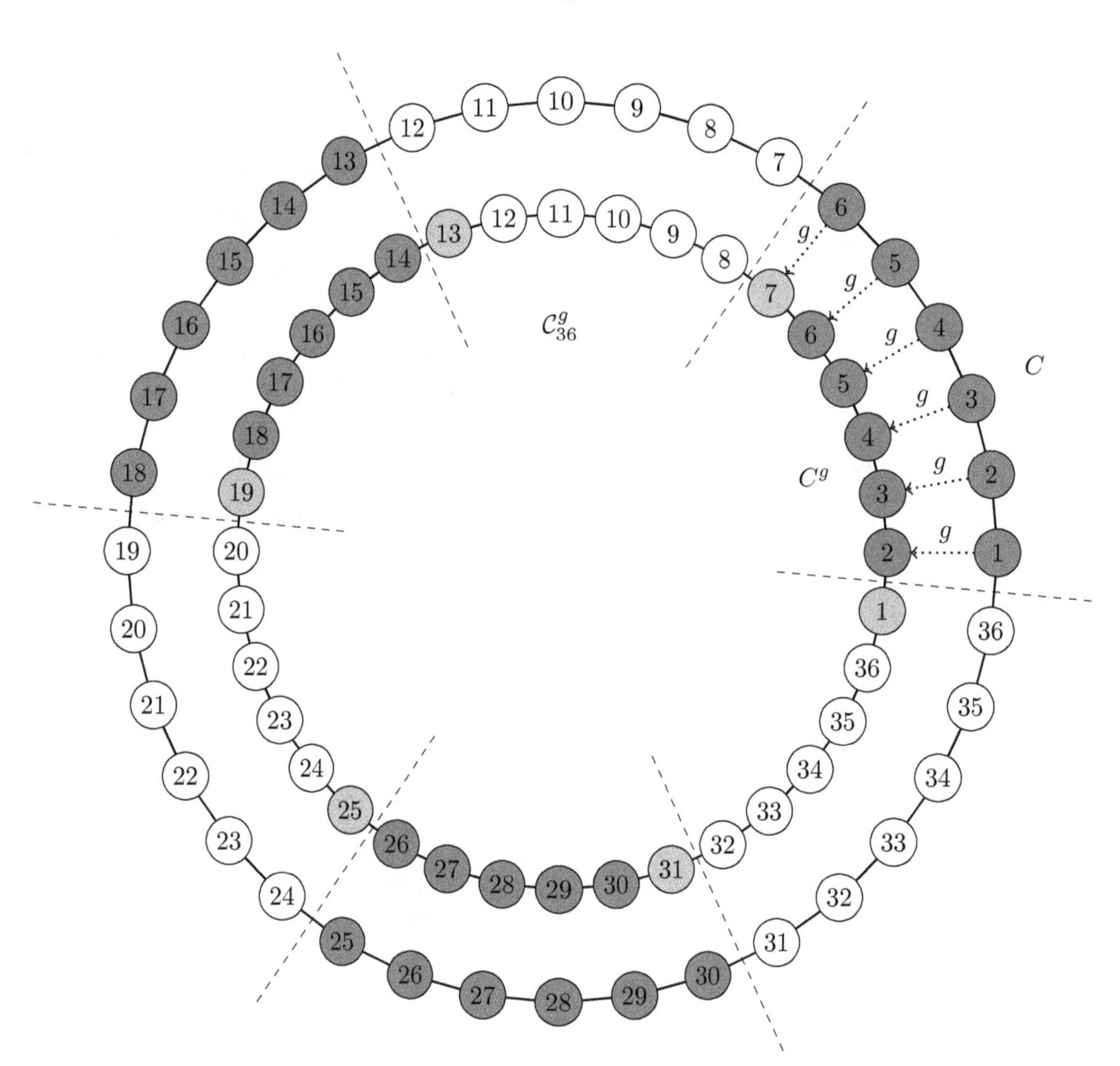

Figure 3. The cycle graph C_{36} (the "outer" cycle) and an initial partition of six clusters (connected nodes of the same color, separated by dashed lines). A single application of $g = (1\,2\,\ldots\,36)$ "rotates" the graph by one node (the "inner" cycle $\mathcal{C}_{36}^{g}$). As a consequence, in each cluster, one node drops out and is added to another cluster: For instance, Node 1 drops out of the "original" cluster $C = \{1,2,3,4,5,6\}$, and Node 7 is added, resulting in $C^{g} = \{2,3,4,5,6,7\}$. All dropped nodes are shown in light gray.

4. The Construction of Invariant Measures for Finite Permutation Groups

The purpose of this section is to construct invariant counterparts for most of the partition comparison measures in Appendix B. We proceed in two steps:

1. We construct a pseudometric space from the images of the actions of $Aut(G)$ on partitions in $P(V)$ (Definition 1).
2. We extend the metrics for partition comparison by constructing invariant metrics on the pseudo-metric space of partitions.

4.1. The Construction of the Pseudometric Space of Equivalence Classes of Graph Partitions

We use a variant of the idea of Doob's concept of a pseudometric space [24] (p. 5). A metric for a space S (with $s,t,u \in S$) is a function $d : S \times S \to \mathbb{R}^{+}$ for which the following holds:

1. Symmetry: $d(s,t) = d(t,s)$.

2. Identity: $d(s,t) = 0$ if and only if $s = t$.
3. Triangle inequality: $d(s,u) \leq d(s,t) + d(t,u)$.

A pseudometric space (S, d^*) relaxes the identity condition to $d^*(s,s) = 0$. The distance between two elements s_1, s_2 of an equivalence class $[s]$ is defined as $d^*(s_1, s_2) = 0$ by Definition 3.

For graphs, S is the finite set of partitions $P(V)$ and S^* is the partition of $P(V)$ into orbits of $Aut(G)$:

$$S^*(V) = P(V)^{Aut(G)} = \left\{\mathcal{P}^{Aut(G)} \mid \mathcal{P} \in P(V)\right\}.$$

A partition $\mathcal{P}$ in S corresponds to its orbit $\mathcal{P}^{Aut(G)}$ in S^*. The relations between the spaces used in the following are:

1. (S, d) is a metric space with $S = P(V)$ and with the function $d : P(V) \times P(V) \to \mathbb{R}$.
2. (S^*, d^*) is a metric space with $S^* = P(V)^{Aut(G)} = \{\mathcal{P}^{Aut(G)} \mid \mathcal{P} \in P(V)\}$ and the function d^*: $P(V)^{Aut(G)} \times P(V)^{Aut(G)} \to \mathbb{R}$. We construct three variants of d^* in Section 4.2.
3. (S, d^*) is the pseudometric space with $S = P(V)$ and with the metric d^*. The partitions in S are mapped to arguments of d^* by the transformation $ec : P(V) \to P(V)^{Aut(G)}$, which is defined as $ec(\mathcal{P}) := \mathcal{P}^{Aut(G)}$.

Table 4 illustrates S^* (the space of equivalence classes) of the pseudometric space (S, d^*) of the butterfly graph (shown in Figure 2). S^* is the partition of $P(\{1,2,3,4,5\})$ into 17 equivalence classes. Only the four classes E_1, E_8, E_{12} and E_{17} are stable because they are trivial orbits. The three partitions from Table 2 are contained in the following equivalence classes: $\mathcal{P}_1 \in E_8$, $\mathcal{P}_2 \in E_{14}$, and $\mathcal{P}_3 \in E_{13}$.

Table 4. The equivalence classes of the pseudometric space (S, d^*) of the butterfly graph (see Figure 2). Classes are grouped by their partition type, which is the corresponding integer partition. k is the number of partitions per type; l is the number of clusters the partitions of a type consists of; dia_{1-RI} is the diameter (see Equation (2)) of the equivalence class computed for the distance d_{RI} computed from the Rand Index (RI) by $1 - RI$.

	$\mathcal{P}^{Aut(G)}$	**Q**	**dia_{1-RI}**
	Partition type $(1,1,1,1,1)$, $k = 1$, $l = 5$		
E_1	$\{1\},\{2\},\{3\},\{4\},\{5\}$	$-\frac{2}{9}$	0.0
	Partition type $(1,1,1,2)$, $k = 10$, $l = 4$		
E_2	$\{1\},\{2\},\{3\},\{4,5\}$ $\{1,2\},\{3\},\{4\},\{5\}$	$-\frac{1}{9}$	0.2
E_3	$\{1\},\{2\},\{3,4\},\{5\}$ $\{1\},\{2\},\{3,5\},\{4\}$ $\{1\},\{2,3\},\{4\},\{5\}$ $\{1,3\},\{2\},\{4\},\{5\}$	$-\frac{1}{6}$	0.2
E_4	$\{1\},\{2,4\},\{3\},\{5\}$ $\{1\},\{2,5\},\{3\},\{4\}$ $\{1,4\},\{2\},\{3\},\{5\}$ $\{1,5\},\{2\},\{3\},\{4\}$	$-\frac{5}{18}$	0.2
	Partition type $(1,1,3)$ $k = 10$, $l = 3$		
E_5	$\{1\},\{2\},\{3,4,5\}$ $\{4\},\{5\},\{1,2,3\}$	0	0.6
E_6	$\{1\},\{3\},\{2,4,5\}$ $\{3\},\{5\},\{1,2,4\}$ $\{3\},\{4\},\{1,2,5\}$ $\{2\},\{3\},\{1,4,5\}$	$-\frac{2}{9}$	0.4
E_7	$\{1\},\{5\},\{2,3,4\}$ $\{1\},\{4\},\{2,3,5\}$ $\{2\},\{5\},\{1,3,4\}$ $\{2\},\{4\},\{1,3,5\}$	$-\frac{1}{6}$	0.6
	Partition type $(1,2,2)$, $k = 15$, $l = 3$		
E_8	$\{3\},\{1,2\},\{4,5\}$	0	0.0
E_9	$\{3\},\{1,4\},\{2,5\}$ $\{3\},\{1,5\},\{2,4\}$	$-\frac{1}{3}$	0.4
E_{10}	$\{1\},\{2,3\},\{4,5\}$ $\{5\},\{1,2\},\{3,4\}$ $\{4\},\{1,2\},\{3,5\}$ $\{2\},\{1,3\},\{4,5\}$	$-\frac{1}{18}$	0.4
E_{11}	$\{1\},\{2,4\},\{3,5\}$ $\{1\},\{2,5\},\{3,4\}$ $\{5\},\{1,3\},\{2,4\}$ $\{4\},\{1,3\},\{2,5\}$ $\{2\},\{1,4\},\{3,5\}$ $\{5\},\{1,4\},\{2,3\}$ $\{2\},\{1,5\},\{3,4\}$ $\{4\},\{1,5\},\{2,3\}$	$-\frac{2}{9}$	0.4

Table 4. *Cont.*

	$\mathcal{P}^{Aut(G)}$			**Q**	**dia$_{1-RI}$**
	Partition type $(1,4)$, $k=5, l=2$				
E_{12}	$\{1,2,4,5\},\{3\}$			$-\frac{2}{9}$	0.0
E_{13}	$\{2,3,4,5\},\{1\}$	$\{1,2,3,4\},\{5\}$	$\{1,2,3,5\},\{4\}$	$-\frac{1}{18}$	0.6
	$\{1,3,4,5\},\{2\}$				
	Partition type $(2,3)$, $k=10, l=2$				
E_{14}	$\{1,2\},\{3,4,5\}$	$\{4,5\},\{1,2,3\}$		$\frac{1}{9}$	0.4
E_{15}	$\{3,5\},\{1,2,4\}$	$\{3,4\},\{1,2,5\}$	$\{1,3\},\{2,4,5\}$	$-\frac{1}{6}$	0.6
	$\{2,3\},\{1,4,5\}$				
E_{16}	$\{2,5\},\{1,3,4\}$	$\{2,4\},\{1,3,5\}$	$\{1,4\},\{2,3,5\}$	$-\frac{2}{9}$	0.6
	$\{1,5\},\{2,3,4\}$				
	Partition type (5), $k=1, l=1$				
E_{17}	$\{1,2,3,4,5\}$			0	0.0

4.2. The Construction of Left-Invariant and Additive Measures on the Pseudometric Space of Equivalence Classes of Graph Partitions

In the following, we consider only partition comparison measures, which are distance functions of a metric space. Note that a normalized similarity measure s can be transformed into a distance by the transformation $d = 1 - s$.

In a pseudometric space (S, d^*), we measure the distance $d^*(\mathcal{P}, \mathcal{Q})$ between equivalence classes (which are sets) of partitions instead of the distance $d(\mathcal{P}, \mathcal{Q})$ between partitions. The partitions $\mathcal{P}$ and $\mathcal{Q}$ are formal arguments of d^*, which are expanded to equivalence classes by $\mathcal{P}^{Aut(G)}$ and $\mathcal{Q}^{Aut(G)}$. The standard construction of a distance measure between sets has been developed for the point set topology and is due to Felix Hausdorff [25] (p. 166) and Kazimierz Kuratowski [26] (p. 209). For finite sets, it requires the computation of the distances for all pairs of the direct product of the two sets. Since for finite permutation groups, we deal with distances between two finite sets of partitions, we use the following definitions for the lower and upper measures, respectively. Both definitions have the form of an optimization problem:

$$d^*_L(\mathcal{P}, \mathcal{Q}) = \min_{\substack{\tilde{\mathcal{P}} \in \mathcal{P}^{Aut(G)}, \\ \tilde{\mathcal{Q}} \in \mathcal{Q}^{Aut(G)}}} d(\tilde{\mathcal{P}}, \tilde{\mathcal{Q}})$$

and:

$$d^*_U(\mathcal{P}, \mathcal{Q}) = \begin{cases} 0 & \text{if } \mathcal{P}^{Aut(G)} = \mathcal{Q}^{Aut(G)} \\ \max_{\substack{\tilde{\mathcal{P}} \in \mathcal{P}^{Aut(G)}, \\ \tilde{\mathcal{Q}} \in \mathcal{Q}^{Aut(G)}}} d(\tilde{\mathcal{P}}, \tilde{\mathcal{Q}}) & \text{else} \end{cases}$$

The diameter of a finite equivalence class of partitions is defined by

$$\text{dia}(\mathcal{P}) = \max_{\substack{\tilde{\mathcal{P}} \in \mathcal{P}^{Aut(G)}, \\ \tilde{\mathcal{Q}} \in \mathcal{P}^{Aut(G)}}} d(\tilde{\mathcal{P}}, \tilde{\mathcal{Q}}). \quad (2)$$

The third option of defining a distance between two finite equivalence classes of partitions of taking the average distance is due to John von Neumann [27]:

$$d^*_{av}(\mathcal{P}, \mathcal{Q}) = \begin{cases} 0 & \text{if } \mathcal{P}^{Aut(G)} = \mathcal{Q}^{Aut(G)} \\ \frac{1}{|\mathcal{P}^{Aut(G)}| \cdot |\mathcal{Q}^{Aut(G)}|} \sum_{\substack{\tilde{\mathcal{P}} \in \mathcal{P}^{Aut(G)}, \\ \tilde{\mathcal{Q}} \in \mathcal{Q}^{Aut(G)}}} d(\tilde{\mathcal{P}}, \tilde{\mathcal{Q}}) & \text{else} \end{cases}$$

Note that the definitions for d_L^*, d_U^* and d_{av}^* require the computation of the minimal, maximal and average distance of all pairs of the direct product $\mathcal{P}^{Aut(G)} \times \mathcal{Q}^{Aut(G)}$. The computational complexity of this is quadratic in the size of the larger equivalence class.

Posed as a measurement problem, we can instead fix one partition in one of the orbits and measure the minimal, maximal and average distance between all pairs of either the direct product of $\{\mathcal{P}\} \times \mathcal{Q}^{Aut(G)}$ or $\{\mathcal{Q}\} \times \mathcal{P}^{Aut(G)}$. The complexity of this is linear in the size of the smaller equivalence class.

Theorems 4 and 5 and their proofs are based on these observations. They are the basis for the development of algorithms for the computation of invariant partition comparison measures of a computational complexity of at most linear order and often of constant order.

Theorem 4. *For all $\mathcal{P}^{Aut(G)} \neq \mathcal{Q}^{Aut(G)}$, the following equations hold:*

$$\begin{aligned} d_L^*(\mathcal{P},\mathcal{Q}) &= \min_{\substack{\tilde{\mathcal{P}}\in\mathcal{P}^{Aut(G)},\\ \tilde{\mathcal{Q}}\in\mathcal{Q}^{Aut(G)}}} d(\tilde{\mathcal{P}},\tilde{\mathcal{Q}}) = \min_{g,h\in Aut(G)} d(\mathcal{P}^h,\mathcal{Q}^g) \\ &= \min_{\tilde{\mathcal{Q}}\in\mathcal{Q}^{Aut(G)}} d(\mathcal{P},\tilde{\mathcal{Q}}) = \min_{g\in Aut(G)} d(\mathcal{P},\mathcal{Q}^g) \\ &= \min_{\tilde{\mathcal{P}}\in\mathcal{P}^{Aut(G)}} d(\tilde{\mathcal{P}},\mathcal{Q}) = \min_{h\in Aut(G)} d(\mathcal{P}^h,\mathcal{Q}) \end{aligned}$$

For $\mathcal{P}^{Aut(G)} \neq \mathcal{Q}^{Aut(G)}$:

$$\begin{aligned} d_U^*(\mathcal{P},\mathcal{Q}) &= \max_{\substack{\tilde{\mathcal{P}}\in\mathcal{P}^{Aut(G)},\\ \tilde{\mathcal{Q}}\in\mathcal{P}^{Aut(G)}}} d(\tilde{\mathcal{P}},\tilde{\mathcal{Q}}) = \max_{g,h\in Aut(G)} d(\mathcal{P}^h,\mathcal{Q}^g) \\ &= \max_{\tilde{\mathcal{Q}}\in\mathcal{Q}^{Aut(G)}} d(\mathcal{P},\tilde{\mathcal{Q}}) = \max_{g\in Aut(G)} d(\mathcal{P},\mathcal{Q}^g) \\ &= \max_{\tilde{\mathcal{P}}\in\mathcal{P}^{Aut(G)}} d(\tilde{\mathcal{P}},\mathcal{Q}) = \max_{h\in Aut(G)} d(\mathcal{P}^h,\mathcal{Q}) \end{aligned}$$

Proof. Let $g, h, f \in Aut(G)$, $\tilde{\mathcal{P}} \in \mathcal{P}^{Aut(G)}$ and $\tilde{\mathcal{Q}} \in \mathcal{Q}^{Aut(G)}$, that is $\tilde{\mathcal{P}} = \mathcal{P}^h$ and $\tilde{\mathcal{Q}} = \mathcal{Q}^g$. Then, since the orbits of both partitions are generated by $Aut(G)$, the following identities between distances hold:

$$\begin{aligned} d(\mathcal{P},\tilde{\mathcal{Q}}) &= d(\mathcal{P},\mathcal{Q}^g) = d(\mathcal{P}^{g^{-1}},\mathcal{Q}), \\ d(\tilde{\mathcal{P}},\mathcal{Q}) &= d(\mathcal{P}^h,\mathcal{Q}) = d(\mathcal{P},\mathcal{Q}^{h^{-1}}) \end{aligned}$$

as well as:

$$d(\tilde{\mathcal{P}},\tilde{\mathcal{Q}}) = d(\mathcal{P}^h,\mathcal{Q}^g) = d(\mathcal{P}^{hg^{-1}},\mathcal{Q}),$$

and:

$$d(\tilde{\mathcal{P}},\tilde{\mathcal{Q}}) = d(\mathcal{P}^h,\mathcal{Q}^g) = d(\mathcal{P},\mathcal{Q}^{gh^{-1}}).$$

Furthermore, let $f = gh^{-1}$.

1. For d_L^*, we have:

$$\begin{aligned} \min_{\tilde{\mathcal{Q}}\in\mathcal{Q}^{Aut(G)}} d(\mathcal{P},\tilde{\mathcal{Q}}) &= \min_{g\in Aut(G)} d(\mathcal{P},\mathcal{Q}^g) \\ &= \min_{g^{-1}\in Aut(G)} d(\mathcal{P}^{g^{-1}},\mathcal{Q}) = \min_{\tilde{\mathcal{P}}\in\mathcal{P}^{Aut(G)}} d(\tilde{\mathcal{P}},\mathcal{Q}) \end{aligned}$$

by switching the reference systems. In the next sequence of equations, we establish that taking the minimum over all reference systems is equivalent to finding the minimum for one arbitrarily fixed reference system.

$$\min_{\substack{\tilde{\mathcal{P}}\in\mathcal{P}^{Aut(G)},\\ \tilde{\mathcal{Q}}\in\mathcal{Q}^{Aut(G)}}} d(\tilde{\mathcal{P}},\tilde{\mathcal{Q}}) = \min_{g,h\in Aut(G)} d(\mathcal{P}^h,\mathcal{Q}^g) = \min_{g,h\in Aut(G)} d(\mathcal{P},\mathcal{Q}^{gh^{-1}})$$
$$= \min_{f\in Aut(G)} d(\mathcal{P},\mathcal{Q}^f) = \min_{\tilde{\mathcal{Q}}\in\mathcal{Q}^{Aut(G)}} d(\mathcal{P},\tilde{\mathcal{Q}})$$

2. For the proof of d^*_U for $\mathcal{P}^{Aut(G)} \neq \mathcal{Q}^{Aut(G)}$ we substitute max for min in the proof of d^*_L.

□

Theorem 5. *For all $\mathcal{P}^{Aut(G)} \neq \mathcal{Q}^{Aut(G)}$, the following equations hold:*

$$d^*_{av}(\mathcal{P},\mathcal{Q}) = \frac{1}{|\mathcal{P}^{Aut(G)}|\cdot|\mathcal{Q}^{Aut(G)}|} \sum_{\substack{\tilde{\mathcal{P}}\in\mathcal{P}^{Aut(G)},\\ \tilde{\mathcal{Q}}\in\mathcal{Q}^{Aut(G)}}} d(\tilde{\mathcal{P}},\tilde{\mathcal{Q}}) \tag{3}$$
$$= \frac{1}{|Aut(G)|^2} \sum_{h,g\in Aut(G)} d(\mathcal{P}^h,\mathcal{Q}^g) \tag{4}$$
$$= \frac{1}{|\mathcal{P}^{Aut(G)}|} \sum_{\tilde{\mathcal{P}}\in\mathcal{P}^{Aut(G)}} d(\tilde{\mathcal{P}},\mathcal{Q}) \tag{5}$$
$$= \frac{1}{|Aut(G)|} \sum_{h\in Aut(G)} d(\mathcal{P}^h,\mathcal{Q}) \tag{6}$$
$$= \frac{1}{|\mathcal{Q}^{Aut(G)}|} \sum_{\tilde{\mathcal{Q}}\in\mathcal{Q}^{Aut(G)}} d(\mathcal{P},\tilde{\mathcal{Q}}) \tag{7}$$
$$= \frac{1}{|Aut(G)|} \sum_{g\in Aut(G)} d(\mathcal{P},\mathcal{Q}^g) \tag{8}$$

Proof. For the proof of the equality of the identities of d^*_{av}, we use the property of an average of n observations $x_{i,j}$ with k identical groups of size m with $i \in 1,\ldots,k$, $j \in 1,\ldots,m$:

$$\frac{1}{n}\sum_{i=1}^{k}\sum_{j=1}^{m} x_{i,j} = \frac{k}{km}\sum_{j=1}^{m} x_{1,j} = \frac{1}{m}\sum_{j=1}^{m} x_{1,j} \tag{9}$$

The computation of an average over the group equals the result of the computation of an average over the orbit, because the orbit stabilizer Theorem 1 implies that each element of the orbit is generated $|Aut(G)_{\mathcal{P}}|$ times, and this means that we average $|Aut(G)_{\mathcal{P}}|$ groups of identical values and that Equation (9) applies. This establishes the equality of Expressions (3) and (4), as well as of Expressions (5) and (6) and of Expressions (7) and (8), respectively.

The two decompositions of the direct product $Aut(G) \times Aut(G)$ establish the equality of Expressions (4) and (6), as well as of Expressions (4) and (8). □

Note that these proofs also show that $d^*_L(\mathcal{P},\mathcal{Q})$, $d^*_U(\mathcal{P},\mathcal{Q})$ and $d^*_{av}(\mathcal{P},\mathcal{Q})$ are invariant. Next, we prove that the three measures $d^*_L(\mathcal{P},\mathcal{Q})$, $d^*_U(\mathcal{P},\mathcal{Q})$ and $d^*_{av}(\mathcal{P},\mathcal{Q})$ are invariant measures.

Theorem 6. *The lower pseudometric space (S, d^*_L) has the following properties:*

1. *Identity: $d^*_L(\mathcal{P},\mathcal{Q}) = 0$, if $\mathcal{P}^{Aut(G)} = \mathcal{Q}^{Aut(G)}$.*
2. *Invariance: $d^*_L(\mathcal{P},\mathcal{Q}) = d^*_L(\tilde{\mathcal{P}},\tilde{\mathcal{Q}})$, for all $\mathcal{P},\mathcal{Q} \in P(V)$ and $\tilde{\mathcal{P}} \in \mathcal{P}^{Aut(G)}$, $\tilde{\mathcal{Q}} \in \mathcal{Q}^{Aut(G)}$.*
3. *Symmetry: $d^*_L(\mathcal{P},\mathcal{Q}) = d^*_L(\mathcal{Q},\mathcal{P})$.*
4. *Triangle inequality: $d^*_L(\mathcal{P},\mathcal{R}) \leq d^*_L(\mathcal{P},\mathcal{Q}) + d^*_L(\mathcal{Q},\mathcal{R})$*

These properties also hold for the upper pseudometric space (S, d^*_U) and the average pseudometric space (S, d^*_{av}).

Proof.

1. Identity holds because of the definition of the distance d^* between two elements in an equivalence class of the pseudometric space (S, d^*).
2. Invariance of $d^*_L(\mathcal{P},\mathcal{Q})$, $d^*_U(\mathcal{P},\mathcal{Q})$ and $d^*_{av}(\mathcal{P},\mathcal{Q})$ is proven by Theorems 4 and 5.
3. Symmetry holds, because d is symmetric, and min, max and the average do not depend on the order of their respective arguments.
4. To proof the triangular inequality, we make use of Theorems 4 and 5 and of the fact that d is a metric for which the triangular inequality holds:

   (a) For d^*_L follows:

$$\begin{aligned} d^*_L(\mathcal{P},\mathcal{R}) &= \min_{\substack{\tilde{\mathcal{P}}\in\mathcal{P}^{Aut(G)},\\ \tilde{\mathcal{R}}\in\mathcal{R}^{Aut(G)}}} d(\tilde{\mathcal{P}},\tilde{\mathcal{R}}) \\ &\leq \min_{\substack{\tilde{\mathcal{P}}\in\mathcal{P}^{Aut(G)},\\ \tilde{\mathcal{Q}}\in\mathcal{Q}^{Aut(G)},\\ \tilde{\mathcal{R}}\in\mathcal{R}^{Aut(G)}}} (d(\tilde{\mathcal{P}},\tilde{\mathcal{Q}}) + d(\tilde{\mathcal{Q}},\tilde{\mathcal{R}})) \\ &= \min_{\substack{\tilde{\mathcal{P}}\in\mathcal{P}^{Aut(G)},\\ \tilde{\mathcal{R}}\in\mathcal{R}^{Aut(G)}}} (d(\tilde{\mathcal{P}},\mathcal{Q}) + d(\mathcal{Q},\tilde{\mathcal{R}})) \\ &= \min_{\tilde{\mathcal{P}}\in\mathcal{P}^{Aut(G)}} d(\tilde{\mathcal{P}},\mathcal{Q}) + \min_{\tilde{\mathcal{R}}\in\mathcal{R}^{Aut(G)}} d(\mathcal{Q},\tilde{\mathcal{R}}) \\ &= d^*_L(\mathcal{P},\mathcal{Q}) + d^*_L(\mathcal{Q},\mathcal{R}) \end{aligned}$$

   (b) For the proof of the triangular inequality for d^*_U, we substitute max for min and d_U for d_L in the proof of the triangular inequality for d^*_L.

   (c) For d^*_{av}, it follows:

$$\begin{aligned} d^*_{av}(\mathcal{P},\mathcal{R}) &= \frac{1}{|\mathcal{P}^{Aut(G)}|\cdot|\mathcal{R}^{Aut(G)}|} \sum_{\tilde{\mathcal{P}}\in\mathcal{P}^{Aut(G)}} \sum_{\tilde{\mathcal{R}}\in\mathcal{R}^{Aut(G)}} d(\tilde{\mathcal{P}},\tilde{\mathcal{R}}) \\ &\leq \frac{1}{|\mathcal{P}^{Aut(G)}|\cdot|\mathcal{R}^{Aut(G)}|} \sum_{\tilde{\mathcal{P}}}\sum_{\tilde{\mathcal{R}}} [d(\tilde{\mathcal{P}},\mathcal{Q}) + d(\mathcal{Q},\tilde{\mathcal{R}})] \\ &= \frac{1}{|\mathcal{P}^{Aut(G)}|\cdot|\mathcal{R}^{Aut(G)}|} \sum_{\tilde{\mathcal{P}}}\sum_{\tilde{\mathcal{R}}} d(\tilde{\mathcal{P}},\mathcal{Q}) + \frac{1}{|\mathcal{P}^{Aut(G)}|\cdot|\mathcal{R}^{Aut(G)}|} \sum_{\tilde{\mathcal{P}}}\sum_{\tilde{\mathcal{R}}} d(\mathcal{Q},\tilde{\mathcal{R}}) \\ &= \frac{1}{|\mathcal{R}^{Aut(G)}|} \sum_{\tilde{\mathcal{R}}} d^*_{av}(\mathcal{P},\mathcal{Q}) + \frac{1}{|\mathcal{P}^{Aut(G)}|} \sum_{\tilde{\mathcal{P}}} d^*_{av}(\mathcal{Q},\mathcal{R}) \\ &= d^*_{av}(\mathcal{P},\mathcal{Q}) + d(\tilde{\mathcal{Q}},\mathcal{R}) \end{aligned}$$

□

5. Decomposition of Partition Comparison Measures

In this section, we assess the structural (dis)similarity between two partitions and the effect of the group actions by combining a partition comparison measure and its invariant counterpart defined in Section 4. The distances $d(\mathcal{P},\mathcal{Q})$, $d^*_L(\mathcal{P},\mathcal{Q})$, $d^*_U(\mathcal{P},\mathcal{Q})$ and $d^*_{av}(\mathcal{P},\mathcal{Q})$ allow the decomposition of a partition comparison measure (transformed into a distance) into a structural component d_{struc} and the effect $d_{Aut(G)}$ of the automorphism group $Aut(G)$:

$$d(\mathcal{P},\mathcal{Q}) = \underbrace{d_L^*(\mathcal{P},\mathcal{Q})}_{d_{struc}} + \underbrace{(d(\mathcal{P},\mathcal{Q}) - d_L^*(\mathcal{P},\mathcal{Q}))}_{d_{Aut(G)}}$$
$$= \underbrace{d_U^*(\mathcal{P},\mathcal{Q})}_{d_{struc}} - \underbrace{(d_U^*(\mathcal{P},\mathcal{Q}) - d(\mathcal{P},\mathcal{Q}))}_{d_{Aut(G)}}$$
$$= \underbrace{d_{av}^*(\mathcal{P},\mathcal{Q})}_{d_{struc}} - \underbrace{(d_{av}^*(\mathcal{P},\mathcal{Q}) - d(\mathcal{P},\mathcal{Q}))}_{d_{Aut(G)}}$$

$\mathrm{dia}(\mathcal{P})$ measures the effect of the automorphism group $Aut(G)$ on the equivalence class $\mathcal{P}^{Aut(G)}$ (see the last column of Table 4). $e_{\max}^{Aut(G)}$ is an upper bound of the automorphism effect on the distance of two partitions $\mathcal{P}$ and $\mathcal{Q}$:

$$e_{\max}^{Aut(G)} = \min(\mathrm{dia}(\mathcal{P}), \mathrm{dia}(\mathcal{Q})).$$

This follows from Theorem 4. Note that $e_{\max}^{Aut(G)} \geq d_U^* - d_L^*$, as Case 1 in Table 5 shows.

Table 5. Measure decomposition for partitions of the butterfly graph for the Rand distance $d_{RI} = 1 - RI$.

Case	$\mathcal{P}$	$\mathcal{Q}$	d_{RI}	d^*	d_{struc}	$d_{Aut(G)}$
1	$\{\{1,2,3,4\}\{5\}\}$	$\{\{4\}\{5\}\{1,2,3\}\}$	0.3	d_L^*	0.3	0.0
	$\in E_{13}$	$\in E_5$	0.3	d_{av}^*	0.5	-0.2
	$\mathrm{dia}(E_{13}) = 0.6$	$\mathrm{dia}(E_5) = 0.6$	0.3	d_U^*	0.7	-0.4
2	$\{\{2,4\}\{1,3,5\}\}$	$\{\{3\}\{1,4\}\{2,5\}\}$	0.6	d_L^*	0.2	0.4
	$\in E_{16}$	$\in E_9$	0.6	d_{av}^*	0.4	0.2
	$\mathrm{dia}(E_{16}) = 0.6$	$\mathrm{dia}(E_9) = 0.4$	0.6	d_U^*	0.6	0.0
3	$\{\{1\}\{2,5\}\{3,4\}\}$	$\{\{1\}\{2,3\}\{4\}\{5\}\}$	0.3	d_L^*	0.1	0.2
	$\in E_{11}$	$\in E_3$	0.3	d_{av}^*	0.25	0.05
	$\mathrm{dia}(E_{11}) = 0.4$	$\mathrm{dia}(E_3) = 0.2$	0.3	d_U^*	0.3	0.0
4	$\{\{3\}\{1,2\}\{4,5\}\}$	$\{\{1\}\{2,3\}\{4,5\}\}$	0.3	d_L^*	0.3	0.0
	$\in E_8$	$\in E_{10}$	0.3	d_{av}^*	0.3	0.0
	$(\mathrm{dia}(E_8) = 0$, stable)	$\mathrm{dia}(E_{10}) = 0.4$	0.3	d_U^*	0.3	0.0

In Table 5, we show a few examples of measure decomposition for partitions of the butterfly graph for the Rand distance d_{RI}:

1. In Case 1, we compare two partitions from nontrivial equivalence classes: the difference of 0.4 between d_U^* and d_L^* indicates that the potential maximal automorphism effect is larger than the lower measure. In addition, it is also smaller (by 0.2) than the automorphism effect in each of the equivalence classes. That $d_{Aut(G)}$ is zero for the lower measure implies that the pair $(\mathcal{P},\mathcal{Q})$ is a pair with the minimal distance between the equivalence classes. The fact that $d_{av}^* = 0.5$ is the mid-point between the lower and upper measures indicates a symmetric distribution of the distances between the equivalence classes.
2. That $d_{Aut(G)}$ is zero for the upper measure in Case 2 means that we have found a pair with the maximal distance between the equivalence classes.
3. In Case 3, we have also found a pair with maximal distance between the equivalence classes. However, the maximal potential automorphism effect is smaller than for Cases 1 and 2. In addition, the distribution of distances between the equivalence classes is asymmetric.
4. Case 4 shows the comparison of a partition from a trivial with a partition from a non-trivial equivalence class. Note, that in this case, all three invariant measures, as well as d_{RI} coincide and that no automorphism effect exists.

A different approach to measure the potential instability in clustering a graph G is the computation of the Kolmogorov–Sinai entropy of the finite permutation group $Aut(G)$ acting on the graph [28].

Note, that the Kolmogorov–Sinai entropy of a finite permutation group is a measure of the uncertainty of the automorphism group. It cannot be used as a measure to compare two graph partitions.

6. Invariant Measures for the Karate Graph

In this section, we illustrate the use of invariant measures for the three partitions $\mathcal{P}_O$, $\mathcal{P}_1$ and $\mathcal{P}_2$ of the Karate graph K [29], which is shown in Figure 4. $Aut(K)$ is of order 480, and it consists of the three subgroups $G_1 = Sym(\Omega_1)$ with $\Omega_1 = \{15, 16, 19, 21, 23\}$, $G_2 = Sym(\Omega_2)$ with $\Omega_2 = \{18, 22\}$ and $G_3 = \{(), (5\,11), (6\,7)\}$. In addition to the modularity optimal partition $\mathcal{P}_O$ (with its clusters separated by longer and dashed lines in Figure 4), we use the partitions $\mathcal{P}_1$ and $\mathcal{P}_2$:

$$\begin{aligned}\mathcal{P}_1 = \{&\{5,6,7,\mathbf{19},\mathbf{21}\},\{1,2,3,4,8,12,13,14,18,20,22\},\\ &\{9,10,\mathit{11},\mathbf{15},\mathbf{16},17,\mathbf{23},27,30,31,33,34\},\{24,25,26,28,29,32\}\}\end{aligned}$$

$$\begin{aligned}\mathcal{P}_2 = \{&\{5,6,7,8,12,\mathbf{19},\mathbf{21}\},\{1,2,3,4,13,14,18,20,22\},\\ &\{9,10,\mathit{11},\mathbf{15},\mathbf{16},17,\mathbf{23},27,30,31,33,34\},\{24,25,26,28,29,32\}\}\end{aligned}$$

Both partitions are affected by the orbits $\{\mathbf{15}, \mathbf{16}, \mathbf{19}, \mathbf{21}, \mathbf{23}\}$ and $\{\mathit{5}, \mathit{11}\}$, each overlapping two clusters. The dissimilarity to $\mathcal{P}_O$ is larger for $\mathcal{P}_2$, which is reflected in Tables 6 and 7.

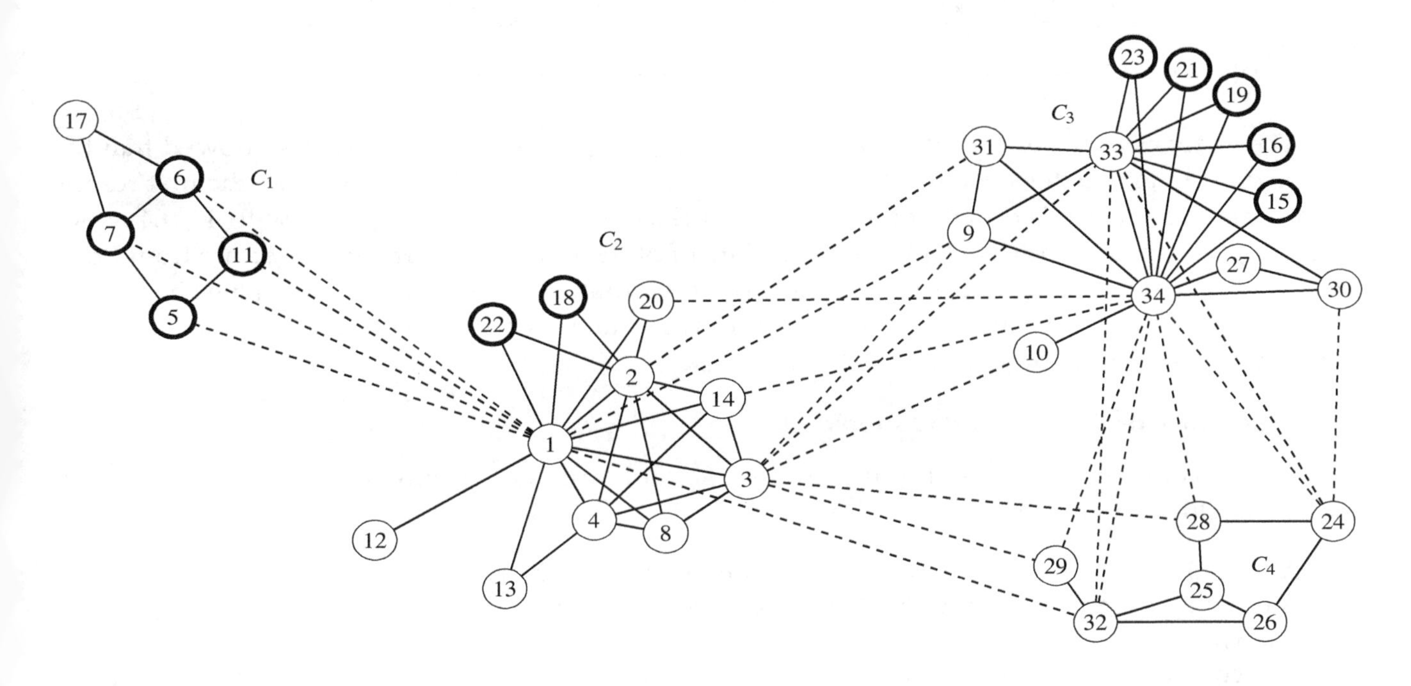

Figure 4. Zachary's Karate graph K with the vertices of the orbits of the three subgroups of $Aut(K)$ in bold and the clusters of P_O separated by dashed edges.

For the optimal partition $\mathcal{P}_O$ of type $(5, 6, 11, 12)$, the upper bound of the size of the equivalence class is 480 [30] (p. 112). The actual size of the equivalence class of $\mathcal{P}_O$ is one, which means the optimal solution is not affected by $Aut(K)$. Partition $\mathcal{P}_1$, which is of the same type as $\mathcal{P}_O$, also has an upper bound of 480 for its equivalence class. The actual size of the equivalence classes of both $\mathcal{P}_1$ and $\mathcal{P}_2$ is 20. Note that the actual size of the equivalence classes that drive the complexity of computing invariant measures is in our example far below the upper bound. Table 6 shows the diameters of the equivalence classes of the partitions.

Table 6. Diameter (computed using d_{RI}), orbit size and stability of partitions $\mathcal{P}_O$, $\mathcal{P}_1$ and $\mathcal{P}_2$.

$\mathcal{X}$	$\mathcal{P}_O$	$\mathcal{P}_1$	$\mathcal{P}_2$
dia($\mathcal{X}$)	0.0000	0.1176	0.1390
$\lvert\mathcal{X}^{Aut(G)}\rvert$	1	20	20
$\mathcal{X}$ stable?	yes	no	no

Table 7 illustrates the decomposition into structural effects and automorphism effects for the three partitions of the Karate graph. We see that for the comparison of a stable partition ($\mathcal{P}_O$) with one of the unstable partitions, the classic partition comparison measures are sufficient. However, when comparing the two unstable partitions $\mathcal{P}_1$ and $\mathcal{P}_2$, the structural effect (0.0499) is dominated by the maximal automorphism effect (0.1176). Furthermore, we note that the distribution of values over the orbit of the automorphism group is asymmetric (by looking at d_L^*, d_U^* and d_{av}^*).

Table 7. Invariant measures and automorphism effects for the Karate graph. The R package `partitionComparison` has been used for the computations [23].

Measure $d = d_{RI}$	$m(\mathcal{P}_O, \mathcal{P}_1)$	$m(\mathcal{P}_O, \mathcal{P}_2)$	$m(\mathcal{P}_1, \mathcal{P}_2)$
d	0.0927	0.1426	0.0499
$d_L^* + d_{Aut(G)}$	0.0927	0.1426	$0.0499 + 0.0000$
$d_U^* - d_{Aut(G)}$	0.0927	0.1426	$0.1676 - 0.1176$
$d_{av}^* - d_{Aut(G)}$	0.0927	0.1426	$0.1280 - 0.0781$
$e_{\max}^{Aut(K)}$	0.0000	0.0000	0.1176

The analysis of the effects of the automorphism group of the Karate network showed that the automorphism group does not affect the stability of the optimal partition. However, the first results show that the situation is different for other networks like the Internet AS graph with 40,164 nodes and 85,123 edges (see Rossi et al. [31], and the data of of the graph `tech-internet-as` are from Rossi and Ahmed [32]): for this graph, several locally optimal solutions with a modularity value above 0.694 exist, all of which are unstable. Further analysis of the structural properties of the solution landscape of this graph is work in progress.

7. Discussion, Conclusions and Outlook

In this contribution, we study the effects of graph automorphisms on partition comparison measures. Our main results are:

1. A formal definition of partition stability, namely $\mathcal{P}$ is stable iff $|\mathcal{P}^{Aut(G)}| = 1$.
2. A proof of the non-invariance of all partition comparison measures if the automorphism group is nontrivial ($|Aut(G)| > 1$).
3. The construction of a pseudometric space of equivalence classes of graph partitions for three classes of invariant measures concerning finite permutation groups of graph automorphisms.
4. The proof that the measures are invariant and that for these measures (after the transformation to a distance), the axioms of a metric space hold.
5. The space of partitions is equipped with a metric (the original partition comparison measure) and a pseudometric (the invariant partition comparison measure).
6. The decomposition of the value of a partition comparison measure into a structural part and a remainder that measures the effect of group actions.

Our definitions of invariant measures have the advantage that any existing partition comparison measure (as long as it is a distance or can be transformed into one) can still be used for the task. Moreover, the decomposition of measures restores the primary purpose of the existing comparison measures, which is to quantify structural difference. However, the construction of these measures leads directly to the classic graph isomorphism problem, whose complexity—despite considerable efforts and hopes to the contrary [33]—is still an open theoretical problem [34,35]. However, from a pragmatic point of view, today, quite efficient and practically usable algorithms exist to tackle the graph isomorphism problem [34]. In addition, for very large and sparse graphs, algorithms for finding generators of the automorphism group exist [11]. Therefore, this dependence on a computationally hard problem in general is not an actual disadvantage and allows one to implement the presented measure decomposition. The efficient implementation of algorithms for the decomposition of graph partition comparison measures is left for further research.

Another constraint is that we have investigated the effects of automorphisms on partition comparison measures in the setting of graph clustering only. The reason for this restriction is that the automorphism group of the graph is already defined by the graph itself and, therefore, is completely contained in the graph data. For arbitrary datasets, the information about the automorphism group is usually not contained in the data, but must be inferred from background theories. However, provided we know the automorphism group, our results on the decomposition of the measures generalize to arbitrary cluster problems.

All in all, this means that this article provides two major assets: first, it provides a theoretic framework that is independent of the preferred measure and the data. Second, we provide insights into a source of possible partition instability that has not yet been discussed in the literature. The downsides (symmetry group must be known and graph clustering only) are in our opinion not too severe, as we discussed above. Therefore, we think that our study indicates that a better understanding of the principle of symmetry is important for future research in data analysis.

Author Contributions: The F.B. implemented the R package mentioned in the Supplementary Materials and conducted the non-invariance proof by counterexample. The more general proof of the non-existence of invariant measures, as well as the idea of creating a pseudometric to repair a measure's deficiency of not being invariant is mainly due to the A.G.-S. Both authors contributed equally to writing the article and revising it multiple times.

Acknowledgments: We thank Andreas Geyer-Schulz (Institute of Analysis, Faculty of Mathematics, KIT, Karlsruhe) for repeated corrections and suggestions for improvement in the proofs.

Appendix A. Modularity

Newman's and Girvan's modularity [14] is defined as:

$$Q = \sum_i \left(e_{ii} - a_i^2 \right)$$

with the edge fractions:

$$e_{ij} = \frac{|\{\{u,v\} \in E | u \in C_i \wedge v \in C_j\}|}{2|E|}, \ i \neq j,$$

and the cluster density:

$$e_{ii} = \frac{|\{\{u,v\} \in E | u,v \in C_i\}|}{|E|}.$$

We have to distinguish e_{ij} and e_{ii} because of the set-based definition E. e_{ij} is the fraction of edges from cluster C_i to cluster C_j and e_{ji}, vice versa. Therefore, the edges are counted twice, and thus, the fraction has to be weighted with $\frac{1}{2}$. The second part of Q is the marginal distribution:

$$a_i^2 = \left(\sum_j e_{ij} \right)^2 .$$

High values of Q indicate good partitions. The range of Q is $[-\frac{1}{2}, 1)$. Even if the modularity has some problems by design (e.g., the resolution limit [36], unbalanced cluster sizes [37], multiple equivalent, but unstable solutions generated by automorphisms [38]), maximization of Q is the de facto standard formal optimization criterion for graph clustering algorithms.

Appendix B. Measures for Comparing Partitions

We classify the measures that are used in the literature to compare object partitions as three categories [39]:

1. Pair-counting measures.
2. Set-based comparison measures.
3. Information theory based measures.

All these measures come from a general context and, therefore, may be used to compare any object partitions, not only graph partitions. The flip side of the coin is that they do not consider any adjacency information from the underlying graph at all.

The column Abbr. of Tables A1–A4 denotes the Abbreviations used throughout this paper; the column $\mathcal{P} = \mathcal{P}$ denotes the value resulting when identical partitions are compared (max stands for some maximum value depending on the partition).

Appendix B.1. Pair-Counting Measures

All the measures within the first class are based on the four coefficients N_{xy} that count pairs of objects (nodes in our context). Let $\mathcal{P}, \mathcal{Q}$ be partitions of the node set V of a graph G. C and C' denote clusters (subsets of vertices $C, C' \subseteq V$). The coefficients are defined as:

$$N_{11} := \left| \{ \{u,v\} \subseteq V \mid (\exists C \in \mathcal{P} : \{u,v\} \subseteq C) \land (\exists C' \in \mathcal{Q} : \{u,v\} \subseteq C') \} \right| ,$$

$$N_{10} := \left| \{ \{u,v\} \subseteq V \mid (\exists C \in \mathcal{P} : \{u,v\} \subseteq C) \land (\forall C' \in \mathcal{Q} : \{u,v\} \not\subseteq C') \} \right| ,$$

$$N_{01} := \left| \{ \{u,v\} \subseteq V \mid (\forall C \in \mathcal{P} : \{u,v\} \not\subseteq C) \land (\exists C' \in \mathcal{Q} : \{u,v\} \subseteq C') \} \right| ,$$

$$N_{00} := \left| \{ \{u,v\} \subseteq V \mid (\forall C \in \mathcal{P} : \{u,v\} \not\subseteq C) \land (\forall C' \in \mathcal{Q} : \{u,v\} \not\subseteq C') \} \right| .$$

Please note that $N_{11} + N_{10} + N_{01} + N_{00} = \binom{n}{2} = \frac{n(n-1)}{2}$. One easily can see that for identical partitions $N_{10} = N_{01} = 0$, because two nodes either occur in a cluster together or not. Completely different partitions result in $N_{11} = 0$. All the measures we examined are given in Tables A1 and A2. The RV coefficient is used by Youness and Saporta [40] for partition comparison, and p and q are the cluster counts (e.g., $p = |\mathcal{P}|$) for the two partitions. For a detailed definition of the Lerman index (especially the definitions of the expectation and standard deviation), see Denœud and Guénoche [41].

Table A1. The pair counting measures used in Table 3 [42]. The above measures are similarity measures. Distance measures and non-normalized measures are listed in Table A2. For brevity: $N_{21} = N_{11} + N_{10}$, $N_{12} = N_{11} + N_{01}$, $N'_{01} = N_{00} + N_{01}$ and $N'_{10} = N_{00} + N_{10}$. Abbr., Abbreviation.

Abbr.	Measure	Formula	$\mathcal{P} = \mathcal{P}$
RI	Rand [43]	$\frac{N_{11}+N_{00}}{\binom{n}{2}}$	1.0
ARI	Hubert and Arabie [44]	$\frac{2(N_{00}N_{11}-N_{10}N_{01})}{N'_{01}N_{12}+N'_{10}N_{21}}$	1.0
H	Hamann [45]	$\frac{(N_{11}+N_{00})-(N_{10}+N_{01})}{\binom{n}{2}}$	1.0
CZ	Czekanowski [46]	$\frac{2N_{11}}{2N_{11}+N_{10}+N_{01}}$	1.0
K	Kulczynski [47]	$\frac{1}{2}\left(\frac{N_{11}}{N_{21}}+\frac{N_{11}}{N_{12}}\right)$	1.0
MC	McConnaughey [48]	$\frac{N_{11}^2-N_{10}N_{01}}{N_{21}N_{12}}$	1.0
P	Peirce [49]	$\frac{N_{11}N_{00}-N_{10}N_{01}}{N_{21}N'_{01}}$	1.0
W_I	Wallace [50]	$\frac{N_{11}}{N_{21}}$	1.0
W_{II}	Wallace [50]	$\frac{N_{11}}{N_{12}}$	1.0
FM	Fowlkes and Mallows [51]	$\sqrt{\frac{N_{11}}{N_{21}}\frac{N_{11}}{N_{12}}}$	1.0
Γ	Yule [52]	$\frac{N_{11}N_{00}-N_{10}N_{01}}{\sqrt{N_{21}N_{12}N'_{10}N'_{01}}}$	1.0
SS1	Sokal and Sneath [53]	$\frac{1}{4}\left(\frac{N_{11}}{N_{21}}+\frac{N_{11}}{N_{12}}+\frac{N_{00}}{N'_{10}}+\frac{N_{00}}{N'_{01}}\right)$	1.0
B1	Baulieu [54]	$\frac{\binom{n}{2}^2-\binom{n}{2}(N_{10}+N_{01})+(N_{10}-N_{01})^2}{\binom{n}{2}^2}$	1.0
GL	Gower and Legendre [55]	$\frac{N_{11}+N_{00}}{N_{11}+\frac{1}{2}(N_{10}+N_{01})+N_{00}}$	1.0
SS2	Sokal and Sneath [53]	$\frac{N_{11}}{N_{11}+2(N_{10}+N_{01})}$	1.0
SS3	Sokal and Sneath [53]	$\frac{N_{11}N_{00}}{\sqrt{N_{21}N_{12}N'_{01}N'_{10}}}$	1.0
RT	Rogers and Tanimoto [56]	$\frac{N_{11}+N_{00}}{N_{11}+2(N_{10}+N_{01})+N_{00}}$	1.0
GK	Goodman and Kruskal [57]	$\frac{N_{11}N_{00}-N_{10}N_{01}}{N_{11}N_{00}+N_{10}N_{01}}$	1.0
J	Jaccard [3]	$\frac{N_{11}}{N_{11}+N_{10}+N_{01}}$	1.0
RV	Robert and Escoufier [58]	$\left(N_{11}-\frac{1}{q}N_{21}-\frac{1}{p}N_{12}+\binom{n}{2}\frac{1}{pq}\right)\left[\left(\frac{p-2}{p}N_{21}+\binom{n}{2}\frac{1}{p^2}\right)\left(\frac{q-2}{q}N_{12}+\binom{n}{2}\frac{1}{q^2}\right)\right]^{-\frac{1}{2}}$	1.0

Table A2. Pair counting measures that are not similarity measures. For brevity: $N_{21} = N_{11} + N_{10}$, $N_{12} = N_{11} + N_{01}$, $N'_{01} = N_{00} + N_{01}$ and $N'_{10} = N_{00} + N_{10}$.

Abbr.	Measure	Formula	$\mathcal{P} = \mathcal{P}$
RR	Russel and Rao [59]	$\frac{N_{11}}{\binom{n}{2}}$	max
M	Mirkin and Chernyi [60]	$2(N_{01} + N_{10})$	0.0
Mi	Hilbert [61]	$\sqrt{\frac{N_{10}+N_{01}}{N_{11}+N_{10}}}$	0.0
Pe	Pearson [62]	$\frac{N_{11}N_{00}-N_{10}N_{01}}{N_{21}N_{12}N'_{01}N'_{10}}$	max
B2	Baulieu [54]	$\frac{N_{11}N_{00}-N_{10}N_{01}}{\binom{n}{2}^2}$	max
LI	Lerman [63]	$\frac{N_{11}-E(N_{11})}{\sqrt{\sigma^2(N_{11})}}$	max
NLI	Lerman [63] (normalized)	$\frac{\mathrm{LI}(P_1,P_2)}{\mathrm{LI}(P_1,P_1)\mathrm{LI}(P_2,P_2)}$	1.0
FMG	Fager and McGowan [64]	$\frac{N_{11}}{\sqrt{N_{21}N_{12}}} - \frac{1}{2\sqrt{N_{21}}}$	max

Appendix B.2. Set-Based Comparison Measures

The second class is based on plain set comparison. We investigate three measures (see Table A3), namely the measure of Larsen and Aone [65], the so-called classification error distance [66] and Dongen's metric [67].

Table A3. References and formulas for the three set-based comparison measures used in Table 3. σ is the result of a maximum weighted matching of a bipartite graph. The bipartite graph is constructed from the partitions that shall be compared: the two node sets are derived from the two partitions, and each cluster is represented by a node. By definition, the two node sets are disjoint. The node sets are connected by edges of weight $w_{ij} = \left|\{C_i \cap C'_j \mid C_i \in \mathcal{P}, C'_j \in \mathcal{Q}\}\right|$. As in our context $|\mathcal{P}| = |\mathcal{Q}|$, the found σ is assured to be a perfect (bijective) matching. n is the number of nodes $|V|$.

Abbr.	Measure	Formula	$\mathcal{P} = \mathcal{P}$
LA	Larsen and Aone [65]	$\frac{1}{\lvert\mathcal{P}\rvert} \sum_{C \in \mathcal{P}} \max_{C' \in \mathcal{Q}} \frac{2\lvert C \cap C'\rvert}{\lvert C\rvert + \lvert C'\rvert}$	1.0
d_{CE}	Meilă and Heckerman [66]	$1 - \frac{1}{n} \max_{\sigma} \sum_{C \in \mathcal{P}} \lvert C \cap \sigma(C)\rvert$	0.0
D	van Dongen [67]	$2n - \sum_{C \in \mathcal{P}} \max_{C' \in \mathcal{Q}} \lvert C \cap C'\rvert - \sum_{C' \in \mathcal{Q}} \max_{C \in \mathcal{P}} \lvert C \cap C'\rvert$	0.0

Appendix B.3. Information Theory-Based Measures

The last class of measures contains those that are rooted in information theory. We show the measures in Table A4, and we recap the fundamentals briefly: the entropy of a random variable X is defined as:

$$H(X) = -\sum_{i=1}^{k} p_i \log p_i$$

with p_i being the probability of a specific incidence. The entropy of a partition can analogously be defined as:

$$H(\mathcal{P}) = -\sum_{C \in \mathcal{P}} \frac{|C|}{n} \log \frac{|C|}{n}.$$

The mutual information of two random variables is:

$$I(X,Y) = \sum_{i=0}^{k}\sum_{j=0}^{l} p_{ij} \log \frac{p_{ij}}{p_i p_j}$$

and again, analogously:

$$MI(\mathcal{P},\mathcal{Q}) = \sum_{C\in\mathcal{P}}\sum_{C'\in\mathcal{Q}} \frac{|C\cap C'|}{n} \log n\frac{|C\cap C'|}{|C||C'|}$$

is the mutual information of two partitions [68]. Meilă [69] introduced the Variation of Information as $VI = H(\mathcal{P}) + H(\mathcal{Q}) - 2MI$.

Table A4. Information theory-based measures used in Table 3. All measures are based on Shannon's definition of entropy. Again, $n = |V|$.

Abbr.	Measure	Formula	$\mathcal{P} = \mathcal{P}$
MI	e.g., Vinh et al. [68]	$\sum_{C\in\mathcal{P}}\sum_{C'\in\mathcal{Q}} \frac{\lvert C\cap C'\rvert}{n} \log n\frac{\lvert C\cap C'\rvert}{\lvert C\rvert\lvert C'\rvert}$	max
NMI_φ	Danon et al. [70]	$\frac{\mathrm{MI}}{\varphi(H(\mathcal{P}),H(\mathcal{Q}))}$, $\varphi \in \{\min, \max\}$	1.0
NMI_Σ	Danon et al. [70]	$\frac{2\cdot \mathrm{MI}}{H(\mathcal{P})+H(\mathcal{Q})}$	1.0
VI	Meilă [69]	$H(\mathcal{P}) + H(\mathcal{Q}) - 2MI$	0.0

Appendix B.4. Summary

As one can see, all three classes of measures rely mainly on set matching between node sets (clusters), as an alternative definition of $N_{11} = \sum_{C\in\mathcal{P}}\sum_{C'\in\mathcal{Q}} \binom{|C\cap C'|}{2}$ shows [42]. The adjacency information of the graph is completely ignored.

References

1. Melnykov, V.; Maitra, R. CARP: Software for fishing out good clustering algorithms. *J. Mach. Learn. Res.* **2011**, *12*, 69–73.
2. Bader, D.A.; Meyerhenke, H.; Sanders, P.; Wagner, D. (Eds.) *10th DIMACS Implementation Challenge—Graph Partitioning and Graph Clustering*; Rutgers University, DIMACS (Center for Discrete Mathematics and Theoretical Computer Science): Piscataway, NJ, USA, 2012.
3. Jaccard, P. Nouvelles recherches sur la distribution florale. *Bull. Soc. Vaud. Sci. Nat.* **1908**, *44*, 223–270.
4. Horta, D.; Campello, R.J.G.B. Comparing hard and overlapping clusterings. *J. Mach. Learn. Res.* **2015**, *16*, 2949–2997.
5. Romano, S.; Vinh, N.X.; Bailey, J.; Verspoor, K. Adjusting for chance clustering comparison measures. *J. Mach. Learn. Res.* **2016**, *17*, 1–32.
6. Von Luxburg, U.; Williamson, R.C.; Guyon, I. Clustering: Science or art? *JMLR Workshop Conf. Proc.* **2011**, *27*, 65–79.
7. Hennig, C. What are the true clusters? *Pattern Recognit. Lett.* **2015**, *64*, 53–62. [CrossRef]
8. Van Craenendonck, T.; Blockeel, H. *Using Internal Validity Measures to Compare Clustering Algorithms*; Benelearn 2015 Poster Presentations (Online); Benelearn: Delft, The Netherlands, 2015; pp. 1–8.
9. Filchenkov, A.; Muravyov, S.; Parfenov, V. Towards cluster validity index evaluation and selection. In Proceedings of the 2016 IEEE Artificial Intelligence and Natural Language Conference, St. Petersburg, Russia, 10–12 November 2016; pp. 1–8.
10. MacArthur, B.D.; Sánchez-García, R.J.; Anderson, J.W. Symmetry in complex networks. *Discret. Appl. Math.* **2008**, *156*, 3525–3531. [CrossRef]

11. Darga, P.T.; Sakallah, K.A.; Markov, I.L. Faster Symmetry Discovery Using Sparsity of Symmetries. In Proceedings of the 2008 45th ACM/IEEE Design Automation Conference, Anaheim, CA, USA, 8–13 June 2008; pp. 149–154.
12. Katebi, H.; Sakallah, K.A.; Markov, I.L. Graph Symmetry Detection and Canonical Labeling: Differences and Synergies. In *Turing-100. The Alan Turing Centenary*; EPiC Series in Computing; Voronkov, A., Ed.; EasyChair: Manchester, UK, 2012; Volume 10, pp. 181–195.
13. Ball, F.; Geyer-Schulz, A. How symmetric are real-world graphs? A large-scale study. *Symmetry* **2018**, *10*, 29. [CrossRef]
14. Newman, M.E.J.; Girvan, M. Finding and evaluating community structure in networks. *Phys. Rev. E* **2004**, *69*, 026113. [CrossRef] [PubMed]
15. Ovelgönne, M.; Geyer-Schulz, A. An Ensemble Learning Strategy for Graph Clustering. In *Graph Partitioning and Graph Clustering*; Bader, D.A., Meyerhenke, H., Sanders, P., Wagner, D., Eds.; American Mathematical Society: Providence, RI, USA, 2013; Volume 588, pp. 187–205.
16. Wielandt, H. *Finite Permutation Groups*; Academic Press: New York, NY, USA, 1964.
17. James, G.; Kerber, A. The Representation Theory of the Symmetric Group. In *Encyclopedia of Mathematics and Its Applications*; Addison-Wesley: Reading, MA, USA, 1981; Volume 16.
18. Coxeter, H.; Moser, W. Generators and Relations for Discrete Groups. In *Ergebnisse der Mathematik und ihrer Grenzgebiete*; Springer: Berlin, Germany, 1965; Volume 14.
19. Dixon, J.D.; Mortimer, B. Permutation Groups. In *Graduate Texts in Mathematics*; Springer: New York, NY, USA, 1996; Volume 163.
20. Beth, T.; Jungnickel, D.; Lenz, H. *Design Theory*; Cambridge University Press: Cambridge, UK, 1993.
21. Erdős, P.; Rényi, A.; Sós, V.T. On a problem of graph theory. *Stud. Sci. Math. Hung.* **1966**, *1*, 215–235.
22. Burr, S.A.; Erdős, P.; Spencer, J.H. Ramsey theorems for multiple copies of graphs. *Trans. Am. Math. Soc.* **1975**, *209*, 87–99. [CrossRef]
23. Ball, F.; Geyer-Schulz, A. *R Package Partition Comparison*; Technical Report 1-2017, Information Services and Electronic Markets, Institute of Information Systems and Marketing; KIT: Karlsruhe, Germany, 2017.
24. Doob, J.L. Measure Theory. In *Graduate Texts in Mathematics*; Springer: New York, NY, USA, 1994.
25. Hausdorff, F. *Set Theory*, 2nd ed.; Chelsea Publishing Company: New York, NY, USA, 1962.
26. Kuratowski, K. *Topology Volume I*; Academic Press: New York, NY, USA, 1966; Volume 1.
27. Von Neumann, J. Construction of Haar's invariant measure in groups by approximately equidistributed finite point sets and explicit evaluations of approximations. In *Invariant Measures*; American Mathematical Society: Providence, RI, USA, 1999; Chapter 6, pp. 87–134.
28. Ball, F.; Geyer-Schulz, A. Weak invariants of actions of the automorphism group of a graph. *Arch. Data Sci. Ser. A* **2017**, *2*, 1–22.
29. Zachary, W.W. An information flow model for conflict and fission in small groups. *J. Anthropol. Res.* **1977**, *33*, 452–473. [CrossRef]
30. Bock, H.H. *Automatische Klassifikation: Theoretische und praktische Methoden zur Gruppierung und Strukturierung von Daten*; Vandenhoeck und Ruprecht: Göttingen, Germany, 1974.
31. Rossi, R.; Fahmy, S.; Talukder, N. A Multi-level Approach for Evaluating Internet Topology Generators. In Proceedings of the 2013 IFIP Networking Conference, Trondheim, Norway, 2–4 June 2013, pp. 1–9.
32. Rossi, R.A.; Ahmed, N.K. The Network Data Repository with Interactive Graph Analytics and Visualization. In Proceedings of the Twenty-Ninth AAAI Conference on Artificial Intelligence, Austin, TX, USA, 25–30 January 2015.
33. Furst, M.; Hopcroft, J.; Luks, E. Polynomial-time Algorithms for Permutation Groups. In Proceedings of the 21st Annual Symposium on Foundations of Computer Science, Syracuse, NY, USA, 13–15 October 1980; pp. 36–41.
34. McKay, B.D.; Piperno, A. Practical graph isomorphism, II. *J. Symb. Comput.* **2014**, *60*, 94–112. [CrossRef]
35. Babai, L. Graph isomorphism in quasipolynomial time. *arXiv* **2015**, arXiv:1512.03547.
36. Fortunato, S.; Barthélemy, M. Resolution limit in community detection. *Proc. Natl. Acad. Sci. USA* **2007**, *104*, 36–41. [CrossRef] [PubMed]
37. Lancichinetti, A.; Fortunato, S. Limits of modularity maximization in community detection. *Phys. Rev. E* **2011**, *84*, 66122. [CrossRef] [PubMed]

38. Geyer-Schulz, A.; Ovelgönne, M.; Stein, M. Modified randomized modularity clustering: Adapting the resolution limit. In *Algorithms from and for Nature and Life*; Lausen, B., Van den Poel, D., Ultsch, A., Eds.; Studies in Classification, Data Analysis, and Knowledge Organization; Springer International Publishing: Heidelberg, Germany, 2013; pp. 355–363.
39. Meilă, M. Comparing clusterings—An information based distance. *J. Multivar. Anal.* **2007**, *98*, 873–895. [CrossRef]
40. Youness, G.; Saporta, G. Some measures of agreement between close partitions. *Student* **2004**, *51*, 1–12.
41. Denœud, L.; Guénoche, A. Comparison of distance indices between partitions. In *Data Science and Classification*; Batagelj, V., Bock, H.H., Ferligoj, A., Žiberna, A., Eds.; Studies in Classification, Data Analysis, and Knowledge Organization; Springer: Berlin/Heidelberg, Germany, 2006; pp. 21–28.
42. Albatineh, A.N.; Niewiadomska-Bugaj, M.; Mihalko, D. On similarity indices and correction for chance agreement. *J. Classif.* **2006**, *23*, 301–313. [CrossRef]
43. Rand, W.M. Objective criteria for the evaluation of clustering algorithms. *J. Am. Stat. Assoc.* **1971**, *66*, 846–850. [CrossRef]
44. Hubert, L.; Arabie, P. Comparing partitions. *J. Classif.* **1985**, *2*, 193–218. [CrossRef]
45. Hamann, U. Merkmalsbestand und Verwandtschaftsbeziehungen der Farinosae: Ein Beitrag zum System der Monokotyledonen. *Willdenowia* **1961**, *2*, 639–768.
46. Czekanowski, J. "Coefficient of Racial Likeness" und "Durchschnittliche Differenz". *Anthropol. Anz.* **1932**, *9*, 227–249.
47. Kulczynski, S. Zespoly roslin w Pieninach. *Bull. Int. Acad. Pol. Sci. Lett.* **1927**, *2*, 57–203.
48. McConnaughey, B.H. The determination and analysis of plankton communities. *Mar. Res.* **1964**, *1*, 1–40.
49. Peirce, C.S. The numerical measure of the success of predictions. *Science* **1884**, *4*, 453–454. [CrossRef] [PubMed]
50. Wallace, D.L. A method for comparing two hierarchical clusterings: Comment. *J. Am. Stat. Assoc.* **1983**, *78*, 569–576. [CrossRef]
51. Fowlkes, E.B.; Mallows, C.L. A method for comparing two hierarchical clusterings. *J. Am. Stat. Assoc.* **1983**, *78*, 553–569. [CrossRef]
52. Yule, G.U. On the association of attributes in statistics: With illustrations from the material of the childhood society. *Philos. Trans. R. Soc. A* **1900**, *194*, 257–319. [CrossRef]
53. Sokal, R.R.; Sneath, P.H.A. *Principles of Numerical Taxonomy*; W. H. Freeman: San Francisco, CA, USA; London, UK, 1963.
54. Baulieu, F.B. A classification of presence/absence based dissimilarity coefficients. *J. Classif.* **1989**, *6*, 233–246. [CrossRef]
55. Gower, J.C.; Legendre, P. Metric and euclidean properties of dissimilarity coefficients. *J. Classif.* **1986**, *3*, 5–48. [CrossRef]
56. Rogers, D.J.; Tanimoto, T.T. A computer program for classifying plants. *Science* **1960**, *132*, 1115–1118. [CrossRef] [PubMed]
57. Goodman, L.A.; Kruskal, W.H. Measures of association for cross classifications. *J. Am. Stat. Assoc.* **1954**, *49*, 732–764.
58. Robert, P.; Escoufier, Y. A unifying tool for linear multivariate statistical methods: The RV-coefficient. *J. R. Stat. Soc. Ser. C* **1976**, *25*, 257–265. [CrossRef]
59. Russel, P.F.; Rao, T.R. On habitat and association of species of anopheline larvae in south-eastern madras. *J. Malar. Inst. India* **1940**, *3*, 153–178.
60. Mirkin, B.G.; Chernyi, L.B. Measurement of the distance between partitions of a finite set of objects. *Autom. Remote Control* **1970**, *31*, 786–792.
61. Hilbert, D. *Gesammelte Abhandlungen von Hermann Minkowski, Zweiter Band*; Number 2; B. G. Teubner: Leipzig, UK; Berlin, Germany, 1911.
62. Pearson, K. On the coefficient of racial likeness. *Biometrika* **1926**, *18*, 105–117. [CrossRef]
63. Lerman, I.C. Comparing Partitions (Mathematical and Statistical Aspects). In *Classification and Related Methods of Data Analysis*; Bock, H.H., Ed.; North-Holland: Amsterdam, The Netherlands, 1988; pp. 121–132.
64. Fager, E.W.; McGowan, J.A. Zooplankton species groups in the north pacific co-occurrences of species can be used to derive groups whose members react similarly to water-mass types. *Science* **1963**, *140*, 453–460. [CrossRef] [PubMed]

65. Larsen, B.; Aone, C. Fast and Effective Text Mining Using Linear-time Document Clustering. In Proceedings of the Fifth ACM SIGKDD International Conference on Knowledge Discovery and Data Mining, San Diego, CA, USA, 15–18 August 1999; ACM: New York, NY, USA, 1999; pp. 16–22.
66. Meilă, M.; Heckerman, D. An experimental comparison of model-based clustering methods. *Mach. Learn.* **2001**, *42*, 9–29. [CrossRef]
67. Van Dongen, S. *Performance Criteria for Graph Clustering and Markov Cluster Experiments*; Technical Report INS-R 0012; CWI (Centre for Mathematics and Computer Science): Amsterdam, The Netherlands, 2000.
68. Vinh, N.X.; Epps, J.; Bailey, J. Information theoretic measures for clusterings comparison: Variants, properties, normalization and correction for chance. *J. Mach. Learn. Res.* **2010**, *11*, 2837–2854.
69. Meilă, M. Comparing clusterings by the variation of information. In *Learning Theory and Kernel Machines*; Schölkopf, B., Warmuth, M.K., Eds.; Number 2777 in Lecture Notes in Computer Science; Springer: Berlin/Heidelberg, Germany, 2003; pp. 173–187.
70. Danon, L.; Díaz-Guilera, A.; Duch, J.; Arenas, A. Comparing community structure identification. *J. Stat. Mech. Theory Exp.* **2005**, *2005*, P09008. [CrossRef]

8

Multi-Granulation Graded Rough Intuitionistic Fuzzy Sets Models based on Dominance Relation

Zhan-Ao Xue [1,2,*], Min-Jie Lv [1,2], Dan-Jie Han [1,2] and Xian-wei Xin [1,2]

[1] College of Computer and Information Engineering, Henan Normal University, Xinxiang 453007, China; lmj2921419592@163.com (M.-j.L.); handanjie2017@163.com (D.-j.H.); Xin_XianWei@163.com (X.-w.X.)

[2] Engineering Lab of Henan Province for Intelligence Business & Internet of Things, Henan Normal University, Xinxiang 453007, China

* Correspondence: 121017@htu.edu.cn

Abstract: From the perspective of the degrees of classification error, we proposed graded rough intuitionistic fuzzy sets as the extension of classic rough intuitionistic fuzzy sets. Firstly, combining dominance relation of graded rough sets with dominance relation in intuitionistic fuzzy ordered information systems, we designed type-I dominance relation and type-II dominance relation. Type-I dominance relation reduces the errors caused by single theory and improves the precision of ordering. Type-II dominance relation decreases the limitation of ordering by single theory. After that, we proposed graded rough intuitionistic fuzzy sets based on type-I dominance relation and type-II dominance relation. Furthermore, from the viewpoint of multi-granulation, we further established multi-granulation graded rough intuitionistic fuzzy sets models based on type-I dominance relation and type-II dominance relation. Meanwhile, some properties of these models were discussed. Finally, the validity of these models was verified by an algorithm and some relative examples.

Keywords: graded rough sets; rough intuitionistic fuzzy sets; dominance relation; logical conjunction operation; logical disjunction operation; multi-granulation

---

1. Introduction

Pawlak proposed a rough set model in 1982, which is a significant method in dealing with uncertain, incomplete, and inaccurate information [1]. Its key strategy is to consider the lower and upper approximations based on precise classification.

As a tool, the classic rough set is based on precise classification. It is too restrictive for some problems in the real world. Considering this defect of classic rough sets, Yao proposed the graded rough sets (GRS) model [2]. Then researchers paid more attention to it and relative literatures began to accumulate on its theory and application. GRS can be defined as the lower approximation being $\underline{R}_k(X) = \{x \mid |[x]_R| - |[x]_R \cap X| \leq k, x \in U\}$ and the upper approximation being $\overline{R}_k(X) = \{x \mid |[x]_R \cap X| > k, x \in U\}$. $|[x]_R \cap X|$ is the absolute number of the elements of $|[x]_R|$ inside X and should be called internal grade, $|[x]_R| - |[x]_R \cap X|$ is the absolute number of the elements of $|[x]_R|$ outside X and should be called external grade. $\overline{R}_k(X)$ means union of the elements whose equivalence class' internal-grade about X is greater than k, $\underline{R}_k(X)$ means union of the elements whose equivalence class' external grade about X is at most k [3].

In the view of granular computing [4], the classic rough set is a single-granulation rough set. However, in the real world, we need multiple granularities to analyze and solve problems and Qian et al. proposed multi-granulation rough sets solving this issue [5]. Subsequently, multi-granulation rough sets were extended in References [6–9]. In addition, in the viewpoint of the degrees of classification error, Hu et al. and Wang et al. established a novel model of multi-granulation graded covering rough sets [10,11]. Simultaneously, Wu et al. constructed graded multi-granulation rough sets [12].

References [13–17] discussed GRS in a multi-granulation environment. Moreover, for GRS, it has been studied that the equivalence relation has been extended to the dominance relation [13,14], the limited tolerance relation [17] and so forth [10,11]. In general, all these aforementioned studies have naturally contributed to the development of GRS.

Inspired by the research reported in References [5,13–17], intuitionistic fuzzy sets (IFS) are also a theory which describe uncertainty [18]. IFS consisting of a membership function and a non-membership function are commonly encountered in uncertainty, imprecision, and vagueness [18]. The notion of IFS, proposed by Atanassov, was initially developed in the framework of fuzzy sets [19]. Furthermore, it can describe the "fuzzy concept" of "not this and not that", that is to say, neutral state or neutral degree, thus it is more precise to portray the ambiguous nature of the objective world. IFS theory is applicable in decision-making, logical planning, medical diagnosis, machine learning, and market forecasting, etc. Applications of IFS have attracted people's attention and achieved fruitful results [20–27].

In recent years, IFS have been a hot research topic in uncertain information systems [6,28–30]. For example, in the development of IFS theory, Ai et al. proposed intuitionistic fuzzy line integrals and gave their concrete values in Reference [31]. Zhang et al. researched the fuzzy logic algebraic system and neutrosophic sets as generalizations of IFS in References [23,26,27]. Furthermore, Guo et al. provided the dominance relation of intuitionistic fuzzy information systems [30].

Both rough sets and IFS not only describe uncertain information but also have strong complementarity in practical problems. As such many researchers have studied the combination of rough sets and IFS, namely, rough intuitionistic fuzzy sets (RIFS) and intuitionistic fuzzy rough sets (IFRS) [32]. For example, Huang et al., Gong et al., Zhang et al., He et al., and Tiwari et al. effectively developed IFRS respectively from uncertainty measures, variable precision rough sets, dominance–based rough sets, interval-valued IFS, and attribute selection [29,30,33–35]. Additionally, Zhang and Chen, Zhang and Yang, Huang et al. studied dominance relation of IFRS [19–21]. With respect to RIFS, Xue et al. provided a multi-granulation covering the RIFS model [9].

The above models did not consider the classification of some degrees of error [6–9,29,30,33,36] in dominance relation on GRS and dominance relation in intuitionistic fuzzy ordered information systems [37]. Therefore, in this paper, firstly, we introduce GRS into RIFS to get graded rough intuitionistic fuzzy sets (GRIFS) solving this problem. Then, considering the need for more precise sequence information in the real world, based on dominance relation of GRS and an intuitionistic fuzzy ordered information system, we respectively perform logical conjunction and disjunction operation to gain type-I dominance relation and type-II dominance relation. After that, we use type-I dominance relation and type-II dominance relation thereby replacing equivalence relation to generalize GRIFS. We design two novel models of GRIFS based on type-I dominance relation and type-II dominance relation. In addition, to accommodate a complex environment, we further extend GRIFS models based on type-I dominance relation and type-II dominance relation, respectively, to multi-granulation GRIFS models based on type-I dominance relation and type-II dominance relation. These models present a new path to extract more flexible and accurate information.

The rest of this paper is organized as follows. In Section 2, some basic concepts of IFS and GRS, RIFS are briefly reviewed, at the same time, we give the definition of GRS based on dominance relation. In Section 3, we respectively propose two novel models of GRIFS models based on type-I dominance relation and type-II dominance relation and verify the validity of these two models. In Section 4, the basic concepts of multi-granulation RIFS are given. Then, we propose multi-granulation GRIFS models based on type-I dominance relation and type-II dominance relation, and provide the concepts of optimistic and pessimistic multi-granulation GRIFS models based on type-I dominance relation and type-II dominance relation, respectively. In Section 5, we use an algorithm and example to study and illustrate the multi-granulation GRIFS models based on type-I dominance relation and type-II dominance relation, respectively. In Section 6, we conclude the paper and illuminate on future research.

2. Preliminaries

Definition 1 ([22]). *Let U be a non-empty classic universe of discourse. U is denoted by:*

$$A = \{< x,\ \mu_A(x), \nu_A(x) > | x \in U\},$$

A can be viewed as IFS on U, where $\mu_A(x) : U \to [0,1]$ and $\nu_A(x) : U \to [0,1]$. $\mu_A(x)$ and $\nu_A(x)$ are denoted as membership and non-membership degrees of the element x in A, satisfying $0 \leq \mu_A(x) + \nu_A(x) \leq 1$. For $\forall\, x \in U$, the hesitancy degree is $\pi_A(x) = 1 - \mu_A(x) - \nu_A(x)$, noticeably, $\pi_A(x) :\ U \to [0,1]$. $\forall\, A,\ B \in IFS(U)$, the basic operations of A and B are given as follows:

(1) $A \subseteq B \Leftrightarrow \mu_A(x) \leq \mu_B(x),\ \nu_A(x) \geq \nu_B(x), \forall x \in U,$
(2) $A = B \Leftrightarrow \mu_A(x) = \mu_B(x),\ \nu_A(x) = \nu_B(x),\ \forall x \in U,$
(3) $A \cup B = \{< x,\ \max\{\mu_A(x),\ \mu_B(x)\},\ \min\{\nu_A(x),\ \nu_B(x)\} > | x \in U\},$
(4) $A \cap B = \{< x,\ \min\{\mu_A(x),\ \mu_B(x)\},\ \max\{\nu_A(x),\ \nu_B(x)\} > | x \in U\},$
(5) $\sim A = \{< x,\ \nu_A(x),\ \mu_A(x) > | x \in U\}.$

Definition 2 ([2]). *Let (U, R) be an approximation space, assume $k \in N$, where N is the natural number set. Then GRS can be defined as follows:*

$$\underline{R}_k(X) = \{x || [x]_R| - |[x]_R \cap X| \leq k,\ x \in U\},$$
$$\overline{R}_k(X) = \{x || [x]_R \cap X| > k,\ x \in U\}.$$

$\underline{R}_k(X)$ and $\overline{R}_k(X)$ can be considered as the lower and upper approximations of X with respect to the graded k. Then we call the pair $(\underline{R}_k(X),\ \overline{R}_k(X))$ GRS. When $k = 0$, $\underline{R}_0(X) = \underline{R}(X)$, $\overline{R}_0(X) = \overline{R}(X)$. However, in general, $\underline{R}_k(X)\overline{R}_k(X)$, $\overline{R}_k(X)\underline{R}_k(X)$.

In Reference [4], the positive and negative domains of X are given as follows:

$$POS(X) = \underline{R}_k(X) \cap \overline{R}_k(X),\ NEG(X) = \neg(\underline{R}_k(X) \cup \overline{R}_k(X)).$$

Definition 3 ([36]). *If we denote $R_{\overline{a}}^{\geq} = \{(x_i, x_j) \in U \times U : f(x_i) \geq f(x_j), \forall a \in A\}$ where A is a subset of the attributes set and $f(x)$ is the value of attribute a, then $[x]_{\overline{a}}^{\geq}$ is referred to as the dominance class of dominance relation $R_{\overline{a}}^{\geq}$. Moreover, we denote approximation space based on dominance relations by $S = (U, R_{\overline{a}}^{\geq})$.*

Definition 4. *Let $(U, R_{\overline{a}}^{\geq})$ be an information approximation. $U/R_{\overline{a}}^{\geq}$ is the set of dominance classes induced by a dominance relation $R_{\overline{a}}^{\geq}$, and $[x]_{\overline{a}}^{\geq}$ is called the dominance class containing x. Assume $k \in N$, where N is the natural number set. GRS based on dominance relation can be defined:*

$$\underline{R}_{\overline{k}}^{\geq}(X) = \{x || [x]_{\overline{a}}^{\geq}| - |[x]_{\overline{a}}^{\geq} \cap X| \leq k,\ x \in U\},$$
$$\overline{R}_{\overline{k}}^{\geq}(X) = \{x || [x]_{\overline{a}}^{\geq} \cap X| > k,\ x \in U\}.$$

When $k = 0$, $(\overline{R}_{\overline{0}}^{\geq}(X),\ \underline{R}_{\overline{0}}^{\geq}(X))$ will be rough sets based on dominance relation.

Example 1. *Suppose there are nine patients $U = \{x_1, x_2, x_3, x_4, x_5, x_6, x_7, x_8, x_9\}$, they may suffer from a cold. According to their fever, we get $U/R_{\overline{a}}^{\geq} = \{\{x_1, x_2, x_4\}, \{x_3, x_8\}, \{x_6, x_8\}, \{x_5, x_7, x_8, x_9\}\}$, $X \subseteq U$. Then suppose $X = \{x_1, x_2, x_4, x_7, x_9\}$, we can obtain GRS based on dominance relation.*

The demonstration process is given as follows:

Suppose $k = 1$, then we can get,

$$[x_1]_{\overline{a}}^{\geq} = [x_2]_{\overline{a}}^{\geq} = [x_4]_{\overline{a}}^{\geq} = \{x_1, x_2, x_4\},\ [x_3]_{\overline{a}}^{\geq} = [x_8]_{\overline{a}}^{\geq} = \{x_3, x_8\},\ [x_6]_{\overline{a}}^{\geq} = [x_8]_{\overline{a}}^{\geq} = \{x_6, x_8\},$$
$$[x_5]_{\overline{a}}^{\geq} = [x_7]_{\overline{a}}^{\geq} = [x_8]_{\overline{a}}^{\geq} = [x_9]_{\overline{a}}^{\geq} = \{x_5, x_7, x_8, x_9\}.$$

Then, we can calculate $\underline{R}_1^{\geq}(X)$, $\overline{R}_1^{\geq}(X)$ *and* $POS(X)$, $NEG(X)$.

$$\underline{R}_1^{\geq}(X) = \{x_1, x_2, x_4\},\ \overline{R}_1^{\geq}(X) = \{x_1, x_2, x_4, x_5, x_7, x_8, x_9\}.$$

$$POS(X) = \underline{R}_1^{\geq}(X) \cap \overline{R}_1^{\geq}(X) = \{x_1, x_2, x_4\} \cap \{x_1, x_2, x_4, x_5, x_7, x_8, x_9\} = \{x_1, x_2, x_4\},$$
$$NEG(X) = \neg(\underline{R}_1^{\geq}(X) \cup \overline{R}_1^{\geq}(X)) = \neg(\{x_1, x_2, x_4\} \cup \{x_1, x_2, x_4, x_5, x_7, x_8, x_9\}) = \{x_3, x_6\}.$$

Through the above analysis, we can see x_1, x_2, *and* x_4 *patients suffering from a cold disease and* x_3 *and* x_6 *patients not having a cold disease.*

When $k = 0$, $(\overline{R}_0^{\geq}(X), \underline{R}_0^{\geq}(X))$ *will be rough sets based on dominance relation.*

Definition 5 ([8,32,35]). *Let X be a non-empty set and R be an equivalence relation on X. Let B be IFS in X with the membership function* $\mu_B(x)$ *and non-membership function* $\nu_B(x)$. *The lower and upper approximations, respectively, of B are IFS of the quotient set* X/R *with*

(1) Membership function defined by

$$\mu_{\underline{R}(B)}(X_i) = \inf\{\mu_B(x) |\ x \in X_i\},\ \mu_{\overline{R}(B)}(X_i) = \sup\{\mu_B(x) |\ x \in X_i\}.$$

(2) Non-membership function defined by

$$\nu_{\underline{R}(B)}(X_i) = \sup\{\nu_B(x) |\ x \in X_i\},\ \nu_{\overline{R}(B)}(X_i) = \inf\{\nu_B(x) |\ x \in X_i\}.$$

In this way, we can prove $\underline{R}(B)$ *and* $\overline{R}(B)$ *are IFS.*

For $\forall x \in X_i$, *we can obtain,*

$$\mu_B(x) + \nu_B(x) \leq 1, \mu_B(x) \leq 1 - \nu_B(x), \sup\{\mu_B(x) |\ x \in X_i\} \leq \sup\{1 - \nu_B(x)$$
$$|\ x \in X_i\}, \sup\{\mu_B(x) |\ x \in X_i\} \leq 1 - \inf\{\nu_B$$

Hence $\overline{R}(B)$ *is IFS. Similarly, we can prove that* $\underline{R}(B)$ *is IFS. The RIFS of* $\underline{R}(B)$ *and* $\overline{R}(B)$ *are given as ollows:*

$$\underline{R}(B) = \{< x, \inf_{y \in [x]_i} \mu_B(y), \sup_{y \in [x]_i} \nu_B(y) > | x \in U\},$$

$$\overline{R}(B) = \{< x, \sup_{y \in [x]_i} \mu_B(y), \inf_{y \in [x]_i} \nu_B(y) > | x \in U\}.$$

3. GRIFS Model Based on Dominance Relation

In this section, we propose a GRIFS model based on dominance relation. Moreover, this model contains a GRIFS model based on type-I dominance relation and GRIFS model based on type-II dominance relation, respectively. Then we employ an example to demonstrate the validity of these two models, and finish by discussing some basic properties of these two models.

Definition 6 ([37]). *If* (U, A, V, f) *is an intuitionistic fuzzy ordered information system, so* $(R')_a^{\geq} = \{(x, y) \in U \times U | f_a(y) \geq f_a(x), \forall a \in A\}$ *can be called dominance relation in the intuitionistic fuzzy ordered information system.*

$$\begin{aligned} [x]_a^{\geq'} &= \{y | (x, y) \in (R')_a^{\geq}, \forall a \in A, y \in U\} \\ &= \{y | \mu_a(y) \geq \mu_a(x),\ \nu_a(y) \leq \nu_a(x), \forall a \in A, y \in U\} \end{aligned}$$

$[x]_a^{\geq'}$ *is dominance class of x in terms of dominance relation* $(R')_a^{\geq}$.

3.1. GRIFS Model Based on Type-I Dominance Relation

Definition 7. *Let $IS^{\geq^{I}} = (U, A, V, f)$ be an intuitionistic fuzzy ordered information system and $R_{\overline{a}}^{\geq}$ be a dominance relation of the attribute set A. Suppose X is the GRS of $R_{\overline{a}}^{\geq}$ on U, $a \in A$, and IFS B on U about attribute a satisfies dominance relation $(R')_{\overline{a}}^{\geq}$. The lower approximation $\underline{R_{\overline{k}}}^{\geq^{I}}(B)$ and the upper approximation $\overline{R_{\overline{k}}}^{\geq^{I}}(B)$ with respect to the graded k are given as follows:*

When $k \geq 1$, we can gain,

$$\underline{R_{\overline{k}}}^{\geq^{I}}(B) = \{< x, \inf_{y \in (\bigwedge_{s=1}^{j}(([x]_{\overline{a}}^{\geq})_s \wedge [x]_{\overline{a}}^{\geq'}))} (\mu_B(y) \wedge \mu'_B(y)), \sup_{y \in (\bigwedge_{s=1}^{j}(([x]_{\overline{a}}^{\geq})_s \wedge [x]_{\overline{a}}^{\geq'}))} (\nu_B(y) \vee \nu'_B(y)) > | x \in U\},$$

$$\overline{R_{\overline{k}}}^{\geq^{I}}(B) = \{< x, \sup_{y \in (\bigwedge_{s=1}^{j}(([x]_{\overline{a}}^{\geq})_s \wedge [x]_{\overline{a}}^{\geq'}))} (\mu_B(y) \vee \mu'_B(y)), \inf_{y \in (\bigwedge_{s=1}^{j}(([x]_{\overline{a}}^{\geq})_s \wedge [x]_{\overline{a}}^{\geq'}))} (\nu_B(y) \wedge \nu'_B(y)) > | x \in U\}.$$

$$\mu'_B(y) = \frac{|\overline{R_{\overline{k}}}^{\geq}(X) \cap \underline{R_{\overline{k}}}^{\geq}(X)|}{|U|}, \quad \nu'_B(y) = \frac{|\neg(\overline{R_{\overline{k}}}^{\geq}(X) \cup \underline{R_{\overline{k}}}^{\geq}(X))|}{|U|}.$$

Obviously, $0 \leq \mu'_B(y) \leq 1$, $0 \leq \nu'_B(y) \leq 1$, $j = 1, 2, \cdots, n$.

When $k = 0$, $\mu'_B(y)$ and $\nu'_B(y)$ degenerate to be calculated by the classical rough set. However, under these circumstances, the model is still valid, we call this model RIFS based on type-I dominance relation.

Note that, in GRIFS model based on type-I dominance relation, we let $[x]_{\overline{a}}^{\geq}$ and $[x]_{\overline{a}}^{\geq'}$ perform a conjunction operation $\wedge$, this is to say $\geq^{I}$ means $\bigwedge_{s=1}^{j}(([x]_{\overline{a}}^{\geq})_s \wedge [x]_{\overline{a}}^{\geq'})$.

Note that, $\bigwedge_{s=1}^{j}(([x]_{\overline{a}}^{\geq})_s \wedge [x]_{\overline{a}}^{\geq'})$ in GRIFS model based on type-I dominance relation, if x have j dominance classes $[x]_{\overline{a}}^{\geq}$ of dominance relation $R_{\overline{a}}^{\geq}$ on GRS, we perform a conjunction operation $\wedge$ of j dominance classes $[x]_{\overline{a}}^{\geq}$ and $[x]_{\overline{a}}^{\geq'}$.

According to Definition 7, the following theorem can be obtained.

Theorem 1. *Let $IS^{\geq^{I}} =< U, A, V, f >$ be an intuitionistic fuzzy ordered information system, and B be IFS on U. Then a GRIFS model based on type-I dominance relation has these following properties:*

$$(1)\ \underline{R_{\overline{k}}}^{\geq^{I}}(B) \subseteq B \subseteq \overline{R_{\overline{k}}}^{\geq^{I}}(B),$$
$$(2)\ A \subseteq B,\ \underline{R_{\overline{k}}}^{\geq^{I}}(A) \subseteq \underline{R_{\overline{k}}}^{\geq^{I}}(B),\ \overline{R_{\overline{k}}}^{\geq^{I}}(A) \subseteq \overline{R_{\overline{k}}}^{\geq^{I}}(B),$$
$$(3)\ \underline{R_{\overline{k}}}^{\geq^{I}}(A \cap B) = \underline{R_{\overline{k}}}^{\geq^{I}}(A) \cap \underline{R_{\overline{k}}}^{\geq^{I}}(B),\ \overline{R_{\overline{k}}}^{\geq^{I}}(A \cup B) = \overline{R_{\overline{k}}}^{\geq^{I}}(A) \cup \overline{R_{\overline{k}}}^{\geq^{I}}(B).$$

Proof. (1) From Definition 7, we can get,

$$\inf_{y \in (\bigwedge_{s=1}^{j}(([x]_{\overline{a}}^{\geq})_s \wedge [x]_{\overline{a}}^{\geq'}))} (\mu_B(y) \wedge \mu'_B(y)) \leq \mu_B(x) \leq \sup_{y \in (\bigwedge_{s=1}^{j}(([x]_{\overline{a}}^{\geq})_s \wedge [x]_{\overline{a}}^{\geq'}))} (\mu_B(y) \vee \mu'_B(y)) \Leftrightarrow \mu_{\underline{R_{\overline{k}}}^{\geq^{I}}(B)}(x) \leq \mu_B(x) \leq \mu_{\overline{R_{\overline{k}}}^{\geq^{I}}(B)}(x),$$

$$\sup_{y \in (\bigwedge_{s=1}^{j}(([x]_{\overline{a}}^{\geq})_s \wedge [x]_{\overline{a}}^{\geq'}))} (\nu_B(y) \vee \nu'_B(y)) \geq \nu_B(x) \geq \inf_{y \in (\bigwedge_{s=1}^{j}(([x]_{\overline{a}}^{\geq})_s \wedge [x]_{\overline{a}}^{\geq'}))} (\nu_B(y) \wedge \nu'_B(y)) \Leftrightarrow \nu_{\underline{R_{\overline{k}}}^{\geq^{I}}(B)}(x) \geq \nu_B(x) \geq \nu_{\overline{R_{\overline{k}}}^{\geq^{I}}(B)}(x),$$

Hence, $\underline{R_{\overline{k}}}^{\geq^{I}}(B) \subseteq B \subseteq \overline{R_{\overline{k}}}^{\geq^{I}}(B)$.

(2) Based on Definition 1 and $A \subseteq B$,

Thus we can get, $\mu_A(x) \leq \mu_B(x)$, $\nu_A(x) \geq \nu_B(x)$.

From Definition 7, we can get, $\mu'_A(y) = \mu'_B(y)$, $\nu'_A(y) = \nu'_B(y)$.

Then, in the GRIFS model based on type-I dominance relation, we can get,

$$\inf_{y \in (\bigwedge_{s=1}^{j}(([x]_a^{\geq})_s \wedge [x]_a^{\geq'}))} (\mu_A(y) \wedge \mu'_A(y)) \leq \inf_{y \in (\bigwedge_{s=1}^{j}(([x]_a^{\geq})_s \wedge [x]_a^{\geq'}))} (\mu_B(y) \wedge \mu'_B(y)) \Leftrightarrow \mu_{\underline{R_k^{\geq}}^{\mathrm{I}}(A)}(x) \leq \mu_{\underline{R_k^{\geq}}^{\mathrm{I}}(B)}(x),$$
$$\sup_{y \in (\bigwedge_{s=1}^{j}(([x]_a^{\geq})_s \wedge [x]_a^{\geq'}))} (\nu_A(y) \vee \nu'_A(y)) \geq \sup_{y \in (\bigwedge_{s=1}^{j}(([x]_a^{\geq})_s \wedge [x]_a^{\geq'}))} (\nu_B(y) \vee \nu'_B(y)) \Leftrightarrow \nu_{\underline{R_k^{\geq}}^{\mathrm{I}}(A)}(x) \geq \nu_{\underline{R_k^{\geq}}^{\mathrm{I}}(B)}(x).$$

Thus we can get, $\underline{R_k^{\geq}}^{\mathrm{I}}(A) \subseteq \underline{R_k^{\geq}}^{\mathrm{I}}(B)$.

In the same way, we can get, $\overline{R_k^{\geq}}^{\mathrm{I}}(A) \subseteq \overline{R_k^{\geq}}^{\mathrm{I}}(B)$.

(3) From Definition 7, we can get,

$$\begin{aligned}
\mu_{\underline{R_k^{\geq}}^{\mathrm{I}}(A \cap B)}(x) &= \inf_{y \in (\bigwedge_{s=1}^{j}(([x]_a^{\geq})_s \wedge [x]_a^{\geq'}))} (\mu_{A \cap B}(y) \wedge \mu'_{A \cap B}(y)) = (\inf_{y \in (\bigwedge_{s=1}^{j}(([x]_a^{\geq})_s \wedge [x]_a^{\geq'}))} (\mu_A(y) \wedge \mu'_A(y))) \wedge (\inf_{y \in (\bigwedge_{s=1}^{j}(([x]_a^{\geq})_s \wedge [x]_a^{\geq'}))} (\mu_B(y) \wedge \mu'_B(y))) \\
&= \mu_{\underline{R_k^{\geq}}^{\mathrm{I}}(A)}(x) \wedge \mu_{\underline{R_k^{\geq}}^{\mathrm{I}}(B)}(x), \\
\nu_{\underline{R_k^{\geq}}^{\mathrm{I}}(A \cap B)}(x) &= \sup_{y \in (\bigwedge_{s=1}^{j}(([x]_a^{\geq})_s \wedge [x]_a^{\geq'}))} (\nu_{A \cap B}(y) \vee \nu'_{A \cap B}(y)) = (\sup_{y \in (\bigwedge_{s=1}^{j}(([x]_a^{\geq})_s \wedge [x]_a^{\geq'}))} (\nu_A(y) \vee \nu'_A(y))) \wedge (\sup_{y \in (\bigwedge_{s=1}^{j}(([x]_a^{\geq})_s \wedge [x]_a^{\geq'}))} (\nu_B(y) \vee \nu'_B(y))) \\
&= \nu_{\underline{R_k^{\geq}}^{\mathrm{I}}(A)}(x) \wedge \nu_{\underline{R_k^{\geq}}^{\mathrm{I}}(B)}(x),
\end{aligned}$$

Thus we can get, $\underline{R_k^{\geq}}^{\mathrm{I}}(A \cap B) = \underline{R_k^{\geq}}^{\mathrm{I}}(A) \cap \underline{R_k^{\geq}}^{\mathrm{I}}(B)$.

In the same way, we can get $\overline{R_k^{\geq}}^{\mathrm{I}}(A \cup B) = \overline{R_k^{\geq}}^{\mathrm{I}}(A) \cup \overline{R_k^{\geq}}^{\mathrm{I}}(B)$. □

Example 2. *In a city, the court administration needs to recruit 3 staff. Applicants who pass the application, preliminary examination of qualifications, written examination, interview, qualification review, political review, and physical examination can be employed. In order to facilitate the calculation, we simplify the enrollment process to qualification review, written test, interview. At present, 12 people have passed the preliminary examination of qualifications, and 9 of them have passed the written examination (administrative professional ability test and application).* $U = \{x_1, x_2, x_3, x_4, x_5, x_6, x_7, x_8, x_9\}$ *is the domain. We can get* $U/R_a^{\geq} = \{\{x_1, x_2, x_4\}, \{x_3, x_8\}, \{x_7\}, \{x_4, x_5, x_6, x_9\}\}$ *according to the "excellent" and "pass" of the two results. In addition, through the interview of 9 people, the following IFS can be obtained, and we suppose* $X = \{x_1, x_4, x_5, x_6, x_9\}$, $X \subseteq U$.

$$B = \left\{ \frac{[0.9, 0]}{x_1}, \frac{[0.8, 0.1]}{x_2}, \frac{[0.65, 0.3]}{x_3}, \frac{[0.85, 0.1]}{x_4}, \frac{[0.95, 0.05]}{x_5}, \frac{[0.7, 0.3]}{x_6}, \frac{[0.5, 0.2]}{x_7}, \frac{[0.87, 0.1]}{x_8}, \frac{[0.75, 0.2]}{x_9} \right\}.$$

To solve the above problems, we can use the model described in References [38,39], which are rough sets based on dominance relation.

First, according to $U/R_a^{\geq}$, we can get,

$$\underline{R^{\geq}}(X) = \{x_4, x_5, x_6, x_9\},\ \overline{R^{\geq}}(X) = \{x_1, x_2, x_4, x_5, x_6, x_9\},$$

Through rough sets based on dominance relation, we can get some applicants with better written test scores. However, regarding IFS B, we cannot use rough sets based on dominance relation to handle the data. Therefore, we are even less able to get the final result with the model. To process the interview data, we need to use another model, described in Reference [40]. Through data processing, we can obtain the dominance classes as follows:

$$[x_1]_a^{\geq'} = \{x_1\},\ [x_2]_a^{\geq'} = \{x_2, x_4, x_5, x_8\},\ [x_3]_a^{\geq'} = \{x_3, x_4, x_5, x_6, x_8, x_9\},\ [x_4]_a^{\geq'} = \{x_4, x_5\},$$
$$[x_5]_a^{\geq'} = \{x_5\},\ [x_6]_a^{\geq'} = \{x_6, x_8, x_9\},\ [x_7]_a^{\geq'} = \{x_7, x_8, x_9\},\ [x_8]_a^{\geq'} = \{x_8\},\ [x_9]_a^{\geq'} = \{x_9\}.$$

From the above analysis, we can get,

$$x_5 \geq x_1 \geq x_8 \geq x_4 \geq x_2 \geq x_9 \geq x_6 \geq x_3 \geq x_7$$

Through dominance relation in the intuitionistic fuzzy ordered information system, we can get some applicants with better interview results, but we still cannot get the final results. To get this result, we need to analyze the applicants who have better written test scores and better written test scores. Based on the above conclusions, we can determine that only x_5 and x_4 applicants meet the requirements. However, the performance of others is not certain. If they only need one or two staff, then this analysis can help us to choose the applicant. However, we need 3 applicants, so we cannot get the result in this way. However, there is a model in Definition 6 that can help us get the results. The calculation process is as follows:

According to Example 1, when $k = 1$, we can get

$$\underline{R_{\overline{1}}^{\geq}}(X) = \{x_1, x_2, x_4, x_5, x_6, x_7, x_9\},\ \overline{R_{\overline{1}}^{\geq}}(X) = \{x_1, x_2, x_4, x_5, x_6, x_7, x_9\},$$

According to Definitions 7 and 8, we can then get,

$$\mu'_B(y) = \frac{|\overline{R_{\overline{1}}^{\geq}}(X) \cap \underline{R_{\overline{1}}^{\geq}}(X)|}{|U|} = \frac{7}{9} \approx 0.78,\ \nu'_B(y) = \frac{|\neg(\overline{R_{\overline{1}}^{\geq}}(X) \cup \underline{R_{\overline{1}}^{\geq}}(X))|}{|U|} = \frac{2}{9} \approx 0.22.$$

So, according to Definition 6 and Example 1, we can compute the conjunction operation of $[x]_{\overline{a}}^{\geq}$ and $[x]_{\overline{a}}^{\geq'}$, and the results are as Table 1.

Table 1. The conjunction operation of $[x]_{\overline{a}}^{\geq}$ and $[x]_{\overline{a}}^{\geq'}$.

x	$[x]_{\overline{a}}^{\geq}$	$[x]_{\overline{a}}^{\geq'}$	$[x]_{\overline{a}}^{\geq} \wedge [x]_{\overline{a}}^{\geq'}$
x_1	$\{x_1, x_2, x_4\}$	$\{x_1\}$	$\{x_1\}$
x_2	$\{x_1, x_2, x_4\}$	$\{x_2, x_4, x_5, x_8\}$	$\{x_2, x_4\}$
x_3	$\{x_3, x_8\}$	$\{x_3, x_4, x_5, x_6, x_8, x_9\}$	$\{x_3, x_8\}$
x_4	$\{x_1, x_2, x_4\}, \{x_4, x_5, x_6, x_9\}$	$\{x_4, x_5\}$	$\{x_4\}$
x_5	$\{x_4, x_5, x_6, x_9\}$	$\{x_5\}$	$\{x_5\}$
x_6	$\{x_4, x_5, x_6, x_9\}$	$\{x_6, x_8, x_9\}$	$\{x_6, x_9\}$
x_7	$\{x_7\}$	$\{x_7, x_8, x_9\}$	$\{x_7\}$
x_8	$\{x_3, x_8\}$	$\{x_8\}$	$\{x_8\}$
x_9	$\{x_4, x_5, x_6, x_9\}$	$\{x_9\}$	$\{x_9\}$

GRIFS model based on type-I dominance relation can be obtained as follows:

$$\underline{R_{\overline{1}}^{\geq^{\mathrm{I}}}}(B) = \left\{\frac{[0.78,0.22]}{x_1}, \frac{[0.78,0.22]}{x_2}, \frac{[0.65,0.3]}{x_3}, \frac{[0.78,0.22]}{x_4}, \frac{[0.78,0.22]}{x_5}, \frac{[0.7,0.3]}{x_6}, \frac{[0.5,0.22]}{x_7}, \frac{[0.78,0.22]}{x_8}, \frac{[0.75,0.22]}{x_9}\right\},$$
$$\overline{R_{\overline{1}}^{\geq^{\mathrm{I}}}}(B) = \left\{\frac{[0.9,0]}{x_1}, \frac{[0.85,0.1]}{x_2}, \frac{[0.78,0.1]}{x_3}, \frac{[0.85,0.1]}{x_4}, \frac{[0.95,0.05]}{x_5}, \frac{[0.87,0.1]}{x_6}, \frac{[0.78,0.1]}{x_7}, \frac{[0.87,0.1]}{x_8}, \frac{[0.78,0.2]}{x_9}\right\}.$$

Comprehensive analysis $\underline{R_{\overline{1}}^{\geq^{\mathrm{I}}}}(B)$ and $\overline{R_{\overline{1}}^{\geq^{\mathrm{I}}}}(B)$, we can conclude that x_5, x_1, x_8, x_2 and x_4 applicants are more suitable for the position in the pessimistic situation. From this example we can see that our model is able to handle more complicated situations than the previous theories, and it can help us get more accurate results.

3.2. GRIFS Model Based on Type-II Dominance Relation

Definition 8. *Let U be a non-empty set and A be the attribute set on U, and $a \in A$, $R_{\overline{a}}^{\geq}$ is a dominance relation of attribute A. Let X be GRS of $R_{\overline{a}}^{\geq}$ on U, and IFS B on U about attribute a satisfies dominance relation $(R')_{\overline{a}}^{\geq}$. The lower and upper approximations of B with respect to the graded k are given as follows:*

When $k \geq 1$, we can get,

$$\underline{R_k^{\geq}}^{\Pi}(B) = \{< x, \inf_{y \in (\overset{j}{\underset{s=1}{\vee}}(([x]_a^{\geq})_s \vee [x]_a^{\geq'}))} (\mu_B(y) \wedge \mu'_B(y)), \sup_{y \in (\overset{j}{\underset{s=1}{\vee}}(([x]_a^{\geq})_s \vee [x]_a^{\geq'}))} (\nu_B(y) \vee \nu'_B(y)) > | x \in U\},$$

$$\overline{R_k^{\geq}}^{\Pi}(B) = \{< x, \sup_{y \in (\overset{j}{\underset{s=1}{\vee}}(([x]_a^{\geq})_s \vee [x]_a^{\geq'}))} (\mu_B(y) \vee \mu'_B(y)), \inf_{y \in (\overset{j}{\underset{s=1}{\vee}}(([x]_a^{\geq})_s \vee [x]_a^{\geq'}))} (\nu_B(y) \wedge \nu'_B(y)) > | x \in U\}.$$

$$\mu'_B(y) = \frac{|\overline{R_k^{\geq}}(X) \cap \underline{R_k^{\geq}}(X)|}{|U|}, \nu'_B(y) = \frac{|\neg(\overline{R_k^{\geq}}(X) \cup \underline{R_k^{\geq}}(X))|}{|U|}.$$

Obviously, $0 \leq \mu'_B(y) \leq 1$, $0 \leq \nu'_B(y) \leq 1$, $j = 1, 2, \cdots, n$.

When $k = 0$, $\mu'_B(y)$ and $\nu'_B(y)$ are calculated from the classical rough set. However, under these circumstances the model is still valid and we call this model RIFS based on type-II dominance relation.

Note that in the GRIFS model based on type-II dominance relation, we perform a disjunction operation $\vee$ on $[x]_a^{\geq}$ and $[x]_a^{\geq'}$, this is to say $\geq^{\Pi}$ means $\overset{j}{\underset{s=1}{\vee}}(([x]_a^{\geq})_s \vee [x]_a^{\geq'})$.

Note that, $\overset{j}{\underset{s=1}{\vee}}(([x]_a^{\geq})_s \vee [x]_a^{\geq'})$ in the GRIFS model based on type-II dominance relation. If x have j dominance classes $[x]_a^{\geq}$ of dominance relation $R_a^{\geq}$ on GRS, we perform a disjunction operation $\vee$ of j dominance classes $[x]_a^{\geq}$ and $[x]_a^{\geq'}$, respectively.

According to Definition 8, the following theorem can be obtained.

Theorem 2. *Let $IS^{\geq^{\Pi}} =< U, A, V, f >$ be an intuitionistic fuzzy ordered information system, and B be IFS on U. Then GRIFS model based on type-II dominance relation will have the following properties:*

$$(1)\ \underline{R_k^{\geq}}^{\Pi}(B) \subseteq B \subseteq \overline{R_k^{\geq}}^{\Pi}(B),$$
$$(2)\ A \subseteq B, \underline{R_k^{\geq}}^{\Pi}(A) \subseteq \underline{R_k^{\geq}}^{\Pi}(B),\ \overline{R_k^{\geq}}^{\Pi}(A) \subseteq \overline{R_k^{\geq}}^{\Pi}(B),$$
$$(3)\ \underline{R_k^{\geq}}^{\Pi}(A \cap B) = \underline{R_k^{\geq}}^{\Pi}(A) \cap \underline{R_k^{\geq}}^{\Pi}(B), \overline{R_k^{\geq}}^{\Pi}(A \cup B) = \overline{R_k^{\geq}}^{\Pi}(A) \cup \overline{R_k^{\geq}}^{\Pi}(B).$$

Proof. The proving process of Theorem 2 is similar to Theorem 1. □

Example 3. *Nine senior university students are going to graduate from a computer department and they want to work for a famous internet company. Let $U = \{x_1, x_2, x_3, x_4, x_5, x_6, x_7, x_8, x_9\}$ be the domain. The company has a campus recruitment at this university. Based on their confidence in programming skills, we get the following IFS B whether they succeed in the campus recruitment or not. At the same time, according to programming skills grades in school, $U/R_a^{\geq} = \{\{x_1, x_2, x_4\}, \{x_4, x_5, x_6, x_9\}, \{x_3, x_8\}, \{x_7\}\}$ can be obtained. We suppose $X = \{x_1, x_4, x_5, x_6, x_9\}$, $X \subseteq U$.*

$$B = \left\{ \frac{[0.9, 0]}{x_1}, \frac{[0.8, 0.1]}{x_2}, \frac{[0.65, 0.3]}{x_3}, \frac{[0.85, 0.1]}{x_4}, \frac{[0.95, 0.05]}{x_5}, \frac{[0.7, 0.3]}{x_6}, \frac{[0.5, 0.2]}{x_7}, \frac{[0.87, 0.1]}{x_8}, \frac{[0.75, 0.2]}{x_9} \right\}.$$

We can try to use rough sets based on dominance relation to solve the above problems, as described in Reference [38].

First, according to $U/R_a^{\geq}$, we can get the result as follows.

$$\underline{R^{\geq}}(X) = \{x_4, x_5, x_6, x_9\},\ \overline{R^{\geq}}(X) = \{x_1, x_2, x_4, x_5, x_6, x_9\},$$

From the upper and lower approximations, we can get that x_4, x_5, x_6 and x_9 students may pass the campus interview. However, we cannot use the rough set based on dominance relation to deal with the data of the test scores of their programming skills. In order to process B, we need to use another model, outlined in Reference [40]. The result is as follows:

$$x_5 \geq x_1 \geq x_8 \geq x_4 \geq x_2 \geq x_9 \geq x_6 \geq x_3 \geq x_7$$

Through IFS, we can get that x_4, x_2, x_1 and x_7 students are better than other students. From the above analysis, we can get student x_4 who can be successful in the interview. However, we are not sure about other students. At the same time, from the process of analysis, we find that different models are built for the examples, and the predicted results will have deviation. Our model is based on GRS based on dominance relation and the dominance relation in intuitionistic fuzzy ordered information system. Thus, we can use the model to predict the campus interview.

Consequently, according to Definition 8 and Example 1, we can compute the disjunction operation of $[x]_{\tilde{a}}^{\geq}$ and $[x]_{\tilde{a}}^{\geq'}$, the results are as Table 2.

Table 2. The disjunction operation of $[x]_{\tilde{a}}^{\geq}$ and $[x]_{\tilde{a}}^{\geq'}$.

x	$[x]_{\tilde{a}}^{\geq}$	$[x]_{\tilde{a}}^{\geq'}$	$[x]_{\tilde{a}}^{\geq} \vee [x]_{\tilde{a}}^{\geq'}$
x_1	$\{x_1, x_2, x_4\}$	$\{x_1\}$	$\{x_1, x_2, x_4\}$
x_2	$\{x_1, x_2, x_4\}$	$\{x_2, x_4, x_5, x_8\}$	$\{x_1, x_2, x_4, x_5, x_8\}$
x_3	$\{x_3, x_8\}$	$\{x_3, x_4, x_5, x_6, x_8, x_9\}$	$\{x_3, x_4, x_5, x_6, x_8, x_9\}$
x_4	$\{x_1, x_2, x_4\}, \{x_4, x_5, x_6, x_9\}$	$\{x_4, x_5\}$	$\{x_1, x_2, x_4, x_5, x_6, x_9\}$
x_5	$\{x_4, x_5, x_6, x_9\}$	$\{x_5\}$	$\{x_4, x_5, x_6, x_9\}$
x_6	$\{x_4, x_5, x_6, x_9\}$	$\{x_6, x_8, x_9\}$	$\{x_4, x_5, x_6, x_8, x_9\}$
x_7	$\{x_7\}$	$\{x_7, x_8, x_9\}$	$\{x_7, x_8, x_9\}$
x_8	$\{x_3, x_8\}$	$\{x_8\}$	$\{x_3, x_8\}$
x_9	$\{x_4, x_5, x_6, x_9\}$	$\{x_9\}$	$\{x_4, x_5, x_6, x_9\}$

GRIFS model based on type-II dominance relation can be obtained as follows:

$$\underline{R}_{\tilde{1}}^{\geq \Pi}(B) = \left\{ \frac{[0.78,0.22]}{x_1}, \frac{[0.78,0.22]}{x_2}, \frac{[0.65,0.3]}{x_3}, \frac{[0.7,0.22]}{x_4}, \frac{[0.7,0.22]}{x_5}, \frac{[0.7,0.3]}{x_6}, \frac{[0.5,0.22]}{x_7}, \frac{[0.65,0.3]}{x_8}, \frac{[0.7,0.3]}{x_9} \right\},$$
$$\overline{R}_{\tilde{1}}^{\geq \Pi}(B) = \left\{ \frac{[0.9,0]}{x_1}, \frac{[0.95,0]}{x_2}, \frac{[0.95,0.05]}{x_3}, \frac{[0.95,0]}{x_4}, \frac{[0.95,0.05]}{x_5}, \frac{[0.95,0.05]}{x_6}, \frac{[0.87,0.1]}{x_7}, \frac{[0.87,0.1]}{x_8}, \frac{[0.95,0.05]}{x_9} \right\}.$$

Through the above analysis, the students' interviews prediction can be obtained. x_4, x_2 and x_1 students are better than others. From this example, the model can help us to analyze the same situation though two kinds of dominance relations. Therefore, this example can be analyzed more comprehensively

4. Multi-Granulation GRIFS Models Based on Dominance Relation

In this section, we give the multi-granulation RIFS conception, and then propose optimistic and pessimistic multi-granulation GRIFS models based on type-I dominance relation and type-II dominance relation, respectively. These four models are constructed by multiple granularities GRIFS models based on type-I and type-II dominance relation. Finally, we discuss some properties of these models.

Definition 9 ([39]). *Let $IS = < U, A, V, f >$ be an information system, $A_1, A_2, \cdots, A_m \subseteq A$, and R_{A_i} is an equivalence relation of x in terms of attribute set A. $[x]_{A_i}$ is the equivalence class of R_{A_i}, $\forall B \subseteq U$, B is IFS. Then the optimistic multi-granulation lower and upper approximations of A_i can be defined as follows:*

$$\underline{\sum_{i=1}^{m} R^{O}_{A_i}}(B) = \{< x, \mu_{\underline{\sum_{i=1}^{m} R^{O}_{A_i}(B)}}(x), \nu_{\underline{\sum_{i=1}^{m} R^{O}_{A_i}(B)}}(x) > | x \in U\},$$

$$\overline{\sum_{i=1}^{m} R^{O}_{A_i}}(B) = \{< x, \mu_{\overline{\sum_{i=1}^{m} R^{O}_{A_i}(B)}}(x), \nu_{\overline{\sum_{i=1}^{m} R^{O}_{A_i}(B)}}(x) > | x \in U\},$$

$$\mu_{\underline{\sum_{i=1}^{m} R^{O}_{A_i}(B)}}(x) = \bigvee_{i=1}^{m} \inf_{y \in [x]_{A_i}} \mu_B(y),\ \nu_{\underline{\sum_{i=1}^{m} R^{O}_{A_i}(B)}}(x) = \bigwedge_{i=1}^{m} \sup_{y \in [x]_{A_i}} \nu_B(y),$$

$$\mu_{\overline{\sum_{i=1}^{m} R^{O}_{A_i}(B)}}(x) = \bigwedge_{i=1}^{m} \sup_{y \in [x]_{A_i}} \mu_B(y),\ \nu_{\overline{\sum_{i=1}^{m} R^{O}_{A_i}(B)}}(x) = \bigvee_{i=1}^{m} \inf_{y \in [x]_{A_i}} \nu_B(y).$$

where $[x]_{A_i}$ is the equivalence class of x in terms of the equivalence relation A_i. $[x]_{A_1}, [x]_{A_2}, \cdots, [x]_{A_m}$ are m equivalence classes, and $\vee$ is a disjunction operation.

Definition 10 ([39]). *Let $IS = < U, A, V, f >$ be an information system, $A_1, A_2, \cdots, A_m \subseteq A$, and R_{A_i} is an equivalence relation of x in terms of attribute set A. $[x]_{A_i}$ is the equivalence class of R_{A_i}, $\forall B \subseteq U$, B is IFS. Then the pessimistic multi-granulation lower and upper approximations of A_i can be easily obtained by:*

$$\underline{\sum_{i=1}^{m} R^{P}_{A_i}}(B) = \{< x, \mu_{\underline{\sum_{i=1}^{m} R^{P}_{A_i}(B)}}(x), \nu_{\underline{\sum_{i=1}^{m} R^{P}_{A_i}(B)}}(x) > | x \in U\},$$

$$\overline{\sum_{i=1}^{m} R^{P}_{A_i}}(B) = \{< x, \mu_{\overline{\sum_{i=1}^{m} R^{P}_{A_i}(B)}}(x), \nu_{\overline{\sum_{i=1}^{m} R^{P}_{A_i}(B)}}(x) > | x \in U\},$$

$$\mu_{\underline{\sum_{i=1}^{m} R^{P}_{A_i}(B)}}(x) = \bigwedge_{i=1}^{m} \inf_{y \in [x]_{A_i}} \mu_B(y),\ \nu_{\underline{\sum_{i=1}^{m} R^{P}_{A_i}(B)}}(x) = \bigvee_{i=1}^{m} \sup_{y \in [x]_{A_i}} \nu_B(y),$$

$$\mu_{\overline{\sum_{i=1}^{m} R^{P}_{A_i}(B)}}(x) = \bigvee_{i=1}^{m} \sup_{y \in [x]_{A_i}} \mu_B(y),\ \nu_{\overline{\sum_{i=1}^{m} R^{P}_{A_i}(B)}}(x) = \bigwedge_{i=1}^{m} \inf_{y \in [x]_{A_i}} \nu_B(y).$$

where $[x]_{A_i}$ is the equivalence class of x in terms of the equivalence relation A_i. $[x]_{A_1}$, $[x]_{A_2}, \cdots$, $[x]_{A_m}$are m equivalence classes, and $\wedge$ is a conjunction operation.

4.1. GRIFS Model Based on Type-I Dominance Relation

Definition 11. *Let $IS^{\geq^{I}} = < U, A, V, f >$ be an intuitionistic fuzzy ordered information system, $A_1, A_2, \cdots, A_m \subseteq A$. $(R^{\geq}_{\overline{a}})_i$ is a dominance relation of x in terms of attribute $A_i, a \in A_i$, where $([x]^{\geq}_{\overline{a}})_i$ is the dominance class of $(R^{\geq}_{\overline{a}})_i$. Suppose X is GRS of $(R^{\geq}_{\overline{a}})_i$ and B is IFS on U. IFS B with respect to attribute a satisfies dominance relation $((R')^{\geq}_{\overline{a}})_i$. Therefore, the lower and upper approximations of B with respect to the graded k are given as follows:*

When $k \geq 1$, we can get,

$$\underline{\sum_{i=1}^{m} R^{O}_{A_i}}^{\geq^{I}}_{(k)}(B) = \{< x, \mu_{\underline{\sum_{i=1}^{m} R^{O}_{A_i}}^{\geq^{I}}_{(k)}(B)}(x), \nu_{\underline{\sum_{i=1}^{m} R^{O}_{A_i}}^{\geq^{I}}_{(k)}(B)}(x) > | x \in U\},$$

$$\overline{\sum_{i=1}^{m} R^{O}_{A_i}}^{\geq^{I}}_{(k)}(B) = \{< x, \mu_{\overline{\sum_{i=1}^{m} R^{O}_{A_i}}^{\geq^{I}}_{(k)}(B)}(x), \nu_{\overline{\sum_{i=1}^{m} R^{O}_{A_i}}^{\geq^{I}}_{(k)}(B)}(x) > | x \in U\},$$

$$\mu'_{B_i}(y) = \frac{|\overline{R}^{\geq}_{k}(X) \cap \underline{R}^{\geq}_{k}(X)|}{|U|},\ \nu'_{B_i}(y) = \frac{|\neg(\overline{R}^{\geq}_{k}(X) \cup \underline{R}^{\geq}_{k}(X))|}{|U|}.$$

We can get GRS in $A_1, A_2, \cdots, A_m$, then there will be $\mu'_{B_1}(y), \mu'_{B_2}(y), \mu'_{B_3}(y), \cdots, \mu'_{B_m}(y)$ and $\nu'_{B_1}(y), \nu'_{B_2}(y), \nu'_{B_3}(y), \cdots, \nu'_{B_m}(y)$. Subsequently, we can obtain,

$$\mu_{\underline{\sum_{i=1}^{m} R^{O}_{A_i}}^{\geq^{I}}_{(k)}(B)}(x) = \bigvee_{i=1}^{m} \inf_{y \in (\bigwedge_{s=1}^{j}(([x]^{\geq}_{a})_s \wedge [x]^{\geq'}_{a}))_i} (\mu_B(y) \wedge \mu'_{B_i}(y)),\ \nu_{\underline{\sum_{i=1}^{m} R^{O}_{A_i}}^{\geq^{I}}_{(k)}(B)}(x) = \bigwedge_{i=1}^{m} \sup_{y \in (\bigwedge_{s=1}^{j}(([x]^{\geq}_{a})_s \wedge [x]^{\geq'}_{a}))_i} (\nu_B(y) \vee \nu'_{B_i}(y)),$$

$$\mu_{\overline{\sum_{i=1}^{m} R^{O}_{A_i}}^{\geq^{I}}_{(k)}(B)}(x) = \bigwedge_{i=1}^{m} \sup_{y \in (\bigwedge_{s=1}^{j}(([x]^{\geq}_{a})_s \wedge [x]^{\geq'}_{a}))_i} (\mu_B(y) \vee \mu'_{B_i}(y)),\ \nu_{\overline{\sum_{i=1}^{m} R^{O}_{A_i}}^{\geq^{I}}_{(k)}(B)}(x) = \bigvee_{i=1}^{m} \inf_{y \in (\bigwedge_{s=1}^{j}(([x]^{\geq}_{a})_s \wedge [x]^{\geq'}_{a}))_i} (\nu_B(y) \wedge \nu'_{B_i}(y)).$$

Obviously, $0 \leq \mu'_B(y) \leq 1,\ 0 \leq \nu'_B(y) \leq 1,\ j = 1, 2, \cdots, n$.

When $\underline{\sum_{i=1}^{m} R^{O}_{A_i}}^{\geq^{I}}_{(k)}(B) \neq \overline{\sum_{i=1}^{m} R^{O}_{A_i}}^{\geq^{I}}_{(k)}(B)$, B is an optimistic multi-granulation GRIFS model based on type-I dominance relation.

When $k = 0$, $\mu'_{B_i}(y)$ and $\nu'_{B_i}(y)$ are calculated through the classical rough set. However, under these circumstances the model is still valid and we call this model an optimistic multi-granulation RIFS based on type-I dominance relation.

Definition 12. *Let $IS^{\geq^{I}} = < U, A, V, f >$ be an intuitionistic fuzzy ordered information system, $A_1, A_2, \cdots, A_m \subseteq A$. $(R^{\geq}_{\bar{a}})_i$ is a dominance relation of x in terms of attribute A_i, where $([x]^{\geq}_{\bar{a}})_i$ is the dominance class of $(R^{\geq}_{\bar{a}})_i$. Suppose X is GRS of $(R^{\geq}_{\bar{a}})_i$ and B is IFS on U. IFS B about attribute a satisfies dominance relation $((R')^{\geq}_{\bar{a}})_i$, $a \in A_i$. Then the lower and upper approximations of B with respect to the graded k are given as follows:*

When $k \geq 1$, we can get,

$$\underline{\sum_{i=1}^{m} R^{p}_{A_i}}^{\geq^{I}}_{(k)}(B) = \{< x, \mu_{\underline{\sum_{i=1}^{m} R^{p}_{A_i}}^{\geq^{I}}_{(k)}(B)}(x), \nu_{\underline{\sum_{i=1}^{m} R^{p}_{A_i}}^{\geq^{I}}_{(k)}(B)}(x) > | x \in U\},$$

$$\overline{\sum_{i=1}^{m} R^{p}_{A_i}}^{\geq^{I}}_{(k)}(B) = \{< x, \mu_{\overline{\sum_{i=1}^{m} R^{p}_{A_i}}^{\geq^{I}}_{(k)}(B)}(x), \nu_{\overline{\sum_{i=1}^{m} R^{p}_{A_i}}^{\geq^{I}}_{(k)}(B)}(x) > | x \in U\},$$

$$\mu'_{B_i}(y) = \frac{|\overline{R}^{\geq}_{k}(X) \cap \underline{R}^{\geq}_{k}(X)|}{|U|},\ \nu'_{B_i}(y) = \frac{|\neg(\overline{R}^{\geq}_{k}(X) \cup \underline{R}^{\geq}_{k}(X))|}{|U|}.$$

We can obtain GRS in $A_1, A_2, \cdots, A_m$, then there will be $\mu'_{B_1}(y), \mu'_{B_2}(y), \mu'_{B_3}(y), \cdots, \mu'_{B_m}(y)$ and $\nu'_{B_1}(y), \nu'_{B_2}(y), \nu'_{B_3}(y), \cdots, \nu'_{B_m}(y)$. Subsequently, we can obtain,

$$\mu_{\underline{\sum_{i=1}^{m} R^{p}_{A_i}}^{\geq^{I}}_{(k)}(B)}(x) = \bigwedge_{i=1}^{m} \inf_{y \in (\bigwedge_{s=1}^{j}(([x]^{\geq}_{a})_s \wedge [x]^{\geq'}_{a}))_i} (\mu_B(y) \wedge \mu'_{B_i}(y)),\ \nu_{\underline{\sum_{i=1}^{m} R^{p}_{A_i}}^{\geq^{I}}_{(k)}(B)}(x) = \bigvee_{i=1}^{m} \sup_{y \in (\bigwedge_{s=1}^{j}(([x]^{\geq}_{a})_s \wedge [x]^{\geq'}_{a}))_i} (\nu_B(y) \vee \nu'_{B_i}(y)),$$

$$\mu_{\overline{\sum_{i=1}^{m} R^{p}_{A_i}}^{\geq^{I}}_{(k)}(B)}(x) = \bigvee_{i=1}^{m} \sup_{y \in (\bigwedge_{s=1}^{j}(([x]^{\geq}_{a})_s \wedge [x]^{\geq'}_{a}))_i} (\mu_B(y) \vee \mu'_{B_i}(y)),\ \nu_{\overline{\sum_{i=1}^{m} R^{p}_{A_i}}^{\geq^{I}}_{(k)}(B)}(x) = \bigwedge_{i=1}^{m} \inf_{y \in (\bigwedge_{s=1}^{j}(([x]^{\geq}_{a})_s \wedge [x]^{\geq'}_{a}))_i} (\nu_B(y) \wedge \nu'_{B_i}(y)).$$

Obviously, $0 \leq \mu'_B(y) \leq 1,\ 0 \leq \nu'_B(y) \leq 1,\ j = 1, 2, \cdots, n$.

When $\underline{\sum_{i=1}^{m} R^{p}_{A_i}}^{\geq^{I}}_{(k)}(B) \neq \overline{\sum_{i=1}^{m} R^{p}_{A_i}}^{\geq^{I}}_{(k)}(B)$, B is a pessimistic multi-granulation GRIFS model based on type-I dominance relation.

When $k = 0$, $\mu'_{B_i}(y)$ and $\nu'_{B_i}(y)$ are calculated through the classical rough set. However, under these circumstances the model is still valid and we call this model a pessimistic multi-granulation RIFS based on type-I dominance relation.

Note that, $(\overset{j}{\underset{s=1}{\wedge}}(([x]_{\bar{a}}^{\geq})_s \wedge [x]_{\bar{a}}^{\geq'}))_i$ *in multi-granulation GRIFS models based on type-I dominance relation. If x have j dominance classes* $[x]_{\bar{a}}^{\geq}$ *of dominance relation* $R_{\bar{a}}^{\geq}$ *on GRS, we perform a conjunction operation* $\wedge$ *of j dominance classes* $[x]_{\bar{a}}^{\geq}$ *and* $[x]_{\bar{a}}^{\geq'}$*, respectively.*

Note that multi-granulation GRIFS models based on type-I dominance relation are formed by combining multiple granularities GRIFS models based on type-I dominance relation.

According to Definitions 11 and 12, the following theorem can be obtained.

Theorem 3. *Let* $IS^{\geq^{\mathrm{I}}} =< U, A, V, f >$ *be an intuitionistic fuzzy ordered information system,* $A_1, A_2, \cdots, A_m \subseteq A$*, and B be IFS on U. Then the optimistic and pessimistic multi-granulation GRIFS models based on type-I dominance relation have the following properties:*

$$\underline{\sum_{i=1}^{m} R_{A_i}^{O}}_{(k)}^{\geq^{\mathrm{I}}}(B) = \bigcup_{i=1}^{m} \underline{R_{A_i}}_{(k)}^{\geq^{\mathrm{I}}}(B),\ \overline{\sum_{i=1}^{m} R_{A_i}^{O}}_{(k)}^{\geq^{\mathrm{I}}}(B) = \bigcap_{i=1}^{m} \overline{R_{A_i}}_{(k)}^{\geq^{\mathrm{I}}}(B).$$

$$\underline{\sum_{i=1}^{m} R_{A_i}^{p}}_{(k)}^{\geq^{\mathrm{I}}}(B) = \bigcap_{i=1}^{m} \underline{R_{A_i}}_{(k)}^{\geq^{\mathrm{I}}}(B),\ \overline{\sum_{i=1}^{m} R_{A_i}^{p}}_{(k)}^{\geq^{\mathrm{I}}}(B) = \bigcup_{i=1}^{m} \overline{R_{A_i}}_{(k)}^{\geq^{\mathrm{I}}}(B).$$

Proof. One can derive them from Definitions 7, 11, and 12. □

4.2. GRIFS Model Based on Type-II Dominance Relation

Definition 13. *Let* $IS^{\geq^{\Pi}} =< U, A, V, f >$ *be an intuitionistic fuzzy ordered information system,* $A_1, A_2, \cdots, A_m \subseteq A$*, and U be the universe of discourse.* $(R_{\bar{a}}^{\geq})_i$ *is a dominance relation of x in terms of attribute* A_i*,* $a \in A_i$*, where* $([x]_{\bar{a}}^{\geq})_i$ *is the dominance class of* $(R_{\bar{a}}^{\geq})_i$*. Suppose X is GRS of* $(R_{\bar{a}}^{\geq})_i$ *and B is IFS on U. IFS B about attribute a satisfies dominance relation* $((R')_{\bar{a}}^{\geq})_i$*. So the lower and upper approximations of B with respect to the graded k are given as follows:*

When $k \geq 1$*, we can get,*

$$\underline{\sum_{i=1}^{m} R_{A_i}^{O}}_{(k)}^{\geq^{\Pi}}(B) = \{< x, \mu_{\underline{\sum_{i=1}^{m} R_{A_i}^{O}}_{(k)}^{\geq^{\Pi}}(B)}(x), \nu_{\underline{\sum_{i=1}^{m} R_{A_i}^{O}}_{(k)}^{\geq^{\Pi}}(B)}(x) > | x \in U\},$$

$$\overline{\sum_{i=1}^{m} R_{A_i}^{O}}_{(k)}^{\geq^{\Pi}}(B) = \{< x, \mu_{\overline{\sum_{i=1}^{m} R_{A_i}^{O}}_{(k)}^{\geq^{\Pi}}(B)}(x), \nu_{\overline{\sum_{i=1}^{m} R_{A_i}^{O}}_{(k)}^{\geq^{\Pi}}(B)}(x) > | x \in U\},$$

$$\mu'_{B_i}(y) = \frac{|\overline{R}_k^{\geq}(X) \cap \underline{R}_k^{\geq}(X)|}{|U|},\ \nu'_{B_i}(y) = \frac{|\neg(\overline{R}_k^{\geq}(X) \cup \underline{R}_k^{\geq}(X))|}{|U|}.$$

We can obtain GRS in $A_1, A_2, \cdots, A_m$*, then there will be* $\mu'_{B_1}(y), \mu'_{B_2}(y), \mu'_{B_3}(y), \cdots, \mu'_{B_m}(y)$ *and* $\nu'_{B_1}(y), \nu'_{B_2}(y), \nu'_{B_3}(y), \cdots, \nu'_{B_m}(y)$*. Subsequently, we can obtain,*

$$\mu_{\underline{\sum_{i=1}^{m} R_{A_i}^{O}}_{(k)}^{\geq^{\Pi}}(B)}(x) = \bigvee_{i=1}^{m} \inf_{y \in (\overset{j}{\underset{s=1}{\vee}}(([x]_{\bar{a}}^{\geq})_s \vee [x]_{\bar{a}}^{\geq'}))_i} (\mu_B(y) \wedge \mu'_{B_i}(y)),\ \nu_{\underline{\sum_{i=1}^{m} R_{A_i}^{O}}_{(k)}^{\geq^{\Pi}}(B)}(x) = \bigwedge_{i=1}^{m} \sup_{y \in (\overset{j}{\underset{s=1}{\vee}}(([x]_{\bar{a}}^{\geq})_s \vee [x]_{\bar{a}}^{\geq'}))_i} (\nu_B(y) \vee \nu'_{B_i}(y)),$$

$$\mu_{\overline{\sum_{i=1}^{m} R_{A_i}^{O}}_{(k)}^{\geq^{\Pi}}(B)}(x) = \bigwedge_{i=1}^{m} \sup_{y \in (\overset{j}{\underset{s=1}{\vee}}(([x]_{\bar{a}}^{\geq})_s \vee [x]_{\bar{a}}^{\geq'}))_i} (\mu_B(y) \vee \mu'_{B_i}(y)),\ \nu_{\overline{\sum_{i=1}^{m} R_{A_i}^{O}}_{(k)}^{\geq^{\Pi}}(B)}(x) = \bigvee_{i=1}^{m} \inf_{y \in (\overset{j}{\underset{s=1}{\vee}}(([x]_{\bar{a}}^{\geq})_s \vee [x]_{\bar{a}}^{\geq'}))_i} (\nu_B(y) \wedge \nu'_{B_i}(y)).$$

Obviously, $0 \leq \mu'_B(y) \leq 1$*,* $0 \leq \nu'_B(y) \leq 1$*,* $j = 1, 2, \cdots, n$*.*

When $\underline{\sum_{i=1}^{m} R^{O}_{A_i}}^{\geq\Pi}_{(k)}(B) \neq \overline{\sum_{i=1}^{m} R^{O}_{A_i}}^{\geq\Pi}_{(k)}(B)$, *B is an optimistic multi-granulation GRIFS model based on type-II dominance relation.*

When $k = 0$, $\mu'_{B_i}(y)$ and $\nu'_{B_i}(y)$ *are calculated from the classical rough set. Under these circumstances, the model is still valid.*

Definition 14. *Let* $IS^{\geq\Pi} =< U, A, V, f >$ *be an intuitionistic fuzzy ordered information system,* $A_1, A_2, \cdots, A_m \subseteq A$. $(R^{\geq}_{\overline{a}})_i$ *is a dominance relation of x in terms of attribute* $A_i, a \in A_i$*, where* $([x]^{\geq}_{\overline{a}})_i$ *is the dominance class of* $(R^{\geq}_{\overline{a}})_i$*. Suppose X is GRS of* $(R^{\geq}_{\overline{a}})_i$ *on U and B is IFS on U. IFS B with respect to attribute a satisfies dominance relation* $((R')^{\geq}_{\overline{a}})_i$*. Then lower and upper approximations of B with respect to the graded k are as follows:*

When $k \geq 1$, *we can get,*

$$\underline{\sum_{i=1}^{m} R^{p}_{A_i}}^{\geq\Pi}_{(k)}(B) = \{< x, \mu_{\underline{\sum_{i=1}^{m} R^{p}_{A_i}}^{\geq\Pi}_{(k)}(B)}(x), \nu_{\underline{\sum_{i=1}^{m} R^{p}_{A_i}}^{\geq\Pi}_{(k)}(B)}(x) > | x \in U\},$$

$$\overline{\sum_{i=1}^{m} R^{p}_{A_i}}^{\geq\Pi}_{(k)}(B) = \{< x, \mu_{\overline{\sum_{i=1}^{m} R^{p}_{A_i}}^{\geq\Pi}_{(k)}(B)}(x), \nu_{\overline{\sum_{i=1}^{m} R^{p}_{A_i}}^{\geq\Pi}_{(k)}(B)}(x) > | x \in U\},$$

$$\mu'_{B_i}(y) = \frac{|\overline{R}^{\geq}_{k}(X) \cap \underline{R}^{\geq}_{k}(X)|}{|U|}, \quad \nu'_{B_i}(y) = \frac{|\neg(\overline{R}^{\geq}_{k}(X) \cup \underline{R}^{\geq}_{k}(X))|}{|U|}.$$

We can obtain GRS in $A_1, A_2, \cdots, A_m$*, then there will be* $\mu'_{B_1}(y), \mu'_{B_2}(y), \mu'_{B_3}(y), \cdots, \mu'_{B_m}(y)$ *and* $\nu'_{B_1}(y), \nu'_{B_2}(y), \nu'_{B_3}(y), \cdots, \nu'_{B_m}(y)$*. Subsequently, we can obtain,*

$$\mu_{\underline{\sum_{i=1}^{m} R^{p}_{A_i}}^{\geq\Pi}_{(k)}(B)}(x) = \bigwedge_{i=1}^{m} \inf_{y \in (\bigvee_{s=1}^{j}(([x]^{\geq}_{\overline{a}})_s \vee [x]^{\geq'}_{\overline{a}}))_i} (\mu_B(y) \wedge \mu'_{B_i}(y)), \quad \nu_{\underline{\sum_{i=1}^{m} R^{p}_{A_i}}^{\geq\Pi}_{(k)}(B)}(x) = \bigvee_{i=1}^{m} \sup_{y \in (\bigvee_{s=1}^{j}(([x]^{\geq}_{\overline{a}})_s \vee [x]^{\geq'}_{\overline{a}}))_i} (\nu_B(y) \vee \nu'_{B_i}(y)),$$

$$\mu_{\overline{\sum_{i=1}^{m} R^{p}_{A_i}}^{\geq\Pi}_{(k)}(B)}(x) = \bigvee_{i=1}^{m} \sup_{y \in (\bigvee_{s=1}^{j}(([x]^{\geq}_{\overline{a}})_s \vee [x]^{\geq'}_{\overline{a}}))_i} (\mu_B(y) \vee \mu'_{B_i}(y)), \quad \nu_{\overline{\sum_{i=1}^{m} R^{p}_{A_i}}^{\geq\Pi}_{(k)}(B)}(x) = \bigwedge_{i=1}^{m} \inf_{y \in (\bigvee_{s=1}^{j}(([x]^{\geq}_{\overline{a}})_s \vee [x]^{\geq'}_{\overline{a}}))_i} (\nu_B(y) \wedge \nu'_{B_i}(y)).$$

Obviously, $0 \leq \mu'_B(y) \leq 1$, $0 \leq \nu'_B(y) \leq 1$, $j = 1, 2, \cdots, n$.

When $\underline{\sum_{i=1}^{m} R^{p}_{A_i}}^{\geq\Pi}_{(k)}(B) \neq \overline{\sum_{i=1}^{m} R^{p}_{A_i}}^{\geq\Pi}_{(k)}(B)$, *B is a pessimistic multi-granulation GRIFS model based on type-II dominance relation.*

When $k = 0$, $\mu'_{B_i}(y)$ and $\nu'_{B_i}(y)$ *are calculated from the classical rough set. Under these circumstances, the model is still valid.*

Note that, in $(\bigvee_{s=1}^{j}(([x]^{\geq}_{\overline{a}})_s \vee [x]^{\geq'}_{\overline{a}}))_i$*, if x have j dominance classes* $[x]^{\geq}_{\overline{a}}$ *of dominance relation* $R^{\geq}_{\overline{a}}$ *on GRS, we perform a disjunction operation* $\vee$ *of j dominance classes* $[x]^{\geq}_{\overline{a}}$ *and* $[x]^{\geq'}_{\overline{a}}$*, respectively.*

Note that multi-granulation GRIFS models based on type-II dominance relation are formed by combining multiple granularities GRIFS models based on type-II dominance relation.

According to Definitions 13 and 14, the following theorem can be obtained.

Theorem 4. *Let $IS^{\geq^{\Pi}} =< U, A, V, f >$ be an intuitionistic fuzzy ordered information system, $A_1, A_2, \cdots, A_m \subseteq A$, and IFS $B \subseteq U$. Then optimistic and pessimistic multi-granulation GRIFS models based on type-II dominance relation have the following properties:*

$$\underline{\sum_{i=1}^{m} R^{O}_{A_i}}^{\geq^{\Pi}}_{(k)}(B) = \bigcup_{i=1}^{m} \underline{R_{A_i}}^{\geq^{\Pi}}_{(k)}(B),\ \overline{\sum_{i=1}^{m} R^{O}_{A_i}}^{\geq^{\Pi}}_{(k)}(B) = \bigcap_{i=1}^{m} \overline{R_{A_i}}^{\geq^{\Pi}}_{(k)}(B),$$

$$\underline{\sum_{i=1}^{m} R^{P}_{A_i}}^{\geq^{\Pi}}_{(k)}(B) = \bigcap_{i=1}^{m} \underline{R_{A_i}}^{\geq^{\Pi}}_{(k)}(B),\ \overline{\sum_{i=1}^{m} R^{P}_{A_i}}^{\geq^{\Pi}}_{(k)}(B) = \bigcup_{i=1}^{m} \overline{R_{A_i}}^{\geq^{\Pi}}_{(k)}(B).$$

Proof. One can derive them from Definitions 7, 13 and 14. □

5. Algorithm and Example Analysis

5.1. Algorithm

Through Examples 1–3, we can conclude that the GRIFS model is effective, and now we use multi-granulation GRIFS models based on dominance relation to predict results under the same situations again as Algorithm 1.

Algorithm 1. Computing multi-granulation GRIFS models based on dominance relation.

Input: $IS =< U, A, V, f >, X \subseteq U$, IFS $B \subseteq U$, k is a natural number
Output: Multi-granulation GRIFS models based on dominance relation
1:**if** $(U \neq \phi$ **and** $A \neq \phi)$
2: **if** can build up GRS
3: **if** $(k \geq 1\ \&\&\ i = 1$ to m $\&\&\ a \in A_i)$
4: compute $\mu'_B(y)$ and $\nu'_B(y)$, $[x]_a^{\geq}$ and $[x]_a^{\geq'}$, for each $A_i \subseteq A$;
5: then compute $\wedge$ and $\vee$ of $\mu_B(y)$ and $\mu'_{B_i}(y)$, $\nu'_{B_i}(y)$ and $\nu_B(y)$ and compute $\wedge$ and $\vee$ of $[x]_a^{\geq}$ and $[x]_a^{\geq'}$;
6: **if** $(x \in U\ \&\&\ \forall y \in (\overset{j}{\underset{s=1}{*}}(([x]_a^{\geq})_s * [x]_a^{\geq'}))_i)$
7: **for** $(i = 1$ to m $\&\&\ 1 \leq j \leq n)$
8: compute $\mu_{\underline{\sum_{i=1}^{m} R^{\Delta}_{A_i}}^{\geq^{\bullet}}_{(k)}(B)}(x)$, $\nu_{\underline{\sum_{i=1}^{m} R^{\Delta}_{A_i}}^{\geq^{\bullet}}_{(k)}(B)}(x)$, $\mu_{\overline{\sum_{i=1}^{m} R^{\Delta}_{A_i}}^{\geq^{\bullet}}_{(k)}(B)}(x)$, $\nu_{\overline{\sum_{i=1}^{m} R^{\Delta}_{A_i}}^{\geq^{\bullet}}_{(k)}(B)}(x)$;
9: **end**
10: compute $\underline{\sum_{i=1}^{m} R^{\Delta}_{A_i}}^{\geq^{\bullet}}_{(k)}(B)$, $\overline{\sum_{i=1}^{m} R^{\Delta}_{A_i}}^{\geq^{\bullet}}_{(k)}(B)$.
11: **end**
12: **end**
13: **end**
14:**else**
return NULL
15:**end**

Note that Δ represents optimistic and pessimistic and $*$ means $\wedge$ or $\vee$ operation, and in $(\overset{j}{\underset{s=1}{*}}(([x]_a^{\geq})_s * [x]_a^{\geq'}))_i$, $(\overset{j}{\underset{s=1}{*}}(([x]_a^{\geq})_s * [x]_a^{\geq'}))_i$ represent $(\overset{j}{\underset{s=1}{\wedge}}(([x]_a^{\geq})_s \wedge [x]_a^{\geq'}))_i$ or $(\overset{j}{\underset{s=1}{\vee}}(([x]_a^{\geq})_s \vee [x]_a^{\geq'}))_i$ in this algorithm. $\bullet$ represents I or II.

Through this algorithm, we next illustrate these models by example again.

5.2. An Illustrative Example

We use this example to illustrate Algorithm 1 of multi-granulation GRIFS models based on type-I and type-II dominance relation. According to Algorithm 1, we will not discuss this case where k is 0. There are 9 patients. Let $U = \{x_1, x_2, x_3, x_4, x_5, x_6, x_7, x_8, x_9\}$ be the domain. Next, we analyzed these 9 patients from these symptoms of fever and salivation. The set of condition attributes are A = {fever, salivation, streaming nose}. For fever, we can get $U/R^{\geq} = \{\{x_1, x_2, x_4\}, \{x_3, x_8\}, \{x_7\}, \{x_4, x_5, x_6, x_9\}\}$, for salivation there is $U/R^{\geq} = \{\{x_1\}, \{x_1, x_4\}, \{x_3, x_5, x_6\}, \{x_5, x_6\}, \{x_6, x_9\}, \{x_2, x_7, x_8\}\}$, and for streaming nose $U/R^{\geq} = \{\{x_1\}, \{x_1, x_2, x_4\}, \{x_3, x_5, x_6\}, \{x_4, x_6, x_7, x_9\}, \{x_2, x_7, x_8\}\}$. According to the cold disease, these patients have the have the following IFS

$$B = \left\{ \frac{[0.9,0]}{x_1}, \frac{[0.8,0.1]}{x_2}, \frac{[0.65,0.3]}{x_3}, \frac{[0.85,0.1]}{x_4}, \frac{[0.95,0.05]}{x_5}, \frac{[0.7,0.3]}{x_6}, \frac{[0.5,0.2]}{x_7}, \frac{[0.87,0.1]}{x_8}, \frac{[0.75,0.2]}{x_9} \right\}.$$

Suppose $X = \{x_1, x_4, x_5, x_6, x_9\}$, $k = 1$. Then we can obtain multi-granulation GRIFS models based on type-I and type-II dominance relation through Definitions 11–14. Results are as follows.

For fever, according to $U/R^{\geq}$, we can get,

$$\underline{R_{\overline{1}}^{\geq}}(X) = \{x_1, x_2, x_4, x_5, x_6, x_7, x_9\},\ \overline{R_{\overline{1}}^{\geq}}(X) = \{x_1, x_2, x_4, x_5, x_6, x_7, x_9\},$$
$$\mu'_{B_1}(y) = \frac{|\overline{R_{\overline{1}}^{\geq}}(X) \cap \underline{R_{\overline{1}}^{\geq}}(X)|}{|U|} = \frac{7}{9} \approx 0.78,\ \nu'_{B_1}(y) = \frac{|\neg(\overline{R_{\overline{1}}^{\geq}}(X) \cup \underline{R_{\overline{1}}^{\geq}}(X))|}{|U|} = \frac{2}{9} \approx 0.22.$$

Similarly, for salivation and streaming nose, the results are as follows:

$$\mu'_{B_2}(y) = \frac{6}{9} \approx 0.67,\ \nu'_{B_2}(y) = \frac{3}{9} \approx 0.33.\ \mu'_{B_2}(y) = \frac{8}{9} \approx 0.89,\ \nu'_{B_2}(y) = \frac{1}{9} \approx 0.11.$$

According to Definitions 11–14, we can obtain multi-granulation GRIFS models based on type-I dominance relation and type-II dominance relation.

For $\mu'_{B_1}(y)$ and $\nu'_{B_1}(y)$, $\mu'_{B_2}(y)$ and $\nu'_{B_2}(y)$ and $\mu'_{B_3}(y)$ and $\nu'_{B_3}(y)$, the results are the followings as Table 3.

Table 3. The conjunction and disjunction operation of $\mu_B(y)$ and $\mu'_{B_1}(y)$.

x	x_1	x_2	x_3	x_4	x_5	x_6	x_7	x_8	x_9
$\mu_B(y)$	0.9	0.8	0.65	0.85	0.95	0.7	0.5	0.87	0.75
$\mu_B(y)$	0	0.1	0.3	0.1	0.05	0.3	0.2	0.1	0.2
$\mu_B(y) \wedge \mu'_{B_1}(y)$	0.78	0.78	0.65	0.78	0.78	0.7	0.5	0.72	0.75
$\nu_B(y) \wedge \nu'_{B_1}(y)$	0.22	0.22	0.3	0.22	0.22	0.3	0.22	0.22	0.22
$\mu_B(y) \vee \mu'_{B_1}(y)$	0.9	0.8	0.78	0.85	0.95	0.78	0.78	0.78	0.78
$\nu_B(y) \wedge \nu'_{B_1}(y)$	0	0.1	0.22	0.1	0.05	0.22	0.2	0.1	0.2
$\mu_B(y) \wedge \mu'_{B_2}(y)$	0.67	0.67	0.65	0.67	0.67	0.67	0.5	0.6	0.67
$\nu_B(y) \vee \nu'_{B_2}(y)$	0.33	0.33	0.33	0.33	0.33	0.33	0.33	0.33	0.33
$\mu_B(y) \vee \mu'_{B_2}(y)$	0.89	0.8	0.67	0.85	0.95	0.7	0.67	0.67	0.75
$\nu_B(y) \wedge \nu'_{B_2}(y)$	0.11	0.1	0.3	0.1	0.05	0.3	0.2	0.1	0.2
$\mu_B(y) \wedge \mu'_{B_3}(y)$	0.9	0.8	0.65	0.85	0.89	0.7	0.5	0.87	0.75
$\nu_B(y) \vee \nu'_{B_3}(y)$	0	0.11	0.3	0.11	0.11	0.3	0.2	0.11	0.2
$\mu_B(y) \vee \mu'_{B_3}(y)$	0.9	0.89	0.89	0.89	0.95	0.89	0.89	0.89	0.89
$\nu_B(y) \wedge \nu'_{B_3}(y)$	0	0.1	0.11	0.1	0.05	0.11	0.11	0.1	0.11

Then, according to Definition 6, for B, we can get $[x]_a^{\geq'}$. Then, the conjunction operation of $[x]_a^{\geq}$ and $[x]_a^{\geq'}$ can be computed as Table 1.

For fever, we can get GRIFS based on type-I dominance relation as follows:

$$\underline{R}_{1}^{\geq^{\mathrm{I}}}(B)=\left\{\frac{[0.78,0.22]}{x_1},\frac{[0.78,0.22]}{x_2},\frac{[0.65,0.3]}{x_3},\frac{[0.78,0.22]}{x_4},\frac{[0.78,0.22]}{x_5},\frac{[0.7,0.3]}{x_6},\frac{[0.5,0.22]}{x_7},\frac{[0.78,0.22]}{x_8},\frac{[0.75,0.22]}{x_9}\right\},$$

$$\overline{R}_{1}^{\geq^{\mathrm{I}}}(B)=\left\{\frac{[0.9,0]}{x_1},\frac{[0.85,0.1]}{x_2},\frac{[0.78,0.1]}{x_3},\frac{[0.85,0.1]}{x_4},\frac{[0.95,0.05]}{x_5},\frac{[0.87,0.1]}{x_6},\frac{[0.78,0.1]}{x_7},\frac{[0.87,0.1]}{x_8},\frac{[0.78,0.2]}{x_9}\right\}.$$

For streaming nose, similar to Table 1, we can obtain $[x]_{\tilde{a}}^{\geq}\wedge[x]_{\tilde{a}}^{\geq'}$ as Table 4.

Table 4. The conjunction operation of $[x]_{\tilde{a}}^{\geq}$ and $[x]_{\tilde{a}}^{\geq'}$.

x	$[x]_{\tilde{a}}^{\geq}$	$[x]_{\tilde{a}}^{\geq'}$	$[x]_{\tilde{a}}^{\geq}\wedge[x]_{\tilde{a}}^{\geq'}$
x_1	$\{x_1\},\{x_1,x_2,x_4\}$	$\{x_1\}$	$\{x_1\}$
x_2	$\{x_1,x_2,x_4\},\{x_2,x_7,x_8\}$	$\{x_2,x_4,x_5,x_8\}$	$\{x_2\}$
x_3	$\{x_3,x_5,x_6\}$	$\{x_3,x_4,x_5,x_6,x_8,x_9\}$	$\{x_3,x_5,x_6\}$
x_4	$\{x_1,x_2,x_4\},\{x_4,x_6,x_7,x_9\}$	$\{x_4,x_5\}$	$\{x_4\}$
x_5	$\{x_3,x_5,x_6\}$	$\{x_5\}$	$\{x_5\}$
x_6	$\{x_3,x_5,x_6\},\{x_4,x_6,x_7,x_9\}$	$\{x_6,x_8,x_9\}$	$\{x_6\}$
x_7	$\{x_2,x_7,x_8\},\{x_4,x_6,x_7,x_9\}$	$\{x_7,x_8,x_9\}$	$\{x_7\}$
x_8	$\{x_2,x_7,x_8\}$	$\{x_8\}$	$\{x_8\}$
x_9	$\{x_4,x_6,x_7,x_9\}$	$\{x_9\}$	$\{x_9\}$

For salivation, similar to Table 1, we can obtain $[x]_{\tilde{a}}^{\geq}\wedge[x]_{\tilde{a}}^{\geq'}$ as Table 5.

Table 5. The conjunction operation of $[x]_{\tilde{a}}^{\geq}$ and $[x]_{\tilde{a}}^{\geq'}$.

x	$[x]_{\tilde{a}}^{\geq}$	$[x]_{\tilde{a}}^{\geq'}$	$[x]_{\tilde{a}}^{\geq}\wedge[x]_{\tilde{a}}^{\geq'}$
x_1	$\{x_1,x_2\}$	$\{x_1\}$	$\{x_1\}$
x_2	$\{x_1,x_2\},\{x_2,x_4\}$	$\{x_2,x_4,x_5,x_8\}$	$\{x_2\}$
x_3	$\{x_3,x_8\}$	$\{x_3,x_4,x_5,x_6,x_8,x_9\}$	$\{x_3,x_8\}$
x_4	$\{x_2,x_4\}$	$\{x_4,x_5\}$	$\{x_4\}$
x_5	$\{x_5,x_6\}$	$\{x_5\}$	$\{x_5\}$
x_6	$\{x_5,x_6\}$	$\{x_6,x_8,x_9\}$	$\{x_6\}$
x_7	$\{x_7,x_9\},\{x_7,x_8,x_9\}$	$\{x_7,x_8,x_9\}$	$\{x_7,x_9\}$
x_8	$\{x_3,x_8\},\{x_7,x_8,x_9\}$	$\{x_8\}$	$\{x_8\}$
x_9	$\{x_7,x_9\},\{x_7,x_8,x_9\}$	$\{x_9\}$	$\{x_9\}$

For streaming nose, we can get GRIFS based on type-I dominance relation as follows:

$$\underline{R}_{1}^{\geq^{\mathrm{I}}}(B)=\left\{\frac{[0.9,0.11]}{x_1},\frac{[0.8,0.11]}{x_2},\frac{[0.65,0.3]}{x_3},\frac{[0.85,0.11]}{x_4},\frac{[0.89,0.11]}{x_5},\frac{[0.7,0.3]}{x_6},\frac{[0.5,0.2]}{x_7},\frac{[0.87,0.11]}{x_8},\frac{[0.75,0.2]}{x_9}\right\},$$

$$\overline{R}_{1}^{\geq^{\mathrm{I}}}(B)=\left\{\frac{[0.9,0]}{x_1},\frac{[0.89,0.1]}{x_2},\frac{[0.95,0.05]}{x_3},\frac{[0.89,0.1]}{x_4},\frac{[0.95,0.05]}{x_5},\frac{[0.89,0.11]}{x_6},\frac{[0.89,0.11]}{x_7},\frac{[0.89,0.1]}{x_8},\frac{[0.89,0.11]}{x_9}\right\}.$$

For salivation, we can get GRIFS based on type-I dominance relation as follows:

$$\underline{R}_{1}^{\geq^{\mathrm{I}}}(B)=\left\{\frac{[0.67,0.33]}{x_1},\frac{[0.67,0.33]}{x_2},\frac{[0.65,0.33]}{x_3},\frac{[0.67,0.33]}{x_4},\frac{[0.67,0.33]}{x_5},\frac{[0.67,0.33]}{x_6},\frac{[0.67,0.33]}{x_7},\frac{[0.67,0.33]}{x_8},\frac{[0.67,0.33]}{x_9}\right\},$$

$$\overline{R}_{1}^{\geq^{\mathrm{I}}}(B)=\left\{\frac{[0.9,0]}{x_1},\frac{[0.8,0.1]}{x_2},\frac{[0.87,0.1]}{x_3},\frac{[0.85,0.1]}{x_4},\frac{[0.95,0.05]}{x_5},\frac{[0.7,0.3]}{x_6},\frac{[0.75,0.2]}{x_7},\frac{[0.87,0.1]}{x_8},\frac{[0.75,0.2]}{x_9}\right\}.$$

For $[x_2]_{\bar{a}}^{\geq} = \{x_1, x_2\}$, $[x_2]_{\bar{a}}^{\geq} = \{x_2, x_4\}$ and $[x_2]_{\bar{a}}^{\geq'} = \{x_2, x_4, x_5, x_8\}$, based on Definitions 11–14, we should perform the conjunction operation of them, respectively.

$$([x_1]_{\bar{a}}^{\geq} \wedge [x_1]_{\bar{a}}^{\geq'}) \wedge ([x_1]_{\bar{a}}^{\geq} \wedge [x_1]_{\bar{a}}^{\geq'}) = (\{x_1\} \wedge \{x_1, x_1, x_4\}) \wedge (\{x_1, x_2\} \wedge \{x_1, x_2, x_4\}) = \{x_1\} \wedge \{x_1, x_2\} = \{x_1\}.$$

Similarly, for x_2, x_4, x_6 and x_7, we can get the results as Tables 4 and 5.

Therefore, according to the Definitions 11 and 12 and the above calculations, we can get multi-granulation GRIFS models based on a type-I dominance relation as follows:

$$\underline{\sum_{i=1}^{3} R_{A_i}^{O}}_{1}^{\geq^{I}}(B) = \left\{ \frac{[0.89,0.11]}{x_1}, \frac{[0.8,0.11]}{x_2}, \frac{[0.65,0.3]}{x_3}, \frac{[0.85,0.11]}{x_4}, \frac{[0.89,0.11]}{x_5}, \frac{[0.7,0.3]}{x_6}, \frac{[0.67,0.2]}{x_7}, \frac{[0.87,0.11]}{x_8}, \frac{[0.75,0.2]}{x_9} \right\},$$

$$\overline{\sum_{i=1}^{3} R_{A_i}^{O}}_{1}^{\geq^{I}}(B) = \left\{ \frac{[0.9,0]}{x_1}, \frac{[0.8,0.1]}{x_2}, \frac{[0.78,0.1]}{x_3}, \frac{[0.85,0.1]}{x_4}, \frac{[0.95,0.05]}{x_5}, \frac{[0.7,0.3]}{x_6}, \frac{[0.75,0.2]}{x_7}, \frac{[0.87,0.1]}{x_8}, \frac{[0.75,0.2]}{x_9} \right\}.$$

$$\underline{\sum_{i=1}^{3} R_{A_i}^{P}}_{1}^{\geq^{I}}(B) = \left\{ \frac{[0.67,0.33]}{x_1}, \frac{[0.67,0.33]}{x_2}, \frac{[0.65,0.33]}{x_3}, \frac{[0.67,0.33]}{x_4}, \frac{[0.67,0.33]}{x_5}, \frac{[0.67,0.33]}{x_6}, \frac{[0.5,0.33]}{x_7}, \frac{[0.67,0.33]}{x_8}, \frac{[0.67,0.33]}{x_9} \right\},$$

$$\overline{\sum_{i=1}^{3} R_{A_i}^{P}}_{1}^{\geq^{I}}(B) = \left\{ \frac{[0.89,0]}{x_1}, \frac{[0.85,0.1]}{x_2}, \frac{[0.95,0.05]}{x_3}, \frac{[0.85,0.1]}{x_4}, \frac{[0.95,0.05]}{x_5}, \frac{[0.87,0.05]}{x_6}, \frac{[0.78,0.1]}{x_7}, \frac{[0.87,0.1]}{x_8}, \frac{[0.78,0.1]}{x_9} \right\}.$$

From the above results, Figures 1 and 2 can be drawn as follows:

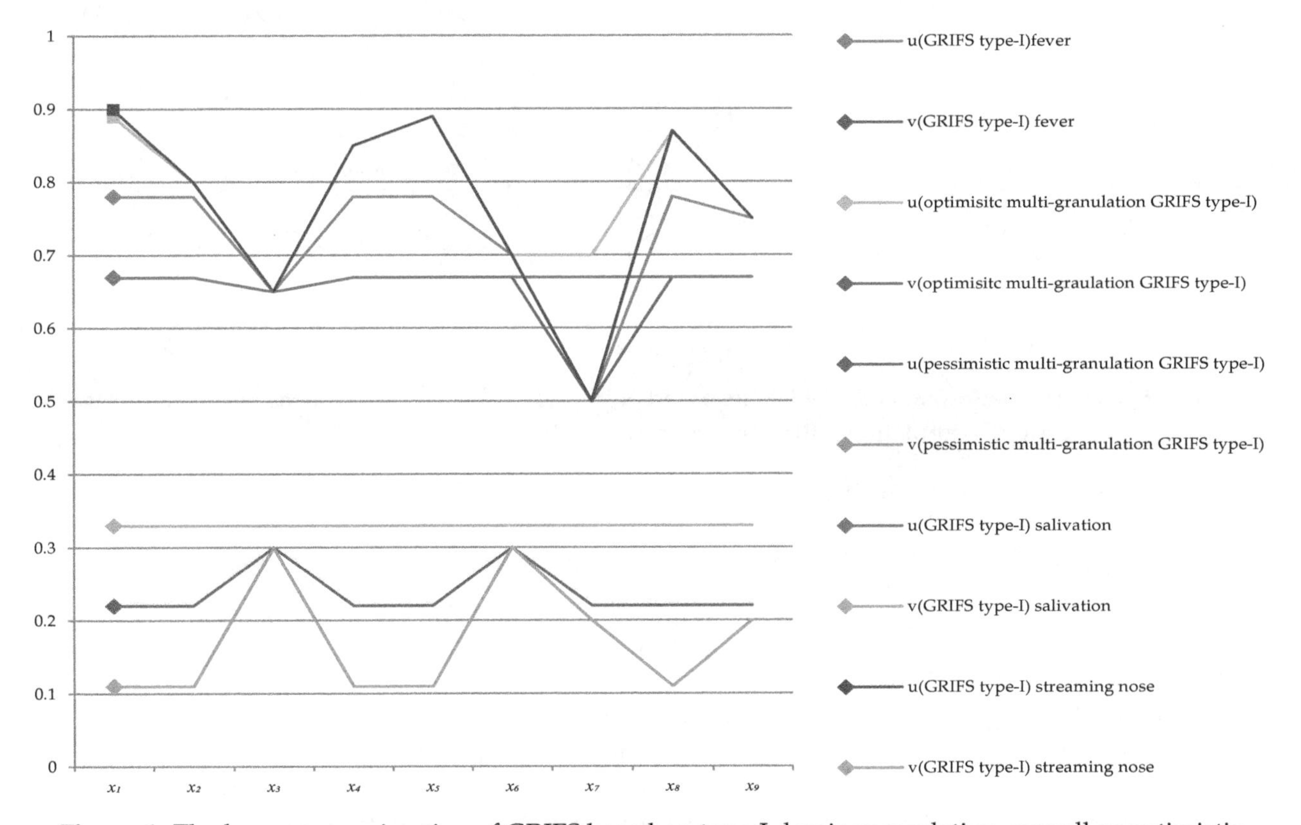

Figure 1. The lower approximation of GRIFS based on type-I dominance relation, as well as optimistic and pessimistic multi-granulation GRIFS based on type-I dominance relation.

For Figure 1, we can obtain,

$$\mu(y)_{OI1} \geq \mu(y)_{GIn1} \Theta\, \mu(y)_{GIf1} \Theta(y)_{GIs1} \geq \mu(y)_{PI1}, \nu(y)_{GIs1} \geq \nu(y)_{PI1} \geq \nu(y)_{OI1} = \nu(y)_{GIf1} \geq \nu(y)_{GIn1};$$

Note:

Θ represents $\leq$ or $\geq$;
$\mu(y)_{GIf1}$ and $\nu(y)_{GIf1}$ represent GRIFS type-I dominance relation (fever);
$\mu(y)_{GIs1}$ and $\nu(y)_{GIs1}$ represent GRIFS type-I dominance relation (salivation);
$\mu(y)_{GIn1}$ and $\nu(y)_{GIn1}$ represent GRIFS type-I dominance relation (streaming nose);
$\mu(y)_{OI1}$ and $\nu(y)_{OI1}$ represent optimistic multi-granulation GRIFS type-I dominance relation;
$\mu(y)_{PI1}$ and $\nu(y)_{PI1}$ represent pessimistic multi-granulation GRIFS type-I dominance relation;

From Figure 1, we can get that x_1, x_2, x_4, x_5 and x_8 patients have the disease, and x_7 patients do not have the disease.

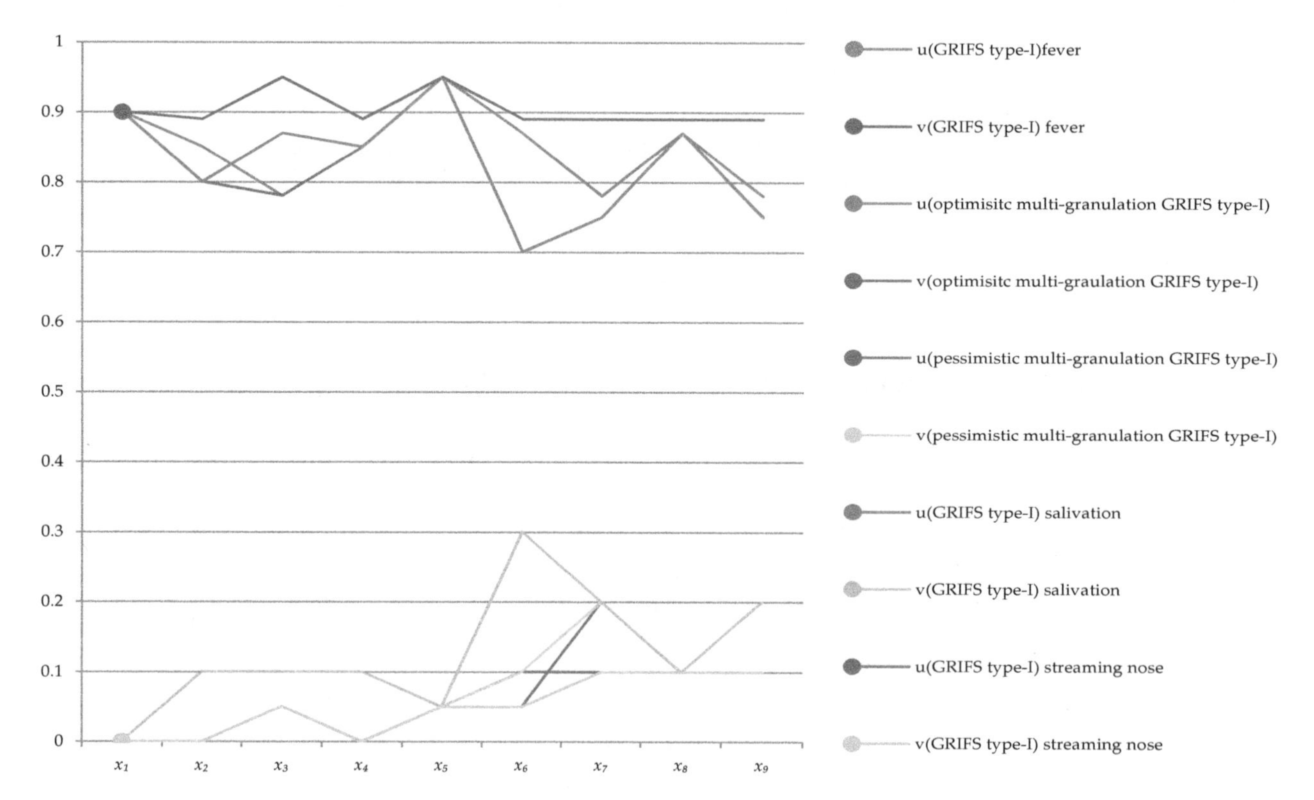

Figure 2. The upper approximation of GRIFS based on type-I dominance relation, as well as optimistic and pessimistic multi-granulation GRIFS based on type-I dominance relation.

Then, from Figure 2, we can obtain,

$$\mu(y)_{OI2} = \mu(y)_{GIn2} \geq \mu(y)_{PI2} \Theta\ \mu(y)_{GIs2} \Theta\ \mu(y)_{GIf2}, \nu(y)_{GIs2} \geq \nu(y)_{OI2} \Theta\ \nu(y)_{PI2} \Theta\ \nu(y)_{GIf2} \geq \nu(y)_{GIn2};$$

Note:

Θ represents $\leq$ or $\geq$;
$\mu(y)_{GIf2}$ and $\nu(y)_{GIf2}$ represent GRIFS type-I dominance relation (fever);
$\mu(y)_{GIs2}$ and $\nu(y)_{GIs2}$ represent GRIFS type-I dominance relation (salivation);
$\mu(y)_{GIn2}$ and $\nu(y)_{GIn2}$ represent GRIFS type-I dominance relation (streaming nose);
$\mu(y)_{OI2}$ and $\nu(y)_{OI2}$ represent optimistic multi-granulation GRIFS type-I dominance relation;
$\mu(y)_{PI2}$ and $\nu(y)_{PI2}$ represent pessimistic multi-granulation GRIFS type-I dominance relation;

From Figure 2, we can get that x_1, x_2, x_3, x_4, x_5, x_6, x_8, and x_9 patients have the disease, and x_6 patients do not have the disease.

For multi-granulation GRIFS models based on type-II dominance relation, the calculations for this model are similar to multi-granulation GRIFS models based on type-I dominance relation.

Firstly, for streaming nose, we can compute the disjunction operation of $[x]_a^{\geq}$ and $[x]_a^{\geq'}$, and the results are as Table 6.

Table 6. The disjunction operation of $[x]_a^{\geq}$ and $[x]_a^{\geq'}$.

x	$[x]_a^{\geq}$	$[x]_a^{\geq'}$	$[x]_a^{\geq} \vee [x]_a^{\geq'}$
x_1	$\{x_1\}, \{x_1, x_2, x_4\}$	$\{x_1\}$	$\{x_1, x_2, x_4\}$
x_2	$\{x_1, x_2, x_4\}, \{x_2, x_7, x_8\}$	$\{x_2, x_4, x_5, x_8\}$	$\{x_1, x_2, x_4, x_5, x_7, x_8\}$
x_3	$\{x_3, x_5, x_6\}$	$\{x_3, x_4, x_5, x_6, x_8, x_9\}$	$\{x_3, x_4, x_5, x_6, x_8, x_9\}$
x_4	$\{x_1, x_2, x_4\}, \{x_4, x_6, x_7, x_9\}$	$\{x_4, x_5\}$	$\{x_1, x_2, x_4, x_5, x_6, x_7, x_9\}$
x_5	$\{x_3, x_5, x_6\}$	$\{x_5\}$	$\{x_3, x_5, x_6\}$
x_6	$\{x_3, x_5, x_6\}, \{x_4, x_6, x_7, x_9\}$	$\{x_6, x_8, x_9\}$	$\{x_3, x_4, x_5, x_6, x_7, x_8, x_9\}$
x_7	$\{x_2, x_7, x_8\}, \{x_4, x_6, x_7, x_9\}$	$\{x_7, x_8, x_9\}$	$\{x_2, x_4, x_6, x_7, x_8, x_9\}$
x_8	$\{x_2, x_7, x_8\}$	$\{x_8\}$	$\{x_2, x_7, x_8\}$
x_9	$\{x_4, x_6, x_7, x_9\}$	$\{x_9\}$	$\{x_4, x_6, x_7, x_9\}$

Next, for salivation, we can compute the disjunction operation of $[x]_a^{\geq}$ and $[x]_a^{\geq'}$, and the results are as Table 7.

Table 7. The disjunction operation of $[x]_a^{\geq}$ and $[x]_a^{\geq'}$.

x	$[x]_a^{\geq}$	$[x]_a^{\geq'}$	$[x]_a^{\geq} \vee [x]_a^{\geq'}$
x_1	$\{x_1, x_2\}$	$\{x_1\}$	$\{x_1, x_2\}$
x_2	$\{x_1, x_2\}, \{x_2, x_4\}$	$\{x_2, x_4, x_5, x_8\}$	$\{x_1, x_2, x_4, x_5, x_8\}$
x_3	$\{x_3, x_8\}$	$\{x_3, x_4, x_5, x_6, x_8, x_9\}$	$\{x_3, x_4, x_5, x_6, x_8, x_9\}$
x_4	$\{x_2, x_4\}$	$\{x_4, x_5\}$	$\{x_2, x_4, x_5\}$
x_5	$\{x_5, x_6\}$	$\{x_5\}$	$\{x_5, x_6\}$
x_6	$\{x_5, x_6\}$	$\{x_6, x_8, x_9\}$	$\{x_5, x_6, x_8, x_9\}$
x_7	$\{x_7, x_9\}, \{x_7, x_8, x_9\}$	$\{x_7, x_8, x_9\}$	$\{x_7, x_8, x_9\}$
x_8	$\{x_3, x_8\}, \{x_7, x_8, x_9\}$	$\{x_8\}$	$\{x_3, x_7, x_8, x_9\}$
x_9	$\{x_7, x_9\}, \{x_7, x_8, x_9\}$	$\{x_9\}$	$\{x_7, x_8, x_9\}$

Then, for fever, we compute the disjunction operation of $[x]_a^{\geq}$ and $[x]_a^{\geq'}$, and these results are shown as Table 8.

Table 8. The disjunction operation of $[x]_a^{\geq}$ and $[x]_a^{\geq'}$.

x	$[x]_a^{\geq}$	$[x]_a^{\geq'}$	$[x]_a^{\geq} \vee [x]_a^{\geq'}$
x_1	$\{x_1, x_2, x_4\}$	$\{x_1\}$	$\{x_1, x_2, x_4\}$
x_2	$\{x_1, x_2, x_4\}$	$\{x_2, x_4, x_5, x_8\}$	$\{x_1, x_2, x_4, x_5, x_8\}$
x_3	$\{x_3, x_8\}$	$\{x_3, x_4, x_5, x_6, x_8, x_9\}$	$\{x_3, x_4, x_5, x_6, x_8, x_9\}$
x_4	$\{x_1, x_2, x_4\}, \{x_4, x_5, x_6, x_9\}$	$\{x_4, x_5\}$	$\{x_1, x_2, x_4, x_5, x_6, x_9\}$
x_5	$\{x_4, x_5, x_6, x_9\}$	$\{x_5\}$	$\{x_4, x_5, x_6, x_9\}$
x_6	$\{x_4, x_5, x_6, x_9\}$	$\{x_6, x_8, x_9\}$	$\{x_4, x_5, x_6, x_8, x_9\}$
x_7	$\{x_7\}$	$\{x_7, x_8, x_9\}$	$\{x_7, x_8, x_9\}$
x_8	$\{x_3, x_8\}$	$\{x_8\}$	$\{x_3, x_8\}$
x_9	$\{x_4, x_5, x_6, x_9\}$	$\{x_9\}$	$\{x_4, x_5, x_6, x_9\}$

For streaming nose, we can get GRIFS based on type-II dominance relation,

$$\underline{R_1}^{\geq^{\Pi}}(B) = \left\{ \frac{[0.8,0.11]}{x_1}, \frac{[0.5,0.2]}{x_2}, \frac{[0.65,0.3]}{x_3}, \frac{[0.5,0.3]}{x_4}, \frac{[0.65,0.3]}{x_5}, \frac{[0.5,0.3]}{x_6}, \frac{[0.5,0.3]}{x_7}, \frac{[0.5,0.2]}{x_8}, \frac{[0.5,0.3]}{x_9} \right\},$$
$$\overline{R_1}^{\geq^{\Pi}}(B) = \left\{ \frac{[0.9,0]}{x_1}, \frac{[0.95,0]}{x_2}, \frac{[0.95,0.05]}{x_3}, \frac{[0.95,0]}{x_4}, \frac{[0.95,0.05]}{x_5}, \frac{[0.95,0.05]}{x_6}, \frac{[0.89,0.1]}{x_7}, \frac{[0.89,0.1]}{x_8}, \frac{[0.89,0.1]}{x_9} \right\}.$$

For salivation, we can get GRIFS based on type-II dominance relation,

$$\underline{R}_1^{\geq\Pi}(B) = \left\{ \frac{[0.67,0.33]}{x_1}, \frac{[0.67,0.33]}{x_2}, \frac{[0.65,0.33]}{x_3}, \frac{[0.67,0.33]}{x_4}, \frac{[0.67,0.33]}{x_5}, \frac{[0.67,0.33]}{x_6}, \frac{[0.5,0.33]}{x_7}, \frac{[0.5,0.33]}{x_8}, \frac{[0.5,0.33]}{x_9} \right\},$$

$$\overline{R}_1^{\geq\Pi}(B) = \left\{ \frac{[0.9,0]}{x_1}, \frac{[0.95,0]}{x_2}, \frac{[0.95,0.05]}{x_3}, \frac{[0.95,0.05]}{x_4}, \frac{[0.95,0.05]}{x_5}, \frac{[0.95,0.05]}{x_6}, \frac{[0.87,0.1]}{x_7}, \frac{[0.87,0.1]}{x_8}, \frac{[0.87,0.1]}{x_9} \right\}.$$

For fever, GRIFS type-II dominance relation can be calculated as follows:

$$\underline{R}_1^{\geq\Pi}(B) = \left\{ \frac{[0.78,0.22]}{x_1}, \frac{[0.78,0.22]}{x_2}, \frac{[0.65,0.3]}{x_3}, \frac{[0.7,0.22]}{x_4}, \frac{[0.7,0.22]}{x_5}, \frac{[0.7,0.3]}{x_6}, \frac{[0.5,0.22]}{x_7}, \frac{[0.65,0.3]}{x_8}, \frac{[0.7,0.3]}{x_9} \right\},$$

$$\overline{R}_1^{\geq\Pi}(B) = \left\{ \frac{[0.9,0]}{x_1}, \frac{[0.95,0]}{x_2}, \frac{[0.95,0.05]}{x_3}, \frac{[0.95,0]}{x_4}, \frac{[0.95,0.05]}{x_5}, \frac{[0.95,0.05]}{x_6}, \frac{[0.87,0.1]}{x_7}, \frac{[0.87,0.1]}{x_8}, \frac{[0.95,0.05]}{x_9} \right\}.$$

Based on Definitions 13 and 14, the condition of these patients based on multi-granulation GRIFS type-II dominance relation can be obtained as follows:

$$\underline{\sum_{i=1}^{3} R_{A_i}^{O}}_{1}^{\geq\Pi}(B) = \left\{ \frac{[0.8,0.11]}{x_1}, \frac{[0.78,0.2]}{x_2}, \frac{[0.65,0.3]}{x_3}, \frac{[0.7,0.22]}{x_4}, \frac{[0.7,0.3]}{x_5}, \frac{[0.7,0.3]}{x_6}, \frac{[0.5,0.22]}{x_7}, \frac{[0.65,0.2]}{x_8}, \frac{[0.7,0.3]}{x_9} \right\},$$

$$\overline{\sum_{i=1}^{3} R_{A_i}^{O}}_{1}^{\geq\Pi}(B) = \left\{ \frac{[0.9,0]}{x_1}, \frac{[0.95,0]}{x_2}, \frac{[0.95,0.05]}{x_3}, \frac{[0.95,0.05]}{x_4}, \frac{[0.95,0.05]}{x_5}, \frac{[0.95,0.05]}{x_6}, \frac{[0.87,0.1]}{x_7}, \frac{[0.87,0.1]}{x_8}, \frac{[0.87,0.1]}{x_9} \right\}.$$

$$\underline{\sum_{i=1}^{3} R_{A_i}^{P}}_{1}^{\geq\Pi}(B) = \left\{ \frac{[0.67,0.33]}{x_1}, \frac{[0.5,0.33]}{x_2}, \frac{[0.65,0.33]}{x_3}, \frac{[0.5,0.4]}{x_4}, \frac{[0.65,0.33]}{x_5}, \frac{[0.5,0.33]}{x_6}, \frac{[0.5,0.33]}{x_7}, \frac{[0.5,0.33]}{x_8}, \frac{[0.5,0.33]}{x_9} \right\},$$

$$\overline{\sum_{i=1}^{3} R_{A_i}^{P}}_{1}^{\geq\Pi}(B) = \left\{ \frac{[0.9,0]}{x_1}, \frac{[0.95,0]}{x_2}, \frac{[0.95,0.05]}{x_3}, \frac{[0.95,0]}{x_4}, \frac{[0.95,0.05]}{x_5}, \frac{[0.95,0.05]}{x_6}, \frac{[0.89,0.1]}{x_7}, \frac{[0.89,0.1]}{x_8}, \frac{[0.95,0.05]}{x_9} \right\}.$$

Then, from Figure 3, we can obtain,

$$\mu(y)_{O\Pi 3} \geq \mu(y)_{G\Pi f3} \Theta\, \mu(y)_{G\Pi n3} \Theta\, \mu(y)_{G\Pi s3} \geq \mu(y)_{P\Pi 3}, \nu(y)_{P\Pi 3} \geq \nu(y)_{G\Pi s3} \Theta\, \nu(y)_{O\Pi 3} \Theta\, \nu(y)_{G\Pi f3} \Theta\, \nu(y)_{G\Pi n3};$$

Note:

Θ represents $\leq$ or $\geq$;
$\mu(y)_{G\Pi f3}$ and $\nu(y)_{G\Pi f3}$ represent GRIFS type-II dominance relation (fever);
$\mu(y)_{G\Pi s3}$ and $\nu(y)_{G\Pi s3}$ represent GRIFS type-II dominance relation (salivation);
$\mu(y)_{G\Pi n3}$ and $\nu(y)_{G\Pi n3}$ represent GRIFS type-II dominance relation (streaming nose);
$\mu(y)_{O\Pi 3}$ and $\nu(y)_{O\Pi 3}$ represent optimistic multi-granulation GRIFS type-II dominance relation;
$\mu(y)_{P\Pi 3}$ and $\nu(y)_{P\Pi 3}$ represent pessimistic multi-granulation GRIFS type-II dominance relation;

From Figure 3, we can see that x_1, x_2, x_4 patients have the disease, and $x_3, x_5, x_6, x_7, x_8, x_9$ patients do not have the disease.

Then, from Figure 4, we can obtain,

$$\mu(y)_{O\Pi 4} \geq \mu(y)_{G\Pi n4} \Theta\, \mu(y)_{G\Pi f4} \Theta\, \mu(y)_{G\Pi s4} \geq \mu(y)_{P\Pi 4}, \nu(y)_{P\Pi 4} \Theta\, \nu(y)_{G\Pi s4} \Theta\, \nu(y)_{G\Pi f4} \Theta\, \nu(y)_{G\Pi n4} \Theta\, \nu(y)_{O\Pi 4};$$

Note:

Θ represents $\leq$ or $\geq$;
$\mu(y)_{G\Pi f4}$ and $\nu(y)_{G\Pi f4}$ represent GRIFS type-II dominance relation (fever);
$\mu(y)_{G\Pi s4}$ and $\nu(y)_{G\Pi s4}$ represent GRIFS type-II dominance relation (salivation);
$\mu(y)_{G\Pi n4}$ and $\nu(y)_{G\Pi n4}$ represent GRIFS type-II dominance relation (streaming nose);
$\mu(y)_{O\Pi 4}$ and $\nu(y)_{O\Pi 4}$ represent optimistic multi-granulation GRIFS type-II dominance relation;
$\mu(y)_{P\Pi 4}$ and $\nu(y)_{P\Pi 4}$ represent pessimistic multi-granulation GRIFS type-II dominance relation;

From Figure 4, we can see that x_1, x_2, x_3, x_4, x_5, x_6, x_7, x_8 and x_9 patients have the disease.

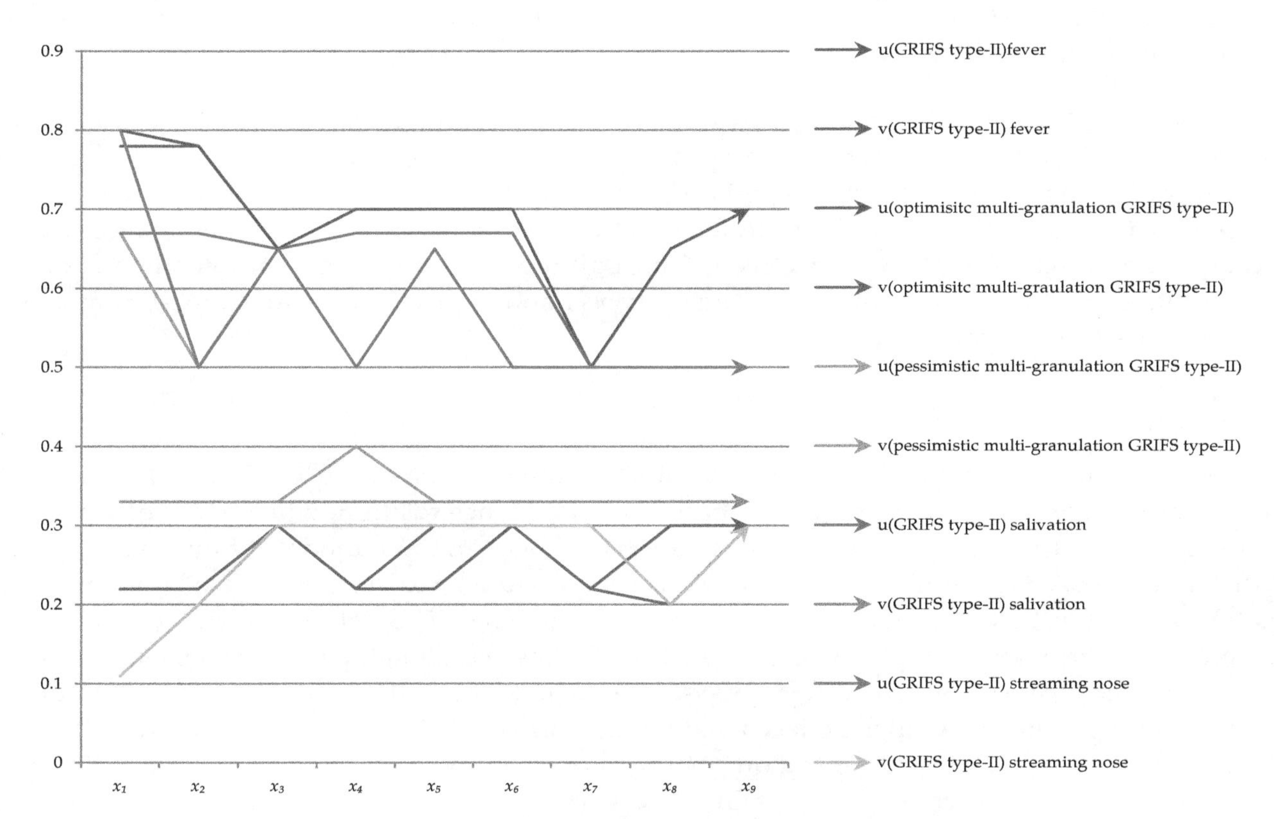

Figure 3. The lower approximation of GRIFS based on type-II dominance relation, as well as optimistic and pessimistic multi-granulation GRIFS based on type-II dominance relation.

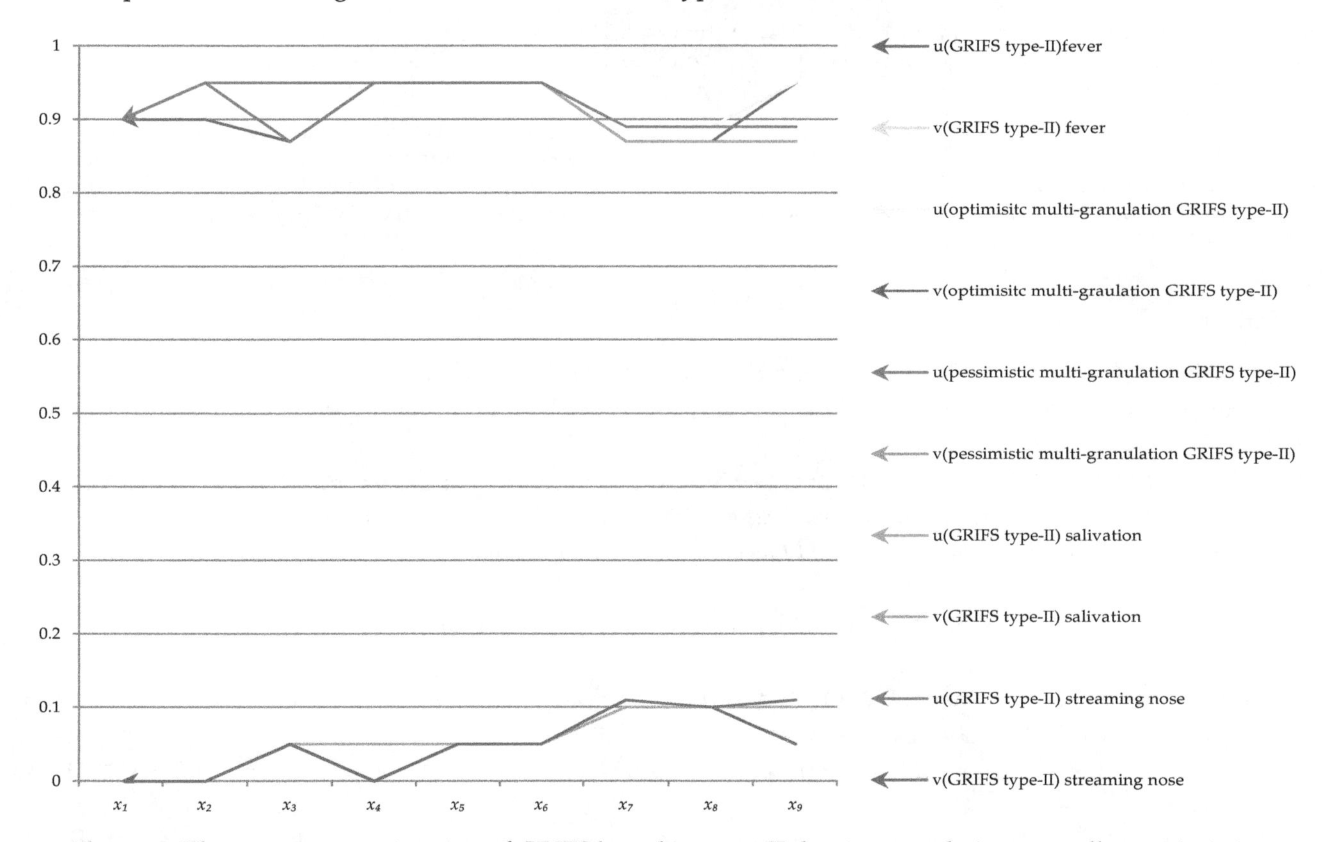

Figure 4. The upper approximation of GRIFS based on type-II dominance relation, as well as optimistic and pessimistic multi-granulation GRIFS based on type-II dominance relation.

From Figures 1 and 2, x_1, x_2, x_4 and x_8 patients have the disease, x_6 and x_7 patient do not have the disease. From Figures 3 and 4, x_1, x_2 and x_4 patients have the disease, x_3, x_5, x_6, x_7, x_8 and x_9 patients do not have the disease. Furthermore, this example proves the accuracy of Algorithm 1.

This example analyzes and discusses multi-granulation GRIFS models based on dominance relation. From conjunction and disjunction operations of two kinds of dominance classes perspective, we analyzed GRIFS models based on type-I dominance relation and type-II dominance relation and also optimistic and pessimistic multi-granulation GRIFS models based on type-I dominance relation and type-II dominance relation, respectively. Through the analysis of this example, the validity of these multi-granulation GRIFS models based on type-I dominance relation and type-II dominance relation models can be obtained.

6. Conclusions

These theories of GRS and RIFS are extensions of the classical rough set theory. In this paper, we proposed a series of models on GRIFS based on dominance relation, which were based on the combination of GRS, RIFS, and dominance relations. Moreover, these models of multi-granulation GRIFS models based on dominance relation were established on GRIFS models based on dominance relation using multiple dominance relations on the universe. The validity of these models was demonstrated by giving examples. Compared with GRS based on dominance relation, GRIFS models based on dominance relation can be more precise. Compared with GRIFS models based on dominance relation, multi-granulation GRIFS models based on dominance relation can be more accurate. It can be demonstrated using the algorithm, and our methods provide a way to combine GRS and RIFS. Our next work is to study the combination of GRS and variable precision rough sets on the basis of our proposed methods.

Author Contributions: Z.-a.X. and M.-j.L. initiated the research and wrote the paper, D.-j.H. participated in some of the search work, and X.-w.X. supervised the research work and provided helpful suggestions.

Acknowledgments: This work is supported by the national natural science foundation of China under Grant Nos. 61772176, 61402153, and the scientific and technological project of Henan Province of China under Grant Nos. 182102210078, 182102210362, and the Plan for Scientific Innovation of Henan Province of China under Grant No. 18410051003, and the key scientific and technological project of Xinxiang City of China under Grant No. CXGG17002.

References

1. Pawlak, Z. Rough sets. *Int. J. Comput. Inf. Sci.* **1982**, *11*, 341–356. [CrossRef]
2. Yao, Y.Y.; Lin, T.Y. Generalization of rough sets using modal logics. *Intell. Autom. Soft Comput.* **1996**, *2*, 103–119. [CrossRef]
3. Zhang, X.Y.; Mo, Z.W.; Xiong, F.; Cheng, W. Comparative study of variable precision rough set model and graded rough set model. *Int. J. Approx. Reason.* **2012**, *53*, 104–116. [CrossRef]
4. Zadeh, L.A. *Fuzzy Sets, Fuzzy Logic, and Fuzzy Systems*; World Scientific Publishing Corporation: Singapore, 1996; pp. 433–448.
5. Qian, Y.H.; Liang, J.Y.; Yao, Y.Y.; Dang, C.Y. MGRS: A multi-granulation rough set. *Inf. Sci.* **2010**, *180*, 949–970. [CrossRef]
6. Xu, W.H.; Wang, Q.R.; Zhang, X.T. Multi-granulation fuzzy rough sets in a fuzzy tolerance approximation space. *Int. J. Fuzzy Syst.* **2011**, *13*, 246–259.
7. Lin, G.P.; Liang, J.Y.; Qian, Y.H. Multigranulation rough sets: From partition to covering. *Inf. Sci.* **2013**, *241*, 101–118. [CrossRef]
8. Liu, C.H.; Pedrycz, W. Covering-based multi-granulation fuzzy rough sets. *J. Intell. Fuzzy Syst.* **2015**, *30*, 303–318. [CrossRef]

9. Xue, Z.A.; Si, X.M.; Xue, T.Y.; Xin, X.W.; Yuan, Y.L. Multi-granulation covering rough intuitionistic fuzzy sets. *J. Intell. Fuzzy Syst.* **2017**, *32*, 899–911.
10. Hu, Q.J. Extended Graded Rough Sets Models Based on Covering. Master's Thesis, Shanxi Normal University, Linfen, China, 21 March 2016. (In Chinese)
11. Wang, H.; Hu, Q.J. Multi-granulation graded covering rough sets. In Proceedings of the International Conference on Machine Learning and Cybernetics, Guangzhou, China, 12–15 July 2015; Institute of Electrical and Electronics Engineers Computer Society: New York, NY, USA, 2015.
12. Wu, Z.Y.; Zhong, P.H.; Hu, J.G. Graded multi-granulation rough sets. *Fuzzy Syst. Math.* **2014**, *28*, 165–172. (In Chinese)
13. Yu, J.H.; Zhang, X.Y.; Zhao, Z.H.; Xu, W.H. Uncertainty measures in multi-granulation with different grades rough set based on dominance relation. *J. Intell. Fuzzy Syst.* **2016**, *31*, 1133–1144. [CrossRef]
14. Yu, J.H.; Xu, W.H. Multigranulation with different grades rough set in ordered information system. In Proceedings of the International Conference on Fuzzy Systems and Knowledge Discovery, Zhangjiajie, China, 15–17 August 2015; Institute of Electrical and Electronics Engineers Incorporated: New York, NY, USA, 2016.
15. Wang, X.Y.; Shen, J.Y.; Shen, J.L.; Shen, Y.X. Graded multi-granulation rough set based on weighting granulations and dominance relation. *J. Shandong Univ.* **2017**, *52*, 97–104. (In Chinese)
16. Zheng, Y. Graded multi-granularity rough sets based on covering. *J. Shanxi Normal Univ.* **2017**, *1*, 5–9. (In Chinese)
17. Shen, J.R.; Wang, X.Y.; Shen, Y.X. Variable grade multi-granulation rough set. *J. Chin. Comput. Syst.* **2016**, *37*, 1012–1016. (In Chinese)
18. Atanassov, K.T.; Rangasamy, P. Intuitionistic fuzzy sets. *Fuzzy Sets Syst.* **1986**, *20*, 87–96. [CrossRef]
19. Zadeh, L.A. Fuzzy sets. *Inf. Control* **1965**, *8*, 338–353. [CrossRef]
20. Huang, B.; Zhuang, Y.L.; Li, H.X. Information granulation and uncertainty measures in interval-valued intuitionistic fuzzy information systems. *Eur. J. Oper. Res.* **2013**, *231*, 162–170. [CrossRef]
21. Slowinski, R.; Vanderpooten, D. A generalized definition of rough approximations based on similarity. *IEEE Trans. Knowl. Data Eng.* **1996**, *12*, 331–336. [CrossRef]
22. Chang, K.H.; Cheng, C.H. A risk assessment methodology using intuitionistic fuzzy set in FMEA. *Int. J. Syst. Sci.* **2010**, *41*, 1457–1471. [CrossRef]
23. Zhang, X.H. Fuzzy anti-grouped filters and fuzzy normal filters in pseudo-BCI algebras. *J. Intell. Fuzzy Syst.* **2017**, *33*, 1767–1774. [CrossRef]
24. Gong, Z.T.; Zhang, X.X. The further investigation of variable precision intuitionistic fuzzy rough set model. *Int. J. Mach. Learn. Cybern.* **2016**, *8*, 1565–1584. [CrossRef]
25. He, Y.P.; Xiong, L.L. Generalized inter-valued intuitionistic fuzzy soft rough set and its application. *J. Comput. Anal. Appl.* **2017**, *23*, 1070–1088.
26. Zhang, X.H.; Bo, C.X.; Smarandache, F.; Dai, J.H. New inclusion relation of neutrosophic sets with applications and related lattice structure. *Int. J. Mach. Learn. Cybern.* **2018**, *9*, 1753–1763. [CrossRef]
27. Zhang, X.H.; Smarandache, F.; Liang, X.L. Neutrosophic duplet semi-group and cancellable neutrosophic triplet groups. *Symmetry* **2017**, *9*, 275. [CrossRef]
28. Huang, B.; Guo, C.X.; Li, H.X.; Feng, G.F.; Zhou, X.Z. An intuitionistic fuzzy graded covering rough set. *Knowl.-Based Syst.* **2016**, *107*, 155–178. [CrossRef]
29. Tiwari, A.K.; Shreevastava, S.; Som, T. Tolerance-based intuitionistic fuzzy-rough set approach for attribute reduction. *Expert Syst. Appl.* **2018**, *101*, 205–212. [CrossRef]
30. Guo, Q.; Ming, Y.; Wu, L. Dominance relation and reduction in intuitionistic fuzzy information system. *Syst. Eng. Electron.* **2014**, *36*, 2239–2243.
31. Ai, A.H.; Xu, Z.S. Line integral of intuitionistic fuzzy calculus and their properties. *IEEE Trans. Fuzzy Syst.* **2018**, *26*, 1435–1446. [CrossRef]
32. Rizvi, S.; Naqvi, H.J.; Nadeem, D. Rough intuitionistic fuzzy sets. In Proceedings of the 6th Joint Conference on Information Sciences, Research Triangle Park, NC, USA, 8–13 March 2002; Duke University/Association for Intelligent Machinery: Durham, NC, USA, 2002.
33. Zhang, X.X.; Chen, D.G.; Tsang, E.C.C. Generalized dominance rough set models for the dominance intuitionistic fuzzy information systems. *Inf. Sci.* **2017**, *378*, 1339–1351. [CrossRef]

34. Zhang, Y.Q.; Yang, X.B. An intuitionistic fuzzy dominance-based rough set. In Proceedings of the 7th International Conference on Intelligent Computing, Zhengzhou, China, 11–14 August 2011; Springer: Berlin, Germany, 2011.
35. Huang, B.; Zhuang, Y.L.; Li, H.X.; Wei, D.K. A dominance intuitionistic fuzzy-rough set approach and its applications. *Appl. Math. Model.* **2013**, *37*, 7128–7141. [CrossRef]
36. Zhang, W.X.; Wu, W.Z.; Liang, J.Y.; Li, D.Y. *Theory and Method of Rough Sets*; Science Press: Beijing, China, 2001. (In Chinese)
37. Wen, X.J. Uncertainty measurement for intuitionistic fuzzy ordered information system. Master's Thesis, Shanxi Normal University, Linfen, China, 21 March 2015. (In Chinese)
38. Lezanski, P.; Pilacinska, M. The dominance-based rough set approach to cylindrical plunge grinding process diagnosis. *J. Intell. Manuf.* **2018**, *29*, 989–1004. [CrossRef]
39. Huang, B.; Guo, C.X.; Zhang, Y.L.; Li, H.X.; Zhou, X.Z. Intuitionistic fuzzy multi-granulation rough sets. *Inf. Sci.* **2014**, *277*, 299–320. [CrossRef]
40. Greco, S.; Matarazzo, B.; Slowinski, R. An algorithm for induction decision rules consistent with the dominance principle. In Proceedings of the 2nd International Conference on Rough Sets and Current Trends in Computing, Banff, AB, Canada, 16–19 October 2000; Springer: Berlin, Germany, 2011.

9

Weak Embeddable Hypernear-Rings

Jelena Dakić [1], Sanja Jančić-Rašović [1] and Irina Cristea [2,*]

[1] Department of Mathematics, Faculty of Natural Science and Mathematics, University of Montenegro, 81000 Podgorica, Montenegro
[2] Centre for Information Technologies and Applied Mathematics, University of Nova Gorica, 5000 Nova Gorica, Slovenia
* Correspondence: irina.cristea@ung.si or irinacri@yahoo.co.uk.

Abstract: In this paper we extend one of the main problems of near-rings to the framework of algebraic hypercompositional structures. This problem states that every near-ring is isomorphic with a near-ring of the transformations of a group. First we endow the set of all multitransformations of a hypergroup (not necessarily abelian) with a general hypernear-ring structure, called the multitransformation general hypernear-ring associated with a hypergroup. Then we show that any hypernear-ring can be weakly embedded into a multitransformation general hypernear-ring, generalizing the similar classical theorem on near-rings. Several properties of hypernear-rings related with this property are discussed and illustrated also by examples.

Keywords: hypernear-ring; multitransformation; embedding

---

1. Introduction

Generally speaking, the embedding of an algebraic structure into another one requires the existence of an injective map between the two algebraic objects, that also preserves the structure, i.e., a monomorphism. The most natural, canonical and well-known embeddings are those of numbers: the natural numbers into integers, the integers into the rational numbers, the rational numbers into the real numbers and the real numbers into the complex numbers. One important type of rings is that one of the endomorphisms of an abelian group under function pointwise addition and composition of functions. It is well known that every ring is isomorphic with a subring of such a ring of endomorphisms. But this result holds only in the commutative case, since the set of the endomorphisms of a non-abelian group is no longer closed under addition. This aspect motivates the interest in studying near-rings, that appear to have applications also in characterizing transformations of a group. More exactly, the set of all transformations of a group G, i.e., $T(G) = \{f : G \to G\}$ can be endowed with a near-ring structure under pointwise addition and composition of mappings, such a near-ring being called the *transformation near-ring* of the group G.

In 1959 Berman and Silverman [1] claimed that every near-ring is isomorphic with a near-ring of transformations. At that time only some hints were presented, while a direct and clear proof of this result appeared in Malone and Heatherly [2] almost ten years later. Since $T(G)$ has an identity, it immediately follows that any near-ring can be embedded in a near-ring with identity. Moreover, in the same paper [2], it was proved that a group $(H, +)$ can be embedded in a group $(G, +)$ if and only if the near-ring $T_0(H)$, consisting of all transformations of H which multiplicatively commute with the zero transformation, can be embedded into the similar near-ring $T_0(G)$ on G under a kernel-preserving monomorphism of near-rings.

Similarly to near-rings, but in the framework of algebraic hyperstructures, Dašić [3] defined the hypernear-rings as hyperstructures with the additive part being a quasicanonical hypergroup [4,5] (called also a polygroup [6,7]), and the multiplicative part being a semigroup with a bilaterally absorbing element, such that the multiplication is distributive with respect to the hyperaddition on the left-hand side. Later on, this algebraic hyperstructure was called a *strongly distributive hypernear-ring*, or a *zero-symmetric hypernear-ring*, while in a hypernear-ring the distributivity property was replaced by the "inclusive distributivity" from the left (or right) side. Moreover, when the additive part is a hypergroup and all the other properties related to the multiplication are conserved, we talk about a *general hypernear-ring* [8]. The distributivity property is important also in other types of hyperstructures, see e.g., [9]. A detailed discussion about the terminology related to hypernear-rings is included in [10]. In the same paper, the authors defined on the set of all transformations of a quasicanonical hypergroup that preserves the zero element a hyperaddition and a multiplication (as the composition of functions) in such a way to obtain a hypernear-ring. More general, the set of all transformations of a hypergroup (not necessarily commutative) together with the same hyperaddition and multiplication is a strongly distributive hypernear-ring [3]. In this note we will extend the study to the set of all multimappings (or multitransformations) of a (non-abelian) hypergroup, defining first a structure of (left) general hypernear-ring, called the multitransformation general hypernear-ring associated with a hypergroup. Then we will show that any hypernear-ring can be weakly embedded into a multitransformation general hypernear-ring, generalizing the similar classical theorem on near-rings [2]. Besides, under same conditions, any additive hypernear-ring is weakly embeddable into the additive hypernear-ring of the transformations of a hypergroup with identity element that commute multiplicatively with the zero-function. The paper ends with some conclusive ideas and suggestions of future works on this topic.

2. Preliminaries

We start with some basic definitions and results in the framework of hypernear-rings and near-rings of group mappings. For further properties of these concepts we refer the reader to the papers [2,3,11,12] and the fundamental books [13–15]. For the consistence of our study, regarding hypernear-rings we keep the terminology established and explained in [8,16].

First we recall the definition introduced by Dašić in 1978.

Definition 1. *[12] A hypernear-ring is an algebraic system* $(R,+,\cdot)$, *where* R *is a non-empty set endowed with a hyperoperation* $+ : R \times R \to P^*(R)$ *and an operation* $\cdot : R \times R \to R$, *satisfying the following three axioms:*

1. $(R,+)$ *is a quasicanonical hypergroup (named also polygroup [6]), meaning that:*
   (a) $x+(y+z) = (x+y)+z$ *for any* $x,y,z \in R$,
   (b) *there exists* $0 \in R$ *such that, for any* $x \in R, x+0 = 0+x = \{x\}$,
   (c) *for any* $x \in R$ *there exists a unique element* $-x \in R$, *such that* $0 \in x+(-x) \cap (-x)+x$,
   (d) *for any* $x,y,z \in R, z \in x+y$ *implies that* $x \in z+(-y), y \in (-x)+z$.
2. $(R,\cdot)$ *is a semigroup endowed with a two-sided absorbing element* 0, *i.e., for any* $x \in R, x \cdot 0 = 0 \cdot x = 0$.
3. *The operation "$\cdot$" is distributive with respect to the hyperoperation "$+$" from the left-hand side: for any* $x,y,z \in R$, *there is* $x \cdot (y+z) = x \cdot y + x \cdot z$.

This kind of hypernear-ring was called by Gontineac [11] a *zero-symmetric hypernear-ring*. In our previous works [10,16], regarding the distributivity, we kept the Vougiouklis' terminology [17], and therefore, we say that a hypernear-ring is a hyperstructure $(R,+,\cdot)$ satisfying the above mentioned axioms 1. and 2., and the new one:

$3'$. The operation "$\cdot$" is inclusively distributive with respect to the hyperoperation "$+$" from the left-hand side: for any $x,y,z \in R, x \cdot (y+z) \subseteq x \cdot y + x \cdot z$. Accordingly, the Dašić 's hypernear-ring (satisfying the axioms 1., 2., and 3.) is called *strongly distributive hypernear-ring*.

Furthermore, if the additive part is a hypergroup (and not a polygroup), then we talk about a more general type of hypernear-rings.

Definition 2. *[8] A general (left) hypernear-ring is an algebraic structure* $(R, +, \cdot)$ *such that* $(R, +)$ *is a hypergroup,* $(R, \cdot)$ *is a semihypergroup and the hyperoperation "$\cdot$" is inclusively distributive with respect to the hyperoperation "$+$" from the left-hand side, i.e.,* $x \cdot (y + z) \subseteq x \cdot y + x \cdot z$, *for any* $x, y, z \in R$. *If in the third condition the equality is valid, then the structure* $(R, +, \cdot)$ *is called strongly distributive general (left) hypernear-ring. Besides, if the multiplicative part* $(R, \cdot)$ *is only a semigroup (instead of a semihypergroup), we get the notion of general (left) additive hypernear-ring.*

Definition 3. *Let* $(R_1, +, \cdot)$ *and* $(R_2, +, \cdot)$ *be two general hypernear-rings. A map* $\rho : R_1 \to R_2$ *is called an inclusion homomorphism if the following conditions are satisfied:*

1. $\rho(x + y) \subseteq \rho(x) + \rho(y)$
2. $\rho(x \cdot y) \subseteq \rho(x) \cdot \rho(y)$ *for all* $x, y \in R_1$.

A map ρ *is called a good (strong) homomorphism if in the conditions* 1. *and* 2. *the equality is valid.*

In the second part of this section we will briefly recall the fundamentals on *near-rings of group mappings*. A *left near-ring* $(N, +, \cdot)$ is a non-empty set endowed with two binary operations, the addition $+$ and the multiplication $\cdot$, such that $(N, +)$ is a group (not necessarily abelian) with the neutral element 0, $(N, \cdot)$ is a semigroup, and the multiplication is distributive with respect to the addition from the left-hand side. Similarly, we have a right near-ring. Several examples of near-rings are obtained on the set of "non-linear" mappings and here we will see two of them.

Let $(G, +)$ be a group (not necessarily commutative) and let $T(G)$ be the set of all functions from G to G. On $T(G)$ define two binary operations: "$+$" is the pointwise addition of functions, while the multiplication "$\cdot$" is the composition of functions. Then $(T(G), +, \cdot)$ is a (left) near-ring, called the *transformation near-ring* on the group G. Moreover, let $T_0(G)$ be the subnear-ring of $T(G)$ consisting of the functions of $T(G)$ that commute multiplicatively with the zero function, i.e., $T_0(G) = \{f \in T(G) \mid f(0) = 0\}$. These two near-rings, $T(G)$ and $T_0(G)$, have a fundamental role in embeddings. Already in 1959, it was claimed by Berman and Silverman [1] that every near-ring is isomorphic with a near-ring of transformations. One year later the proof was given by the same authors, but using an elaborate terminology and methodology. Here below we recall this result together with other related properties, as presented by Malone and Heatherly [2].

Theorem 1. *[2] Let* $(R, +, \cdot)$ *be a near-ring. If* $(G, +)$ *is any group containing* $(R, +)$ *as a proper subgroup, then* $(R, +, \cdot)$ *can be embedded in the transformation near-ring* $T(G)$.

Corollary 1. *[2] Every near-ring can be embedded in a near-ring with identity.*

Theorem 2. *[2] A group* $(H, +)$ *can be embedded in a group* $(G, +)$ *if and only if* $T_0(H)$ *can be embedded in* $T_0(G)$ *by a near-ring monomorphism which is kernel-preserving.*

Theorem 3. *[2] A group* $(H, +)$ *can be embedded in a group* $(G, +)$ *if and only if the near-ring* $T(H)$ *can be embedded in the near-ring* $T(G)$.

3. Weak Embeddable Hypernear-Rings

In this section we aim to extend the results related to embeddings of near-rings to the case of hypernear-rings. In this respect, instead of a group $(G, +)$ we will consider a hypergroup $(H, +)$ and then the set of all multimappings on H, which we endow with a structure of general hypernear-ring.

Theorem 4. *Let* $(H, +)$ *be a hypergroup (not necessarily abelian) and* $T^*(H) = \{h : H \to P^*(H)\}$ *the set of all multimappings of the hypergroup* $(H, +)$. *Define, for all* $(f, g) \in T^*(H) \times T^*(H)$, *the following hyperoperations:*

$$f \oplus g = \{h \in T^*(H) \mid (\forall x \in H)\, h(x) \subseteq f(x) + g(x)\}$$

$$f \odot g = \{h \in T^*(H) \mid (\forall x \in H)\ h(x) \subseteq g(f(x)) = \bigcup_{u \in f(x)} g(u)\}.$$

The structure $(T^*(H), \oplus, \odot)$ is a (left) general hypernear-ring.

Proof. For any $f, g \in T^*(H)$ it holds: $f \oplus g \neq \emptyset$. Indeed, for any $x \in H$, it holds $f(x) \neq \emptyset$ and $g(x) \neq \emptyset$ and thus, $f(x) + g(x) \neq \emptyset$. Therefore, for the map $h : H \to P^*(H)$ defined by: $h(x) = f(x) + g(x)$ for all $x \in H$, it holds $h \in f \oplus g$. Now, we prove that the hyperoperation $\oplus$ is associative. Let $f, g, h \in T^*(H)$ and set

$$L = (f \oplus g) \oplus h = \bigcup\{h' \oplus h \mid h' \in f \oplus g\} =$$

$$= \bigcup\{h' \oplus h \mid (\forall x \in H)\ h'(x) \subseteq f(x) + g(x)\}.$$

Thus, if $h'' \in L$, then, for all $x \in H$, it holds: $h''(x) \subseteq h'(x) + h(x) \subseteq (f(x) + g(x)) + h(x)$. Conversely, if h'' is an element of $T^*(H)$ such that: $h''(x) \subseteq (f(x) + g(x)) + h(x)$, for all $x \in H$, and if we choose h' such that $h'(x) = f(x) + g(x)$ for all $x \in H$, then $h' \in f \oplus g$ and $h'' \in h' \oplus h$ i.e., $h'' \in L$. So, $L = \{h'' \in T^*(H) | (\forall x \in H) h''(x) \subseteq (f(x) + g(x)) + h(x)\}$. On the other side, take $D = f \oplus (g \oplus h)$. Then, $D = \{h'' \in T^*(H) | (\forall x \in H) h''(x) \subseteq f(x) + (g(x) + h(x))\}$. By the associativity of the hyperoperation $+$ we obtain that $L = D$, meaning that the hyperoperation $\oplus$ is associative.

Let $f, g \in T^*(H)$. We prove that the equation $f \in g \oplus a$ has a solution $a \in T^*(H)$. If we set $a(x) = H$, for all $x \in H$, then $a \in T^*(H)$ and for all $x \in H$ it holds $g(x) + a(x) = H \supseteq f(x)$. So, $f \in g \oplus a$. Similarly, the equation $f \in a \oplus g$ has a solution in $T^*(H)$. Thus, $(T^*(H), \oplus)$ is a hypergroup.

Now, we show that $(T^*(H), \odot)$ is a semihypergroup. Let $f, g \in T^*(H)$. For all $x \in H$ it holds $g(x) \neq \emptyset$ and so $g(f(x)) \neq \emptyset$. Let $h : H \to P^*(H)$ be a multimapping defined by $h(x) = g(f(x))$, for all $x \in H$. Obviously, $h \in f \odot g$ and so $f \odot g \neq \emptyset$. Let us prove that $\odot$ is a associative. Let $f, g, h \in T^*(H)$. Set:

$$L = (f \odot g) \odot h = \bigcup\{h' \odot h \mid h' \in f \odot g\} = \{h' \odot h \mid (\forall x \in H)\ h'(x) \subseteq g(f(x))\} =$$

$$= \{h'' \mid (\forall x \in H)\ h''(x) \subseteq h(h'(x)) \wedge h'(x) \subseteq g(f(x))\}.$$

So, if $h'' \in L$, then $h''(x) \subseteq h(g(f(x)))$, for all $x \in H$. On the other side, if $h'' \in T^*(H)$ and $h''(x) \subseteq h(g(f(x)))$ for all $x \in H$, then we choose $h' \in T^*(H)$ such that $h'(x) = g(f(x))$ and consequently we obtain that $h'' \subseteq h(h'(x))$. Thus, $h'' \in L$. So, $L = \{h'' \in T^*(H) \mid (\forall x \in H)\ h''(x) \subseteq h(g(f(x)))\}$.

Similarly, $D = f \odot (g \odot h) = \{h'' \mid (\forall x \in H)\ h''(x) \subseteq h(g(f(x)))\}$. Thus, $L = D$.

It remains to prove that the hyperoperation $\oplus$ is inclusively distributive with respect to the hyperoperation $\odot$ on the left-hand side. Let $f, g, h \in T^*(H)$. Set $L = f \odot (g \oplus h) = \bigcup\{f \odot h' | h' \in g \oplus h\} = \bigcup\{f \odot h' | h' \in T^*(H) \wedge (\forall x) h'(x) \subseteq g(x) + h(x)\}$. So, if $k \in L$ then for all $x \in H$ it holds: $k(x) \subseteq h'(f(x)) \subseteq g(f(x)) + h(f(x))$.

On the other hand, $D = (f \odot g) \oplus (f \odot h) = \bigcup\{k_1 \oplus k_2 | k_1 \in f \odot g, k_2 \in f \odot h\}$. Let $k \in L$. Choose, $k_1, k_2 \in T^*(H)$ such that $k_1(x) = g(f(x))$ and $k_2(x) = h(f(x))$ for all $x \in H$. Then $k_1 \in f \odot g$ and $k_2 \in f \odot h$. Thus, $k(x) \subseteq k_1(x) + k_2(x)$ for all $x \in H$, i.e., $k \in k_1 \oplus k_2$ and $k_1 \in f \odot g, k_2 \in f \odot h$. So, $k \in D$. Therefore, $L \subseteq D$. □

Definition 4. *$T^*(H)$ is called the multitransformations general hypernear-ring on the hypergroup H.*

Remark 1. *Let $(G, +)$ be a group and $T(G)$ be the transformations near-ring on G. Obviously, $T(G) \subset T^*(G) = \{f : G \to P^*(G)\}$ and, for all $f, g \in T(G)$, it holds: $f \oplus g = f + g$, $f \odot g = f \cdot g$, meaning that the hyperoperations defined in Theorem 4 are the same as the operations in Theorem 1. It follows that $T(G)$ is a sub(hyper)near-ring of $(T^*(G), \oplus, \odot)$.*

Definition 5. *We say that the hypernear-ring $(R_1, +, \cdot)$ is weak embeddable (by short $W-$ embeddable) in the hypernear-ring $(R_2, +, \cdot)$ if there exists an injective inclusion homomorphism $\mu : R_1 \to R_2$.*

The next theorem is a generalization of Theorem 1 [5].

Theorem 5. *For every general hypernear-ring $(R, +, \cdot)$ there exists a hypergroup $(H, +)$ such that R is $W-$ embeddable in the associated hypernear-ring $T^*(H)$.*

Proof. Let $(R, +, \cdot)$ be a hypernear-ring and let $(H, +)$ be a hypergroup such that $(R, +)$ is a proper subhypergroup of $(H, +)$. For a fixed element $r \in R$ we define a multimapping $f_r : H \to P^*(H)$ as follows

$$f_r(g) = \begin{cases} g \cdot r, & \text{if } g \in R \\ r, & \text{if } g \in H \setminus R. \end{cases}$$

Let us define now the mapping $\mu : R \to T^*(H)$ as $\mu(r) = f_r$, which is an inclusion homomorphism. Indeed, if $a, b \in R$ then we have $\mu(a + b) = \{f_c \mid c \in a + b\}$ and $\mu(a) \oplus \mu(b) = f_a \oplus f_b = \{h \mid (\forall g \in H)\ h(g) \subseteq f_a(g) + f_b(g)\}$.

Consider $c \in a + b$ and $g \in H$. If $g \in R$, then $f_c(g) = g \cdot c \subseteq g \cdot (a + b) \subseteq g \cdot a + g \cdot b = f_a(g) + f_b(g)$. If $g \in H \setminus R$, then $f_c(g) = c \in a + b = f_a(g) + f_b(g)$. It follows that, for all $g \in H$, we have $f_c(g) \subseteq f_a(g) + f_b(g)$ and therefore $f_c \in \mu(a) \oplus \mu(b)$, meaning that $\mu(a + b) \subseteq \mu(a) \oplus \mu(b)$.

Similarly, there is $\mu(a \cdot b) = \{f_c \mid c \in a \cdot b\}$ and $\mu(a) \odot \mu(b) = f_a \odot f_b = \{h \in T^*(H) \mid (\forall g \in H)\ h(g) \subseteq f_b(f_a(g))\}$. Let $c \in a \cdot b$. Then, for $g \in R$, it holds: $f_c(g) = g \cdot c \subseteq g \cdot (a \cdot b) = (g \cdot a) \cdot b = f_b(f_a(g))$. If $g \in H \setminus R$, then there is $f_c(g) = c \in a \cdot b = f_b(a) = f_b(f_a(g))$. Thus, $f_c \in \mu(a) \odot \mu(b)$ and so $\mu(a \odot b) \subseteq \mu(a) \odot \mu(b)$.

Based on Definition 3, we conclude that μ is an inclusive homomorphism. It remains to show that μ is injective. If $\mu(a) = \mu(b)$, then for all $g \in H$, it holds $f_a(g) = f_b(g)$. So, if we choose $g \in H \setminus R$, then we get that $a = f_a(g) = f_b(g) = b$.

These all show that the general hypernear-ring R is W-embeddable in $T^*(H)$. □

Remark 2. *If $(R, +, \cdot)$ is a near-ring such that $(R, +)$ is a proper subgroup of a group $(G, +)$, then for a fixed $r \in R$ the multimapping f_r constructed in the proof of Theorem 5 is in fact a map from G to G, since in this case the multiplication $\cdot$ is an ordinary operation, i.e., $g \cdot r \in G$, for all $g \in R$. Thus $f_r : G \to G$ and thereby $\mu(R) \subseteq T(G)$. By consequence $\mu : R \to T(G)$ is an ordinary monomorphism. In other words, Theorem 5 is a generalization of Theorem 1.*

Example 1. *Let $(R, +, \cdot)$ be a left near-ring. Let P_1 and P_2 be non-empty subsets of R such that $R \cdot P_1 \subseteq P_1$ and $P_1 \subseteq Z(R)$, where $Z(R)$ is the center of R, i.e., $Z(R) = \{x \in R \mid (\forall y \in R) x + y = y + x\}$. For any $(x, y) \in R^2$ define:*

$$x \oplus_{P_1} y = x + y + P_1, \quad x \odot_{P_2} y = xP_2y.$$

Then the structure $(R, \oplus_{P_1}, \odot_{P_2})$ is a general left hypernear-ring [8,18]. Let $H = R \cup \{a\}$ and define on H the hyperoperation $\oplus'_P$ as follows:

$$x \oplus'_{P_1} y = \begin{cases} x \oplus_{P_1} y, & \text{if } x, y \in R \\ H, & \text{if } x = a \vee y = a. \end{cases}$$

It is clear that H is a hypergroup such that $(R, +)$ is a proper subhypergroup of $(H, +)$. Besides, based on Theorem 5, for every $r \in R$ the multimapping $f_r : H \to P^*(H)$ is defined as

$$f_r(g) = \begin{cases} g \odot_{P_2} r, & \text{if } g \in R \\ r, & \text{if } g = a \end{cases} = \begin{cases} gP_2r, & \text{if } g \in R \\ r, & \text{if } g = a. \end{cases}$$

Clearly it follows that $\mu : R \to P^*(H)$, defined by $\mu(r) = f_r$, is an inclusive homomorphism, so the general left hypernear-ring $(R, \oplus_{P_1}, \odot_{P_2})$ is W-embeddable in $T^*(H)$.

Example 2. *Consider the semigroup $(\mathbb{N}, \cdot)$ of natural numbers with the standard multiplication operation and the order "$\leq$". Define on it the hyperoperations $+_{\leq}$ and $\cdot_{\leq}$ as follows:*

$$x +_{\leq} y = \{z \mid x \leq z \vee y \leq z\}$$

$$x \cdot_{\leq} y = \{z \mid x \cdot y \leq z\}.$$

Then the structure $(\mathbb{N}, +_{\leq}, \cdot_{\leq})$ is a strongly distributive general hypernear-ring (in fact it is a hyperring). This follows from Theorem 4.3 [19]. Furthermore, for any $a \notin \mathbb{N}$, it can be easily verified that $(\mathbb{N}, +_{\leq})$ is a proper subhypergroup of $(\mathbb{N} \cup \{a\}, +'_{\leq})$, where the hyperoperation $+'_{\leq}$ is defined by:

$$x +'_{\leq} y = \begin{cases} x +_{\leq} y, & \text{if } x, y \in \mathbb{N} \\ \mathbb{N} \cup \{a\}, & \text{if } x = a \vee y = a. \end{cases}$$

In this case, for a fixed $n \in \mathbb{N}$, we can define the multimapping $f_n : \mathbb{N} \cup \{a\} \to P^*(\mathbb{N} \cup \{a\})$ as follows:

$$f_n(g) = \begin{cases} g \cdot_{\leq} n, & \text{if } g \in \mathbb{N} \\ n, & \text{if } g = a \end{cases} = \begin{cases} \{k \in \mathbb{N} \mid g \cdot n \leq k\}, & \text{if } g \in \mathbb{N} \\ n, & \text{if } g = a \end{cases}$$

and therefore the mapping $\mu : \mathbb{N} \to P^*(\mathbb{N} \cup \{a\})$ is an inclusive homomorphism. Again this shows that the general hypernear-ring $(\mathbb{N}, +_{\leq}, \cdot_{\leq})$ is W-embeddable in $T^*(\mathbb{N} \cup \{a\})$.

Example 3. *Let $R = \{0, 1, 2, 3\}$. Consider now the semigroup $(R, \cdot)$ defined by Table 1:*

Table 1. The Cayley table of the semigroup $(R, \cdot)$

$\cdot$	**0**	**1**	**2**	**3**
0	0	0	0	0
1	0	1	2	3
2	0	1	2	3
3	0	1	2	3

Define on R the hyperoperation $+_{\leq}$ as follows: $x +_{\leq} y = \{z \mid x \leq z \vee y \leq z\}$, so its Cayley table is described in Table 2:

Table 2. The Cayley table of the hypergroupoid $(R, +_{\leq})$

$+_{\leq}$	**0**	**1**	**2**	**3**
0	R	R	R	R
1	R	{1,2,3}	{1,2,3}	{1,2,3}
2	R	{1,2,3}	{2,3}	{2,3}
3	R	{1,2,3}	{2,3}	{3}

Obviously, the relation $\leq$ is reflexive and transitive and, for all $x, y, z \in R$, it holds: $x \leq y \Rightarrow z \cdot x \leq z \cdot y$. Thus, $(R, +_{\leq}, \cdot)$ is an (additive) hypernear-ring. Let $H = R \cup \{4\}$ and define the hyperoperation $+_{\leq}$ as follows:

$$x +_{\leq} y = \begin{cases} x +_{\leq} y, & \text{if } x, y \in \{0, 1, 2, 3\} \\ H, & \text{if } x = 4 \vee y = 4 \end{cases}$$

It follows that $(R,+)$ is a proper subhypergroup of $(H,+)$ and for a fixed $r \in R$ it holds $f_r(x) = r$, for all $x \in H$. This implies that the mapping $\mu : H \to P^*(H)$, defined by $\mu(r) = f_r$ for any $r \in R$, is an inclusive homomorphism.

Now we will construct a left general additive hypernear-ring associated with an arbitrary hypergroup.

Theorem 6. *Let $(H,+)$ be a hypergroup and $T(H) = \{f : H \to H\}$. On the set $T(H)$ define the hyperoperation $\oplus_T$ and the operation $\odot_T$ as follows:*

$$f \oplus_T g = \{h \in T(H) \mid (\forall x \in H)\ \ h(x) \in f(x) + g(x)\},$$

$$(f \odot_T g)(x) = g(f(x)), \ \text{for all} \ \ x \in H.$$

The obtained structure $(T(H), \oplus_T, \odot_T)$ is a (left) general additive hypernear-ring.

Proof. Let $f, g \in T(H)$. We prove that there exists $h \in T(H)$ such that $h(x) \in f(x) + g(x)$ for all $x \in H$. Let $x \in H$. Since $f(x) + g(x) \neq \emptyset$ we can choose $h_x \in f(x) + g(x)$ and define $h(x) = h_x$. Obviously, $h \in f \oplus_T g$. Now we prove that the hyperoperation $\oplus_T$ is associative. Let $f, g, h \in T(H)$. Set $L = (f \oplus_T g) \oplus_T h = \{h'' \mid (\forall x)\ h''(x) \in h'(x) + h(x) \wedge h'(x) \in f(x) + g(x)\}$ and $D = f \oplus_T (g \oplus_T h) = \{f'' \mid (\forall x)\ f''(x) \in f(x) + f'(x) \wedge f'(x) \in g(x) + h(x)\}$. Thus, if $h'' \in L$, then $h''(x) \in (f(x) + g(x)) + h(x) = f(x) + (g(x) + h(x))$. Thereby, for any $x \in H$, there exists $a_x \in g(x) + h(x)$ such that $h''(x) \in f(x) + a_x$. Define $f'(x) = a_x$. Then, $f' \in g \oplus_T h$ and for all $x \in H$ it holds $h''(x) \in f(x) + f'(x)$. Therefore, $h'' \in D$. So, $L \subseteq D$. Similarly, we obtain that $D \subseteq L$. Now, let $f, g \in T(H)$. We prove that the equation $f \in g \oplus_T h$ has a solution $h \in T(H)$. Since $(H,+)$ is a hypergroup, it follows that, for any $x \in H$, there exists $b_x \in H$ such that $f(x) \in g(x) + b_x$. Define $h : H \to H$ by $h(x) = b_x$. Then $h \in T(H)$ and $f \in g \oplus_T h$. Similarly, we obtain that the equation $f \in h \oplus_T g$ has a solution in $T(H)$. We may conclude that $(T(H), \oplus_T)$ is a hypergroup.

Obviously, $(T(H), \odot_T)$ is a semigroup, because the composition of functions is associative. Now we prove that the hyperoperation $\oplus_T$ is left inclusively distributive with respect to the operation $\odot_T$. Let $f, g, h \in T(H)$. Set $L = f \odot_T (g \oplus_T h) = \{f \odot_T k \mid k \in g \oplus_T h\}$ and $D = (f \odot_T g) \oplus_T (f \odot_T h) = \{h' \mid (\forall x \in H)\ h'(x) \in g(f(x)) + h(f(x))\}$. Let $k \in g \oplus h$. Then, for all $x \in H$, it holds $(f \odot k)(x) = k(f(x)) \subseteq g(f(x)) + h(f(x))$. Thus, $f \odot k \in D$, meaning that $L \subseteq D$. $\square$

For an arbitrary group G, Malone and Heatherly [2] denote by $T_0(G)$ the subset of $T(G)$ consisting of the functions which commute multiplicatively with the zero-function, i.e., $T_0(G) = \{f : G \to G \mid f(0) = 0\}$. Obviously, $T_0(G)$ is a sub-near-ring of $(T(G), +, \cdot)$. The next result extends this property to the case of hyperstructures.

Theorem 7. *Let $(H,+)$ be a hypergroup with the identity element 0 (i.e., for all $x \in H$, it holds $x \in x + 0 \cap 0 + x$), such that $0 + 0 = \{0\}$. Let $T_0(H) = \{f : H \to H \mid f(0) = 0\}$. Then, $T_0(H)$ is a subhypernear-ring of the general additive hypernear-ring $(T(H), \oplus_T, \odot_T)$.*

Proof. Let $f, g \in T_0(H)$. If $h \in f \oplus_T g$, then $h(0) \in f(0) + g(0) = 0 + 0 = \{0\}$, i.e., $h(0) = 0$. Thus, $h \in T_0(H)$. Let $f, g \in T_0(H)$. We prove now that the equation $f \in g \oplus a$ has a solution $a \in T_0(H)$. If we set $a(0) = 0$ and $a(x) = a_x$, where $f(x) \in g(x) + a_x$, for $x \neq 0$ and $a_x \in H$, then $a \in T_0(H)$ and $f \in g + a$. Similarly the equation $f \in a \oplus g$ has a solution $a \in T_0(H)$. Thus, $(T_0(H), \oplus_T)$ is a subhypergroup of $(T(H), \oplus_T)$. Obviously, if $f, g \in T_0(H)$, then it follows that $(f \odot_T g)(0) = g(f(0)) = g(0) = 0$, i.e., $f \odot_T g \in T_0(H)$. So, $(T_0(H), \odot_T)$ is a subsemihypergroup of $(T(H), \odot_T)$, implying that $T_0(H)$ is a subsemihypernear-ring of $T(H)$. $\square$

Theorem 8. *Let $(R, +, \cdot)$ be an additive hypernear-ring such that $(R, +)$ is a proper subhypergroup of the hypergroup $(H, +)$, having an identity element 0 satisfying the following properties:*

1. $0 + 0 = \{0\}$ *and*
2. $0 \cdot r = 0$, *for all* $r \in R$.

Then the hypernear-ring $(R, +, \cdot)$ is $W-$embeddable in the additive hypernear-ring $T_0(H)$.

Proof. For a fixed $r \in R$, define a map $f : H \to H$ as follows

$$f_r(g) = \begin{cases} g \cdot r, & \text{if } g \in R \\ r, & \text{if } g \in H \setminus R. \end{cases}$$

Obviously, $f_r(0) = 0 \cdot r = 0$. So, $f_r \in T_0(H)$ and, similarly as in the proof of Theorem 5, we obtain that the map $\rho : (R, +, \cdot) \to (T_0(H), \oplus_T, \odot_T)$ defined by $\rho(r) = f_r$ is an injective inclusion homomorphism. □

Example 4. *On the set $H = \{0, 1, 2, 3, 4, 5, 6\}$ define an additive hyperoperation and a multiplicative operation having the Cayley tables described in Tables 3 and 4, respectively:*

Table 3. The Cayley table of the hypergroupoid $(H, +)$

+	0	1	2	3	4	5	6
0	0	1	2	3	4	5	6
1	1	2	3	4	5	{0, 6}	1
2	2	3	4	5	{0, 6}	1	2
3	3	4	5	{0, 6}	1	2	3
4	4	5	{0, 6}	1	2	3	4
5	5	{0, 6}	1	2	3	4	5
6	6	1	2	3	4	5	0

Table 4. The Cayley table of the semigroup $(H, \cdot)$

·	0	1	2	3	4	5	6
0	0	0	0	0	0	0	0
1	0	5	4	3	2	1	0
2	0	1	2	3	4	5	0
3	0	0	0	0	0	0	0
4	0	5	4	3	2	1	0
5	0	1	2	3	4	5	0
6	0	0	0	0	0	0	0

The structure $(H, +, \cdot)$ is an (additive) hypernear-ring [16].

Let $R = \{0, 3, 6\}$. Then $(R, +, \cdot)$ is a hypernear-ring (in particular it is a subhypernear-ring of $(H, +, \cdot)$). Obviously, $(R, +)$ is a proper subhypergroup of the hypergroup $(H, +)$, which has the identity 0 such that $0 + 0 = \{0\}$ and $0 \cdot r = 0$, for all $r \in R$. It follows that, for each $r \in \{0, 3, 6\}$, $f_r : H \to H$ is a map such that $f_0(g) = 0$, for all $g \in H$,

$$f_3(g) = \begin{cases} g \cdot 3, & \text{if } g \in \{0,3,6\} \\ 3, & \text{if } g \in \{1,2,4\} \end{cases} = \begin{cases} 0, & \text{if } g \in \{0,3,6\} \\ 3, & \text{if } g \in \{1,2,4\}, \end{cases}$$

while

$$f_6(g) = \begin{cases} g \cdot 6, & \text{if } g \in \{0,3,6\} \\ 6, & \text{if } g \in \{1,2,4\} \end{cases} = \begin{cases} 0, & \text{if } g \in \{0,3,6\} \\ 6, & \text{if } g \in \{1,2,4\}. \end{cases}$$

Clearly, the map $\rho : (R, +, \cdot) \to (T_0(H), \oplus_T, \odot_T)$, *defined by* $\rho(r) = f_r$, *is an injective inclusion homomorphism, so the hypernear-ring* R *is W-embeddable in* $T_0(H)$.

Remark 3. *If* $(G, +)$ *is a group, then, for any* $f, g \in T(G) = \{f : G \to G\}$, *it holds* $f \oplus_T g = f + g$ *and* $f \odot_T g = f \cdot g$, *meaning that the transformation near-ring* $(T(G), +, \cdot)$ *of a group* G *is in fact the structure* $(T(G), \oplus_T, \odot_T)$. *Furthermore, if* $(R, +, \cdot)$ *is a zero-symmetric near-ring, i.e., a near-ring in which any element* x *satisfies the relation* $x \cdot 0 = 0 \cdot x = 0$, *then the map* ρ *constructed in the proof of Theorem 8 is the injective homomorphism* $\rho : R \to T_0(G)$. *Thus, according with Theorem 8, it follows that the zero-symmetric near-ring* $(R, +, \cdot)$ *is W-embeddable in the near-ring* $T_0(G)$, *where* $(G, +)$ *is any group containing* $(R, +)$ *as a proper subgroup.*

Remark 4. *If* $(G, +)$ *is a group, then the following inclusions hold:* $T_0(G) \subseteq T(G) \subseteq T^*(G)$, *where both* $T(G)$ *and* $T_0(G)$ *are sub-(hyper)near-rings of the hypernear-ring* $T^*(G)$. *Considering now* $(H, +)$ *a hypergroup, the same inclusions exist:* $T_0(H) \subseteq T(H) \subseteq T^*(H)$, *but generally* $T(H)$ *and* $T_0(H)$ *are not subhypernear-rings of* $T^*(H)$.

Proposition 1. *Let* $(H, +)$ *be a hypergroup with the identity element* 0 *(i.e., for all* $x \in H$ *it holds* $x \in x + 0 \cap 0 + x$*) such that* $0 + 0 = \{0\}$. *Let* $T_0^*(H) = \{f : H \to P^*(H) \mid f(0) = 0\}$. *Then,* $T_0^*(H)$ *is a subhypernear-ring of the general hypernear-ring* $(T^*(H), \oplus, \odot)$.

Proof. Let $f, g \in T_0^*(H)$. If $h \in f \oplus g$, then it holds $h(0) \subseteq f(0) + g(0) = 0 + 0 = \{0\}$. Since $h(0) \neq \emptyset$, it follows that $h(0) = \{0\}$. Thus, $h \in T_0^*(H)$. Let $f, g \in T_0^*(H)$. We prove that the equation $f \in g \oplus a$ has a solution $a \in T_0^*(H)$. If we set $a(0) = 0$ and $a(x) = H$, for all $x \neq 0$, then $a \in T_0^*(H)$ and, for all $x \neq 0$, it holds $g(x) + a(x) = H \supseteq f(x)$ and $g(0) + a(0) = \{0\} = f(0)$, meaning that $f \in g \oplus a$. Similarly, the equation $f \in a \oplus g$ has a solution in $T_0^*(H)$. So, $(T_0^*(H))$ is a subhypergroup of $(T^*(H), \oplus)$. Obviously, if $h \in f \odot g$, then $h(0) \subseteq g(f(0)) = \{0\}$. So, $h \in T_0^*(H)$. Thus $T_0^*(H)$ is a subsemihypergroup of $(T^*(H), \odot)$. Therefore, $T_0^*(H)$ is a subhypernear-ring of $(T^*(H), \oplus, \odot)$. □

4. Conclusions

Distributivity property plays a fundamental role in the ring-like structures, i.e., algebraic structures endowed with two operations, usually denoted by addition and multiplication, where the multiplication distributes over the addition. If this happens only from one-hand side, then we talk about near-rings. Similarly, in the framework of algebraic hypercompositional structures, a general hypernear-ring has the additive part an arbitrary hypergroup, the multiplicative part is a semihypergroup, and the multiplication hyperoperation inclusively distributes over the hyperaddition from the left or right-hand side, i.e., for three arbitrary elements x, y, z, there is $x \cdot (y + z) \subseteq x \cdot y + x \cdot z$ for the left-hand side, and respectively, $(y + z) \cdot x \subseteq y \cdot x + z \cdot x$ for the right-hand side. If the inclusion is substituted by equality, then the general hypernear-ring is called *strongly distributive*. We also recall here that there exist also hyperrings having the additive part a group, while the multiplicative one is a semihypergroup, being called *multiplicative hyperrings* [20].

The set of all transformations of a group G, i.e., $T(G) = \{g : G \to G\}$, can be endowed with a near-ring structure, while similarly, on the set of all multitransformations of a hypergroup H, i.e., $T^*(H) = \{h : H \to \mathcal{P}^*(H)\}$, can be defined a general hypernear-ring structure, called the *multitransformations general hypernear-ring* associated with the hypergroup H. We have shown that for every general hypernear-ring R there exists a hypergroup H such that R is weakly embeddable in the associated multitransformations general hypernear-ring $T^*(H)$ (see Theorem 5). Moreover, considering the set $T(H) = \{f : H \to H\}$ of all transformations of a hypergroup H, we have defined on it a hyperaddition and a multiplication such that $T(H)$ becomes a general additive hypernear-ring. We have determined conditions under which the set $T_0(H)$, formed with the transformations of H that multiplicatively commute with the zero function on H, is a subhypernear-ring of $T(H)$. Besides, an additive hypernear-ring satisfying certain conditions can be weakly embedded in the additive hypernear-ring $T_0(H)$ (see Theorem 8).

In our future work, we intend to introduce and study properties of $\Delta-$endomorphisms and $\Delta-$multiendomorphisms of hypernear-rings as generalizations of similar notions on near-rings.

Author Contributions: The authors contributed equally to this paper.

References

1. Berman, G.; Silverman, R.J. Near-rings. *Am. Math. Mon.* **1959**, *66*, 23–34. [CrossRef]
2. Malone, J.J.; Heatherly, H.E., Jr. Some Near-Ring Embeddings. *Quart. J. Math. Oxf. Ser.* **1969**, *20*, 81–85. [CrossRef]
3. Dašić, V. Hypernear-rings. In *Algebraic Hyperstructures and Applications (Xanthi, 1990)*; World Scientific Publishing: Teaneck, NJ, USA, 1991; pp. 75–85.
4. Bonansinga, P. Sugli ipergruppi quasicanonici. *Atti Soc. Peloritana Sci. Fis. Mat. Natur* **1981**, *27*, 9–17.
5. Massouros, C.G. Quasicanonical hypergroups. In *Algebraic Hyperstructures and Applications (Xanthi, 1990)*; World Scientific Publishing: Teaneck, NJ, USA, 1991; pp. 129–136.
6. Comer, S.D. Polygroups derived from cogroups. *J. Algebra* **1984**, *89*, 387–405. [CrossRef]
7. Davvaz, B. *Polygroup Theory and Related Systems*; World Scientific Publishing, Co. Pte. Ltd.; Hackensack, NJ, USA, 2013.
8. Jančić-Rašović, S.; Cristea, I. A note on near-rings and hypernear-rings with a defect of distributivity. *AIP Conf. Proc.* **1978**, *1978*, 34007.
9. Ameri, R.; Amiri-Bideshki, M.; Hoskova-Mayerova, S.; Saeid, A.B. Distributive and Dual Distributive Elements in Hyperlattices. *Ann. Univ. Ovidius Constanta Ser. Mat.* **2017**, *25*, 25–36. [CrossRef]
10. Jančić-Rašović, S.; Cristea, I. Division hypernear-rings. *Ann. Univ. Ovidius Constanta Ser. Mat.* **2018**, *26*, 109–126. [CrossRef]
11. Gontineac, M. On Hypernear-ring and H-hypergroups. In *Algebraic Hyperstructures and Applications (Lasi, 1993)*; Hadronic Press: Palm Harbor, FL, USA, 1994; pp. 171–179.
12. Dašić, V. A defect of distributivity of the near-rings. *Math. Balk.* **1978**, *8*, 63–75.
13. Clay, J. *Nearrings: Geneses and Application*; Oxford University Press: Oxford, UK, 1992.
14. Meldrum, J. *Near-Rings and Their Links with Groups*; Pitman: London, UK, 1985.
15. Pilz, G. *Near-Rings: The theory and Its Applications*; North-Holland Publication Co.: New York, NY, USA, 1983.
16. Jančić-Rašović, S.; Cristea, I. Hypernear-rings with a defect of distributivity. *Filomat* **2018**, *32*, 1133–1149. [CrossRef]
17. Vougiouklis, T. *Hyperstructures and Their Representations*; Hadronic Press: Palm Harbor, FL, USA, 1994.
18. Jančić-Rašović, S. On a class of $P_1 - P_2$ hyperrings and hypernear-rings. *Set-Val. Math. Appl.* **2008**, *1*, 25–37.
19. Jančić-Rašović, S.; Dasic, V. Some new classes of (m,n)-hyperrings. *Filomat* **2012**, *26*, 585–596. [CrossRef]
20. Ameri, R.; Kordi, A.; Hoškova-Mayerova, S. Multiplicative hyperring of fractions and coprime hyperideals. *Ann. Univ. Ovidius Constanta Ser. Mat.* **2017**, *25*, 5–23. [CrossRef]

10

Enumeration of Strongly Regular Graphs on up to 50 Vertices having S_3 as an Automorphism Group

Marija Maksimović

Department of Mathematics, University of Rijeka, Rijeka 51000, Croatia; mmaksimovic@math.uniri.hr;

Abstract: One of the main problems in the theory of strongly regular graphs (SRGs) is constructing and classifying SRGs with given parameters. Strongly regular graphs with parameters $(37, 18, 8, 9)$, $(41, 20, 9, 10)$, $(45, 22, 10, 11)$, $(49, 24, 11, 12)$, $(49, 18, 7, 6)$ and $(50, 21, 8, 9)$ are the only strongly regular graphs on up to 50 vertices that still have to be classified. In this paper, we give the enumeration of SRGs with these parameters having S_3 as an automorphism group. The construction of SRGs in this paper is a step in the classification of SRGs on up to 50 vertices.

Keywords: strongly regular graph; automorphism group; orbit matrix

---

1. Introduction

We assume that the reader is familiar with the basic notions of the theory of finite groups. For basic definitions and properties of strongly regular graphs, we refer the reader to [1–3].

A graph is regular if all its vertices have the same valency; a simple regular graph $\Gamma = (\mathcal{V}, \mathcal{E})$ is strongly regular with parameters (v, k, λ, μ) if it has $|\mathcal{V}| = v$ vertices, valency k, and if any two adjacent vertices are together adjacent to λ vertices, while any two nonadjacent vertices are together adjacent to μ vertices. A strongly regular graph with parameters (v, k, λ, μ) is usually denoted by SRG(v, k, λ, μ). An automorphism of a strongly regular graph Γ is a permutation of vertices of Γ, such that every two vertices are adjacent if and only if their images are adjacent.

By $S(V)$, we denote the symmetric group on the nonempty set V. If $G \leq S(V)$ and $x \in V$, then the set $xG = \{xg | g \in G\}$ is called a G-orbit of x. The set $G_x = \{g \in G | xg = x\}$ is called a stabilizer of x in G. If G is finite, then $|xG| = \frac{|G|}{|G_x|}$. By G_x^g, we denote a conjugate subgroup $g^{-1}G_x g$ of G_x.

One of the main problems in the theory of strongly regular graphs (SRGs) is constructing and classifying SRGs with given parameters. A frequently-used method of constructing combinatorial structures is the construction of combinatorial structures with a prescribed automorphism group. Orbit matrices of block designs have been used for such a construction of combinatorial designs since the 1980s. However, orbit matrices of strongly regular graphs had not been introduced until 2011. Namely, Majid Behbahani and Clement Lam introduced the concept of orbit matrices of strongly regular graphs in [4]. They developed an algorithm for the construction of orbit matrices of strongly regular graphs with an automorphism group of prime order and the construction of corresponding strongly regular graphs.

A method of constructing strongly regular graphs admitting an automorphism group of composite order using orbit matrices is introduced and presented in [5]. Using this method, we classify strongly regular graphs with parameters $(37, 18, 8, 9)$, $(41, 20, 9, 10)$, $(45, 22, 10, 11)$, $(49, 18, 7, 6)$, $(49, 24, 11, 12)$ and $(50, 21, 8, 9)$ having S_3 as an automorphism group. These graphs are the only strongly regular graphs with up to 50 vertices that still have to be classified. Enumeration of SRGs with these parameters having a non-abelian automorphism group of order six, i.e., the construction of SRGs with these parameters in this paper, is a step in that classification. Using this construction, we show that there is no SRG(37, 18, 8, 9) having S_3 as an automorphism group. Furthermore, we show that there are 80 SRGs(41, 20, 9, 10), 288 SRGs(45, 22, 10, 11), 72 SRGs(49, 24, 11, 12), 34 SRGs(49, 18, 7, 6) and 45 SRGs(50, 21, 8, 9) having a non-abelian automorphism group of order six.

The paper is organized as follows: After a brief description of the terminology and some background results, in Section 2, we describe the concept of orbit matrices, based on the work of Behbahani and Lam [4]. In Section 3, we explain the method of construction of strongly regular graphs from their orbit matrices presented in [5]. In Section 4, we apply this method to construct strongly regular graphs with parameters (37, 18, 8, 9), (41, 20, 9, 10), (45, 22, 10, 11), (49, 18, 7, 6), (49, 24, 11, 12) and (50, 21, 8, 9) having a non-abelian automorphism group of order six.

For the construction of orbit matrices and graphs, we have used our own computer programs written for GAP [6]. Isomorphism testing for the obtained graphs and the analysis of their full automorphism groups are conducted using the Grape package for GAP [7].

2. Orbit Matrices of Strongly Regular Graphs

Orbit matrices of block designs have been frequently used for the construction of block designs with a presumed automorphism group, see, e.g., [8–11]. In 2011, Behbahani and Lam introduced the concept of orbit matrices of SRGs (see [4]). While Behbahani and Lam were mostly focused on orbit matrices of strongly regular graphs admitting an automorphism of prime order, a general definition of an orbit matrix of a strongly regular graph is given in [12].

Let Γ be an SRG(v, k, λ, μ) and A be its adjacency matrix. Suppose an automorphism group G of Γ partitions the set of vertices V into b orbits $O_1, \ldots, O_b$, with sizes $n_1, \ldots, n_b$, respectively. The orbits divide A into submatrices $[A_{ij}]$, where A_{ij} is the adjacency matrix of vertices in O_i versus those in O_j. We define matrices $C = [c_{ij}]$ and $R = [r_{ij}]$, $1 \leq i, j \leq b$, such that c_{ij} is the column sum of A_{ij} and r_{ij} is the row sum of A_{ij}. The matrix R is related to C by:

$$r_{ij} n_i = c_{ij} n_j. \tag{1}$$

Since the adjacency matrix is symmetric, it follows that:

$$R = C^T. \tag{2}$$

The matrix R is the row orbit matrix of the graph Γ with respect to G, and the matrix C is the column orbit matrix of the graph Γ with respect to G.

Behbahani and Lam showed that orbit matrices $R = [r_{ij}]$ and $R^T = C = [c_{ij}]$ satisfy the condition:

$$\sum_{s=1}^{b} c_{is} r_{sj} n_s = \delta_{ij}(k - \mu) n_j + \mu n_i n_j + (\lambda - \mu) c_{ij} n_j.$$

Since $R = C^T$, it follows that:

$$\sum_{s=1}^{b} \frac{n_s}{n_j} c_{is} c_{js} = \delta_{ij}(k - \mu) + \mu n_i + (\lambda - \mu) c_{ij} \tag{3}$$

and:

$$\sum_{s=1}^{b} \frac{n_s}{n_j} r_{si} r_{sj} = \delta_{ij}(k - \mu) + \mu n_i + (\lambda - \mu) r_{ji}.$$

Therefore, in [12], we introduced the following definition of orbit matrices of strongly regular graphs.

Definition 1. *A $(b \times b)$-matrix $R = [r_{ij}]$ with entries satisfying conditions:*

$$\sum_{j=1}^{b} r_{ij} = \sum_{i=1}^{b} \frac{n_i}{n_j} r_{ij} = k \tag{4}$$

$$\sum_{s=1}^{b} \frac{n_s}{n_j} r_{si} r_{sj} = \delta_{ij}(k - \mu) + \mu n_i + (\lambda - \mu) r_{ji} \tag{5}$$

where $0 \le r_{ij} \le n_j$, $0 \le r_{ii} \le n_i - 1$ and $\sum_{i=1}^{b} n_i = v$, is called a row orbit matrix for a strongly regular graph with parameters (v, k, λ, μ) and the orbit length distribution $(n_1, \dots, n_b)$.

Definition 2. *A $(b \times b)$-matrix $C = [c_{ij}]$ with entries satisfying conditions:*

$$\sum_{i=1}^{b} c_{ij} = \sum_{j=1}^{b} \frac{n_j}{n_i} c_{ij} = k \tag{6}$$

$$\sum_{s=1}^{b} \frac{n_s}{n_j} c_{is} c_{js} = \delta_{ij}(k - \mu) + \mu n_i + (\lambda - \mu) c_{ij} \tag{7}$$

where $0 \le c_{ij} \le n_i$, $0 \le c_{ii} \le n_i - 1$ and $\sum_{i=1}^{b} n_i = v$, is called a column orbit matrix for a strongly regular graph with parameters (v, k, λ, μ) and the orbit length distribution $(n_1, \dots, n_b)$.

3. The Method of Construction

A method of constructing strongly regular graphs admitting an automorphism group of composite order using orbit matrices is introduced and presented in [5]. In this section, we will give a brief overview of this method.

For the construction of strongly regular graphs with parameters (v, k, λ, μ), we first check whether these parameters are feasible (see [2]). Then, we select the group G and assume that it acts as an automorphism group of an SRG(v, k, λ, μ). The construction of strongly regular graphs admitting an action of a presumed automorphism group, using orbit matrices, consists of the following two basic steps:

- Construction of orbit matrices for the presumed automorphism group
- Construction of strongly regular graphs from the obtained orbit matrices (indexing of orbit matrices)

We could use row or column orbit matrices, but since we are constructing matrices row by row, it is more convenient for us to use column orbit matrices. For the construction of orbit matrices for the presumed automorphism group, we need to determine all possible orbit length distributions $(n_1, n_2, \dots, n_b)$ for an action of the group G. Suppose an automorphism group G of Γ partitions the set of vertices V into b orbits $O_1, \dots, O_b$, with sizes $n_1, \dots, n_b$. Obviously, n_i is a divisor of $|G|$, $i = 1, \dots, b$, and:

$$\sum_{i=1}^{b} n_i = v.$$

When determining the orbit length distribution, we also use the following result that can be found in [13].

Theorem 1. *Let $s < r < k$ be the eigenvalues of an SRG(v, k, λ, μ), then:*

$$\phi \le \frac{\max(\lambda, \mu)}{k - r} v,$$

where ϕ is the number of fixed points for a nontrivial automorphism group G.

For each orbit length distribution we construct column orbit matrices. For the construction of orbit matrices, we first need to find prototypes.

3.1. Prototypes for a Row of a Column Orbit Matrix

A prototype for a row of a column orbit matrix C gives us information about the number of occurrences of each integer as an entry of a particular row of C. Behbahani and Lam [4,13] introduced the concept of a prototype for a row of a column orbit matrix C of a strongly regular graph with a presumed automorphism group of prime order. We will generalize this concept and describe a prototype for a row of a column orbit matrix C of a strongly regular graph under a presumed automorphism group of composite order. Prototypes are useful in the first step of the construction of strongly regular graphs, namely the construction of column orbit matrices.

Suppose an automorphism group G of a strongly regular graph Γ with parameters (v, k, λ, μ) partitions the set of vertices V into b orbits $O_1, \ldots, O_b$, of sizes $n_1, \ldots, n_b$. With $l_i, i = 1, \ldots, \rho$, we denote all divisors of $|G|$ in ascending order ($l_1 = 1, \ldots, l_\rho = |G|$).

3.1.1. Prototypes for a Fixed Row

Consider the row r of a column orbit matrix C. We say that it is a fixed row of a matrix C if $n_r = 1$, i.e., if it corresponds to an orbit of length one. The entries in this row are either zero or one. Let d_{l_i} denote the number of orbits whose length are l_i, $i = 1, \ldots, \rho$.

Let x_e denote the number of occurrences of an element $e \in \{0, 1\}$ at the positions of the row r that correspond to the orbits of length one. It follows that:

$$x_0 + x_1 = d_1, \tag{8}$$

where d_1 is the number of orbits of length one. Since the diagonal elements of the adjacency matrix of a strongly regular graph are equal to zero, it follows that $x_0 \geq 1$.

Let $y_e^{(l_i)}$ denote the number of occurrences of an element $e \in \{0, 1\}$ at the positions of the row r that correspond to the orbits of length l_i ($i = 2, \ldots, \rho$). We have:

$$y_0^{(l_i)} + y_1^{(l_i)} = d_{l_i}, \quad i = 2, \ldots, \rho \tag{9}$$

Because the row sum of an adjacency matrix of Γ is equal to k, it follows that:

$$x_1 + \sum_{i=2}^{\rho} l_i \cdot y_1^{(l_i)} = k. \tag{10}$$

The vector:

$$p_1 = (x_0, x_1; y_0^{(l_2)}, y_1^{(l_2)}; \ldots; y_0^{(l_\rho)}, y_1^{(l_\rho)})$$

whose components are nonnegative integer solutions of the equalities (8), (9) and (10) is called a prototype for a fixed row.

3.1.2. Prototypes for a Non-Fixed Row

Let us consider the row r of a column orbit matrix C, where $n_r \neq 1$. Let d_{l_i} denote the number of orbits whose length is l_i, $i = 1, \ldots, \rho$.

If a fixed vertex is adjacent to a vertex from an orbit $O_i, 1 \leq i \leq b$, then it is adjacent to all vertices from the orbit O_i. Therefore, the entries at the positions corresponding to fixed columns are either zero or n_r. Let x_e denote the number of occurrences of an element $e \in \{0, n_r\}$ at those positions of the row r, which correspond to the orbits of length one. We have:

$$x_0 + x_{n_r} = d_1. \tag{11}$$

The entries at the positions corresponding to the orbits whose lengths are greater than one are $0, 1, \ldots, n_r - 1$ or n_r. The entry at the position (r, r) is $0 \leq c_{rr} \leq n_r - 1$, since the diagonal elements of the adjacency matrix of strongly regular graphs are zero.

Let $y_e^{(l_i)}$ denote the number of occurrences of an element $e \in \{0, \ldots, n_r\}$ of row r at the positions that correspond to the orbits of length l_i $(i = 2, \ldots, \rho)$. From (1) and (2), we conclude that:

$$c_{ri} n_i = c_{ir} n_r, \tag{12}$$

where $c_{ir} \in \{0, \ldots, n_i\}$. If $c_{ri} \cdot \frac{n_i}{n_r} \notin \{0, \ldots, n_i\}$, then $y_{c_{ri}}^{(n_i)} = 0$. It follows that:

$$\sum_{e=0}^{n_r} y_e^{(l_i)} = d_{l_i}, \quad i = 2, \ldots, \rho. \tag{13}$$

Since the row sum of an adjacency matrix is equal to k, we have that:

$$x_{n_r} + \sum_{i=2}^{\rho} \sum_{h=1}^{n_r} y_h^{(l_i)} \cdot h \cdot \frac{n_{l_i}}{n_r} = k, \tag{14}$$

From (3), we conclude that:

$$\sum_{s=1}^{b} c_{rs} c_{rs} n_s = (k - \mu) n_r + \mu n_r^2 + (\lambda - \mu) c_{rr} n_r,$$

where $c_{rr} \in \{0, \ldots, n_r - 1\}$. It follows that:

$$n_r^2 x_{n_r} + \sum_{i=2}^{\rho} \sum_{h=1}^{n_r} y_h^{(l_i)} \cdot h^2 \cdot n_{l_i} = (k - \mu) n_r + \mu n_r^2 + (\lambda - \mu) c_{rr} n_r, \tag{15}$$

The vector:

$$p_{n_r} = (x_0, x_{n_r}; y_0^{l_2}, \ldots, y_{n_r}^{l_2}; \ldots; y_0^{l_\rho}, \ldots, y_{n_r}^{l_\rho}),$$

whose components are nonnegative integer solutions of Equalities (11), (13), (14) and (15) is called a prototype for a row corresponding to the orbit of length n_r.

Using prototypes, we construct an orbit matrix row by row.

Not every orbit matrix gives rise to a strongly regular graph, while, on the other hand, a single orbit matrix may produce several nonisomorphic strongly regular graphs. Further, nonisomorphic orbit matrices may produce isomorphic graphs. Therefore, the constructed graphs need to be checked for isomorphism.

Theorem 2. *Let $\Gamma = (V, E)$ be a strongly regular graph, $G \leq Aut(\Gamma)$, and let $(b \times b)$-matrix C be a column orbit matrix of the graph Γ with respect to the group G. Further, let α be an element of $S(V)$ with the following property: if $\alpha(i) = j$, then the stabilizer G_{x_i} is conjugate to G_{x_j}, where $x_i, x_j \in V$ and $O_i = x_i G, O_j = x_j G$. Then, there exists permutation $g^* \in C_{S(V)}(G)$ such that $\alpha(i) = j \iff g^*(O_i) = O_j$.*

Definition 3. *Let $A = (a_{ij})$ be an $(b \times b)$-matrix and $\alpha \in S_b$. The matrix $B = A\alpha$ is the $(b \times b)$-matrix $B = (b_{ij})$, where $b_{\alpha(i)\alpha(j)} = a_{ij}$. If $A\alpha = A$, then α is called an automorphism of the matrix A.*

Definition 4. *Let an $(b \times b)$-matrix $A = (a_{ij})$ be the orbit matrix of a strongly regular graph Γ with respect to the group $G \leq Aut(\Gamma)$. A mapping $\alpha \in S_b$ is called an isomorphism from A to $B = A\alpha$ if the following condition holds: if $\alpha(i) = j$, then the stabilizer G_{x_i} is conjugate to G_{x_j}. We say that the orbit matrices A and B are isomorphic. If $A\alpha = A$, then α is called an automorphism of the orbit matrix A. All automorphisms of an orbit matrix A form the full automorphism group of A, denoted by $Aut(A)$.*

During the construction of orbit matrices, for the elimination of isomorphic structures, we use permutations that satisfy the conditions from Theorem 2, i.e., isomorphisms from Definition 4.

The next big step of the construction of graphs, called indexing, often cannot be performed in a reasonable amount of time. To make such a construction possible, for a refinement of constructed orbit matrices, we use the composition series:

$$\{1\} = H_0 \trianglelefteq H_1 \trianglelefteq \cdots \trianglelefteq H_n = G,$$

of a solvable automorphism group G of a strongly regular graph. Let Γ be a strongly regular graph and $H \trianglelefteq G \leq \mathrm{Aut}(\Gamma)$. Each G-orbit of Γ decomposes to H-orbits of the same size (see [9]). Therefore, each orbit matrix for the group G decomposes to orbit matrices for the group H, and the following theorem holds [5].

Theorem 3. *Let Ω be a finite nonempty set, $H \triangleleft G \leq S(\Omega)$, $x \in \Omega$ and $xG = \bigsqcup_{i=1}^{h} x_i H$. Then, a group G/H acts transitively on the set $\{x_i H \mid i = 1, 2, \ldots, h\}$.*

Therefore, after we have constructed corresponding orbit matrices for the group G, we continue until we find all refinements for the normal subgroup $H_{n-1} \trianglelefteq G$. In the next step, we obtain orbit matrices for the group H_{n-2}, H_{n-3}, and so on. Our last step is the construction of the corresponding orbit matrices for the subgroup $H_0 = \{1\}$, i.e., construction of adjacency matrices of the strongly regular graphs. The concept of the G-isomorphism of two-block designs was introduced in [14]. For the elimination of mutually-isomorphic structures, we use the concept of G-isomorphism.

Definition 5. *Let $\Gamma_1 = (V, E_1)$ and $\Gamma_2 = (V, E_2)$ be strongly regular graphs, and let $G \leq Aut(\Gamma_1) \cap Aut(\Gamma_2) \leq S(V)$. An isomorphism $\alpha : \Gamma_1 \to \Gamma_2$ is called a G-isomorphism from Γ_1 onto Γ_2 if there is an automorphism $\tau : G \to G$ such that for each $x, y \in V$ and each $g \in G$, the following holds:*

$$(\alpha x)(\tau g) = \alpha y \Leftrightarrow xg = y.$$

If α is a $G-$isomorphism from Γ_1 to Γ_2, then the vertices x_i and x_j are in the same G-orbit if and only if the vertices $\alpha(x_i)$ and $\alpha(x_j)$ are in the same G-orbit.

Lemma 1. *Let $\Gamma_1 = (V, E_1)$ and $\Gamma_2 = (V, E_2)$ be strongly regular graphs, and let $G \leq Aut(\Gamma_1) \cap Aut(\Gamma_2) \leq S = S(V)$. A permutation $\alpha \in S$ is a G-isomorphism from Γ_1 onto Γ_2 if and only if α is an isomorphism from Γ_1 to Γ_2 and $\alpha \in N_S(G)$, where $N_S(G)$ is the normalizer of G in S.*

In each step of refinement of an orbit matrix A, we eliminate isomorphic orbit matrices using the automorphisms from $Aut(A)$, because each automorphism of an orbit matrix determines an G-isomorphism.

4. SRGs with up to 50 Vertices Having S_3 as an Automorphism Group

SRGs with parameters $(37, 18, 8, 9)$, $(41, 20, 9, 10)$, $(45, 22, 10, 11)$, $(49, 24, 11, 12)$, $(49, 18, 7, 6)$ and $(50, 21, 8, 9)$ are the only strongly regular graphs on up to 50 vertices that still have to be classified [2,15]. According to [2], it is known that strongly regular graphs with these parameters exist, but their final enumeration result is not known. In this section, we present the results of the constructed strongly regular graphs with parameters $(37, 18, 8, 9)$, $(41, 20, 9, 10)$, $(45, 22, 10, 11)$, $(49, 18, 7, 6)$, $(49, 24, 11, 12)$ and $(50, 21, 8, 9)$ having $S_3 \cong Z_3 : Z_2 \cong \langle \rho\phi | \rho^3 = 1, \phi^2 = 1, \phi\rho\phi = \rho^{-1} \rangle$ as an automorphism group. In each case, we construct strongly regular graphs by using the algorithm described in Section 3. The orbit lengths for an action of the group G at the set of points of a graph can get values from the set $\{1, 2, 3, 6\}$. Using the program Mathematica [16], we get all possible orbit length distributions

(d_1, d_2, d_3, d_6) for the action of S_3 on a particular SRG that satisfy Theorem 1. For each orbit length distribution, we find the corresponding prototypes using Mathematica. Using our own programs, which are written for GAP [6], we construct all orbit matrices for a given orbit length distribution. Having in mind the action of the whole group, we refine the constructed orbit matrices. For the refinement, we use the composition series

$$\{1\} \trianglelefteq \langle \rho \rangle \trianglelefteq S_3$$

and obtain orbit matrices for the action of the subgroup $Z_3 \triangleleft S_3$. In this step, each orbit of length two and six decomposes to two orbits of length one and three, respectively. In the final step of the construction, we obtain adjacency matrices of the strongly regular graphs with particular parameters admitting a non-abelian automorphism group of order six. Finally, we check isomorphisms of strongly regular graphs and determine orders of the full automorphism groups using the Grape package for GAP [7].

4.1. SRGs(37,18,8,9)

In this section, we present the results of SRGs(37,18,8,9) having S_3 as an automorphism group. According to [17], there are at least 6760 SRGs(37,18,8,9), and none of them have S_3 as an automorphism group. We show that there are no strongly regular graphs with parameters (37,18,8,9) having a non-abelian automorphism group of order six.

We get 176 possibilities for orbit length distributions, but only three give rise to orbit matrices. In Table 1, we present the number of mutually-nonisomorphic orbit matrices for each orbit length distribution, the number of orbit matrices for Z_3 (obtained by the refinement) and the number of constructed SRGs with parameters (37,18,8,9). These calculations prove Theorem 4.

Table 1. Number of orbit matrices and SRGs(37,18,8,9) for the automorphism group S_3.

Distribution	**#OM-S_3**	**#OM-Z_3**	**#SRGs**
(1,0,0,6)	3	6	0
(1,0,4,4)	3	3	0
(1,0,8,2)	3	3	0

Theorem 4. *There are no strongly regular graphs with parameters* $(37, 18, 8, 9)$ *having an automorphism group isomorphic to the symmetric group* S_3.

4.2. SRGs(41,20,9,10)

In this section, we present the results of SRGs$(41, 20, 9, 10)$ having S_3 as an automorphism group. We show that there are exactly 80 strongly regular graphs with parameters $(41, 20, 9, 10)$ having a non-abelian automorphism group of order six.

We get 216 possibilities for orbit length distributions, but only one gives rise to any orbit matrices. In Table 2, we present the number of mutually-nonisomorphic orbit matrices for each orbit length distribution, the number of orbit matrices for Z_3 (obtained by the refinement) and the number of constructed SRGs with parameters $(41, 20, 9, 10)$. These calculations prove Theorem 5. Information about the orders of the full automorphism groups is presented in Table 3.

Table 2. Number of orbit matrices and SRGs(41,20,9,10) for the automorphism group S_3.

Distribution	**#OM-S_3**	**#OM-Z_3**	**#SRGs**
(1,2,4,4)	10	10	80

Theorem 5. *Up to isomorphism, there are exactly 80 strongly regular graphs with parameters* $(41, 20, 9, 10)$ *having an automorphism group isomorphic to the symmetric group* S_3.

Table 3. SRGs with parameters (41,20,9,10) having S_3 as an automorphism group.

$\|\mathbf{Aut}(\Gamma)\|$	**#SRGs**
6	80

The adjacency matrices of the constructed SRGs can be found at [18].

4.3. SRGs(45,22,10,11)

In this section, we present the results of SRGs$(45, 22, 10, 11)$ having S_3 as an automorphism group. We show that there are exactly 288 strongly regular graphs with parameters $(45, 22, 10, 11)$ having a non-abelian automorphism group of order six.

We get 309 possibilities for orbit length distributions, but only one gives rise to any orbit matrices. In Table 4, we present the number of mutually-nonisomorphic orbit matrices for each orbit length distribution, the number of orbit matrices for Z_3 (obtained by the refinement) and the number of constructed SRGs with parameters $(45, 22, 10, 11)$. These calculations prove Theorem 6. Information about orders of the full automorphism groups is presented in Table 5.

Table 4. Number of orbit matrices and SRGs(45,22,10,11) for the automorphism group S_3.

Distribution	**#OM-S_3**	**#OM-Z_3**	**#SRGs**
(1,4,4,4)	7	7	288

Table 5. SRGs with parameters (45,22,10,11) having S_3 as an automorphism group

$\|\mathbf{Aut}(\Gamma)\|$	**#SRGs**
6	288

Theorem 6. *Up to isomorphism, there are exactly 288 strongly regular graphs with parameters* $(45, 22, 10, 11)$ *having an automorphism group isomorphic to the symmetric group* S_3.

The adjacency matrices of the constructed SRGs can be found at [19].

4.4. SRGs(49,18,7,6)

In the paper [5], we proved the following theorem.

Theorem 7. *Up to isomorphism, there are exactly* 36 *strongly regular graphs with parameters* $(49, 18, 7, 6)$ *having an automorphism group isomorphic to the symmetric group* S_3.

Two of these graphs have not been constructed in [4,20]. The adjacency matrices of the constructed SRGs can be found at [21].

4.5. SRGs(49,24,11,12)

In this section, we present the results of SRGs$(49, 24, 11, 12)$ having S_3 as an automorphism group. We show that there are exactly 72 strongly regular graphs with parameters $(49, 24, 11, 12)$ having a non-abelian automorphism group of order six.

We get 435 possibilities for orbit length distributions, but only a few give rise to orbit matrices. In Table 6, we present the number of mutually-nonisomorphic orbit matrices for each orbit length distribution, the number of orbit matrices for Z_3 (obtained by the refinement) and the number of constructed SRGs with parameters $(49, 24, 11, 12)$. Thus, we prove Theorem 8. Information about orders of the full automorphism groups is presented in Table 7.

Table 6. Number of orbit matrices and SRGs$(49, 24, 11, 12)$ for the automorphism group S_3.

Distribution	#OM-S_3	#OM-Z_3	#SRGs	Distribution	#OM-S_3	#OM-Z_3	#SRGs
(0,2,3,6)	8	16	6	(1,3,0,7)	6	6	0
(0,2,5,5)	4	0	0	(1,3,2,6)	10	2	0
(0,2,7,4)	8	0	0	(1,3,6,4)	2	2	0
(1,0,0,8)	2	15	2	(1,6,0,6)	1	0	0
(1,0,2,7)	20	32	0	(3,2,0,7)	4	10	0
(1,0,8,4)	26	24	12	(3,2,6,4)	6	16	0
(1,0,10,3)	2	0	0	(4,0,9,3)	6	0	0
(1,0,12,2)	16	0	0	(5,1,0,7)	2	4	0
(1,0,14,1)	12	0	0	(5,1,6,4)	2	2	12
				(7,0,0,7)	2	2	40

Table 7. SRGs with parameters $(49, 24, 11, 12)$ having S_3 as an automorphism group.

$\|\mathrm{Aut}(\Gamma)\|$	#SRGs
6	42
18	22
24	4
126	4

Theorem 8. *Up to isomorphism, there are exactly 72 strongly regular graphs with parameters* $(49, 24, 11, 12)$ *having an automorphism group isomorphic to the symmetric group* S_3.

The adjacency matrices of the constructed SRGs can be found at [22].

4.6. SRGs(50,21,8,9)

In this section, we present the results of SRGs$(50, 21, 8, 9)$ having S_3 as an automorphism group. According to [17], there are 18 graphs obtained from the 18 Steiner (2,4,25) systems, and three of them have S_3 as an automorphism group. We show that there are exactly 45 strongly regular graphs with parameters $(50, 21, 8, 9)$ having a non-abelian automorphism group of order six. Hence, to our best knowledge, 42 of the constructed strongly regular graphs are new.

We get 340 possibilities for orbit length distributions, but only a few give rise to orbit matrices. In Table 8, we present the number of mutually-nonisomorphic orbit matrices for each orbit length distribution, the number of orbit matrices for Z_3 (obtained by the refinement) and the number of constructed SRGs with parameters $(50, 21, 8, 9)$. Thus, we prove Theorem 9. Information about the orders of the full automorphism groups is presented in Table 9.

Theorem 9. *Up to isomorphism, there are exactly 45 strongly regular graphs with parameters* $(50, 21, 8, 9)$ *having an automorphism group isomorphic to the symmetric group* S_3.

The adjacency matrices of the constructed SRGs can be found at [23].

In Table 10, we summarize the obtained results, i.e., give a list of all the obtained strongly regular graphs and orders of their full automorphism groups.

Table 8. Number of orbit matrices and SRGs(50, 21, 8, 9) for the automorphism group S_3.

Distribution	#OM-S_3	#OM-Z_3	#SRGs	Distribution	#OM-S_3	#OM-Z_3	#SRGs
(0,1,2,7)	10	3	2	(2,0,2,7)	10	16	0
(0,1,4,6)	10	4	6	(2,0,8,4)	22	24	12
(0,1,6,5)	12	21	6	(2,0,10,3)	2	0	0
(0,1,8,4)	8	8	1	(2,0,12,2)	27	0	0
(0,4,2,6)	2	2	0	(2,0,14,1)	14	0	0
(0,4,4,5)	4	3	16	(2,3,0,7)	2	3	0
(0,4,6,4)	3	4	0	(2,3,2,6)	6	1	0
(0,4,8,2)	4	6	0	(2,3,6,4)	2	4	0
(1,2,3,6)	10	20	5	(4,2,6,4)	6	12	0
(1,2,5,5)	2	0	0	(5,0,9,3)	2	0	0
(1,2,7,4)	4	0	0	(6,1,6,4)	1	1	4

Table 9. SRGs with parameters (50, 21, 8, 9) having S_3 as an automorphism group.

$\|Aut(\Gamma)\|$	#SRGs
6	35
18	6
72	1
150	1
336	1
504	1

Table 10. SRGs on up to 50 vertices having S_3 as an automorphism group.

(v, k, λ, μ)	$\|Aut(\Gamma)\|$	#SRGs
(41, 20, 9, 10)	6	80
(45, 22, 10, 11)	6	288
(49, 18, 7, 6)	6	18
(49, 18, 7, 6)	12	2
(49, 18, 7, 6)	18	2
(49, 18, 7, 6)	24	4
(49, 18, 7, 6)	48	1
(49, 18, 7, 6)	72	4
(49, 18, 7, 6)	126	1
(49, 18, 7, 6)	144	2
(49, 18, 7, 6)	1008	1
(49, 18, 7, 6)	1764	1
(49, 24, 11, 12)	6	42
(49, 24, 11, 12)	18	22
(49, 24, 11, 12)	24	4
(49, 24, 11, 12)	126	4
(50, 21, 8, 9)	6	35
(50, 21, 8, 9)	18	6
(50, 21, 8, 9)	72	1
(50, 21, 8, 9)	150	1
(50, 21, 8, 9)	336	1
(50, 21, 8, 9)	504	1

Acknowledgments: Special thanks to Dean Crnković.

Abbreviations

The following abbreviations are used in this manuscript:

SRG Strongly regular graph

References

1. Beth, T.; Jungnickel, D.; Lenz, H. *Design Theory Volume I*, 2nd ed.; Cambridge University Press: Cambridge, UK, 1999.
2. Brouwer, A.E. Strongly Regular Graphs. In *Handbook of Combinatorial Designs*, 2nd ed.; Colbourn, C.J., Dinitz, J.H., Eds.; Chapman & Hall/CRC: Boca Raton, FL, USA, 2007; pp. 852–868.
3. Tonchev, V.D. *Combinatorial Configurations: Designs, Codes, Graphs*; Longman Scientific & Technical: New York, NY, USA, 1988.
4. Behbahani, M.; Lam, C. Strongly regular graphs with non-trivial automorphisms. *Discret. Math.* **2011**, *311*, 132–144. [CrossRef]
5. Crnković, D.; Maksimović, M. Construction of strongly regular graphs having an automorphism group of composite order. under review.
6. GAP Groups. Algorithms, Programming—A System for Computational Discrete Algebra, Version 4.7.2. 2013. Available online: http://www.gap-system.org (accessed on 20 January 2017).
7. Soicher, L.H. The GRAPE Package for GAP, Version 4.6.1. 2012. Available online: http://www.maths.qmul.ac.uk/~leonard/grape/ (accessed on 4 April 2018).
8. Janko, Z. Coset enumeration in groups and constructions of symmetric designs. *Ann. Discret. Math.* **1992**, *52*, 275–277.
9. Crnković, D.; Rukavina, S. Construction of block designs admitting an abelian automorphism groups. *Metrika* **2005**, *62*, 175–183. [CrossRef]
10. Crnković, D.; Rukavina, S.; Schmidt, M. A Classification of all Symmetric Block Designs of Order Nine with an Automorphism of Order Six. *J. Combin. Des.* **2006**, *14*, 301–312. [CrossRef]
11. Crnković, D.; Pavčević, M.O. Some new symmetric designs with parameters (64,28,12). *Discret. Math.* **2001**, *237*, 109–118. [CrossRef]
12. Crnković, D.; Maksimović, M.; Rodrigues, B.G.; Rukavina, S. Self-orthogonal codes from the strongly regular graphs on up to 40 vertices. *Adv. Math. Commun.* **2016**, *10*, 555–582. [CrossRef]
13. Behbahani, M. On Strongly Regular Graphs. Ph.D. Thesis, Concordia University, Montreal, QC, Canada, May 2009.
14. Ćepulić, V. On Symmetric Block Designs (40,13,4) with Automorphisms of Order 5. *Discret. Math.* **1994**, *128*, 45–60. [CrossRef]
15. Brouwer, A.E. Parameters of Strongly Regular Graphs. Available online: http://www.win.tue.nl/~aeb/graphs/srg/srgtab1-50.html (accessed on 1 April 2018).
16. Wolfram Mathematica, Version 7.0.0. 2008. Available online: http://www.wolfram.com/mathematica/ (accessed on 4 April 2018).
17. Spence, E. Strongly Regular Graphs on at Most 64 Vertices. Available online: http://www.maths.gla.ac.uk/~es/srgraphs.php (accessed on 10 April 2018).
18. Maksimović, M. SRGs(41,20,9,10) having S_3 as an automorphism group. Available online: http://www.math.uniri.hr/~mmaksimovic/srg41.txt (accessed on 10 June 2018).
19. Maksimović, M. SRGs(45,22,10,11) having S_3 as an automorphism group. Available online: http://www.math.uniri.hr/~mmaksimovic/srg45.txt (accessed on 10 June 2018).
20. Behbahani, M.; Lam, C.; Östergård, P.R.J. On triple systems and strongly regular graphs. *J. Combin. Theory Ser. A* **2012**, *119*, 1414–1426. [CrossRef]
21. Maksimović, M. SRGs(49,18,7,6) having S_3 as an automorphism group. Available online: http://www.math.uniri.hr/~mmaksimovic/srg49s3.txt (accessed on 10 June 2018).
22. Maksimović, M. SRGs(49,24,11,12) having S_3 as an automorphism group. Available online: http://www.math.uniri.hr/~mmaksimovic/srgs49.txt (accessed on 10 June 2018).
23. Maksimović, M. SRGs(50,21,8,9) having S_3 as an automorphism group. Available online: http://www.math.uniri.hr/~mmaksimovic/srg50.txt (accessed on 10 June 2018).

11

A Novel Edge Detection Method based on the Regularized Laplacian Operation

Huilin Xu * and Yuhui Xiao

College of Mathematics and Computer Science, Gannan Normal University, Ganzhou 341000, China; 15707978628@163.com
* Correspondence: xuhuilin@163.com.

Abstract: In this paper, an edge detection method based on the regularized Laplacian operation is given. The Laplacian operation has been used extensively as a second-order edge detector due to its variable separability and rotation symmetry. Since the image data might contain some noises inevitably, regularization methods should be introduced to overcome the instability of Laplacian operation. By rewriting the Laplacian operation as an integral equation of the first kind, a regularization based on partial differential equation (PDE) can be used to compute the Laplacian operation approximately. We first propose a novel edge detection algorithm based on the regularized Laplacian operation. Considering the importance of the regularization parameter, an unsupervised choice strategy of the regularization parameter is introduced subsequently. Finally, the validity of the proposed edge detection algorithm is shown by some comparison experiments.

Keywords: edge detection; Laplacian operation; regularization; parameter selection; performance evaluation

---

1. Introduction

In a digital image, edges can be defined as abrupt changes of the image intensity. Edge is one of the most essential features contained in an image. The result of edge detection not only retains the main information of an image, but also reduces the amount of data to be processed drastically. Therefore, edge detection has been used as a front-end step in many image processing and computer vision applications [1].

Since the abrupt changes in an image can be reflected by their derivatives, differentiation-based methods are widely used in edge detection. Generally, edges can be detected by finding the maximum of first-order derivatives or the zero-crossing of second-order derivatives of the image intensity. From the original contribution of Roberts in 1965, there have been a large number of works concerning this topic. Some researchers have paid attention to constructing optimal filters according to some reasonable hypotheses and criteria (see [2–5]), while some others are interested in designing discrete masks, such as the well-known Prewitt, Sobel and Laplacian of Gaussian (LoG) operators. Some recently developed methods can be found in [6–8].

The differentiation-based edge detection methods need to calculate derivatives numerically. As we know, numerical differentiations are unstable since a small perturbation of the data may cause huge errors in its derivatives [9]. In real applications, the image is often corrupted by noise during the processes of collection, acquisition and transmission. In order to calculate derivatives of the noisy data stably, some regularization methods should be introduced. There have been much work into this over the past years, such as the Tikhonov regularization [10], the Lavrentiev regularization [11], the Lanczos method [12], the mollification method [9] and the total variation method [13]. Some of the regularization methods for computing the first-order numerical differentiation have been applied to detecting image edges (see [10,13]).

Compared with the first-order numerical differentiation, the computation of second-order derivatives is more unstable and more likely to be influenced by noises. However, the edge detection based on second-order derivatives has higher localization accuracy and a stronger response to final details [14]. The most common second-order derivative used in edge detection is the Laplacian operation due to its variable separability and rotation invariance. In order to overcome the instability of Laplacian operation, one of the existing works is the LoG [2]. Since the image data is discrete, the sampled representation of the LoG and some related issues have been discussed in [15]. The performance of a LoG detector depends mainly on the choice of the scale parameter. For larger scales, the zero-crossings deviate from the true edges, which may cause poor localization. For small scales, there would be many false zero-crossings produced by noises. Besides the LoG detector, a model for designing a discrete mask of the Laplacian operator is introduced in [7].

In view of the above-mentioned facts, a natural idea is to compute the Laplacian operation by the regularization method and construct a novel edge detection algorithm based on this. By rewriting the Laplacian operation as an integral equation of the first kind, a PDE-based regularization for computing the Laplacian operation has been proposed in [16]. In this paper, the PDE-based regularization method will be generalized to edge detection. Based on the objective parameter selection for edge detection given in [17], we will introduce a new choice strategy of the regularization parameter. Comparative experiments with the LoG detector and the Laplacian-based mask given in [7] are considered.

The paper is organized as follows. In Section 2, the PDE-based regularization method for computing the Laplacian operation of image data is given. The novel edge detection algorithm based on the regularized Laplacian operation is given in Section 3. Comparative experiments are shown in Section 4. Finally, the main conclusions are summarized in Section 5.

2. Regularized Laplacian Operation

Considering the image intensity as a function $f(r)$, $r = (x, y)$ of two variables, the Laplacian operation can be defined as

$$u = \Delta f = \frac{\partial^2 f}{\partial x^2} + \frac{\partial^2 f}{\partial y^2}, \ (x,y) \in \Omega := [0,a] \times [0,b].$$

Without loss of generality, we assume the value of $f(x,y)$ on the boundary of Ω is zero, i.e., $f|_{\partial\Omega} \equiv 0$. Otherwise, denote f_0 as the solution of

$$\begin{cases} \Delta f_0 = 0, & \text{in } \Omega \\ f_0 = f, & \text{on } \partial\Omega \end{cases},$$

and replace f by $f - f_0$. Since the latter satisfies

$$\Delta(f - f_0) = \Delta f = u, \ (f - f_0)|_{\partial\Omega} \equiv 0,$$

it has

$$\begin{cases} \Delta f = u, & \text{in } \Omega \\ f = 0, & \text{on } \partial\Omega \end{cases}. \tag{1}$$

Problem (1) is the Dirichlet problem of the Poisson equation. According to the classic theory of the Poisson equation, the relationship between f and u can be expressed as

$$A[u] := \int_{\Omega} G(r,r')\, u(r')\mathrm{d}r' = -f, \tag{2}$$

where $G(r,r')$ is the Green function of the Dirichlet problem (see [18]). Since Ω is a rectangular domain, the Green function has the explicit expression

$$G(r,r') = \sum_{k_1,k_2=1}^{\infty} p(k_1,k_2)\, u(r;k_1,k_2)\, u(r';k_1,k_2),$$

where

$$u(r;k_1,k_2) = \sin\frac{k_1\pi x}{a}\sin\frac{k_2\pi y}{b},\ p(k_1,k_2) = \frac{4ab}{\pi^2(k_1^2b^2+k_2^2a^2)}.$$

The calculation of Laplacian operation $u = \Delta f$ is equivalent to solving the integral Equation (2), which can be simplified in the following.

Denote f^δ as the noise data of f; the calculation of the Laplacian operation Δf^δ is unstable, which means the noise may be amplified. A stabilized strategy is to solve the equivalent Equation (2) by the regularization method. Solving the integral Equation (2) by the Lavrentiev regularization method, an efficient method is given in [16]. The Laplacian operation can be computed approximately by solving the regularization equation

$$\alpha u^{\alpha,\delta} + A[u^{\alpha,\delta}] = -f^\delta, \tag{3}$$

where $\alpha > 0$ is the regularization parameter, and $u^{\alpha,\delta}$ is the regularized Laplacian operation. Assuming that $h^{\alpha,\delta}$ is a function satisfying

$$\begin{cases} \Delta h^{\alpha,\delta} = u^{\alpha,\delta}, & \text{in } \Omega \\ h^{\alpha,\delta} = 0, & \text{on } \partial\Omega \end{cases},$$

then it has $A[u^{\alpha,\delta}] = -h^{\alpha,\delta}$. Equation (3) can be rewritten as

$$\begin{cases} \alpha\Delta h^{\alpha,\delta} - h^{\alpha,\delta} = -f^\delta, & \text{in } \Omega \\ h^{\alpha,\delta} = 0, & \text{on } \partial\Omega \end{cases} \tag{4}$$

This boundary value problem of PDE can be solved by classic numerical methods, and then the regularized Laplacian operation $u^{\alpha,\delta}$ can be expressed as

$$u^{\alpha,\delta}(r) = \Delta h^{\alpha,\delta}(r) = \frac{1}{\alpha}[h^{\alpha,\delta}(r) - f^\delta(r)],\ r \in \Omega. \tag{5}$$

From the above rewriting, we can see that (4) and (5) are equivalent to the integral Equation (3). Compared with solving the regularization Equation (3) directly, the computational burden of solving (4) and (5) is reduced drastically.

The work of [16] mainly focuses on the choice of the regularization parameter α and the error estimate of the regularized Laplacian operation $u^{\alpha,\delta}$. Unfortunately, the choice strategy given in [16] depends on the noise level of the noise data, which is unknown in practice. Since the choice strategy of the regularization parameter plays an important role in the regularization method, as the authors stated in [16], the selection of parameter α in the edge detection algorithm should be considered carefully.

3. The Edge Detection Algorithm

In this section, we will construct the novel edge detection algorithm based on the regularized Laplacian operation given in Section 2.

The first thing we are concerned with is the weakness of the Lavrentiev regularization. Notice that $h^{\alpha,\delta}(r) = 0,\ r \in \partial\Omega$, it has $u^{\alpha,\delta}(r) = -\frac{1}{\alpha}f^\delta(r),\ r \in \partial\Omega$. The parameter $\alpha > 0$ is usually a small number, which means the error of the regularized Laplacian operation on the boundary can be amplified $\frac{1}{\alpha}$ times. Thus, the computation is meaningless on $\partial\Omega$. In fact, the validity of the regularized Laplacian operation $u^{\alpha,\delta}(r)$ has been weakened when r is close to the boundary. Experiments in [16] have shown that the weakness only affects the points very close to the boundary. Hence, except a few pixels which are as close as possible to the boundary of the image domain, the edge detection results will be acceptable.

The second thing we are concerned with is the choice strategy of the regularization parameter α. Since the noise level of an image data is unknown, the choice strategy given in [16] cannot be carried out. Considering only the edge detection problem, the objective parameter selection given in [17] can be adopted to choose the regularization parameter.

Once the regularization parameter α is chosen, the regularized Laplacian operation $u^{\alpha,\delta}$ can be obtained by solving Equations (4) and (5), where Equation (4) can be solved by the standard finite difference method or finite element method.

Combined with the objective parameter selection given in [17], the main framework of the choice strategy is summarized as follows:

Step 1: Regularization parameters $\alpha_j,\ j \in \{1,2,\ldots,n\}$ are used to generate N different edge maps $\mathrm{D}_j,\ j \in \{1,2,\ldots,n\}$ by the proposed edge detection algorithm. Then, N potential ground truths (PGTs) are constructed, and each PGT_i includes pixels which have been identified as edges by at least i different edge maps.

Step 2: Each PGT_i is compared with each edge map D_j, and it generates four different probabilities:$\mathrm{TP}_{\mathrm{PGT}_i,\,\mathrm{D}_j}$, $\mathrm{FP}_{\mathrm{PGT}_i,\,\mathrm{D}_j}$, $\mathrm{TN}_{\mathrm{PGT}_i,\,\mathrm{D}_j}$, $\mathrm{FN}_{\mathrm{PGT}_i,\,\mathrm{D}_j}$. Among them, $\mathrm{TP}_{A,B}$ (true positive) means the probability of pixels which have been determined as edges in both edge maps A and B; $\mathrm{FP}_{A,B}$ (false positive) means the probability of pixels determined as edges in A, but non-edges in B; $\mathrm{TN}_{A,B}$ (true negative) means the probability of pixels determined as non-edges in both A and B; and $\mathrm{FN}_{A,B}$ (false negative) means the probability of pixels determined as edges in B, but non-edges in A.

Step 3: For each PGT_i, we average the four probabilities over all edge maps D_j, and get $\overline{\mathrm{TP}}_{\mathrm{PGT}_i}$, $\overline{\mathrm{FP}}_{\mathrm{PGT}_i}$, $\overline{\mathrm{TN}}_{\mathrm{PGT}_i}$, $\overline{\mathrm{FN}}_{\mathrm{PGT}_i}$, where $\overline{\mathrm{TP}}_{\mathrm{PGT}_i} = \frac{1}{N}\sum_{j=1}^{N} \mathrm{TP}_{\mathrm{PGT}_i,\mathrm{D}_j}$, and the expressions of other probabilities are similar. Then, a statistical measurement of each PGT_i is given by the Chi-square test:

$$\chi^2_{\mathrm{PGT}_i} = \frac{\mathrm{TPR}-\mathrm{Q}}{1-\mathrm{Q}} \cdot \frac{(1-\mathrm{FPR})-(1-\mathrm{Q})}{\mathrm{Q}}, \tag{6}$$

where

$$\mathrm{Q} = \overline{\mathrm{TP}}_{\mathrm{PGT}_i} + \overline{\mathrm{FP}}_{\mathrm{PGT}_i},\ \mathrm{TPR} = \frac{\overline{\mathrm{TP}}_{\mathrm{PGT}_i}}{\overline{\mathrm{TP}}_{\mathrm{PGT}_i} + \overline{\mathrm{FN}}_{\mathrm{PGT}_i}},\ \mathrm{FPR} = \frac{\overline{\mathrm{FP}}_{\mathrm{PGT}_i}}{\overline{\mathrm{FP}}_{\mathrm{PGT}_i} + \overline{\mathrm{TN}}_{\mathrm{PGT}_i}}.$$

The PGT_i with the highest $\chi^2_{\mathrm{PGT}_i}$ is considered as the estimated ground truth (EGT).

Step 4: Each edge map's D_j is then matched to the EGT by four new probabilities: $\mathrm{TP}_{\mathrm{D}_j,\mathrm{EGT}}$, $\mathrm{FP}_{\mathrm{D}_j,\mathrm{EGT}}$, $\mathrm{TN}_{\mathrm{D}_j,\mathrm{EGT}}$, $\mathrm{FN}_{\mathrm{D}_j,\mathrm{EGT}}$. The Chi-square measurements $\chi^2_{\mathrm{D}_j}$ are obtained by the same way as in Step 3. Then, the best edge map is the one which gives the highest $\chi^2_{\mathrm{D}_j}$, and the corresponding regularization parameter α_j is the one we want.

The Chi-square measure (6) can reflect the similarity of two edge maps, and the bigger the value of the Chi-square measurement, the better. As Lopez-Molina et al. stated in [19], the Chi-square measurement can evaluate the errors caused by spurious responses (false positives, FPs) and missing edges (false negatives, FNs), but it cannot work on the localization error when the detected edges deviate from their true position. For example, a reference edge image and three polluted edge maps are given in Figure 1. Compared with the reference edge (Figure 1a), the Chi-square measurements of the three polluted edge maps are the same, yet their localization accuracies are different. In order to reflect the localization error in these edge maps, distance-based error measures should be introduced.

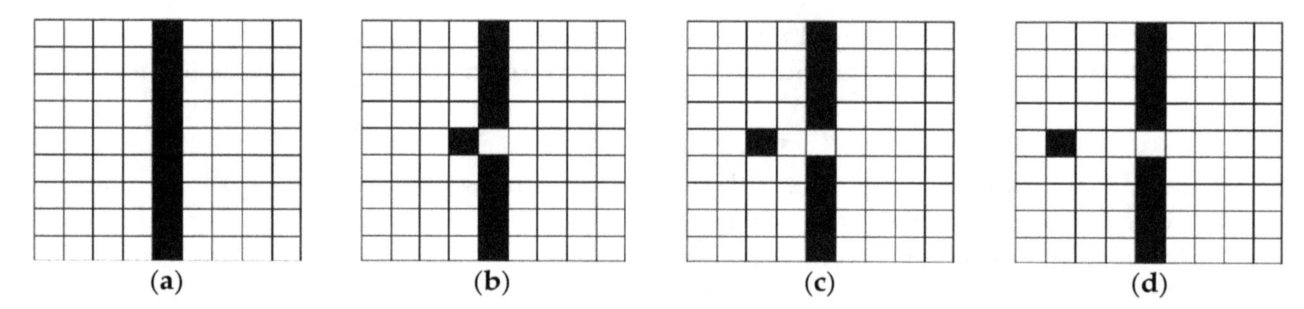

Figure 1. The reference edge image and three polluted edge maps: (**a**) reference edge E_R; (**b**) polluted edge map E_1; (**c**) polluted edge map E_2; (**d**) polluted edge map E_3.

The Baddeley's delta metric (BDM) is one of the most common distance-based measures [20]. It has been proven to be an ideal measure for the comparison of edge detection algorithms [19,21]. Let A and B be two edge maps with the same resolution $M \times N$, and $P = \{1, \ldots, M\} \times \{1, \ldots, N\}$ be the set of pixels in the image. The value of BDM between A and B is defined as

$$\Delta^k(A, B) = \left[\frac{1}{MN} \sum_{p \in P} |w(d(p, A)) - w(d(p, B))|^k \right]^{1/k}, \tag{7}$$

where $d(p, A)$ is the Euclidean distance from $p \in P$ to the closest edge points in A, the parameter k is a given positive integer and $w(d(p, A)) = \min(d(p, A), c)$ for a given constant $c > 0$. Compared with the reference edge E_R in Figure 1, the BDMs of the three polluted edge maps E_i $(i = 1, 2, 3)$ are given in Table 1 with different parameters c and k. The smaller the value of BDM, the better. As we can see from Table 1, localization errors of the three edge maps are apparently distinguished. Therefore, the Chi-square measure (6) will be replaced by the BDM (7) in the choice strategies of the regularization parameter.

Table 1. The Baddeley's delta metrics (BDMs) between the reference edge image E_R and the polluted edge maps E_i $(i = 1, 2, 3)$ with the different choices of parameters c and k.

Parameter Sets	$\Delta^k(E_R, E_1)$	$\Delta^k(E_R, E_2)$	$\Delta^k(E_R, E_3)$
$k = 1, c = 2$	0.0566	0.0937	0.1256
$k = 1, c = 3$	0.0950	0.1879	0.2461
$k = 1, c = 4$	0.1397	0.2614	0.3305
$k = 2, c = 2$	0.2182	0.3307	0.3637
$k = 2, c = 3$	0.2753	0.4925	0.6313
$k = 2, c = 4$	0.3317	0.6159	0.8021

4. Experiments and Results

In order to show the validity of the proposed edge detection algorithm, some comparative experiments are given in this section. In the experiments, our regularized edge detector (RED) will be compared with the LoG detector and the Laplacian-based edge detector (LED) proposed in [7].

As Yitzhaky and Peli said in [17], the parameter selection for edge detection depends mainly on the set of parameters used to generate the initial detection results. In order to reduce this influence properly, the range of the parameter is set to be large enough that instead of forming a very sparse edge map it forms a very dense one. The scale parameter of the LoG detector is set from 1.5 to 4 in steps of 0.25. The regularization parameter of the regularized edge detector is set from 0.01 ($\approx$0) to 0.1 in steps of 0.01. The images we used are taken from [22], and some of them are shown in Figure 2. The optimal edge maps given in [22] will be seen as the ground truth in our quantitative comparisons.

Let us first consider the choice strategy of the regularization parameter α, where the parameters in BDM are set as $k = 1$, $c = 2$. Taking the airplane image as an example, the BDM of each PGT_i, $i \in \{1, 2, \ldots 11\}$ is shown in Figure 3a, from which we can see the EGT is PGT_6. Compared with the EGT, the BDM of each edge map D_j is shown in Figure 3b, from which we can see the best edge map is D_6. Hence, the regularization parameter is chosen as $\alpha = 0.05$. The choice of the scale parameter in the LoG detector is carried out similarly. It does not need any parameters in the LED.

For the airplane image, the ground truth and edges detected by the three edge detectors are shown in Figure 4. From Figure 4b, we can see that the influence of the Lavrentiev regularization's weakness on the RED is negligible. From Figure 4b,c, we can see that the RED is better than the LoG detector for noise suppression and maintaining continuous edges. Comparing Figure 4d with Figure 4b,c, we can see the superiority of the parameter-dependent edge detector. Similar results for the elephant image are shown in Figure 5. For some images taken from [22], quantitative comparisons of the edges detected by the LoG detector, the RED and the LED against the ground truth are given in Table 2. Since the smaller the value of BDM, the better, this shows that the RED has better performance than the LoG detector and the LED in most cases.

(**a**) (**b**)

Figure 2. Some images taken from [22]: (**a**) airplane; (**b**) elephant.

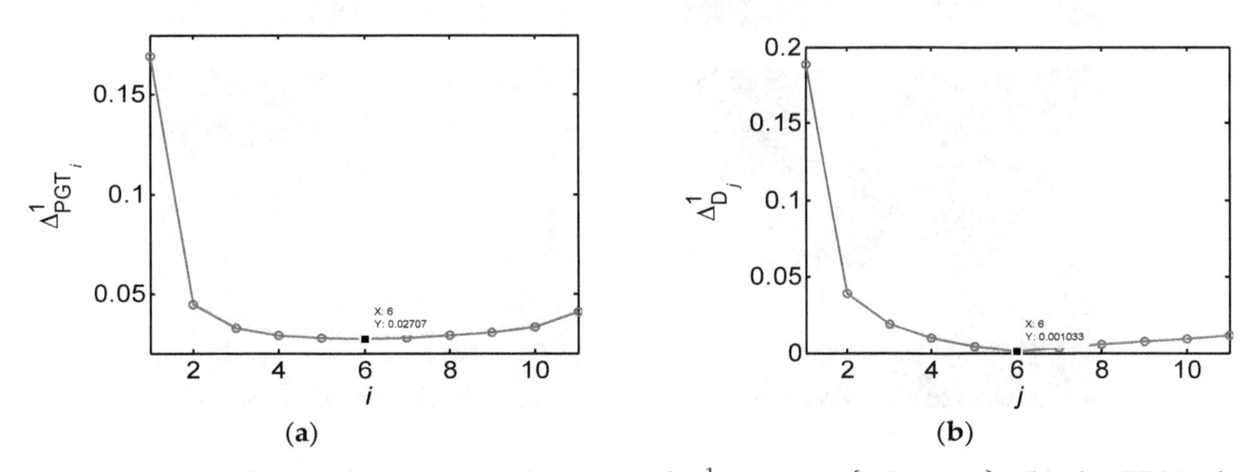

(**a**) (**b**)

Figure 3. The figure of BDMs: (**a**) the BDM of $\Delta^1_{\mathrm{PGT}_i}$, $i \in \{1, 2, \ldots, 11\}$; (**b**) the BDM of $\Delta^1_{\mathrm{D}_j}$, $i \in \{1, 2, \ldots, 11\}$.

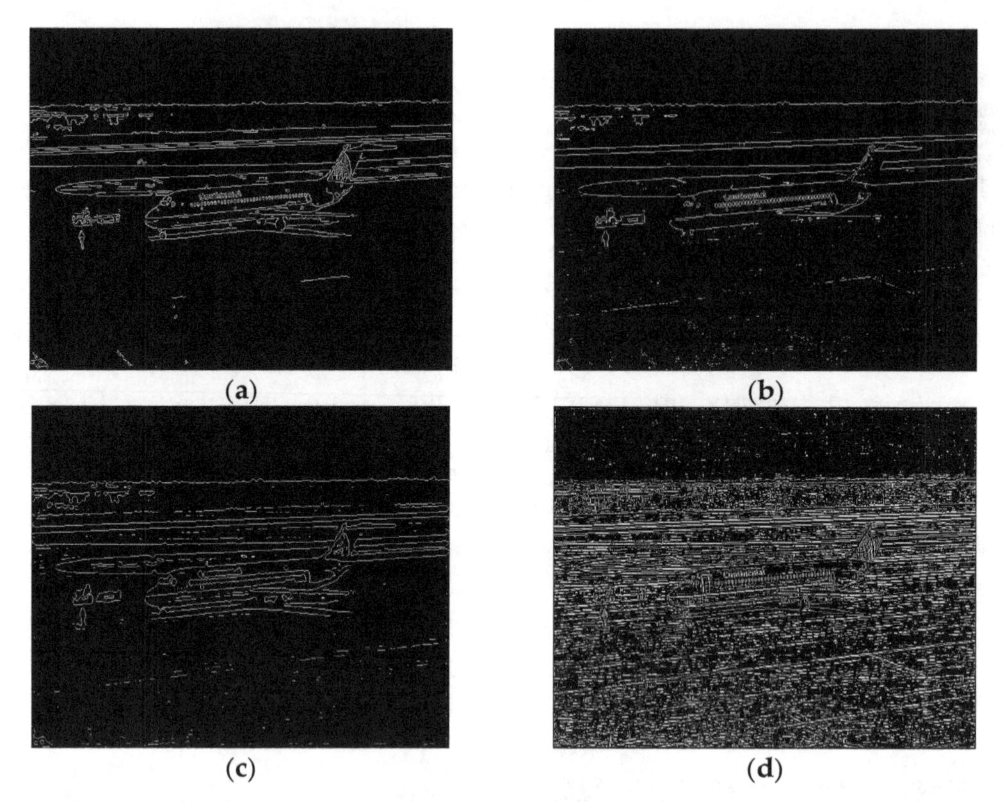

Figure 4. Edge detection results of the airplane image: (**a**) the ground truth; (**b**) the edge detected by the regularized edge detector (RED); (**c**) the Laplacian of Gaussian (LoG); (**d**) the Laplacian-based edge detector (LED).

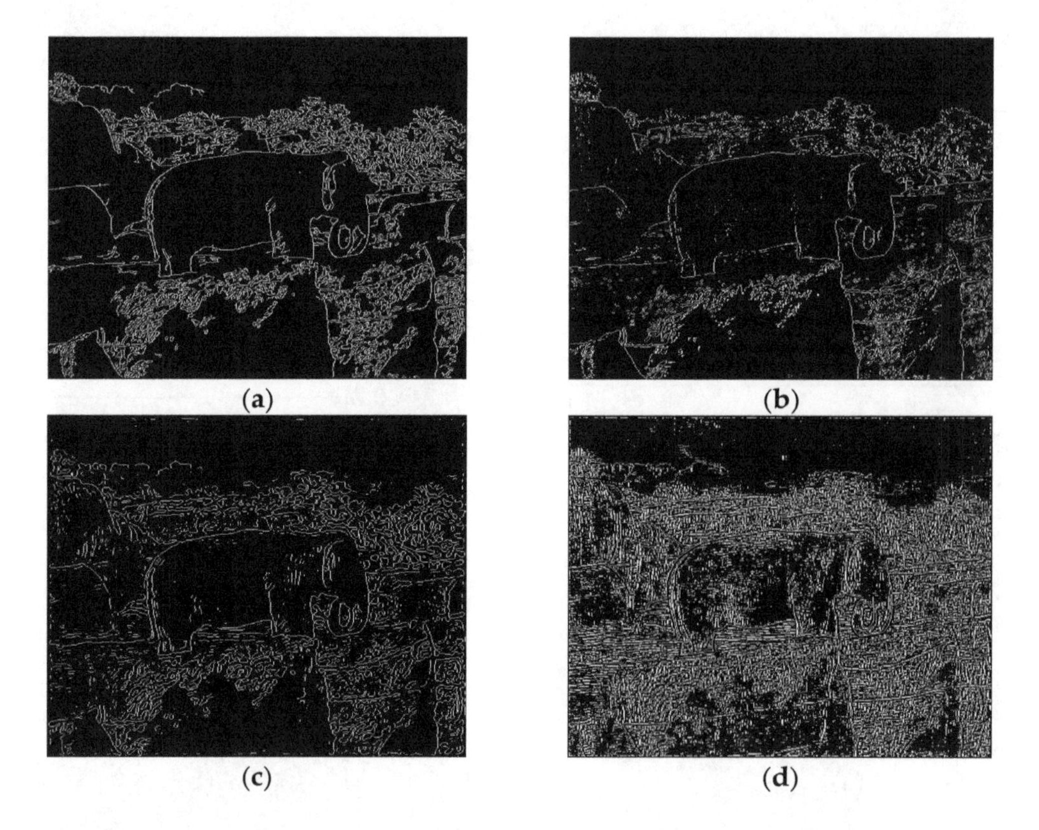

Figure 5. Edge detection results of the elephant image: (**a**) the ground truth; (**b**) the edge detected by the RED; (**c**) the LoG; (**d**) the LED.

Table 2. Quantitative comparison of the edges detected by the LoG, the RED and the LED.

Images	LED	LoG	RED
Airplane	0.7515	0.1270	0.1232
Elephant	0.6619	0.3041	0.2593
Turtle	0.4430	0.1226	0.1323
Brush	0.5790	0.1883	0.1673
Tiger	0.9239	0.2854	0.2748
Grater	0.5537	0.2353	0.2143
Pitcher	0.5032	0.2584	0.2296

5. Conclusions

In this paper, a novel edge detection algorithm is proposed based on the regularized Laplacian operation. The PDE-based regularization enables us to compute the regularized Laplacian operation in a direct way. Considering the importance of the regularization parameter, an objective choice strategy of the regularization parameter is proposed. Numerical implementations of the regularization parameter and the edge detection algorithm are also given. Based on the image database and ground truth edges taken from [22], the superiority of the RED against the LED and the LoG detector has been shown by the edge images and quantitative comparison.

Author Contributions: All the authors inferred the main conclusions and approved the current version of this manuscript.

References

1. Basu, M. Gaussian-based edge-detection methods—A survey. *IEEE Trans. Syst. Man Cybern. Part C Appl. Rev.* **2002**, *32*, 252–260. [CrossRef]
2. Marr, D.; Hildreth, E. Theory of edge detection. *Proc. R. Soc. Lond. B* **1980**, *207*, 187–217. [CrossRef] [PubMed]
3. Canny, J. A computational approach to edge detection. *IEEE Trans. Pattern Anal. Mach. Intell.* **1986**, *8*, 679–698. [CrossRef] [PubMed]
4. Sarkar, S.; Boyer, K. Optimal infinite impulse response zero crossing based edge detectors. *CVGIP Image Underst.* **1991**, *54*, 224–243. [CrossRef]
5. Demigny, D. On optimal linear filtering for edge detection. *IEEE Trans. Image Process.* **2002**, *11*, 728–737. [CrossRef] [PubMed]
6. Kang, C.C.; Wang, W.J. A novel edge detection method based on the maximizing objective function. *Pattern Recognit.* **2007**, *40*, 609–618. [CrossRef]
7. Wang, X. Laplacian operator-based edge detectors. *IEEE Trans. Pattern Anal. Mach. Intell.* **2007**, *29*, 886–890. [CrossRef] [PubMed]
8. Lopez-Molina, C.; Bustince, H.; Fernandez, J.; Couto, P.; De Baets, B. A gravitational approach to edge detection based on triangular norms. *Pattern Recognit.* **2010**, *43*, 3730–3741. [CrossRef]
9. Murio, D.A. *The Mollification Method and the Numerical Solution of Ill-Posed Problems*; Wiley-Interscience: New York, NY, USA, 1993; pp. 1–5, ISBN 0-471-59408-3.
10. Wan, X.Q.; Wang, Y.B.; Yamamoto, M. Detection of irregular points by regularization in numerical differentiation and application to edge detection. *Inverse Probl.* **2006**, *22*, 1089–1103. [CrossRef]
11. Xu, H.L.; Liu, J.J. Stable numerical differentiation for the second order derivatives. *Adv. Comput. Math.* **2010**, *33*, 431–447. [CrossRef]
12. Huang, X.; Wu, C.; Zhou, J. Numerical differentiation by integration. *Math. Comput.* **2013**, *83*, 789–807. [CrossRef]
13. Wang, Y.C.; Liu, J.J. On the edge detection of an image by numerical differentiations for gray function. *Math. Methods Appl. Sci.* **2018**, *41*, 2466–2479. [CrossRef]
14. Gonzalez, R.C.; Woods, R.E. *Digital Image Processing*, 3rd ed.; Pearson: London, UK, 2007; pp. 158–162, ISBN 978-0-13-168728-8.
15. Gunn, S.R. On the discrete representation of the Laplacian of Gaussian. *Pattern Recognit.* **1999**, *32*, 1463–1472. [CrossRef]

16. Xu, H.L.; Liu, J.J. On the Laplacian operation with applications in magnetic resonance electrical impedance imaging. *Inverse Probl. Sci. Eng.* **2013**, *21*, 251–268. [CrossRef]
17. Yitzhaky, Y.; Peli, E. A method for objective edge detection evaluation and detector parameter selection. *IEEE Trans. Pattern Anal. Mach. Intell.* **2003**, *25*, 1027–1033. [CrossRef]
18. Gu, C.H.; Li, D.Q.; Chen, S.X.; Zheng, S.M.; Tan, Y.J. *Equations of Mathematical Physics*, 2nd ed.; Higher Education Press: Beijing, China, 2002; pp. 80–86, ISBN 7-04-010701-5. (In Chinese)
19. Lopez-Molina, C.; Baets De, B.; Bustince, H. Quantitative error measures for edge detection. *Pattern Recognit.* **2013**, *46*, 1125–1139. [CrossRef]
20. Baddeley, A.J. An error metric for binary images. In Proceedings of the IEEE Workshop on Robust Computer Vision, Bonn, Germany, 9–11 March 1992; Wichmann Verlag: Karlsruhe, Germany, 1992; pp. 59–78.
21. Fernández-García, N.L.; Medina-Carnicer, R.; Carmona-Poyato, A.; Madrid-Cuevas, F.J.; Prieto-Villegas, M. Characterization of empirical discrepancy evaluation measures. *Pattern Recognit. Lett.* **2004**, *25*, 35–47. [CrossRef]
22. Heath, M.D.; Sarkar, S.; Sanocki, T.A.; Bowyer, K.W. A robust visual method for assessing the relative performance of edge detection algorithms. *IEEE Trans. Pattern Anal. Mach. Intell.* **1997**, *19*, 1338–1359. [CrossRef]

12

Involution Abel–Grassmann's Groups and Filter Theory of Abel–Grassmann's Groups

Xiaohong Zhang * and Xiaoying Wu

Department of Mathematics, Shaanxi University of Science & Technology, Xi'an 710021, China; 46018@sust.edu.cn
* Correspondence: zhangxiaohong@sust.edu.cn

Abstract: In this paper, some basic properties and structure characterizations of AG-groups are further studied. First, some examples of infinite AG-groups are given, and weak commutative, alternative and quasi-cancellative AG-groups are discussed. Second, two new concepts of involution AG-group and generalized involution AG-group are proposed, the relationships among (generalized) involution AG-groups, commutative groups and AG-groups are investigated, and the structure theorems of (generalized) involution AG-groups are proved. Third, the notion of filter of an AG-group is introduced, the congruence relation is constructed from arbitrary filter, and the corresponding quotient structure and homomorphism theorems are established.

Keywords: Abel–Grassmann's groupoid (AG-groupoid); Abel–Grassmann's group (AG-group); involution AG-group; commutative group; filter

---

1. Introduction

Nowadays, the theories of groups and semigroups [1–5] are attracting increasing attention, which can be used to express various symmetries and generalized symmetries in the real world. Every group or semigroup has a binary operation that satisfies the associative law. On the other hand, non-associative algebraic structures have great research value. Euclidean space $\mathbf{R}^3$ with multiplication given by the vector cross product is an example of an algebra that is not associative, at the same time; Jordan algebra and Lie algebra are non-associative.

For the generalization of commutative semigroup, the notion of an AG-groupoid (Abel–Grassmann's groupoid) is introduced in [6], which is also said to be a left almost semigroup (LA-semigroup). Moreover, a class of non-associative ring with condition $x(yz) = z(yx)$ is investigated in [7]; in fact, the condition $x(yz) = z(yx)$ is a dual distortion of the operation law in AG-groupoids.

An AG-groupoid is a non-associative algebraic structure, but it is a groupoid $(N, *)$ satisfying the left invertive law:

$$(a * b) * c = (c * b) * a, \text{ for any } a, b, c \in N.$$

Now, many characterizations of AG-groupoids and various special subclasses are investigated in [8–13]. As a generalization of commutative group (Abelian group) and a special case of quasigroup, Kamran extended the concept of AG-groupoid to AG-group in [14]. An AG-groupoid is called AG-group if there exists left identity and inverse, and its many properties (similar to the properties of groups) have been revealed successively in [15,16].

In this paper, we further analyze and study the structural characteristics of AG-groups, reveal the relationship between AG-groups and commutative groups, and establish filter and quotient algebra theories of AG-groups. The paper is organized as follows. Section 2 presents several basic concepts and results. Some new properties of AG-groups are investigated in Section 3, especially some examples of infinite AG-groups, and the authors prove that every weak commutative or alternative AG-group is a commutative group (Abelian group) and every AG-group is quasi-cancellative. In Section 4, two special classes of AG-groups are studied and the structure theorems are proved. In Section 5, the filter theory of AG-groups is established, the quotient structures induced by filters are constructed, and some homomorphism theorems are proved. Finally, the main results of this paper are systematically summarized via a schematic figure.

2. Preliminaries

First, we present some basic notions and properties.

A groupoid $(N, *)$ is called an AG-groupoid (Abel–Grassmann's groupoid), if for any $a, b, c \in N$, $(a*b)*c = (c*b)*a$. It is easy to verify that in an AG-groupoid $(N, *)$, the medial law holds:

$$(a * b) * (c * d) = (a * c) * (b * d), \text{ for any } a, b, c, d \in N.$$

Let $(N, *)$ be an AG-groupoid with left identity e, we have

$$a * (b * c) = b * (a * c), \text{ for any } a, b, c \in N;$$

$$(a * b) * (c * d) = (d * b) * (c * a), \text{ for any } a, b, c, d \in N.$$

$$NN = N, N*e = N = e*N.$$

An AG-groupoid $(N, *)$ is called a locally associative AG-groupoid, if it satisfies $a*(a*a) = (a*a)*a, \forall a \in N$.

An AG-groupoid $(N, *)$ is called an AG-band, if it satisfies $a*a = a$ $(\forall a \in N)$.

Definition 1. *([9,10]) Let $(N, *)$ be an AG-groupoid. Then, N is called to be quasi-cancellative if for any $a, b \in N$,*

$$a = a*b \text{ and } b^2 = b*a \text{ imply that } a = b; \text{ and} \tag{1}$$

$$a = b*a \text{ and } b^2 = a*b \text{ imply that } a = b. \tag{2}$$

Proposition 1. *([9,10]) Every AG-band is quasi-cancellative.*

Definition 2. *([14,15]) An AG-groupoid $(N, *)$ is called an AG-group or a left almost group (LA-group), if there exists left identity $e \in N$ (that is $e*a = a$, for all $a \in N$), and there exists $a^{-1} \in N$ such that $a^{-1}*a = a* a^{-1} = e$ $(\forall a \in N)$.*

Proposition 2. *([15]) Assume that $(N, *)$ is an AG-group. We get that $(N, *)$ is a commutative Abel–Grassmann's Group if and only if it is an associative AG-Group.*

Proposition 3. *([15]) Let $(N, *)$ be an AG-group with right identity e. Then, $(N, *)$ is an Abelian group.*

Proposition 4. *([15]) Let $(N, *)$ be an AG-group. Then, $(N, *)$ has exactly one idempotent element, which is the left identity.*

Proposition 5. *([11]) Let $(N, *)$ be an AG-groupoid with a left identity e. Then, the following conditions are equivalent,*

(1) *N is an AG-group.*
(2) *Every element of N has a right inverse.*
(3) *Every element a of N has a unique inverse a^{-1}.*
(4) *The equation $x*a = b$ has a unique solution for all $a, b \in N$.*

Proposition 6. *([16]) Let (N, *) be an AG-group. Define a binary operation ∘ as follows:*

$$x \circ y = (x^*e)^*y, \text{ for any } x, y \in N.$$

Then, (N, ∘) is an Abelian group, denote it by ret(N, *) = (N, ∘).

3. Some Examples and New Results of AG-Groups

In this section, we give some examples of AG-groups (including some infinite examples), and investigate the characterizations of weak commutative AG-groups, alternative AG-groups and quasi-cancellative AG-groups. Moreover, we obtain two subalgebras from arbitrary AG-group.

Example 1. *Let us consider the rotation transformations of a square. A square is rotated 90°, 180° and 270° to the right (clockwise) and they are denoted by φ_a, φ_b and φ_c, respectively (see Figure 1). There is of course the movement that does nothing, which is denoted by φ_e. The following figure gives an intuitive description of these transformations. Denote N = $\{\varphi_e, \varphi_a, \varphi_b, \varphi_c\}$.*

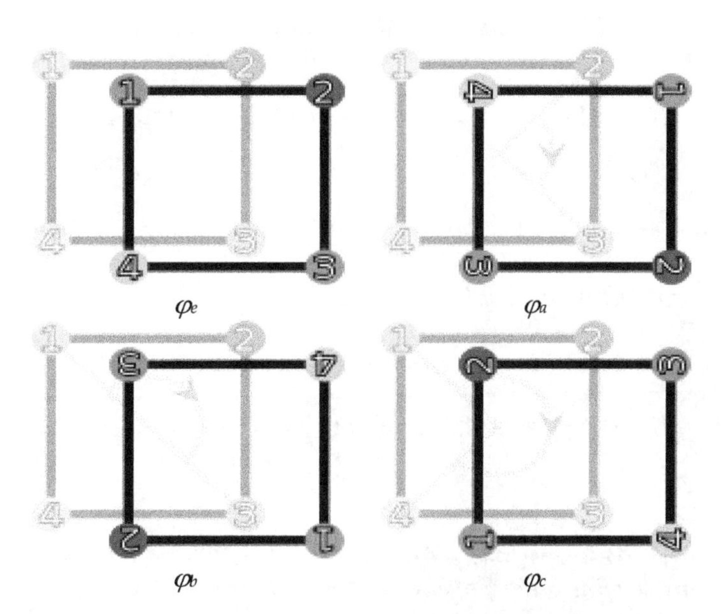

Figure 1. The rotation transformations of a square.

*Obviously, two consecutive rotations have the following results: $\varphi_e\varphi_e = \varphi_e$, $\varphi_a\varphi_c = \varphi_c\varphi_a = \varphi_e$, $\varphi_b\varphi_b = \varphi_e$. That is, ${\varphi_e}^{-1} = \varphi_e$, ${\varphi_a}^{-1} = \varphi_c$, ${\varphi_b}^{-1} = \varphi_b$, ${\varphi_c}^{-1} = \varphi_a$. Now, we define operations * on N as follows:*

$$\varphi_x{}^*\varphi_y = {\varphi_x}^{-1}\ \varphi_y, \forall x, y \in \{e, a, b, c\}.$$

*Then, (N, *) satisfies the left invertive law, and the operation * is as follows in Table 1. We can verify that (N, *) is an AG-Group.*

Table 1. AG-group generated by rotation transformations of a square.

*	φ_e	φ_a	φ_b	φ_c
φ_e	φ_e	φ_a	φ_b	φ_c
φ_a	φ_c	φ_e	φ_a	φ_b
φ_b	φ_b	φ_c	φ_e	φ_a
φ_c	φ_a	φ_b	φ_c	φ_e

Example 2. *Let $X = \{(a, b)|a, b \in \mathbf{R}-\{0\}\}$, where* **R** *represents the set of all real numbers. Define binary operation * as follows:*

$$(a, b) * (c, d) = (ac, d/b), \text{ for any } (a, b), (c, d) \in X.$$

Then,

$$[(a, b) * (c, d)] * (e, f) = (ac, d/b) * (e, f) = (ace, fb/d);$$

$$[(e, f) * (c, d)] * (a, b) = (ec, d/f) * (a, b) = (ace, fb/d).$$

*Therefore, $[(a, b) * (c, d)] * (e, f) = [(e, f) * (c, d)] * (a, b)$, that is, the operation * satisfies left invertive law. For any $(a, b) \in X$, $(1, 1)$ is the left identity of (a, b) and $(1/a, b)$ is the left inverse of (a, b):*

$$(1,1) * (a, b) = (a, b); (1/a, b) * (a, b) = (1, 1).$$

Therefore, $(X, *)$ is an AG-Group.

Example 3. *Let $Y = \{(a, b)|a \in \mathbf{R}, b = 1 \text{ or } -1\}$, where* **R** *represents the set of all real numbers. Define binary operation * as follows:*

$$(a, b) * (c, d) = (ac, b/d), \text{ for any } (a, b), (c, d) \in Y.$$

Then,

$$[(a, b) * (c, d)] * (e, f) = (ac, b/d) * (e, f) = (ace, b/df);$$

$$[(e, f) * (c, d)] * (a, b) = (ec, f/d) * (a, b) = (ace, f/bd).$$

*Because $b, f \in \{1, -1\}$, $b^2 = f^2$, and $b/f = f/b$. We can get $b/df = f/bd$. Therefore, $[(a, b) * (c, d)] * (e, f) = [(e, f) * (c, d)] * (a, b)$, that is, the operation * satisfies left invertive law. Moreover, we can verify that $(1, 1)$ is the left identity and $(1/a, \pm 1)$ is the left inverse of $(a, \pm 1)$, since*

$$(1, 1) * (a, b) = (a,1/b) = (a, b); \text{ (because } b=1 \text{ or } -1)$$

$$(1/a, 1) * (a, 1) = (1, 1) \text{ and } (1/a, -1) * (a, -1) = (1, 1).$$

Therefore, $(Y, *)$ is an AG-group.

Example 4. *Let $Z = \{(a, b)|a \in \mathbf{R}, b = 1, -1, i, \text{ or } -i\}$, where* **R** *represents the set of all real numbers and I represents the imaginary unit. Define binary operation * as follows:*

$$(a, b) * (c, d) = (ac, b/d), \text{ for any } (a, b), (c, d) \in Z$$

Then,

$$[(a, b) * (c, d)] * (e, f) = (ac, b/d) * (e, f) = (ace, b/df);$$

$$[(e, f) * (c, d)] * (a, b) = (ec, f/d) * (a, b) = (ace, f/bd).$$

*Because $b, f \in \{1, -1, i, -i\}$, hence $b^2 = f^2$, and $b/f = f/b$. We can get $b/df = f/bd$. Therefore, $[(a, b) * (c, d)] * (e, f) = [(e, f) * (c, d)] * (a, b)$, that is, the operation * satisfies left invertive law. Therefore, $(Z, *)$ is an AG-groupoid. However, it is not an AG-group, since*

$$(1, 1) * (a, 1) = (a, 1), (1, 1) * (a, -1) = (a, -1);$$

$$(1, -1) * (a, i) = (a, i), (1, -1) * (a, -i) = (a, -i).$$

That is, $(1, 1)$ and $(1, -1)$ are locally identity, not an identity.

Definition 3. *Assume that $(N, *)$ is an AG-group. $(N, *)$ is said to be a weak commutative Abel– Grassmann's group (AG-group), if one of the following conditions holds:*

(1) $e^*x^{-1} = x^{-1}*e$, *for all x in N;*
(2) $e^*x = x^*e$, *for all x in N; or*
(3) $x^{-1}*y^{-1} = y^{-1}*x^{-1}$, *for all x, y in N.*

Theorem 1. *Let (N, *) be an AG-group. We can get that N is a weak commutative AG-group if and only if it is an Abelian group.*

Proof. First, we prove that the Conditions (1)–(3) in Definition 3 are equivalent for an AG-group (N, *).

(1)→(2): Suppose thatCondition (1) holds in the AG-group (N, *). For all x in N, by $(x^{-1})^{-1} = x$, we have $e*(x^{-1})^{-1} = (x^{-1})^{-1}*e$, that is, $e*x = x*e$.

(2)→(3): Suppose that Condition (2) holds in the AG-group (N, *). For all x, y in N, by Proposition 3, we know that N is an Abelian group, that is, $x*y = y*x$, it follows that $x^{-1}*y^{-1} = y^{-1}*x^{-1}$.

(3)→(1): Suppose that Condition (3) holds in the AG-group (N, *). Then, for all x in N, we have $(e^{-1})^{-1}*x^{-1} = x^{-1}*(e^{-1})^{-1}$, that is, $e*x^{-1} = x^{-1}*e$.

Now, we prove that an AG-group (N, *) satisfying Condition (2) in Definition 3 is an Abelian group. Through Condition (2), $e*a = a*e$ for any $a\in N$. Then, $a*e = e*a = a$, which means that e is right identity. Applying Proposition 3, we get that $(N, *)$ is an Abelian group. Moreover, obviously, every Abelian group is a weak commutative AG-group. Therefore, the proof is completed. □

Theorem 2. *Assume that (N, *) is an AG-group, we have that (N, *) is quasi-cancellative AG-groupoid, that is, if it satisfies the following conditions, for any x, y∈N,*

(1) $x = x * y$ *and* $y^2 = y*x$ *imply that* $x = y$; *and*
(2) $x = y * x$ *and* $y^2 = x * y$ *imply that* $x = y$.

Proof. (1) Suppose that $x = x*y$ and $y^2 = y*x$, where $x, y\in N$. Then,

$$x = x*y = (e*x)*y = (y*x)*e = y^2*e = (e*y)*y = y^2. \quad \text{(a)}$$

That is, $x = y^2$; it follows that $x*y = y*x$. Moreover, we have

$$y*e = y*(x^{-1}*x) = (e*y)*(x^{-1}*x) = (e*x^{-1})*(y*x) = x^{-1}*(y*x) = x^{-1}*x = e. \quad \text{(b)}$$

$$\begin{aligned} x*e = (x*y)*e &= (x*y)*(x^{-1}*x) = (x*x^{-1})*(y*x) = (x*x^{-1})*y^2 = (x*x^{-1})*(y*y) \\ &= (x*y)*(x^{-1}*y) = x*(x^{-1}*y) = (e*x)*(x^{-1}*y) = (e*x^{-1})*(x*y) \\ &= x^{-1}*(x*y) = x^{-1}*x = e. \end{aligned} \quad \text{(c)}$$

Combining Equations (b) and (c), we can get

$$x = e*x = (y*e)*x = (x*e)*y = e*y = y.$$

(2) Suppose that $x=y*x$ and $y^2=x*y$, where $x, y\in N$. Then,

$$x = y*x = (e*y)*(x*y)*e = y^2*e = (e*y)*y = y^2*e = (e*y)*y = y^2. \quad \text{(d)}$$

That is, $x = y^2$; it follows that $x*y = y*x$. Then, we have

$$y*e = y*(x^{-1}*x) = (e*y)*(x^{-1}*x) = (e*x^{-1})*(y*x) = (e*x^{-1})*x = e. \quad \text{(e)}$$

$$x*e = x*(y^{-1}*y) = (e*x)*(y^{-1}*y) = y^{-1}*(x*y) = y^{-1}*y^2 = y^{-1}*(y*y) = y*(y^{-1}*y) = y*e = e. \quad \text{(f)}$$

Combining Equations (e) and (f), we can get

$$x = e*x = (y*e)*x = (x*e)*y = e*y = y.$$

Hence, $(N, *)$ is quasi-cancellative AG-groupoid. □

Definition 4. *Let (N, *) be an AG-group. Then, (N, *) is called to be alternative, if it satisfies one of the following conditions,*

(1) $a*(a*b) = (a*a)*b, \forall a, b \in N$; *or*
(2) $a*(b*b) = (a*b)*b, \forall a, b \in N$.

Theorem 3. *Let (N, *) be an AG-group. Then, (N, *) is alternative if and only if it is an Abelian group.*

Proof. (1) Suppose that $(N, *)$ is an alternative AG-group, then Condition (2) in Definition 4 holds. Then, for any $a, b \in N$, $a*(b*b) = (a*b)*b$. Putting $b = e$ and applying left invertive law, we get that $a*e = a*(e*e) = (a*e)*e = (e*e)*a = e*a = a$; by Proposition 3, we know that $(N, *)$ is an Abelian group.

(2) Suppose that $(N, *)$ is an alternative AG-group, then Condition (1) in Definition 4 holds. For any $a, b \in N$, it satisfies $a*(a*b) = (a*a)*b$. Putting $b = e$, we have $(a*a)*e = a*(a*e)$. According to the arbitrariness of a, we can get that

$$((a*e)*(a*e))*e = (a*e)*((a*e)*e).$$

Then,

$$a*a = (e*a)*a = (a*a)*e = ((a*a)*(e*e))*e = ((a*e)*(a*e))*e = (a*e)*((a*e)*e) = (a*e)*a.$$

Let $b*a = e$, using Condition (1) in Definition 4, $(a*a)*b = a*(a*b)$. It follows that $(a*a)*b = ((a*e)*a)*b$. Thus,

$$a = e*a = (b*a)*a = (a*a)*b = ((a*e)*a)*b = (b*a)*(a*e) = e*(a*e) = a*e.$$

Applying Proposition (3), we know that $(N, *)$ is an Abelian group.

Conversely, it is obvious that every Abelian group is an alternative AG-group. Therefore, the proof is completed. □

Theorem 4. *Let (N, *) be an AG-group. Denote*

$$U(N) = \{x \in N | \ x = x*e\}.$$

Then,

(1) *U(N) is sub-algebra of N.*
(2) *U(N) is maximal subgroup of N with identity e.*

Proof. (1) Obviously, $e \in U(N)$, that is, $U(N)$ is not empty. Suppose $x, y \in U(N)$, then $x*e = x$ and $y*e = y$. Thus, $x*y = (x*e)*(y*e) = (x*y)*e \in U(N)$. This means that $U(N)$ is a subalgebra of N.

(2) For any $x \in U(N)$, that is, $x*e = x$. Assume that y is the left inverse of x in N, then $y*x = e$. Thus,

$$x*y = (e*x)*y = ((y*x)*x)*y = (y*x)*(y*x) = e*e = e,$$

$$y = e*y = (x*y)*y = ((x*e)*y)*y = ((y*e)*x)*y = (y*x)*(y*e) = e*(y*e) = y*e.$$

It follows that $y \in U(N)$. Therefore, $U(N)$ is a group, and it is a subgroup of N with identity e. If M is a subgroup of N with identity e, and $U(N) \subseteq M$, then M is an Abelian group (by Proposition (3)) and satisfies $x*e = e*x = x$, for any $x \in M$. Thus, $M \subseteq U(N)$, it follows that $M = U(N)$. Therefore, $U(N)$ is maximal subgroup of N with identity e. □

Theorem 5. *Let (N, *) be an AG-group. Denote* $P(N) = \{x \in N | \exists a \in N, s.t\ x = a*a\}$. *Then*

(1) *P(N) is the subalgebra of N;*
(2) *f is a homomorphism mapping from N to P(N), where* $f: N \rightarrow P(N), f(x)=x*x \in P(N)$.

Proof. (1) Obviously, $e \in P(N)$, that is, P(N) is not empty. Suppose $x, y \in P(N)$ and $a, b \in N$. Then, $a*a = x$ and $b*b = y$. Thus, $x*y = (a*a)*(b*b) = (a*b)*(a*b) \in P(N)$. This means that P(N) is a subalgebra of N.

(2) For any $x, y \in N$, we have

$$f(x*y) = (x*y)*(x*y) = (x*x)*(y*y) = f(x)*f(y).$$

Therefore, f is a homomorphism mapping from N to P(N). □

4. Involution AG-Groups and Generalized Involution AG-Groups

In this section, we discuss two special classes of AG-groups, that is, involution AG-groups and generalized involution AG-groups. Some research into the involutivity in AG-groupoids is presented in [16,17] as the foundation, and further results are given in this section, especially the close relationship between these algebraic structures and commutative groups (Abelian groups), and their structural characteristics.

Definition 5. *Let (N, *) be an AG-group. If (N, *) satisfies* $a*a = e$, *for any* $a \in N$, *then (N, *) is called an involution AG-Group.*

*We can verify that (N, *) in Example 1 is an involution AG-Group.*

Example 5. *Denote N = {a, b, c, d}, define operations * on N as shown in Table 1. We can verify that (N, *) is an involution AG-group (Table 2).*

Table 2. Involution AG-group (N, *).

*	a	b	c	d
a	a	b	c	d
b	b	a	d	c
c	d	c	a	b
d	c	d	b	a

Example 6. *Let (G, +) be an Abelian group. Define operations * on G as follows:*

$$x*y = (-x) + y,\ \forall x, y \in G$$

*where (−x) is the inverse of x in G. Then, (G, *) is an involution AG-group. Denote (G, *) by der (G, +) (see [15]), and call it derived AG-group by Abelian group (G, +).*

Theorem 6. *Let (N, *) be an AG-group. Then, (N, *) is an involution AG-Group if and only if it satisfies one of the following conditions:*

(1) $P(N) = \{e\}$, *where P(N) is defined as Theorem 5.*
(2) $(x*x)*x = x$ *for any* $x \in N$.

Proof. Obviously, (N, *) is an involution AG-group if and only if $P(N) = \{e\}$.

If (N, *) is an involution AG-group, then apply Definition 5, for any $x \in N$,

$$(x*x)*x = e*x = x.$$

Conversely, if $(N, *)$ satisfies the Condition (2), then for any $x \in N$,

$$(x^*x)^*(x^*x) = ((x^*x)^*x)^*x = x^*x.$$

This means that (x^*x) is an idempotent element. Using Proposition 4, we have $x^*x = e$. Thus, $(N, *)$ is an involution AG-group. □

Theorem 7. *Let $(N, *)$ be an involution AG-group. Then, $(N, \circ) = ret\ (N, *)$ defined in Proposition 6 is an Abelian group, and the derived AG-group der $(N, \circ)$ by ret $(N, *)$ (see Example 5) is equal to $(N, *)$, that is,*

$$der(ret(N, *)) = (N, *).$$

Proof. (1) By Proposition 6 and Definition 5, $\forall x, y, z \in N$, we can get that

$$x \circ y = y \circ x;\ x \circ e = e \circ x = x;\ (x \circ y) \circ z = x \circ (y \circ z);\ x \circ x^{-1} = x^{-1} \circ x = e.$$

This means that $(N, \circ) = \text{ret}(N, *)$ is an Abelian group.

(2) For any $x, y \in \text{der}(\text{ret}(N, *)) = \text{der}(N, \circ) = (N, \bullet)$,

$$x \bullet y = (-x) \circ y = ((-x) * e)^* y = ((x * e) * e)^* y = ((e * e) * x)^* y = (e * x)^* y = x^* y.$$

That is, $\text{der}(\text{ret}(N, *)) = (N, \bullet) = (N, *)$. □

Definition 6. *Let $(N, *)$ be an AG-group. Then, $(N, *)$ is called a generalized involution AG-group if it satisfies: for any $x \in N$, $(x^*x)^*(x^*x) = e$.*

Obviously, every involution AG-group is a generalized involution AG-group. The inverse is not true, see the following example.

Example 7. *Denote $N = \{e, a, b, c\}$, and define the operations $*$ on N as shown in Table 3. We can verify that $(N, *)$ is a generalized involution AG-group, but it is not an involution AG-group.*

Table 3. Generalized involution AG-group $(N, *)$.

*	e	a	b	c
e	e	a	b	c
a	a	e	c	b
b	c	b	a	e
c	b	c	e	a

Theorem 8. *Let $(N, *)$ be a generalized involution AG-group. Define binary relation $\approx$ on N as follows:*

$$x \approx y \Leftrightarrow x * x = y * y,\ for\ any\ x, y \in N.$$

Then,

(1) *$\approx$ is an equvalent relation on N, and we denote the equivalent class contained x by $[x]_{\approx}$.*
(2) *The equivalent class contained e by $[e]_{\approx}$ is an involution sub-AG-group.*
(3) *For any $x, y, z \in N$, $x \approx y$ implies $x^*z \approx y^*z$ and $z^*x \approx z^*y$.*
(4) *The quotient $(N/\approx, *)$ is an involution AG-group.*

Proof. (1) For any $a \in N$, we have $a^*a = a^*a$, thus $a \approx a$.

If $a \approx b$, then $a^*a = b^*b$; it is obvious that $b \approx a$.

If $a \approx b$ and $b \approx c$, then $a^*a = b^*b$ and $b^*b = c^*c$; it is obvious that $a^*a = c^*c$, that is, $a \approx c$.

Therefore, $\approx$ is an equivalent relation on N.

(2) $\forall x, y \in [e]_{\approx}$, we have $x^*x = y^*y = e^*e = e$, thus

$$(x^*y)^*(x^*y) = (x^*x)^*(y^*y) = e^*e = e.$$

This means that $[e]_{\approx}$ is a subalgebra of N. Thus, $[e]_{\approx}$ is an involution sub-AG-group of N.

(3) Assume that $x \approx y$, then $x^*x = y^*y$. Thus,

$$(x^*z)^*(x^*z) = (x^*x)^*(z^*z) = (y^*y)^*(z^*z) = (y^*z)^*(y^*z);$$

$$(z^*x)^*(z^*x) = (z^*z)^*(x^*x) = (z^*z)^*(y^*y) = (z^*y)^*(z^*y).$$

It follows that $x^*z \approx y^*z$ and $z^*x \approx z^*y$.

(4) By (3), we know that $(N/\approx, *)$ is an AG-group. Moreover, for any

$$x \in [a]_{\approx} * [a]_{\approx} = [a*a]_{\approx},\ x*x = (a*a)*(a*a)$$

By Definition 6,

$$(a^*a)^*(a^*a) = e.$$

Then,

$$x*x = e \text{ for any } x \in [a*a]_{\approx}.$$

From this, we have $x \in [e]_{\approx}$, $[a^*a]_{\approx} \subseteq [e]_{\approx}$. Hence, $[a^*a]_{\approx} = [e]_{\approx}$. That is, $[a]_{\approx}{}^*[a]_{\approx} = [e]_{\approx}$. Therefore, $(N/\approx, *)$ is an involution AG-group. □

Theorem 9. *Let (N, *) be an AG-group, denote*

$$I(N) = \{x \in N | x^*x = e\},\ GI(N) = \{x \in N | (x^*x)^*(x^*x) = e\}.$$

Then, I(N) and GI(N) are sub-algebra of N. I(N) is an involution AG-group and GI(N) is a generalized involution AG-group.

Proof. (1) It is obvious that $e \in I(N)$. For any $x, y \in I(X)$, we have $x^*x = e$ and $y^*y = e$. By medial law, $(x^*y)^*(x^*y) = (x^*x)^*(y^*y) = e^*e = e$. Hence, $I(N)$ is a sub-algebra of N and $I(N)$ is an involution AG-group.

(2) Obviously, $e \in GI(N)$. Assume that $x, y \in GI(X)$, then

$$(x^*x)^*(x^*x) = (y^*y)^*(y^*y) = e.$$

Thus,

$$((x^*y)^*(x^*y))^*((x^*y)^*(x^*y)) = ((x^*x)^*(y^*y))^*((x^*x)^*(y^*y)) = ((x^*x)^*(x^*x))^*((y^*y)^*(y^*y)) = e^*e = e.$$

It follows that $x^*y \in GI(N)$, and $GI(N)$ is a subalgebra of N. Moreover, from $((x^*x)^*x)^*x = (x^*x)^*(x^*x) = e$, we get that $a = (x^*x)^*x$ is the left inverse of x, and

$$(a^*a)^*(a^*a) = (((x^*x)^*x)^*((x^*x)^*x))^*(a^*a) = ((x^*x)^*(x^*x))^*(x^*x))^*(a^*a) = (e^*(x^*x))^*(a^*a) = (x^*x)^*(a^*a) = (x^*x)^*(x^*x) = e.$$

That is, $a = (x^*x)^*x \in GI(N)$. It follows that $GI(N)$ is an AG-group. By the definition of $GI(N)$, we get that $GI(N)$ is a generalized involution AG-group. □

5. Filter of AG-Groups and Homomorphism Theorems

Definition 7. *Let (N, *) be an AG-group. A non-empty subset F of N is called a filter of N if, for all x, y∈N, F satisfies the following properties,*

(1) $e \in F$;
(2) $x^*x \in F$; and
(3) $x \in F$ and $x^*y \in F$ imply that $y \in F$.

If F is a filter and subalgebra of N, then F will be called a closed filter of N.

Theorem 10. *Let (N, *) be a generalized involution AG-group,* $I(N) = \{x \in N | e = x^*x\}$ *be the involution part of N (see Theorem 9). Then, I(N) is a closed filter of N.*

Proof. It is obvious that $e \in I(N)$. $\forall x \in N$, since

$$(x^*x)^*(x^*x) = e,$$

then $x^*x \in I(N)$. Moreover, assuming that $x \in I(N)$ and $x^*y \in I(N)$, then

$$e = x^*x,\ (x^*y)^*(x^*y) = e.$$

Thus,

$$y^*y = e^*(y^*y) = (x^*x)^*(y^*y) = (x^*y)^*(x^*y) = e.$$

Hence, $y \in I(N)$, and $I(N)$ is a filter of N. By Theorem 9, $I(N)$ is a subalgebra of N. Therefore, $I(N)$ is a closed filter of N. □

Theorem 11. *Let (N, *) be an AG-group and F be a closed filter of N. Define binary relation* $\approx_F$ *on N as follows:*

$$x \approx_F y \Leftrightarrow (x * y \in F, y * x \in F),\ for\ any\ x, y \text{ in } N.$$

Then,

(1) $\approx_F$ *is an equivalent relation on N.*
(2) $x \approx_F y$ *and* $a \approx_F b$ *imply* $x^*a \approx_F y^*b$.
(3) *f: N→N/F is a homomorphism mapping, where* $N/F = \{[x]_F: x \in N\}$, $[x]_F$ *denote the equivalent class contained x.*

Proof. (1) $\forall x \in N$, by Definition 7(2), $x^*x \in F$. Thus, $x \approx_F x$.

Assume $x \approx_F y$, then $x^*y \in F$, $y^*x \in F$. It follows that $y \approx_F x$.

Suppose that $x \approx_F y$ and $y \approx_F z$. We have $x^*y \in F$, $y^*x \in F$, $y^*z \in F$ and $z^*y \in F$. By medial law and Definition 7,

$$(y * y) * (z * x) = (y * z) * (y * x) \in F, \text{ then } (z * x) \in F;$$

$$(y * y) * (x * z) = (y * x) * (y * z) \in F, \text{ then } (x * z) \in F.$$

It follows that $x \approx_F z$.

Therefore, $\approx_F$ is an equivalent relation on N.

(2) Suppose that $x \approx_F y$ and $a \approx_F b$. We have $x^*y \in F$, $y^*x \in F$, $a^*b \in F$ and $b^*a \in F$. By medial law and Definition 7,

$$(x * a) * (y * b) = (x * y) * (a * b) \in F;\ (y * b) * (x * a) = (y * x) * (b * a) \in F.$$

It follows that $x*a \approx_F y*b$.

(3) Combining (1) and (2), we can obtain (3).
The proof complete. □

Theorem 12. *Let (N, *) be a generalized involution AG-group, I(N) the involution part of N (defined as Theorem 9). Then, f: N→N/I(N) is a homomorphism mapping, and N/I(N) is involutive, where N/I(N) = {[x]| x∈N}, [x] is the equivalent class contained x by closed filter I(N).*

Proof. It follows from Theorem 10 and Theorem 11. □

Theorem 13. *Let (N, *) be an AG-group, P(N) = {x∈N|∃a∈N, s.t x =a*a} be the power part of N (see Theorem 5). Then, P(N) is a closed filter of N.*

Proof. It is obvious that $e = e*e \in P(N)$. For any $x \in N$, $x*x \in P(N)$.

Moreover, assume that $x \in P(N)$ and $x*y \in P(N)$, then there exists $a, b \in N$ such that

$$x = a*a,\ x*y = b*b.$$

Denote $c = a^{-1}*b$, where a^{-1} is the left inverse of a in N. Then,

$$c*c = (a^{-1}*b)*(a^{-1}*b) = (a^{-1}*a^{-1})*(b*b) = (a^{-1}*a^{-1})*(x*y) = (a^{-1}*a^{-1})*((a*a)*y) = (a^{-1}*a^{-1})*((y*a)*a) = (a^{-1}*(y*a))*(a^{-1}*a) = (a^{-1}*(y*a))*e = (e*(y*a))*a^{-1} = (y*a)*a^{-1} = (a^{-1}*a)*y = e*y = y.$$

Thus, $y \in P(N)$. It follows that $P(N)$ is a filter of N. By Theorem 5, $P(N)$ is a subalgebra of N, therefore, $P(N)$ is a closed filter of N. □

Theorem 14. *Let (N, *) be an AG-group, P(N) the power part of N (defined as Theorem 13). Then, f: N→N/P(N) is a homomorphism mapping, where N/P(N) = {[x]| x∈N}, [x] is the equivalent class contained x by closed filter P(N).*

Proof. It follows from Theorems 11 and 13. □

6. Conclusions

In the paper, we give some examples of AG-groups, and obtain some new properties of AG-groups: an AG-group is weak commutative (or alternative) if and only if it is an Abelian group; every AG-group is a quasi-cancellative AG-groupoid. We introduce two new concepts of involution AG-group and generalized involution AG-group, establish a one-to-one correspondence between involution AG-groups and Abelian groups, and construct a homomorphism mapping from generalized involution AG-groups to involution AG-groups. Moreover, we introduce the notion of filter in AG-groups, establish quotient algebra by every filter, and obtain some homomorphism theorems. Some results in this paper are expressed in Figure 2. In the future, we can investigate the combination of some uncertainty set theories (fuzzy set, neutrosophic set, etc.) and algebra systems (see [18–22]).

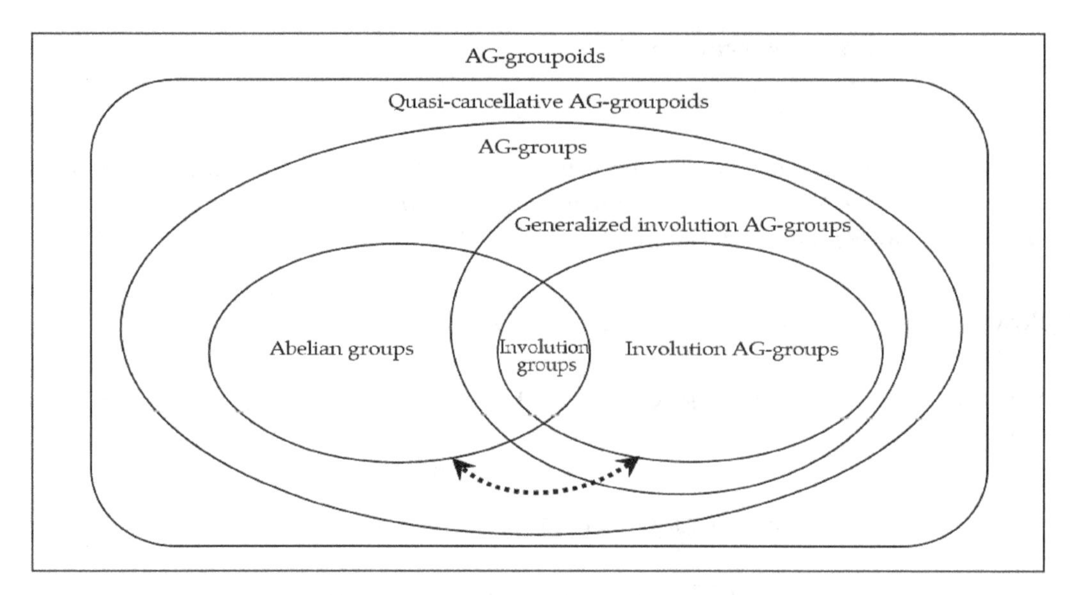

Figure 2. Some results in this paper.

Author Contributions: X.Z. and X.W. initiated the research; and X.Z. wrote final version of the paper.

References

1. Hall, G.G. *Applied Group Theory*; merican Elsevier Publishing Co., Inc.: New York, NY, USA, 1967.
2. Howie, J.M. *Fundamentals of Semigroup Theory*; Clarendon Press: Oxford, UK, 1995.
3. Lawson, M.V. *Inverse Semigroups: The Theory of Partial Symmetries*; World Scientific: Singapore, 1998.
4. Akinmoyewa, T.J. A study of some properties of generalized groups. *Octogon Math. Mag.* **2009**, *7*, 599–626.
5. Smarandache, F.; Ali, M. Neutrosophic triplet group. *Neural Comput. Appl.* **2018**, *29*, 595–601. [CrossRef]
6. Kazim, M.A.; Naseeruddin, M. On almost semigroups. *Port. Math.* **1977**, *36*, 41–47.
7. Kleinfeld, M.H. Rings with $x(yz) = z(yx)$. *Commun. Algebra* **1978**, *6*, 1369–1373. [CrossRef]
8. Mushtaq, Q.; Iqbal, Q. Decomposition of a locally associative LA-semigroup. *Semigroup Forum* **1990**, *41*, 155–164. [CrossRef]
9. Ahmad, I.; Rashad, M. Constructions of some algebraic structures from each other. *Int. Math. Forum* **2012**, *7*, 2759–2766.
10. Ali, A.; Shah, M.; Ahmad, I. On quasi-cancellativity of AG-groupoids. *Int. J. Contemp. Math. Sci.* **2012**, *7*, 2065–2070.
11. Dudek, W.A.; Gigon, R.S. Completely inverse AG**-groupoids. *Semigroup Forum* **2013**, *87*, 201–229. [CrossRef]
12. Zhang, X.H.; Wu, X.Y.; Mao, X.Y.; Smarandache, F.; Park, C. On neutrosophic extended triplet groups (loops) and Abel-Grassmann's groupoids (AG-groupoids). *J. Intell. Fuzzy Syst.* **2019**, in press.
13. Wu, X.Y.; Zhang, X.H. The decomposition theorems of AG-neutrosophic extended triplet loops and strong AG-(l,l)-loops. *Mathematics* **2019**, *7*, 268. [CrossRef]
14. Kamran, M.S. Conditions for LA-Semigroups to Resemble Associative Structures. Ph.D. Thesis, Quaid-i-Azam University, Islamabad, Pakistan, 1993.
15. Shah, M.; Ali, A. Some structure properties of AG-groups. *Int. Math. Forum* **2011**, *6*, 1661–1667.
16. Protić, P.V. Some remarks on Abel-Grassmann's groups. *Quasigroups Relat. Syst.* **2012**, *20*, 267–274.
17. Mushtaq, Q. Abelian groups defined by LA-semigroups. *Stud. Sci. Math. Hung.* **1983**, *18*, 427–428.
18. Ma, Y.C.; Zhang, X.H.; Yang, X.F.; Zhou, X. Generalized neutrosophic extended triplet group. *Symmetry* **2019**, *11*, 327. [CrossRef]
19. Zhang, X.H.; Bo, C.X.; Smarandache, F.; Park, C. New operations of totally dependent-neutrosophic sets and totally dependent-neutrosophic soft sets. *Symmetry* **2018**, *10*, 187. [CrossRef]

20. Zhang, X.H.; Mao, X.Y.; Wu, Y.T.; Zhai, X.H. Neutrosophic filters in pseudo-BCI algebras. *Int. J. Uncertain. Quan.* **2018**, *8*, 511–526. [CrossRef]
21. Zhang, X.H.; Borzooei, R.A.; Jun, Y.B. Q-filters of quantum B-algebras and basic implication algebras. *Symmetry* **2018**, *10*, 573. [CrossRef]
22. Zhan, J.M.; Sun, B.Z.; Alcantud, J.C.R. Covering based multigranulation (I, T)-fuzzy rough set models and applications in multi-attribute group decision-making. *Inf. Sci.* **2019**, *476*, 290–318. [CrossRef]

Some Results on Multigranulation Neutrosophic Rough Sets on a Single Domain

Hu Zhao [1,*] and Hong-Ying Zhang [2]

[1] School of Science, Xi'an Polytechnic University, Xi'an 710048, China
[2] School of Mathematics and Statistics, Xi'an Jiaotong University, Xi'an 710049, China; Zhyemily@mail.xjtu.edu.cn
* Correspondence: zhaohu@xpu.edu.cn or zhaohu2007@yeah.net.

Abstract: As a generalization of single value neutrosophic rough sets, the concept of multi-granulation neutrosophic rough sets was proposed by Bo et al., and some basic properties of the pessimistic (optimistic) multigranulation neutrosophic rough approximation operators were studied. However, they did not do a comprehensive study on the algebraic structure of the pessimistic (optimistic) multigranulation neutrosophic rough approximation operators. In the present paper, we will provide the lattice structure of the pessimistic multigranulation neutrosophic rough approximation operators. In particular, in the one-dimensional case, for special neutrosophic relations, the completely lattice isomorphic relationship between upper neutrosophic rough approximation operators and lower neutrosophic rough approximation operators is proved.

Keywords: neutrosophic set; neutrosophic rough set; pessimistic (optimistic) multigranulation neutrosophic approximation operators; complete lattice

---

1. Introduction

In order to deal with imprecise information and inconsistent knowledge, Smarandache [1,2] first introduced the notion of neutrosophic set by fusing a tri-component set and the non-standard analysis. A neutrosophic set consists of three membership functions, where every function value is a real standard or non-standard subset of the nonstandard unit interval $]0^-, 1^+[$. Since then, many authors have studied various aspects of neutrosophic sets from different points of view, for example, in order to apply the neutrosophic idea to logics, Rivieccio [3] proposed neutrosophic logics which is a generalization of fuzzy logics and studied some basic properties. Guo and Cheng [4] and Guo and Sengur [5] obtained good applications in cluster analysis and image processing by using neutrosophic sets. Salama and Broumi [6] and Broumi and Smarandache [7] first introduced the concept of rough neutrosophic sets, handled incomplete and indeterminate information, and studied some operations and their properties.

In order to apply neutrosophic sets conveniently, Wang et al. [8] proposed single valued neutrosophic sets by simplifying neutrosophic sets. Single valued neutrosophic sets can also be viewed as a generalization of intuitionistic fuzzy sets (Atanassov [9]). Single valued neutrosophic sets have become a new majorly research issue. Ye [10–12] proposed decision making based on correlation coefficients and weighted correlation coefficient of single valued neutrosophic sets, and gave an application of proposed methods. Majumdar and Samant [13] studied similarity, distance and entropy of single valued neutrosophic sets from a theoretical aspect.

Şahin and Küçük [14] gave a subsethood measure of single valued neutrosophic sets based on distance and showed its effectiveness through an example. We know that there's a certain connection among fuzzy rough approximation operators and fuzzy relations (resp., fuzzy topologies, information systems [15–17]). Hence, Yang et al. [18] firstly proposed neutrosophic relations and studied some kinds of kernels and closures of neutrosophic relations. Subsequently they proposed single valued neutrosophic rough sets [19] by fusing single valued neutrosophic sets and rough sets (Pawlak, [20]), and they studied some properties of single value neutrosophic upper and lower approximation operators. As a generalization of single value neutrosophic rough sets, Bao and Yang [21] introduced p-dimension single valued neutrosophic refined rough sets, and they also gave some properties of p-dimension single valued neutrosophic upper and lower approximation operators.

As another generalization of single value neutrosophic rough sets, Bo et al. [22] proposed the concept of multi-granulation neutrosophic rough sets and obtained some basic properties of the pessimistic (optimistic) multigranulation neutrosophic rough approximation operators. However, the lattice structures of those rough approximation operators in references [19,21,22], were not well studied. Following this idea, Zhao and Zhang [23] gave the supremum and infimum of the p-dimension neutrosophic upper and lower approximation operators, but they did not study the relationship between the p-dimension neutrosophic upper approximation operators and the p-dimension neutrosophic lower approximation operators, especially in the one-dimensional case. Inspired by paper [23], a natural problem is: Can the lattice structure of pessimistic (optimistic) multigranulation neutrosophic approximation operators be given?

In the present paper, we study the algebraic structure of optimistic (pessimistic) multigranulation single valued neutrosophic approximation operators.

The structure of the paper is organized as follows. The next section reviews some basic definitions of neutrosophic sets and one-dimensional multi-granulation rough sets. In Section 3, the lattice structure of the pessimistic multigranulation neutrosophic rough approximation operators are studied. In Section 4, for special neutrosophic relations, a one-to-one correspondence relationship between neutrosophic upper approximation operators and lower approximation operators is given. Finally, Section 5 concludes this article and points out the deficiencies of the current research.

2. Preliminaries

In this section, we briefly recall several definitions of neutrosophic set (here "neutrosophic set" refers exclusively to "single value neutrosophic set") and one-dimensional multi-granulation rough set.

Definition 1 ([8]). *A neutrosophic set B in X is defined as follows: $\forall a \in X$,*

$$B = (T_A(a), I_A(a), F_A(a)),$$

where $T_A(a)$, $I_A(a)$, $F_A(a) \in [0,1]$, $0 \leq \sup T_A(a) + \sup I_A(a) + \sup F_A(a) \leq 3$. The set of all neutrosophic sets on X will be denoted by $\mathbb{SVNS}(X)$.

Definition 2 ([11]). *Let C and D be two neutrosophic sets in X, if*

$$T_C(a) \leq T_D(a),\ I_C(a) \geq I_D(a) \text{ and } F_C(a) \geq F_D(a)$$

for each $a \in X$, then we called C is contained in D, i.e., *$C \Subset D$. If $C \Subset D$ and $D \Subset C$, then we called C is equal to D, denoted by $C = D$.*

Definition 3 ([18]). *Let A and B be two neutrosophic sets in X,*

(1) The union of A and B is a s neutrosophic set C, denoted by $A \Cup B$, where $\forall x \in X$,

$$T_C(a) = max\{T_A(a), T_B(a)\},\ I_C(a) = min\{I_A(a), I_B(a)\}, \text{ and}$$
$$F_C(a) = min\{F_A(a), F_B(a)\}.$$

(2) *The intersection of A and B is a neutrosophic set D, denoted by $A \Cap B$, where $\forall x \in X$,*

$$T_D(a) = min\{T_A(a), T_B(a)\}, \ I_D(a) = max\{I_A(a), I_B(a)\}, \text{ and}$$
$$F_D(a) = max\{F_A(a), F_B(a)\}.$$

Definition 4 ([18]). *A neutrosophic relation R in X is defined as follows:*

$$R = \{< (a,b), T_R(a,b), I_R(a,b), F_R(a,b) >| \ (a,b) \in X \times X\},$$

where $T_R : X \times X \to [0,1], I_R : X \times X \to [0,1], F_R : X \times X \to [0,1]$, and

$$0 \leq \sup T_R(a,b) + \sup I_R(a,b) + \sup F_R(a,b) \leq 3.$$

The family of all neutrosophic relations in X will be denoted by **SVNR**(X)*, and the pair (X,R) is called a neutrosophic approximation space.*

Definition 5 ([19]). *Let (X,R) be a neutrosophic approximation space, $\forall A \in \mathbb{SVNS}(X)$, the lower and upper approximations of A with respect to (X,R), denoted by $\underline{R}(A)$ and $\overline{R}(A)$, are two neutrosophic sets whose membership functions are defined as: $\forall a \in X$,*

$$T_{\underline{R}(A)}(a) = \underset{b \in X}{\wedge}[F_R(a,b) \vee T_A(b)], \ I_{\underline{R}(A)}(a) = \underset{b \in X}{\vee}[(1 - I_R(a,b)) \wedge I_A(b)],$$

$$F_{\underline{R}(A)}(a) = \underset{b \in X}{\vee}[T_R(a,b) \wedge F_A(b)], \ T_{\overline{R}(A)}(a) = \underset{b \in X}{\vee}[T_R(a,b) \wedge T_A(b)],$$

$$I_{\overline{R}(A)}(a) = \underset{b \in X}{\wedge}[I_R(a,b) \vee I_A(b)], \ F_{\overline{R}(A)}(a) = \underset{b \in X}{\wedge}[F_R(a,b) \vee F_A(b)].$$

The pair $(\underline{R}(A), \overline{R}(A))$ is called the one-dimensional multi-granulation rough set (also called single value neutrosophic rough set or one-dimension single valued neutrosophic refined rough set) of A with respect to (X,R). $\underline{R}$ and $\overline{R}$ are referred to as the neutrosophic lower and upper approximation operators,respectively.

Lemma 1 ([19]). *Let R_1 and R_2 be two neutrosophic relations in X, $\forall A \in \mathbb{SVNS}(X)$, we have*

(1) $\underline{R_1 \Cup R_2(A)} = \underline{R_1}(A) \Cap \underline{R_2}(A)$;
(2) $\overline{R_1 \Cup R_2(A)} = \overline{R_1}(A) \Cup \overline{R_2}(A)$;
(3) $\underline{R_1 \Cap R_2(A)} \Supset \underline{R_1}(A) \Cup \underline{R_2}(A) \Supset \underline{R_1}(A) \Cap \underline{R_2}(A)$;
(4) $\overline{R_1 \Cap R_2(A)} \Subset \overline{R_1}(A) \Cap \overline{R_2}(A)$.

3. The Lattice Structure of the Pessimistic Multigranulation Neutrosophic Rough Approximation Operators

In this section, set $M = \{R_1, R_2, \cdots, R_n\} = \{R_i\}_{i=\overline{1,n}}$ is called a multigranulation neutrosophic relations set on X if each R_i is a neutrosophic relation on X. In this case, the pair (X,M) will be called an n-dimensional multigranulation neutrosophic apptoximation space.

Definition 6 ([22]). *Let (X,M) be an n-dimensional multigranulation neutrosophic apptoximation space. We define two pairs of approximation operators as follows, for all $\forall A \in \mathbb{SVNS}(X)$ and $a \in X$,*

$$M^O(A) = (\underline{M}^O(A), \overline{M}^O(A)), \ M^P(A) = (\underline{M}^P(A), \overline{M}^P(A)),$$

where

$$T_{\underline{M}^O(A)}(a) = \vee_{i=1}^n T_{\underline{R_i}(A)}(a), \ I_{\underline{M}^O(A)}(a) = \wedge_{i=1}^n I_{\underline{R_i}(A)}(a), \ F_{\underline{M}^O(A)}(a) = \wedge_{i=1}^n F_{\underline{R_i}(A)}(a).$$

$$T_{\overline{M}^O(A)}(a) = \wedge_{i=1}^n T_{\overline{R_i}(A)}(a), \ I_{\overline{M}^O(A)}(a) = \vee_{i=1}^n I_{\overline{R_i}(A)}(a), \ F_{\overline{M}^O(A)}(a) = \vee_{i=1}^n F_{\overline{R_i}(A)}(a).$$

$$T_{\underline{M}^P(A)}(a) = \wedge_{i=1}^{n} T_{\underline{R_i}(A)}(a),\ I_{\underline{M}^P(A)}(a) = \vee_{i=1}^{n} I_{\underline{R_i}(A)}(a),\ F_{\underline{M}^P(A)}(a) = \vee_{i=1}^{n} F_{\underline{R_i}(A)}(a).$$
$$T_{\overline{M}^P(A)}(a) = \vee_{i=1}^{n} T_{\overline{R_i}(A)}(a),\ I_{\overline{M}^P(A)}(a) = \wedge_{i=1}^{n} I_{\overline{R_i}(A)}(a),\ F_{\overline{M}^P(A)}(a) = \wedge_{i=1}^{n} F_{\overline{R_i}(A)}(a).$$

Then the pair $M^O(A) = (\underline{M}^O(A), \overline{M}^O(A))$ is called an optisмistic multigranulation neutrosophic rough set, and the pair $M^P(A) = (\underline{M}^P(A), \overline{M}^P(A))$ is called an pessimistic multigranulation neutrosophic rough set $\overline{M}^O$ and $\overline{M}^P$ are referred to as the optimistic and pessimistic multigranulation neutrosophic upper approximation operators, respectively. Similarly, $\underline{M}^O$ and $\underline{M}^P$ are referred to as the optimistic and pessimistic multigranulation neutrosophic lower approximation operators, respectively.

Remark 1. *If $n = 1$, then the multigranulation neutrosophic rough set will degenerated to a one-dimensional multi-granulation rough set (see Definition 5). In the following, the family of all multigranulation neutrosophic relations set on X will be denoted by $n - \mathbf{SVNR}(X)$. Defined a relation $\sqsubseteq$ on $n - \mathbf{SVNR}(X)$ as follows: $M \sqsubseteq N$ if and only if $M_i \Subset N_i$, then $(n - \mathbf{SVNR}(X), \sqsubseteq)$ is a poset, where $M = \{M_i\}_{i=\overline{1,n}}$ and $N = \{N_i\}_{i=\overline{1,n}}$.*

$\forall \{M^j\}_{j\in\Lambda} \subseteq n - \mathbf{SVNR}(X)$, where $M^j = \left\{M_i^j\right\}_{i=\overline{1,n}}$ and Λ be a index set, we can define union and intersection of M^j as follows:

$$\underset{j\in\Lambda}{\vee} M^j = \left\{\Cup_{j\in\Lambda} M_i^j\right\}_{i=\overline{1,n}},\ \underset{j\in\Lambda}{\wedge} M^j = \left\{\Cap_{j\in\Lambda} M_i^j\right\}_{i=\overline{1,n}},$$

where

$$T_{\Cup_{j\in\Lambda} M_i^j}(a,b) = \underset{j\in\Lambda}{\vee} T_{M_i^j}(a,b),\ I_{\Cup_{j\in\Lambda} M_i^j}(a,b) = \underset{j\in\Lambda}{\wedge} I_{M_i^j}(a,b),$$
$$F_{\Cup_{j\in\Lambda} M_i^j}(a,b) = \underset{j\in\Lambda}{\wedge} F_{M_i^j}(a,b), T_{\Cap_{j\in\Lambda} M_i^j}(a,b) = \underset{j\in\Lambda}{\wedge} T_{M_i^j}(a,b),$$
$$I_{\Cap_{j\in\Lambda} M_i^j}(a,b) = \underset{j\in\Lambda}{\vee} I_{M_i^j}(a,b),\ F_{\Cap_{j\in\Lambda} M_i^j}(a,b) = \underset{j\in\Lambda}{\vee} F_{M_i^j}(a,b).$$

Then $\underset{j\in\Lambda}{\vee} M^j$ and $\underset{j\in\Lambda}{\wedge} M^j$ are two multigranulation neutrosophic relations on X, and we easily show that $\underset{j\in\Lambda}{\vee} M^j$ and $\underset{j\in\Lambda}{\wedge} M^j$ are infimum and supremum of $\{M^j\}_{j\in\Lambda}$, respectively. Hence we can easily obtain the following theorem:

Theorem 1. *$(n - \mathbf{SVNR}(X), \sqsubseteq, \wedge, \vee)$ is a complete lattice, $\widetilde{X_n} = \{\underbrace{X_n, X_n, \cdots, X_n}_{n}\}$ and $\widetilde{\varnothing_N} = \{\underbrace{\varnothing_N, \varnothing_N, \cdots, \varnothing_N}_{n}\}$ are its top element and bottom element, respectively, where X_n and $\varnothing_N$ are two neutrosophic relations in X and defined as follows: $\forall (a,b) \in X \times X$, $T_{X_N}(a,b) = 1$, $I_{X_N}(a,b) = 0$, $F_{X_N}(a,b) = 0$ and $T_{\varnothing_N}(a,b) = 0$, $I_{\varnothing_N}(a,b) = 1$, $F_{\varnothing_N}(a,b) = 1$. In particular, $(\mathbf{SVNR}(X), \Subset, \Cup, \Cap)$ is a complete lattice.*

Theorem 2. *Let $M = \{R_i\}_{i=\overline{1,n}}$ and $N = \{Q_i\}_{i=\overline{1,n}}$ be two multigranulation neutrosophic relations set on X, $\forall A \in \mathbb{SVNS}(X)$, we have*

(1) $\underline{M \vee N}^O(A) \Subset \underline{M}^O(A) \Cap \underline{N}^O(A)$, $\underline{M \vee N}^P(A) = \underline{M}^P(A) \Cap \underline{N}^P(A)$;
(2) $\overline{M \vee N}^O(A) \Supset \overline{M}^O(A) \Cup \overline{N}^O(A)$, $\overline{M \vee N}^P(A) = \overline{M}^P(A) \Cup \overline{N}^P(A)$;
(3) $\underline{M \wedge N}^O(A) \Supset \underline{M}^O(A) \Cup \underline{N}^O(A) \Supset \underline{M}^O(A) \Cap \underline{N}^O(A)$, $\underline{M \wedge N}^P(A) \Supset \underline{M}^P(A) \Cup \underline{N}^P(A) \Supset \underline{M}^P(A) \Cap \underline{N}^P(A)$;
(4) $\overline{M \wedge N}^O(A) \Subset \overline{M}^O(A) \Cap \overline{N}^O(A)$, $\overline{M \wedge N}^P(A) \Subset \overline{M}^P(A) \Cap \overline{N}^P(A)$.

Proof. We only show that the case of the optimistic multigranulation neutrosophic approximation operators.

(1) $\forall a \in X$, by Lemma 1 and Definition 6, we have the following:

$$\begin{aligned}
&T_{\underline{M \vee N}^O(A)}(a) \\
&= \vee_{i=1}^{n} T_{\underline{R_i \Cup Q_i}(A)}(a) = \vee_{i=1}^{n} T_{\underline{R_i}(A) \Cap \underline{Q_i}(A)}(a) \\
&= \vee_{i=1}^{n} \left[T_{\underline{R_i}(A)}(a) \wedge T_{\underline{Q_i}(A)}(a) \right] \\
&\leq \left[\vee_{i=1}^{n} T_{\underline{R_i}(A)}(a) \right] \wedge \left[\vee_{i=1}^{n} T_{\underline{Q_i}(A)}(a) \right] \\
&= T_{\underline{M}^O(A)}(a) \wedge T_{\underline{N}^O(A)}(a) \\
&= T_{\underline{M}^O(A) \Cap \underline{N}^O(A)}(a),
\end{aligned}$$

$$\begin{aligned}
&I_{\underline{M \vee N}^O(A)}(a) \\
&= \wedge_{i=1}^{n} I_{\underline{R_i \Cup Q_i}(A)}(a) = \wedge_{i=1}^{n} I_{\underline{R_i}(A) \Cap \underline{Q_i}(A)}(a) \\
&= \wedge_{i=1}^{n} \left[I_{\underline{R_i}(A)}(a) \vee I_{\underline{Q_i}(A)}(a) \right] \\
&\geq \left[\wedge_{i=1}^{n} I_{\underline{R_i}(A)}(a) \right] \vee \left[\wedge_{i=1}^{n} I_{\underline{Q_i}(A)}(a) \right] \\
&= I_{\underline{M}^O(A)}(a) \vee I_{\underline{N}^O(A)}(a) \\
&= I_{\underline{M}^O(A) \Cap \underline{N}^O(A)}(a),
\end{aligned}$$

$$\begin{aligned}
&F_{\underline{M \vee N}^O(A)}(a) \\
&= \wedge_{i=1}^{n} F_{\underline{R_i \Cup Q_i}(A)}(a) = \wedge_{i=1}^{n} F_{\underline{R_i}(A) \Cap \underline{Q_i}(A)}(a) \\
&= \wedge_{i=1}^{n} \left[F_{\underline{R_i}(A)}(a) \vee F_{\underline{Q_i}(A)}(a) \right] \\
&\geq \left[\wedge_{i=1}^{n} F_{\underline{R_i}(A)}(a) \right] \vee \left[\wedge_{i=1}^{n} F_{\underline{Q_i}(A)}(a) \right] \\
&= F_{\underline{M}^O(A)}(a) \vee F_{\underline{N}^O(A)}(a) \\
&= F_{\underline{M}^O(A) \Cap \underline{N}^O(A)}(a).
\end{aligned}$$

Hence, $\underline{M \vee N}^O(A) \Subset \underline{M}^O(A) \Cap \underline{N}^O(A)$.

(2) $\forall a \in X$, by Lemma 1 and Definition 6, we have the following:

$$\begin{aligned}
&T_{\overline{M \vee N}^O(A)}(a) \\
&= \wedge_{i=1}^{n} T_{\overline{R_i \Cup Q_i}(A)}(a) = \wedge_{i=1}^{n} T_{\overline{R_i}(A) \Cup \overline{Q_i}(A)}(a) \\
&= \wedge_{i=1}^{n} \left[T_{\overline{R_i}(A)}(a) \vee T_{\overline{Q_i}(A)}(a) \right] \\
&\geq \left[\wedge_{i=1}^{n} T_{\overline{R_i}(A)}(a) \right] \vee \left[\wedge_{i=1}^{n} T_{\overline{Q_i}(A)}(a) \right] \\
&= T_{\overline{M}^O(A)}(a) \vee T_{\overline{N}^O(A)}(a) = T_{\overline{M}^O(A) \Cup \overline{N}^O(A)}(a),
\end{aligned}$$

$$\begin{aligned}
&I_{\overline{M \vee N}^O(A)}(a) \\
&= \vee_{i=1}^{n} I_{\overline{R_i \Cup Q_i}(A)}(a) = \vee_{i=1}^{n} I_{\overline{R_i}(A) \Cup \overline{Q_i}(A)}(a) \\
&= \vee_{i=1}^{n} \left[I_{\overline{R_i}(A)}(a) \wedge I_{\overline{Q_i}(A)}(a) \right] \\
&\leq \left[\vee_{i=1}^{n} I_{\overline{R_i}(A)}(a) \right] \wedge \left[\vee_{i=1}^{n} I_{\overline{Q_i}(A)}(a) \right] \\
&= I_{\overline{M}^O(A)}(a) \wedge I_{\overline{N}^O(A)}(a) = I_{\overline{M}^O(A) \Cup \overline{N}^O(A)}(a),
\end{aligned}$$

$$\begin{aligned}
&F_{\overline{M \vee N}^O(A)}(a) \\
&= \vee_{i=1}^{n} F_{\overline{R_i \Cup Q_i}(A)}(a) = \vee_{i=1}^{n} F_{\overline{R_i}(A) \Cup \overline{Q_i}(A)}(a) \\
&= \vee_{i=1}^{n} \left[F_{\overline{R_i}(A)}(a) \wedge F_{\overline{Q_i}(A)}(a) \right] \\
&\leq \left[\vee_{i=1}^{n} F_{\overline{R_i}(A)}(a) \right] \wedge \left[\vee_{i=1}^{n} F_{\overline{Q_i}(A)}(a) \right] \\
&= F_{\overline{M}^O(A)}(a) \wedge F_{\overline{N}^O(A)}(a) = F_{\overline{M}^O(A) \Cup \overline{N}^O(A)}(a).
\end{aligned}$$

Hence, $\overline{M \vee N}^O(A) \Supset \overline{M}^O(A) \Cup \overline{N}^O(A)$.

(3) $\forall a \in X$, by Lemma 1 and Definition 6, we have the following:

$$\begin{aligned} & T_{\underline{M \wedge N}^O(A)}(a) \\ & = \vee_{i=1}^{n} T_{\underline{R_i \Cap Q_i}(A)}(a) \geq \vee_{i=1}^{n} T_{\underline{R_i}(A) \Cup \underline{Q_i}(A)}(a) \\ & = \vee_{i=1}^{n} \left[T_{\underline{R_i}(A)}(a) \vee T_{\underline{Q_i}(A)}(a) \right] \\ & = \left[\vee_{i=1}^{n} T_{\underline{R_i}(A)}(a) \right] \vee \left[\vee_{i=1}^{n} T_{\underline{Q_i}(A)}(a) \right] \\ & = T_{\underline{M}^O(A)}(a) \vee T_{\underline{N}^O(A)}(a) \geq T_{\underline{M}^O(A)}(a) \wedge T_{\underline{N}^O(A)}(a), \end{aligned}$$

$$\begin{aligned} & I_{\underline{M \wedge N}^O(A)}(a) \\ & = \wedge_{i=1}^{n} I_{\underline{R_i \Cap Q_i}(A)}(a) \leq \wedge_{i=1}^{n} I_{\underline{R_i}(A) \Cup \underline{Q_i}(A)}(a) \\ & = \wedge_{i=1}^{n} \left[I_{\underline{R_i}(A)}(a) \wedge I_{\underline{Q_i}(A)}(a) \right] \\ & = \left[\wedge_{i=1}^{n} I_{\underline{R_i}(A)}(a) \right] \wedge \left[\wedge_{i=1}^{n} I_{\underline{Q_i}(A)}(a) \right] \\ & = I_{\underline{M}^O(A)}(a) \wedge I_{\underline{N}^O(A)}(a) \leq I_{\underline{M}^O(A)}(a) \vee I_{\underline{N}^O(A)}(a), \end{aligned}$$

$$\begin{aligned} & F_{\underline{M \wedge N}^O(A)}(a) \\ & = \wedge_{i=1}^{n} F_{\underline{R_i \Cap Q_i}(A)}(a) \leq \wedge_{i=1}^{n} F_{\underline{R_i}(A) \Cup \underline{Q_i}(A)}(a) \\ & = \wedge_{i=1}^{n} \left[F_{\underline{R_i}(A)}(a) \wedge F_{\underline{Q_i}(A)}(a) \right] \\ & = \left[\wedge_{i=1}^{n} F_{\underline{R_i}(A)}(a) \right] \wedge \left[\wedge_{i=1}^{n} F_{\underline{Q_i}(A)}(a) \right] \\ & = F_{\underline{M}^O(A)}(a) \wedge F_{\underline{N}^O(A)}(a) \leq F_{\underline{M}^O(A)}(a) \vee F_{\underline{N}^O(A)}(a). \end{aligned}$$

Hence, $\underline{M \wedge N}^o(A) \Supset \underline{M}^o(A) \Cup \underline{N}^o(A) \Supset \underline{M}^o(A) \Cap \underline{N}^o(A)$.

(4) $\forall a \in X$, by Lemma 1 and Definition 6, we have the following:

$$\begin{aligned} & T_{\overline{M \wedge N}^O(A)}(a) \\ & = \wedge_{i=1}^{n} T_{\overline{R_i \Cap Q_i}(A)}(a) \leq \wedge_{i=1}^{n} T_{\overline{R_i}(A) \Cap \overline{Q_i}(A)}(a) \\ & = \wedge_{i=1}^{n} \left[T_{\overline{R_i}(A)}(a) \wedge T_{\overline{Q_i}(A)}(a) \right] \\ & = \left[\wedge_{i=1}^{n} T_{\overline{R_i}(A)}(a) \right] \wedge \left[\wedge_{i=1}^{n} T_{\overline{Q_i}(A)}(a) \right] \\ & = T_{\overline{M}^O(A)}(a) \wedge T_{\overline{N}^O(A)}(a) = T_{\overline{M}^O(A) \Cap \overline{N}^O(A)}(a), \end{aligned}$$

$$\begin{aligned} & I_{\overline{M \wedge N}^O(A)}(a) \\ & = \vee_{i=1}^{n} I_{\overline{R_i \Cap Q_i}(A)}(a) \geq \vee_{i=1}^{n} I_{\overline{R_i}(A) \Cap \overline{Q_i}(A)}(a) \\ & = \vee_{i=1}^{n} \left[I_{\overline{R_i}(A)}(a) \vee I_{\overline{Q_i}(A)}(a) \right] \\ & = \left[\vee_{i=1}^{n} I_{\overline{R_i}(A)}(a) \right] \vee \left[\vee_{i=1}^{n} I_{\overline{Q_i}(A)}(a) \right] \\ & = I_{\overline{M}^O(A)}(a) \vee T_{\overline{N}^O(A)}(a) = I_{\overline{M}^O(A) \Cap \overline{N}^O(A)}(a), \end{aligned}$$

$$\begin{aligned} & F_{\overline{M \wedge N}^O(A)}(a) \\ & = \vee_{i=1}^{n} F_{\overline{R_i \Cap Q_i}(A)}(a) \geq \vee_{i=1}^{n} F_{\overline{R_i}(A) \Cap \overline{Q_i}(A)}(a) \\ & = \vee_{i=1}^{n} \left[F_{\overline{R_i}(A)}(a) \vee F_{\overline{Q_i}(A)}(a) \right] \\ & = \left[\vee_{i=1}^{n} F_{\overline{R_i}(A)}(a) \right] \vee \left[\vee_{i=1}^{n} F_{\overline{Q_i}(A)}(a) \right] \\ & = F_{\overline{M}^O(A)}(a) \vee F_{\overline{N}^O(A)}(a) = F_{\overline{M}^O(A) \Cap \overline{N}^O(A)}(a). \end{aligned}$$

Hence, $\overline{M \wedge N}^O(A) \Subset \overline{M}^O(A) \Cap \overline{N}^O(A)$. □

From Theorem 2, we can easily obtain the following corollary:

Corollary 1. *Let $M = \{R_i\}_{i=\overline{1,n}}$ and $N = \{Q_i\}_{i=\overline{1,n}}$ be two multigranulation neutrosophic relations set on X. If $M \sqsubseteq N$, then $\forall A \in \mathbb{SVNS}(X)$,*

$$\underline{N}^O(A) \Subset \underline{M}^O(A),\ \underline{N}^P(A) \Subset \underline{M}^P(A)),\ \overline{M}^O(A) \Subset \overline{N}^O(A),\ \overline{M}^P(A) \Subset \overline{N}^P(A).$$

Let $H_n^P = \left\{\overline{M}^P \mid M \in n - \mathbf{SVNR}(X)\right\}$ and $L_n^P = \left\{\underline{M}^P \mid M \in n - \mathbf{SVNR}(X)\right\}$ be the set of pessimistic multigranulation neutrosophic upper and lower approximation operators in X, respectively.

- Defined a relation $\hat{\lesssim}$ on H_n^P as follows: $\overline{M}^P \hat{\lesssim} \overline{N}^P$ if and only if $\overline{M}^P(A) \Subset \overline{N}^P(A)$ for each $A \in \mathbb{SVNS}(X)$. Then $(H_n^P, \hat{\lesssim})$ is a poset.
- Defined a relation $\hat{\lesssim}$ on L_n^P as follows: $\underline{M}^P \hat{\lesssim} \underline{N}^P$ if and only if $\underline{N}^P(A) \Subset \underline{M}^P(A)$ for each $A \in \mathbb{SVNS}(X)$. Then $(L_n^P, \hat{\lesssim})$ is a poset.

Let $H_n^O = \left\{\overline{M}^O \mid M \in n - \mathbf{SVNR}(X)\right\}$ and $L_n^O = \left\{\underline{M}^O \mid M \in n - \mathbf{SVNR}(X)\right\}$ be the set of optimistic multigranulation neutrosophic upper and lower approximation operators in X, respectively.

- Defined a relation $\hat{\lesssim}$ on H_n^O as follows: $\overline{M}^O \hat{\lesssim} \overline{N}^O$ if and only if $\overline{M}^O(A) \Subset \overline{N}^O(A)$ for each $A \in \mathbb{SVNS}(X)$. Then $(H_n^O, \hat{\lesssim})$ is a poset.
- Defined a relation $\hat{\lesssim}$ on L_n^O as follows: $\underline{M}^O \hat{\lesssim} \underline{N}^O$ if and only if $\underline{N}^O(A)) \Subset \underline{M}^O(A)$ for each $A \in \mathbb{SVNS}(X)$. Then $(L_n^O, \hat{\lesssim})$ is a poset.

Theorem 3. *(1) $\forall \left\{\overline{M_i}^P\right\}_{i \in I} \subseteq (H_n^P, \hat{\lesssim})$ and I be a index set, we can define union and intersection of $\overline{M_i}^P$ as follows:*

$$\underset{i \in I}{\hat{\vee}} \overline{M_i}^P = \overline{\underset{i \in I}{\vee} M_i}^P,\ \underset{i \in I}{\hat{\wedge}} \overline{M_i}^P = \overline{[\underset{i \in I}{\wedge} M_i]}^P,$$

where $[\underset{i \in I}{\wedge} M_i] = \vee\left\{M \in n - \mathbf{SVNR}(X) \mid \forall A \in \mathbb{SVNS}(X), \overline{M}^P(A) \Subset \Cap_{i \in I} \overline{M_i}^P(A)\right\}$. Then $\underset{i \in I}{\hat{\vee}} \overline{M_i}^P$ and $\underset{i \in I}{\hat{\wedge}} \overline{M_i}^P$ are supremum and infimum of $\left\{\overline{M_i}^P\right\}_{i \in I}$, respectively.

(2) $\forall \left\{\underline{M_i}^P\right\}_{i \in I} \subseteq (L_n^P, \hat{\lesssim})$ and I be a index set, we can define union and intersection of $\underline{M_i}^P$ as follows:

$$\underset{i \in I}{\hat{\vee}} \underline{M_i}^P = \underline{\underset{i \in I}{\vee} M_i}^P,\ \underset{i \in I}{\hat{\wedge}} \underline{M_i}^P = \underline{[\underset{i \in I}{\vee} M_i]}^P,$$

where $[\underset{i \in I}{\vee} M_i] = \vee\left\{M \in n - \mathbf{SVNR}(X) \mid \forall A \in \mathbb{SVNS}(X), \Cup_{i \in I} \underline{M_i}^P(A) \Subset \underline{M}^P(A)\right\}$. Then $\underset{i \in I}{\hat{\vee}} \underline{M_i}^P$ and $\underset{i \in I}{\hat{\wedge}} \underline{M_i}^P$ are supremum and infimum of $\left\{\underline{M_i}^P\right\}_{i \in I}$, respectively.

Proof. We only show (1).

Let $M = \underset{i \in I}{\vee} M_i$, then $M_i \sqsubseteq M$ for each $i \in I$. By Corollary 1, we have $\overline{M_i}^P(A) \Subset \overline{M}^P(A)$ for any $A \in \mathbb{SVNS}(X)$. Thus $\overline{M_i}^P \hat{\lesssim} \overline{M}^P$. If $M^\star$ is a multigranulation neutrosophic relations set such that $\overline{M_i}^P \hat{\lesssim} \overline{M^\star}^P$ for each $i \in I$, then $A \in \mathbb{SVNS}(X)$, $\overline{M_i}^P(A) \Subset \overline{M^\star}^P(A)$. Hence,

$$\overline{M}^P(A) = \overline{\underset{i \in I}{\vee} M_i}^P(A) = \Cup_{i \in I} \overline{M_i}^P(A) \Subset \overline{M^\star}^P(A).$$

Thus $\overline{M}^P \hat{\lesssim} \overline{M^\star}^P$. So $\underset{i \in I}{\hat{\vee}} \overline{M_i}^P = \overline{\underset{i \in I}{\vee} M_i}^P$ is the supremum of $\left\{\overline{M_i}^P\right\}_{i \in I}$.

Let $Q = [\underset{i \in I}{\wedge} M_i]$, then $\forall B \in \mathbb{SVNS}(X)$, we have

$$\overline{Q}^P(B) = \overline{[\underset{i \in I}{\wedge} M_i]}^P(B) \Subset \Cap_{i \in I} \overline{M_i}^P(B) \Subset \overline{M_i}^P(B).$$

Thus $\overline{Q}^P \hat{\leq} \overline{M_i}^P$ for each $i \in I$. If M^* is a multigranulation neutrosophic relations set such that $\overline{M^*}^P \hat{\leq} \overline{M_i}^P$ for each $i \in I$, then

$$\overline{M^*}^P(A) \Subset \Cap_{i\in I}\overline{M_i}^P(A).$$

By the construction of $[\underset{i\in I}{\wedge} M_i]$, we can easily obtain $M^* \sqsubseteq [\underset{i\in I}{\wedge} M_i] = Q$. Hence,

$$\overline{M^*}^P \hat{\leq} \overline{[\underset{i\in I}{\wedge} M_i]}^P = \overline{Q}^P,$$

So $\underset{i\in I}{\hat{\wedge}} \overline{M_i}^P = \overline{[\underset{i\in I}{\wedge} M_i]}^P$ is the infimum of $\left\{\overline{M_i}^P\right\}_{i\in I}$. □

Remark 2. *(1) $\forall A \in \mathbb{SVNS}(X)$, $\forall a \in X$, we can calculate that the following formula holds.*

$$T_{\overline{\overline{\varnothing_N}}^P(A)}(a) = 0,\ I_{\overline{\overline{\varnothing_N}}^P(A)}(a) = 1,\ F_{\overline{\overline{\varnothing_N}}^P(A)}(a) = 1,$$

$$T_{\underline{\widetilde{\varnothing_N}}^P(A)}(a) = 1,\ I_{\underline{\widetilde{\varnothing_N}}^P(A)}(a) = 0,\ F_{\underline{\widetilde{\varnothing_N}}^P(A)}(a) = 0.$$

Hence, $\forall M \in n-\mathbf{SVNR}(X)$, $\overline{\overline{\varnothing_N}}^P(A) \Subset \overline{M}^P(A)$ and $\underline{M}^P(A) \Subset \underline{\widetilde{\varnothing_N}}^P(A)$. It shows that $\overline{\overline{\varnothing_N}}^P \hat{\leq} \overline{M}^P$ and $\underline{\widetilde{\varnothing_N}}^P \hat{\leq} \underline{M}^P$, i.e., $\overline{\overline{\varnothing_N}}^P$ is the bottom element of $(H_n^P, \hat{\leq})$ and $\underline{\widetilde{\varnothing_N}}^P$ is the bottom element of $(L_n^P, \hat{\leq})$. By Theorem 3, we have the following result: Both $(H_n^P, \hat{\leq}, \hat{\wedge}, \hat{\vee})$ and $(L_n^P, \hat{\leq}, \hat{\wedge}, \hat{\vee})$ are complete lattices.

(2) Similarly, we can prove that both $(H_n^O, \hat{\leq}, \hat{\wedge}, \hat{\vee})$ and $(L_n^O, \hat{\leq}, \hat{\wedge}, \hat{\vee})$ are complete lattices if we can use the generalization formula of

$$\overline{M \vee N}^O(A) \Subset \overline{M}^O(A) \Cup \overline{N}^O(A) \text{ and } \underline{M \vee N}^O(A) \Supset \underline{M}^O(A) \Cap \underline{N}^O(A),$$

However, by Theorem 2, we known that

$$\overline{M \vee N}^O(A) \Supset \overline{M}^O(A) \Cup \overline{N}^O(A) \text{ and } \underline{M \vee N}^O(A) \Subset \underline{M}^O(A) \Cap \underline{N}^O(A).$$

So, naturally, there is the following problem:

How to give the supremum and infimum of the optimistic multigranulation neutrosophic rough approximation operators?

In the one-dimensional case, for convenience, we will use $H = \{\overline{R} \mid R \in \mathbf{SVNR}(X)\}$ and $L = \{\underline{R} \mid R \in \mathbf{SVNR}(X)\}$ to denote the set of neutrosophic upper and lower approximation operators in X, respectively. According to Lemma 1, Remark 2 and Theorem 3, we have the following result: both $(H, \leq, \wedge, \vee)$ and $(L, \leq, \wedge, \vee)$ are complete lattices (it is also the one-dimensional case of Reference [23]).

4. The Relationship between Complete Lattices $(H,\leq,\wedge,\vee)$ and $(L,\leq,\wedge,\vee)$

In this section, we will study the relationship between complete lattices $(H, \leq, \wedge, \vee)$ and $(L, \leq, \wedge, \vee)$. Set

$$\mathcal{A} = \{\mathbf{SVNR}(X) \mid \forall R_1, R_2 \in \mathbf{SVNR}(X), \overline{R_1} \leq \overline{R_2} \Leftrightarrow R_1 \Subset R_2 \Leftrightarrow \underline{R_1} \leq \underline{R_2}\}.$$

Firstly, we will give an example to illustrate that$\mathcal{A}$ is not an empty family.

Example 1. *Let $X = \{a\}$ be a single point set, R_1 and R_2 are two single valued neutrosophic relations in X.*

(1) If $\overline{R_1} \leq \overline{R_2}$, then $R_1 \Subset R_2$. In fact, if $\overline{R_1} \leq \overline{R_2}$, then $\overline{R_1}(A) \Subset \overline{R_2}(A)$ for each $A \in \mathbb{SVNS}(\{a\})$.

Thus, $\forall a \in X$,

$$T_{\overline{R_1(A)}}(a) \leq T_{\overline{R_2(A)}}(a),\ I_{\overline{R_1(A)}}(a) \geq I_{\overline{R_2(A)}}(a), \text{ and } F_{\overline{R_1(A)}}(a) \geq F_{\overline{R_2(A)}}(a).$$

Moreover, $T_{R_1}(a,a) \wedge T_A(a) \leq T_{R_2}(a,a) \wedge T_A(a)$, $I_{R_1}(a,a) \vee I_A(a) \geq I_{R_2}(a,a) \vee I_A(a)$, *and* $F_{R_1}(a,a) \vee F_A(a) \geq F_{R_2}(a,a) \vee F_A(a)$. *Considering the arbitrariness of A, in particular, take* $A = \{< a, (1,0,0) >\}$, *we have* $T_{R_1}(a,a) \leq T_{R_2}(a,a)$, $I_{R_1}(a,a) \geq I_{R_2}(a,a)$ *and* $F_{R_1}(a,a) \geq F_{R_2}(a,a)$.

Hence, $R_1 \Subset R_2$.

Similarly, we also can show that the following result:

(2) *If* $\underline{R_1} \leq \underline{R_2}$, *then* $R_1 \Subset R_2$. *So, by (1), (2) and Corollary 1, we have* $\mathbf{SVNR}(\{a\}) \in \mathcal{A}$, *i.e.,* $\mathcal{A}$ *is not an empty family.*

Now, we will give the relationship between complete lattices $(H, \leq, \wedge, \vee)$ and $(L, \leq, \wedge, \vee)$.

Proposition 1. *If* $\mathbf{SVNR}(X) \in \mathcal{A}$, *then* $[\Cap_{i\in I} R_i] = \Cap_{i\in I} R_i = [\Cup_{i\in I} R_i]$, *where I is a index set, and* $R_i \in \mathbf{SVNR}(X)$ *for each* $i \in I$.

Proof. We first show that $[\Cap_{i\in I} R_i] = \Cap_{i\in I} R_i$. Let R be a neutrosophic relation in X such that $\Cap_{i\in I} \overline{R}_i(A) \Supset \overline{R}(A)$ for each $A \in \mathbb{SVNS}(X)$, then $\overline{R}_i \geq \overline{R}$, this is equivalent to $R_i \Supset R$ since $\mathbf{SVNR}(X) \in \mathcal{A}$. Thus $\Cap_{i\in I} R_i \Supset R$. Moreover, by the construction of $[\Cap_{i\in I} R_i]$, we have $\Cap_{i\in I} R_i \Supset [\Cap_{i\in I} R_i]$. On the other hand, we can show that $\Cap_{i\in I} \overline{R}_i(A) \Supset \overline{\Cap_{i\in I} R_i}(A)$ for each $A \in \mathbb{SVNS}(X)$. So

$$[\Cap_{i\in I} R_i] = \Cup\{R \in \mathbf{SVNR}(X) \mid \forall A \in \mathbb{SVNS}(X), \Cap_{i\in I} \overline{R}_i(A) \Supset \overline{R}(A)\} \Supset \Cap_{i\in I} R_i.$$

Hence $[\Cap_{i\in I} R_i] = \Cap_{i\in I} R_i$.

Now, we show that $\Cap_{i\in I} R_i = [\Cup_{i\in I} R_i]$. Let R be a single valued neutrosophic relation in such that $\Cup_{i\in I} \underline{R}_i(A) \Subset \underline{R}(A)$ for each $A \in \mathbb{SVNS}(X)$, then $\underline{R}_i \geq \underline{R}$, this is equivalent to $R_i \Supset R$ since $\mathbf{SVNR}(X) \in \mathcal{A}$. Thus $\Cap_{i\in I} R_i \Supset R$. Moreover, by the construction of $[\Cup_{i\in I} R_i]$. We have $\Cap_{i\in I} R_i \Supset [\Cup_{i\in I} R_i]$.

On the other hand, we can show that $\Cup_{i\in I} \underline{R}_i(A) \Subset \underline{\Cap_{i\in I} R_i}(A)$ for each $A \in \mathbb{SVNS}(X)$. So

$$[\Cup_{i\in I} R_i] = \Cup\{R \in \mathbf{SVNR}(X) \mid \forall A \in \mathbb{SVNS}(X), \Cup_{i\in I} \underline{R}_i(A) \Subset \underline{R}(A)\} \Supset \Cap_{i\in I} R_i.$$

Hence, $[\Cup_{i\in I} R_i] = \Cap_{i\in I} R_i$.

From above proved, we know that $[\Cap_{i\in I} R_i] = \Cap_{i\in I} R_i = [\Cup_{j\in J} R_j]$. □

Theorem 4. *If* $\mathbf{SVNR}(X) \in \mathcal{A}$, *then* $(\mathbf{SVNR}(X), \Subset, \Cup, \Cap)$ *and* $(H, \leq, \wedge, \vee)$ *are complete lattice isomorphism.*

Proof. Define a mapping $\phi_{12} : \mathbf{SVNR}(X) \to H$ as follows: $\forall R \in \mathbf{SVNR}(X)$, $\phi_{12}(R) = \overline{R}$. Obviously, ϕ_{12} is surjective. If $\overline{R_1} = \overline{R_2}$, notice that $\mathbf{SVNR}(X) \in \mathcal{A}$, we know that $R_1 = R_2$. So ϕ_{12} is one-one. $\forall \{R_i\}_{i\in I} \subseteq \mathbf{SVNR}(X)$ and I be a index set. By Theorem 3 and Proposition 1, we have

$$\phi_{12}(\Cup_{i\in I} R_i) = \overline{\Cup_{i\in I} R_i} = \bigvee_{i\in I} \overline{R_i} = \bigvee_{i\in I} \phi_{12}(R_i),$$

and

$$\phi_{12}(\Cap_{i\in I} R_i) = \overline{\Cap_{i\in I} R_i} = \overline{[\Cap_{i\in I} R_i]} = \bigwedge_{i\in I} \overline{R_i} = \bigwedge_{i\in I} \phi_{12}(R_i).$$

Hence, ϕ_{12} preserves arbitrary union and arbitrary intersection. □

From above proved, we know that $(\mathbf{SVNR}(X), \Subset, \Cup, \Cap)$ and $(H, \leq, \wedge, \vee)$ are complete lattice isomorphism.

Theorem 5. *If* $\mathbf{SVNR}(X) \in \mathcal{A}$, *then* $(\mathbf{SVNR}(X), \Subset, \Cup, \Cap)$ *and* $(L, \leq, \wedge, \vee)$ *are complete lattice isomorphism.*

Proof. Define a mapping $\phi_{13} : \mathbf{SVNR}(X) \to L$ as follows:$\forall R \in \mathbf{SVNR}(X)$, $\phi_{12}(R) = \underline{R}$. Obviously, ϕ_{13} is surjective. If $\underline{R_1} = \underline{R_2}$, notice that $\mathbf{SVNR}(X) \in \mathcal{A}$, we know that $R_1 = R_2$. So ϕ_{13} is one-one. $\forall \{R_i\}_{i \in I} \subseteq \mathbf{SVNR}(X)$ and I be an index set. By Theorem 3 and Proposition 1, we have

$$\phi_{13}(\Cup_{i \in I} R_i) = \underline{\Cup_{i \in I} R_i} = \bigvee_{i \in I} \underline{R_i} = \bigvee_{i \in I} \phi_{13}(R_i),$$

and

$$\phi_{13}(\Cap_{i \in I} R_i) = \underline{\Cap_{i \in I} R_i} = \underline{[\Cup_{i \in I} R_i]} = \bigwedge_{i \in I} \underline{R_i} = \bigwedge_{i \in I} \phi_{13}(R_i).$$

Hence, ϕ_{13} preserves arbitrary union and arbitrary intersection. □

From the above proof, we know that $(\mathbf{SVNR}(X), \Subset, \Cup, \Cap)$ and $(L, \leq, \wedge, \vee)$ are complete lattice isomorphism.

Theorem 6. *If* $\mathbf{SVNR}(X) \in \mathcal{A}$, *then* $(H, \leq, \wedge, \vee)$ *and* $(L, \leq, \wedge, \vee)$ *are complete lattice isomorphism.*

Proof. Through Theorems 4 and 5, we immediately know that the conclusion holds. We can also prove it by the following way:

Define a mapping $\phi_{23} : H \to L$ as follows: $\forall \overline{R} \in H$, $\phi_{23}(\overline{R}) = \underline{R}$. Through Theorems 4 and 5, there must be one and only one $R \in \mathbf{SVNR}(X)$ such that $\phi_{23}(\overline{R}) = \underline{R}$ for each $\underline{R} \in L$. This shows ϕ_{23} is surjective. If $\underline{R_1} = \underline{R_2}$, notice that $\mathbf{SVNR}(X) \in \mathcal{A}$, we know that $\overline{R_1} = \overline{R_2}$. So ϕ_{23} is one-one. $\forall \{\overline{R_i}\}_{i \in I} \subseteq H$ and I be a index set. Through Theorem 3 and Proposition 1, we have

$$\phi_{23}(\bigvee_{i \in I} \overline{R_i}) = \phi_{23}(\overline{\Cup_{i \in I} R_i}) = \underline{\Cup_{i \in I} R_i} = \bigvee_{i \in I} \underline{R_i} = \bigvee_{i \in I} \phi_{13}(\overline{R_i}),$$

and

$$\phi_{13}(\bigwedge_{i \in I} \overline{R_i}) = \phi_{13}(\overline{[\Cap_{i \in I} R_i]}) = \underline{[\Cap_{i \in I} R_i]} = \underline{[\Cup_{i \in I} R_i]} = \bigwedge_{i \in I} \underline{R_i} = \bigwedge_{i \in I} \phi_{23}(\overline{R_i}).$$

Hence, ϕ_{23} preserves arbitrary union and arbitrary intersection. So, $(H, \leq, \wedge, \vee)$ and $(L, \leq, \wedge, \vee)$ are complete lattice isomorphism. □

Remark 3. *Through Theorems 4–6, we can ascertain that* ϕ_{12},ϕ_{13} *and* ϕ_{23} *are isomorphic mappings among complete lattices. Moreover, the following diagram can commute, i.e.,* $\phi_{23} \circ \phi_{12} = \phi_{13}$ *(see Figure 1).*

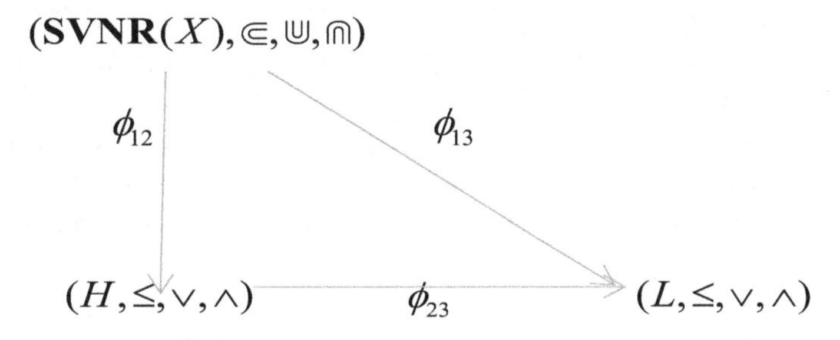

Figure 1. Correspondence relationship among three complete lattices.

5. Conclusions

Following the idea of multigranulation neutrosophic rough sets on a single domain as introduced by Bo et al. (2018), we gave the lattice structure of the pessimistic multigranulation neutrosophic rough approximation operators. In the one-dimensional case, for each special $\mathbf{SVNR}(X)$, we gave a one-to-one correspondence relationship between complete lattices $(H, \leq)$ and $(L, \leq)$. Unfortunately, at the moment, we haven't solved the following problems:

(1) Can the supremum and infimum of the optimistic multigranulation neutrosophic rough approximation operators be given?

(2) For any set , are $(H, \leq)$ and $(L, \leq)$ isomorphic between complete lattices?

Author Contributions: H.Z. provided the idea of the paper and proved the theorems. H.-Y.Z. provided helpful suggestions.

Acknowledgments: The work is partly supported by the National Natural Science Foundation of China (Grant No. 61473181, 11771263 and 11671007), the Doctoral Scientific Research Foundation of Xi'an Polytechnic University (Grant No. BS1426), the Construction Funding of Xi'an Polytechnic University for Mathematics (Grant No. 107090701), and the Scientific Research Program Funded by Shaanxi Provincial Education Department (2018).

References

1. Smarandache, F. *Neutrosophy: Neutrosophic Probability, Set, and Logic*; American Research Press: Rehoboth, MA, USA, 1998.
2. Smarandache, F. Neutrosophic set—-A generialization of the intuitionistics fuzzy sets. *Int. J. Pure Appl. Math.* **2005**, *24*, 287–297.
3. Rivieccio, U. Neutrosophic logics: Prospects and problems. *Fuzzy Sets Syst.* **2008**, *159*, 1860–1868. [CrossRef]
4. Guo, Y.; Cheng, H.D. A new neutrosophic approach to image segmentation. *Pattern Recogn.* **2009**, *42*, 587–595. [CrossRef]
5. Guo, Y.; Sengur, A. NCM: Neutrosophic c-means clustering algorithm. *Pattern Recogn.* **2015**, *48*, 2710–2724. [CrossRef]
6. Salama, A.A.; Broumi, S. Roughness of neutrosophic sets. *Elixir Appl. Math.* **2014**, *74*, 26833–26837.
7. Broumi, S.; Smarandache, F. Rough neutrosophic sets. *Ital. J. Pure Appl. Math.* **2014**, *32*, 493–502.
8. Wang, H.; Smarandache, F.; Zhang, Y.Q.; Sunderraman, R. Single valued neutrosophic sets, Multispace Multistruct. *Google Sch.* **2010**, *4*, 410–413.
9. Atanassov, K. Intuitionistic fuzzy sets. *Fuzzy Sets Syst.* **1986**, *20*, 87–96. [CrossRef]
10. Ye, J. Multicriteria decision-making method using the correlation coefficient under single-valued neutrosophic environment. *Int. J. Gen. Syst.* **2013**, *42*, 386–394. [CrossRef]
11. Ye, J. Improved correlation coefficients of single valued neutrosophic sets and interval neutrosophic sets for multiple attribute decision making. *J. Intell. Fuzzy Syst.* **2014**, *27*, 2453–2462.
12. Ye, S.; Ye, J. Dice similarity measure among single valued neutrosophic multisets and its applcation in medical diagnosis. *Neutrosophic Sets Syst.* **2014**, *6*, 48–53.
13. Majumdar, P.; Samant, S.K. On similarity and entropy of neutrosophic sets. *J. Intell. Fuzzy Syst.* **2014**, *26*, 1245–1252.
14. Şahin, R.; Küçük, A. Subsethood measure for single valued neutrosophic sets. *J. Intell. Fuzzy Syst.* **2015**, *29*, 525–530. [CrossRef]
15. Li, Z.W.; Cui, R.C. *T*-similarity of fuzzy relations and related algebraic structures. *Fuzzy Sets Syst.* **2015**, *275*, 130–143. [CrossRef]
16. Li, Z.W.; Cui, R.C. Similarity of fuzzy relations based on fuzzy topologies induced by fuzzy rough approximation operators. *Inf. Sci.* **2015**, *305*, 219–233. [CrossRef]
17. Li, Z.W.; Liu, X.F.; Zhang, G.Q.; Xie, N.X.; Wang, S.C. A multi-granulation decision-theoretic rough set method for distributed fc-decision information systems: An application in medical diagnosis. *Appl. Soft Comput.* **2017**, *56*, 233–244. [CrossRef]
18. Yang, H.L.; Guo, Z.L.; She, Y.H.; Liao, X.W. On single valued neutrosophic relations. *J. Intell. Fuzzy Syst.* **2016**, *30*, 1045–1056. [CrossRef]

19. Yang, H.L.; Zhang, C.L.; Guo, Z.L.; Liu, Y.L.; Liao, X.W. A hybrid model of single valued neutrosophic sets and rough sets: Single valued neutrosophic rough set model. *Soft Comput.* **2017**, *21*, 6253–6267. [CrossRef]
20. Pawlak, Z. Rough sets. *Int. J. Comput. Inf. Sci.* **1982**, *11*, 341–356. [CrossRef]
21. Bao, Y.L.; Yang, H.L. On single valued neutrosophic refined rough set model and its applition. *J. Intell. Fuzzy Syst.* **2017**, *33*, 1235–1248. [CrossRef]
22. Bo, C.X.; Zhang, X.H.; Shao, S.T.; Smarandache, F. Multi-Granulation Neutrosophic Rough Sets on a Single Domain and Dual Domains with Applications. *Symmetry* **2018**, *10*, 296. [CrossRef]
23. Zhao, H.; Zhang, H.Y. A result on single valued neutrosophic refined rough approximation operators. *J. Intell. Fuzzy Syst.* **2018**, 1–8. [CrossRef]

14

Q-Filters of Quantum B-Algebras and Basic Implication Algebras

Xiaohong Zhang [1,2,*], Rajab Ali Borzooei [3] and Young Bae Jun [3,4]

1 Department of Mathematics, Shaanxi University of Science and Technology, Xi'an 710021, China
2 Department of Mathematics, Shanghai Maritime University, Shanghai 201306, China
3 Department of Mathematics, Shahid Beheshti University, Tehran 1983963113, Iran; borzooei@sbu.ac.ir (R.A.B.); skywine@gmail.com (Y.B.J.)
4 Department of Mathematics Education, Gyeongsang National University, Jinju 52828, Korea
* Correspondence: zhangxiaohong@sust.edu.cn or zhangxh@shmtu.edu.cn

Abstract: The concept of quantum B-algebra was introduced by Rump and Yang, that is, unified algebraic semantics for various noncommutative fuzzy logics, quantum logics, and implication logics. In this paper, a new notion of q-filter in quantum B-algebra is proposed, and quotient structures are constructed by q-filters (in contrast, although the notion of filter in quantum B-algebra has been defined before this paper, but corresponding quotient structures cannot be constructed according to the usual methods). Moreover, a new, more general, implication algebra is proposed, which is called basic implication algebra and can be regarded as a unified frame of general fuzzy logics, including nonassociative fuzzy logics (in contrast, quantum B-algebra is not applied to nonassociative fuzzy logics). The filter theory of basic implication algebras is also established.

Keywords: fuzzy implication; quantum B-algebra; q-filter; quotient algebra; basic implication algebra

---

1. Introduction

For classical logic and nonclassical logics (multivalued logic, quantum logic, t-norm-based fuzzy logic [1–6]), logical implication operators play an important role. In the study of fuzzy logics, fuzzy implications are also the focus of research, and a large number of literatures involve this topic [7–16]. Moreover, some algebraic systems focusing on implication operators are also hot topics. Especially with the in-depth study of noncommutative fuzzy logics in recent years, some related implication algebraic systems have attracted the attention of scholars, such as pseudo-basic-logic (BL) algebras, pseudo- monoidal t-norm-based logic (MTL) algebras, and pseudo- B, C, K axiom (BCK)/ B, C, I axiom (BCI) algebras [17–23] (see also References [5–7]).

For formalizing the implication fragment of the logic of quantales, Rump and Yang proposed the notion of quantum B-algebras [24,25], which provide a unified semantic for a wide class of nonclassical logics. Specifically, quantum B-algebras encompass many implication algebras, like pseudo-BCK/BCI algebras, (commutative and noncommutative) residuated lattices, pseudo- MV/BL/MTL algebras, and generalized pseudo-effect algebras. New research articles on quantum B-algebras can be found in References [26–28]. Note that all hoops and pseudo-hoops are special quantum B-algebras, and they are closely related to L-algebras [29].

Although the definition of a filter in a quantum B-algebra is given in Reference [30], quotient algebraic structures are not established by using filters. In fact, filters in special subclasses of quantum B-algebras are mainly discussed in Reference [30], and these subclasses require a unital element. In this paper, by introducing the concept of a q-filter in quantum B-algebras, we establish the quotient structures using q-filters in a natural way. At the same time, although quantum B-algebra has generality, it cannot include the implication structure of non-associative fuzzy logics [31,32], so we propose a wider concept, that is, basic implication algebra that can include a wider range of implication operations, establish filter theory, and obtain quotient algebra.

2. Preliminaries

Definition 1. *Let $(X, \leq)$ be partially ordered set endowed with two binary operations $\rightarrow$ and $\rightsquigarrow$ [24,25]. Then, $(X, \rightarrow, \rightsquigarrow, \leq)$ is called a quantum B-algebra if it satisfies: $\forall x, y, z \in X$,*

(1) $y \rightarrow z \leq (x \rightarrow y) \rightarrow (x \rightarrow z)$;
(2) $y \rightsquigarrow z \leq (x \rightsquigarrow y) \rightsquigarrow (x \rightsquigarrow z)$;
(3) $y \leq z \Rightarrow x \rightarrow y \leq x \rightarrow z$;
(4) $x \leq y \rightarrow z \Longleftrightarrow y \leq x \rightsquigarrow z$.

If $u \in X$ exists, such that $u \rightarrow x = u \rightsquigarrow x = x$ for all x in X, then u is called a unit element of X. Obviously, the unit element is unique. When a unit element exists in X, we call X unital.

Proposition 1. *An algebra structure $(X, \rightarrow, \rightsquigarrow, \leq)$ endowed with a partially order $\leq$ and two binary operations $\rightarrow$ and $\rightsquigarrow$ is a quantum B-algebra if and only if it satisfies [4]: $\forall x, y, z \in X$,*

(1) $x \rightarrow (y \rightsquigarrow z) = y \rightsquigarrow (x \rightarrow z)$;
(2) $y \leq z \Rightarrow x \rightarrow y \leq x \rightarrow z$;
(3) $x \leq y \rightarrow z \Longleftrightarrow y \leq x \rightsquigarrow z$.

Proposition 2. *Let $(X, \rightarrow, \rightsquigarrow, \leq)$ be a quantum B-algebra [24–26]. Then, $(\forall\, x, y, z \in X)$*

(1) $y \leq z \Rightarrow x \rightsquigarrow y \leq x \rightsquigarrow z$;
(2) $y \leq z \Rightarrow z \rightsquigarrow x \leq y \rightsquigarrow x$;
(3) $y \leq z \Rightarrow z \rightarrow x \leq y \rightarrow x$;
(4) $x \leq (x \rightsquigarrow y) \rightarrow y$, $x \leq (x \rightarrow y) \rightsquigarrow y$;
(5) $x \rightarrow y = ((x \rightarrow y) \rightsquigarrow y) \rightarrow y$, $x \rightsquigarrow y = ((x \rightsquigarrow y) \rightarrow y) \rightsquigarrow y$;
(6) $x \rightarrow y \leq (y \rightarrow z) \rightsquigarrow (x \rightarrow z)$;
(7) $x \rightsquigarrow y \leq (y \rightsquigarrow z) \rightarrow (x \rightsquigarrow z)$;
(8) *assume that u is the unit of X, then $u \leq x \rightsquigarrow y \Longleftrightarrow x \leq y \Longleftrightarrow u \leq x \rightarrow y$;*
(9) *if $0 \in X$ exists, such that $0 \leq x$ for all x in X, then $0 = 0 \rightsquigarrow 0 = 0 \rightarrow 0$ is the greatest element (denote by 1), and $x \rightarrow 1 = x \rightsquigarrow 1 = 1$ for all $x \in X$;*
(10) *if X is a lattice, then $(x \vee y) \rightarrow z = (x \rightarrow z) \wedge (y \rightarrow z)$, $(x \vee y) \rightsquigarrow z = (x \rightsquigarrow z) \vee (y \rightsquigarrow z)$.*

Definition 2. *Let $(X, \leq)$ be partially ordered set and $Y \subseteq X$ [24]. If $x \geq y \in Y$ implies $x \in Y$, then Y is called to be an upper set of X. The smallest upper set containing a given $x \in X$ is denoted by $\uparrow x$. For quantum B-algebra X, the set of upper sets is denoted by $U(X)$. For $A, B \in U(X)$, define*

$$A \cdot B = \{x \in X \mid \exists b \in B: b \rightarrow x \in A\}.$$

We can verify that $A \cdot B = \{x \in X \mid \exists a \in A: a \rightsquigarrow x \in B\} = \{x \in X \mid \exists a \in A, b \in B: a \leq b \rightarrow x\} = \{x \in X \mid \exists a \in A, b \in B: b \leq a \rightsquigarrow x\}$.

Definition 3. *Let A be an empty set, $\leq$ be a binary relation on A [17,18], $\rightarrow$ and $\rightsquigarrow$ be binary operations on A, and 1 be an element of A. Then, structure $(A, \rightarrow, \rightsquigarrow, \leq, 1)$ is called a pseudo-BCI algebra if it satisfies the following axioms: $\forall\, x, y, z \in A$,*

(1) $x \rightarrow y \leq (y \rightarrow z) \rightsquigarrow (x \rightarrow z)$, $x \rightsquigarrow y \leq (y \rightsquigarrow z) \rightarrow (x \rightsquigarrow z)$,
(2) $x \leq (x \rightsquigarrow y) \rightarrow y$, $x \leq (x \rightarrow y) \rightsquigarrow y$;

(3) $x \leq x$;
(4) $x \leq y, y \leq x \Rightarrow x = y$;
(5) $x \leq y \Longleftrightarrow x \rightarrow y = 1 \Longleftrightarrow x \rightsquigarrow y = 1$.

If pseudo-BCI algebra A satisfies: $x \rightarrow 1 = 1$ *(or* $x \rightsquigarrow 1 = 1$*) for all* $x \in A$*, then A is called a pseudo-BCK algebra.*

Proposition 3. *Let* $(A, \rightarrow, \rightsquigarrow, \leq, 1)$ *be a pseudo-BCI algebra [18–20]. We have* $(\forall x, y, z \in A)$

(1) *if* $1 \leq x$, *then* $x = 1$;
(2) *if* $x \leq y$, *then* $y \rightarrow z \leq x \rightarrow z$ *and* $y \rightsquigarrow z \leq x \rightsquigarrow z$;
(3) *if* $x \leq y$ *and* $y \leq z$, *then* $x \leq z$;
(4) $x \rightarrow (y \rightsquigarrow z) = y \rightsquigarrow (x \rightarrow z)$;
(5) $x \leq y \rightarrow z \Longleftrightarrow y \leq x \rightsquigarrow z$;
(6) $x \rightarrow y \leq (z \rightarrow x) \rightarrow (z \rightarrow y)$; $x \rightsquigarrow y \leq (z \rightsquigarrow x) \rightsquigarrow (z \rightsquigarrow y)$;
(7) *if* $x \leq y$, *then* $z \rightarrow x \leq z \rightarrow y$ *and* $z \rightsquigarrow x \leq z \rightsquigarrow y$;
(8) $1 \rightarrow x = 1 \rightsquigarrow x = x$;
(9) $y \rightarrow x = ((y \rightarrow x) \rightsquigarrow x) \rightarrow x$, $y \rightsquigarrow x = ((y \rightsquigarrow x) \rightarrow x) \rightsquigarrow x$;
(10) $x \rightarrow y \leq (y \rightarrow x) \rightsquigarrow 1$, $x \rightsquigarrow y \leq (y \rightsquigarrow x) \rightarrow 1$;
(11) $(x \rightarrow y) \rightarrow 1 = (x \rightarrow 1) \rightsquigarrow (y \rightarrow 1)$, $(x \rightsquigarrow y) \rightsquigarrow 1 = (x \rightsquigarrow 1) \rightarrow (y \rightsquigarrow 1)$;
(12) $x \rightarrow 1 = x \rightsquigarrow 1$.

Proposition 4. *Let* $(A, \rightarrow, \rightsquigarrow, \leq, 1)$ *be a pseudo-BCK algebra* [17], *then* $(\forall x, y \in A)$: $x \leq y \rightarrow x$, $x \leq y \rightsquigarrow x$.

Definition 4. *Let X be a unital quantum B-algebra [24]. If there exists* $x \in X$*, such that* $x \rightarrow u = x \rightsquigarrow u = u$*, then we call that x integral. The subset of integral element in X is denoted by I(X).*

Proposition 5. *Let X be a quantum B-algebra [24]. Then, the following assertions are equivalent:*

(1) *X is a pseudo-BCK algebra;*
(2) *X is unital, and every element of X is integral;*
(3) *X has the greatest element, which is a unit element.*

Proposition 6. *Every pseudo-BCI algebra is a unital quantum B-algebra [25]. And, a quantum B-algebra is a pseudo-BCI algebra if and only if its unit element u is maximal.*

Definition 5. *Let* $(A, \rightarrow, \rightsquigarrow, \leq, 1)$ *be a pseudo-BCI algebra [20,21]. When the following identities are satisfied, we call X an antigrouped pseudo-BCI algebra:*

$$\forall x \in A, (x \rightarrow 1) \rightarrow 1 = x \text{ or } (x \rightsquigarrow 1) \rightsquigarrow 1 = x.$$

Proposition 7. *Let* $(A, \rightarrow, \rightsquigarrow, \leq, 1)$ *be a pseudo-BCI algebra [20]. Then, A is antigrouped if and only if the following conditions are satisfied:*

(G1) *for all* $x, y, z \in A$, $(x \rightarrow y) \rightarrow (x \rightarrow z) = y \rightarrow z$, *and*
(G2) *for all* $x, y, z \in A$, $(x \rightsquigarrow y) \rightsquigarrow (x \rightsquigarrow z) = y \rightsquigarrow z$.

Definition 6. *Let* $(A, \rightarrow, \rightsquigarrow, \leq, 1)$ *be a pseudo-BCI algebra and* $F \subseteq X$ *[19,20]. When the following conditions are satisfied, we call F a pseudo-BCI filter (briefly, filter) of X:*

(F1) $1 \in F$;

(F2) $x \in F$, $x \rightarrow y \in F \Longrightarrow y \in F$;
(F3) $x \in F$, $x \rightsquigarrow y \in F \Longrightarrow y \in F$.

Definition 7. *Let $(A, \rightarrow, \rightsquigarrow, \leq, 1)$ be a pseudo-BCI algebra and F be a filter of X [20,21]. When the following condition is satisfied, we call F an antigrouped filter of X:*

$$(GF)\ \forall x \in X,\ (x \rightarrow 1) \rightarrow 1 \in F \text{ or } (x \rightsquigarrow 1) \rightsquigarrow 1 \in F \Longrightarrow x \in F.$$

Definition 8. *A subset F of pseudo-BCI algebra X is called a p-filter of X if the following conditions are satisfied [20,21]:*

(P1) $1 \in F$,
(P2) $(x \rightarrow y) \rightsquigarrow (x \rightarrow z) \in F$ *and* $y \in F$ *imply* $z \in F$,
(P3) $(x \rightsquigarrow y) \rightarrow (x \rightsquigarrow z) \in F$ *and* $y \in F$ *imply* $z \in F$.

3. Q-Filters in Quantum B-Algebra

In Reference [30], the notion of filter in quantum B-algebra is proposed. If X is a quantum B-algebra and F is a nonempty set of X, then F is called the filter of X if $F \in U(X)$ and $F \cdot F \subseteq F$. That is, F is a filter of X, if and only if: (1) F is a nonempty upper subset of X; (2) $(z \in X, y \in F, y \rightarrow z \in F) \Rightarrow z \in F$. We denote the set of all filters of X by $F(X)$.

In this section, we discuss a new concept of q-filter in quantum B-algebra; by using q-filters, we construct the quotient algebras.

Definition 9. *A nonempty subset F of quantum B-algebra X is called a q-filter of X if it satisfies:*

(1) *F is an upper set of X, that is,* $F \in U(X)$;
(2) *for all* $x \in F$, $x \rightarrow x \in F$ *and* $x \rightsquigarrow x \in F$;
(3) $x \in F$, $y \in X$, $x \rightarrow y \in F \Longrightarrow y \in F$.
(4) *A q-filter of X is normal if* $x \rightarrow y \in F \Longleftrightarrow x \rightsquigarrow y \in F$.

Proposition 8. *Let F be a q-filter of quantum B-algebra X. Then,*

(1) $x \in F$, $y \in X$, $x \rightsquigarrow y \in F \Longrightarrow y \in F$.
(2) $x \in F$ *and* $y \in X \Longrightarrow (x \rightsquigarrow y) \rightarrow y \in F$ *and* $(x \rightarrow y) \rightsquigarrow y \in F$.
(3) *if X is unital, then Condition (2) in Definition 9 can be replaced by* $u \in F$, *where u is the unit element of X.*

Proof. (1) Assume that $x \in F$, $y \in X$, and $x \rightsquigarrow y \in F$. Then, by Proposition 2 (4), $x \leq (x \rightsquigarrow y) \rightarrow y$. Applying Definition 9 (1) and (3), we get that $y \in F$.

(2) Using Proposition 2 (4) and Definition 9 (1), we can get (2).

(3) If X is unital with unit u, then $u \rightarrow u = u$. Moreover, applying Proposition 2 (8), $u \leq x \rightsquigarrow x$ and $u \leq x \rightarrow x$ from $x \leq x$, for all $x \in X$. Therefore, for unital quantum B-algebra X, Condition (2) in Definition 8 can be replaced by condition "$u \in F$". □

By Definition 6, and Propositions 6 and 8, we get the following result (the proof is omitted).

Proposition 9. *Let $(A, \rightarrow, \rightsquigarrow, \leq, 1)$ be a pseudo-BCI algebra. Then, an empty subset of A is a q-filter of A (as a quantum B-algebra) if and only if it is a filter of A (according to Definition 6).*

Example 1. *Let $X = \{a, b, c, d, e, f\}$. Define operations $\rightarrow$ and $\rightsquigarrow$ on X as per the following Cayley Tables 1 and 2; the order on X is defined as follows: $b \leq a \leq f$; $e \leq d \leq c$. Then, X is a quantum B-algebra (we can verify* *it with the Matlab software (The MathWorks Inc., Natick, MA, USA)), but it is not a pseudo-BCI algebra. Let* $F_1 = \{f\}$, $F_2 = \{a, b, f\}$; *then,* F_1 *is a filter but not a q-filter of X, and* F_2 *is a normal q-filter of X.*

Table 1. Cayley table of operation $\rightarrow$.

$\rightarrow$	**a**	**b**	**c**	**d**	**e**	**f**
a	f	a	c	c	c	f
b	f	a	c	c	c	f
c	c	c	f	a	b	c
d	c	c	f	f	a	c
e	c	c	f	f	a	c
f	a	b	c	d	e	f

Table 2. Cayley table of operation $\rightsquigarrow$.

$\rightsquigarrow$	**a**	**b**	**c**	**d**	**e**	**f**
a	f	a	c	c	d	f
b	f	f	c	c	c	f
c	c	c	f	a	a	c
d	c	c	f	a	a	c
e	c	c	f	f	a	c
f	a	b	c	d	e	f

Theorem 1. *Let X be a quantum B-algebra and F a normal q-filter of X. Define the binary* $\approx_F$ *on X as follows:*

$$x \approx_F y \Longleftrightarrow x \rightarrow y \in F \text{ and } y \rightarrow x \in F, \text{ where } x, y \in X.$$

Then,

(1) $\approx_F$ *is an equivalent relation on X;*

(2) $\approx_F$ *is a congruence relation on X, that is,* $x \approx_F y \Longrightarrow (z \rightarrow x) \approx_F (z \rightarrow y)$, $(x \rightarrow z) \approx_F (y \rightarrow z)$, $(z \rightsquigarrow x) \approx_F (z \rightsquigarrow y)$, $(x \rightsquigarrow z) \approx_F (y \rightsquigarrow z)$, *for all* $z \in X$.

Proof. (1) For any $x \in X$, by Definition 9 (2), $x \rightarrow x \in F$, it follows that $x \approx_F x$.

For all $x, y \in X$, if $x \approx_F y$, we can easily verify that $y \approx_F x$.
Assume that $x \approx_F y$, $y \approx_F z$. Then, $x \rightarrow y \in F$, $y \rightarrow x \in F$, $y \rightarrow z \in F$, and $z \rightarrow y \in F$, since

$$y \rightarrow z \leq (x \rightarrow y) \rightarrow (x \rightarrow z) \text{ by Definition 1 (1).}$$

From this and Definition 9, we have $x \rightarrow z \in F$. Similarly, we can get $z \rightarrow x \in F$. Thus, $x \approx_F z$.
Therefore, $\approx_F$ is an equivalent relation on X.
(2) If $x \approx_F y$, then $x \rightarrow y \in F$, $y \rightarrow x \in F$. Since

$$x \rightarrow y \leq (z \rightarrow x) \rightarrow (z \rightarrow y), \text{ by Definition 1 (1).}$$

$$y \rightarrow x \leq (z \rightarrow y) \rightarrow (z \rightarrow x), \text{ by Definition 1 (1).}$$

Using Definition 9 (1), $(z \rightarrow x) \rightarrow (z \rightarrow y) \in F$, $(z \rightarrow y) \rightarrow (z \rightarrow x) \in F$. It follows that $(z \rightarrow x) \approx_F (z \rightarrow y)$.
Moreover, since

$$x \rightarrow y \leq (y \rightarrow z) \rightsquigarrow (x \rightarrow z), \text{ by Proposition 2 (6).}$$

$$y \rightarrow x \leq (x \rightarrow z) \rightsquigarrow (y \rightarrow z), \text{ by Proposition 2 (6).}$$

Then, form $x \rightarrow y \in F$ and $y \rightarrow x \in F$; using Definition 9 (1), we have $(y \rightarrow z) \rightsquigarrow (x \rightarrow z) \in F$, $(x \rightarrow z) \rightsquigarrow (y \rightarrow z) \in F$. Since F is normal, by Definition 9 we get $(y \rightarrow z) \rightarrow (x \rightarrow z) \in F$, $(x \rightarrow z) \rightarrow (y \rightarrow z) \in F$. Thus, $(x \rightarrow z) \approx_F (y \rightarrow z)$.

Similarly, we can get that $x \approx_F y \Longrightarrow (z \rightsquigarrow x) \approx_F (z \rightsquigarrow y)$ and $(x \rightsquigarrow z) \approx_F (y \rightsquigarrow z)$. □

Definition 10. *A quantum B-algebra X is considered to be perfect, if it satisfies:*

(1) *for any normal q-filter F of X, x, y in X, (there exists an* $\in$*X, such that* $[x \rightarrow y]_F = [a \rightarrow a]_F$ *)* $\Longleftrightarrow$ *(there exists b*$\in$*X, such that* $[x \rightsquigarrow y]_F = [b \rightsquigarrow b]_F$ *).*

(1) *for any normal q-filter F of X,* $(X/\approx_F \rightarrow, \rightsquigarrow, \leq)$ *is a quantum B-algebra, where quotient operations* $\rightarrow$ *and* $\rightsquigarrow$ *are defined in a canonical way, and* $\leq$ *is defined as follows:*

$$[x]_F \leq [y]_F \Longleftrightarrow (\text{there exists } a \in X \text{ such that } [x]_F \rightarrow [y]_F = [a \rightarrow a]_F)$$
$$\Longleftrightarrow (\text{there exists } b \in X \text{ such that } [x]_F \rightsquigarrow [y]_F = [b \rightsquigarrow b]_F).$$

Theorem 2. *Let* $(A, \rightarrow, \rightsquigarrow, \leq, 1)$ *be a pseudo-BCI algebra, then A is a perfect quantum B-algebra.*

Proof. By Proposition 6, we know that A is a quantum B-algebra.

(1) For any normal q-filter F of A, $x, y \in A$, if there exists $a \in A$, such that $[x \rightarrow y]_F = [a \rightarrow a]_F$, then

$$[x \rightarrow y]_F = [a \rightarrow a]_F = [1]_F.$$

It follows that $(x \rightarrow y) \rightarrow 1 \in F$, $1 \rightarrow (x \rightarrow y) = x \rightarrow y \in F$. Applying Proposition 3 (11) and (12), we have

$$(x \rightarrow 1) \rightsquigarrow (y \rightarrow 1) = (x \rightarrow y) \rightarrow 1 \in F.$$

Since F is normal, from $(x \rightarrow 1) \rightsquigarrow (y \rightarrow 1) \in F$ and $x \rightarrow y \in F$ we get that

$$(x \rightarrow 1) \rightarrow (y \rightarrow 1) \in F \text{ and } x \rightsquigarrow y \in F.$$

Applying Proposition 3 (11) and (12) again, $(x \rightsquigarrow y) \rightarrow 1 = (x \rightarrow 1) \rightarrow (y \rightarrow 1)$. Thus,

$$(x \rightsquigarrow y) \rightarrow 1 = (x \rightarrow 1) \rightarrow (y \rightarrow 1) \in F \text{ and } 1 \rightarrow (x \rightsquigarrow y) = x \rightsquigarrow y \in F.$$

This means that $[x \rightsquigarrow y]_F = [1]_F = [1 \rightsquigarrow 1]_F$. Similarly, we can prove that the inverse is true. That is, Definition 10 (1) holds for A.

(2) For any normal q-filter F of pseudo-BCI algebra A, binary $\leq$ on $A/\approx_F$ is defined as the following:

$$[x]_F \leq [y]_F \Longleftrightarrow [x]_F \rightarrow [y]_F = [1]_F.$$

We verify that $\leq$ is a partial binary on $A/\approx_F$.

Obviously, $[x]_F \leq [x]_F$ for any $x \in A$.

If $[x]_F \leq [y]_F$ and $[y]_F \leq [x]_F$, then $[x]_F \rightarrow [y]_F = [x \rightarrow y]_F = [1]_F$, $[y]_F \rightarrow [x]_F = [y \rightarrow x]_F = [1]_F$. By the definition of equivalent class, $x \rightarrow y = 1 \rightarrow (x \rightarrow y) \in F$, $y \rightarrow x = 1 \rightarrow (y \rightarrow x) \in F$. It follows that $x \approx_F y$; thus, $[x]_F = [y]_F$.

If $[x]_F \leq [y]_F$ and $[y]_F \leq [z]_F$, then $[x]_F \rightarrow [y]_F = [x \rightarrow y]_F = [1]_F$, $[y]_F \rightarrow [z]_F = [y \rightarrow z]_F = [1]_F$. Thus,

$$x \rightarrow y = 1 \rightarrow (x \rightarrow y) \in F, (x \rightarrow y) \rightarrow 1 \in F;$$

$$y \rightarrow z = 1 \rightarrow (y \rightarrow z) \in F, (y \rightarrow z) \rightarrow 1 \in F.$$

Applying Definition 3 and Proposition 3,

$$y \rightarrow z \leq (x \rightarrow y) \rightarrow (x \rightarrow z),$$

$$(x \to y) \to 1 = (x \to 1) \rightsquigarrow (y \to 1) \leq ([(y \to 1) \rightsquigarrow (z \to 1)] \to [(x \to 1) \rightsquigarrow (z \to 1)]) = [(y \to z) \to 1] \to [(x \to z) \to 1].$$

By Definition 9,

$$1 \to (x \to z) = x \to z \in F, (x \to z) \to 1 \in F.$$

This means that $(x \to z) \approx_F 1$, $[x \to z]_F = [1]_F$. That is, $[x]_F \to [z]_F = [x \to z]_F = [1]_F$, $[x]_F \leq [z]_F$.

Therefore, applying Theorem 1, we know that $(A/\approx_F \to, \rightsquigarrow, [1]_F)$ is a quantum B-algebra and pseudo-BCI algebra. That is, Definition 10 (2) holds for A.

Hence, we know that A is a perfect quantum B-algebra. □

The following examples show that there are some perfect quantum B-algebras that may not be a pseudo-BCI algebra.

Example 2. *Let $X = \{a, b, c, d, e, 1\}$. Define operations $\to$ and $\rightsquigarrow$ on X as per the following Cayley Tables 3 and 4, the order on X is defined as the following: $b \leq a \leq 1$; $e \leq d \leq c$. Then, X is a pseudo-BCI algebra (we can verify it with Matlab). Denote $F_1 = \{1\}$, $F_2 = \{a, b, 1\}$, $F_3 = X$, then F_i ($i = 1, 2, 3$) are all normal q-filters of X, and quotient algebras $(X/\approx_{Fi} \to, \rightsquigarrow, [1]_{Fi})$ are pseudo-BCI algebras. Thus, X is a perfect quantum B-algebra.*

Table 3. Cayley table of operation $\to$.

$\to$	**a**	**b**	**c**	**d**	**e**	**1**
a	1	a	c	c	c	1
b	1	1	c	c	c	1
c	c	c	1	a	b	c
d	c	c	1	1	a	c
e	c	c	1	1	1	c
1	a	b	c	d	e	1

Table 4. Cayley table of operation $\rightsquigarrow$.

$\rightsquigarrow$	**a**	**b**	**c**	**d**	**e**	**1**
a	1	a	c	c	d	1
b	1	1	c	c	c	1
c	c	c	1	a	a	c
d	c	c	1	1	a	c
e	c	c	1	1	1	c
1	a	b	c	d	e	1

Example 3. *Let $X = \{a, b, c, d, e, f\}$. Define operations $\to$ and $\rightsquigarrow$ on X as per the following Cayley Tables 5 and 6, the order on X is defined as follows: $b \leq a \leq f$; $e \leq d \leq c$. Then, X is a quantum B-algebra (we can verify it with Matlab), but it is not a pseudo-BCI algebra, since $e \rightsquigarrow e \neq e \to e$. Denote $F = \{a, b, f\}$, then F, X are all normal q-filters of X, quotient algebras $(X/\approx_F \to, \rightsquigarrow, \leq)$, $(X/\approx_X \to, \rightsquigarrow, \leq)$ are quantum B-algebras, and X is a perfect quantum B-algebra.*

Table 5. Cayley table of operation $\to$.

$\to$	**a**	**b**	**c**	**d**	**e**	***f***
a	f	a	c	c	c	f
b	f	f	c	c	c	f
c	c	c	f	a	b	c
d	c	c	f	f	a	c
e	c	c	f	f	f	c
f	a	b	c	d	e	f

Table 6. Cayley table of operation $\rightsquigarrow$.

$\rightsquigarrow$	**a**	**b**	**c**	**d**	**e**	**f**
a	f	a	c	c	d	f
b	f	f	c	c	c	f
c	c	c	f	a	a	c
d	c	c	f	f	a	c
e	c	c	f	f	a	c
f	a	b	c	d	e	f

4. Basic Implication Algebras and Filters

Definition 11. *Let $(A, \vee, \wedge, \otimes, \to, 0, 1)$ be a type-(2, 2, 2, 2, 0, 0) algebra [32]. A is called a nonassociative residuated lattice, if it satisfies:*

(A1) $(A, \vee, \wedge, 0, 1)$ *is a bounded lattice;*
(A2) $(A, \otimes, 1)$ *is a commutative groupoid with unit element 1;*
(A3) $\forall x, y, z \in A,\ x \otimes y \leq z \Longleftrightarrow x \leq y \to z.$

Proposition 10. *Let $(A, \vee, \wedge, \otimes, \to, 0, 1)$ be a nonassociative residuated lattice [32]. Then, $(\forall\, x, y, z \in A)$*

(1) $x \leq y \Longleftrightarrow x \to y = 1;$
(2) $x \leq y \Rightarrow x \otimes z \leq y \otimes z;$
(3) $x \leq y \Rightarrow y \to z \leq x \to z;$
(4) $x \leq y \Rightarrow z \to x \leq z \to y;$
(5) $x \otimes (y \vee z) = (x \otimes y) \vee (x \otimes z);$
(6) $x \to (y \wedge z) = (x \to y) \wedge (x \to z);$
(7) $(y \vee z) \to x = (y \to x) \wedge (z \to x);$
(8) $(x \to y) \otimes x \leq x, y;$
(9) $(x \to y) \to y \geq x, y.$

Example 4. *Let $A = [0, 1]$, operation $\otimes$ on A is defined as follows:*

$$x \otimes y = 0.5xy + 0.5max\{0, x + y - 1\},\ x, y \in A.$$

Then, $\otimes$ is a nonassociative t-norm on A (see Example 1 in Reference [32]). Operation $\to$ is defined as follows:

$$x \to y = max\{z \in [0, 1] \mid z \otimes x \leq y\},\ x, y \in A.$$

Then, $(A, max, min, \otimes, \to, 0, 1)$ is a nonassoiative residuated lattice (see Theorem 5 in Reference [32]). Assume that $x = 0.55$, $y = 0.2$, $z = 0.1$, then

$$y \to z = 0.2 \to 0.1 = \max\{a \in [0,1|a \otimes 0.2 \leq 0.1\} = \frac{5}{6}.$$

$$x \to y = 0.55 \to 0.2 = \max\{a \in [0,1|a \otimes 0.55 \leq 0.2\} = \frac{17}{31}.$$

$$x \to z = 0.55 \to 0.1 = \max\{a \in [0,1|a \otimes 0.55 \leq 0.1\} = \frac{4}{11}.$$

$$(x \to y) \to (x \to z) = \frac{17}{31} \to \frac{4}{11} = \max\{a \in [0,1|a \otimes \frac{17}{31} \leq \frac{4}{11}\} = \frac{67}{88}.$$

Therefore,

$$y \to z \nleq (x \to y) \to (x \to z).$$

Example 4 shows that Condition (1) in Definition 1 is not true for general non-associative residuated lattices, that is, quantum B-algebras are not common basic of non-associative fuzzy logics. So, we discuss more general implication algebras in this section.

Definition 12. *A basic implication algebra is a partially ordered set $(X, \leq)$ with binary operation $\to$, such that the following are satisfied for x, y, and z in X:*

(1) $x \leq y \Rightarrow z \to x \leq z \to y$;
(2) $x \leq y \Rightarrow y \to z \leq x \to z$.

A basic implication algebra is considered to be normal, if it satisfies:
(3) *for any* $x, y \in X$, $x \to x = y \to y$;
(4) *for any* $x, y \in X$, $x \leq y \Longleftrightarrow x \to y = e$, *where* $e = x \to x = y \to y$.

We can verify that the following results are true (the proofs are omitted).

Proposition 11. *Let $(X, \to, \leq)$ be a basic implication algebra. Then, for all $x, y, z \in X$,*

(1) $x \leq y \Rightarrow y \to x \leq x \to x \leq x \to y$;
(2) $x \leq y \Rightarrow y \to x \leq y \to y \leq x \to y$;
(3) $x \leq y$ *and* $u \leq v \Rightarrow y \to u \leq x \to v$;
(4) $x \leq y$ *and* $u \leq v \Rightarrow v \to x \leq u \to y$.

Proposition 12. *Let $(X, \to, \leq, e)$ be a normal basic implication algebra. Then for all $x, y, z \in X$,*

(1) $x \to x = e$;
(2) $x \to y = y \to x = e \Rightarrow x = y$;
(3) $x \leq y \Rightarrow y \to x \leq e$;
(4) *if e is unit (that is, for all x in X, $e \to x = x$), then e is a maximal element (that is, $e \leq x \Rightarrow e = x$).*

Proposition 13. *(1) If $(X, \to, \rightsquigarrow, \leq)$ is a a quantum B-algebra, then $(X, \to, \leq)$ and $(X, \rightsquigarrow, \leq)$ are basic implication algebras; (2) If $(A, \to, \rightsquigarrow, \leq, 1)$ is a pseudo-BCI algebra, then $(A, \to, \leq, 1)$ and $(A, \rightsquigarrow, \leq, 1)$ are normal basic implication algebras with unit 1; (3) If $(A, \vee, \wedge, \otimes, \to, 0, 1)$ is a non-associative residuated lattice, then $(A, \to, \leq, 1)$ is a normal basic implication algebra.*

The following example shows that element *e* may not be a unit.

Example 5. *Let $X = \{a, b, c, d, 1\}$. Define $a \leq b \leq c \leq d \leq 1$ and operation $\to$ on X as per the following Cayley Table 7. Then, X is a normal basic implication algebra in which element 1 is not a unit. $(X, \to, \leq)$ is not a commutative quantum B-algebra, since*

$$c = 1 \to c \nleq b = (c \to d) \to (1 \to d).$$

Table 7. Cayley table of operation $\to$.

$\to$	*a*	*b*	*c*	*d*	**1**
a	1	1	1	1	1
b	d	1	1	1	1
c	d	d	1	1	1
d	b	c	d	1	1
1	b	b	c	b	1

The following example shows that element *e* may be not maximal.

Example 6. *Let X = {a, b, c, d, 1}. Define $a \leq b \leq c \leq d$, $a \leq b \leq c \leq 1$ and operation → on X as per the following Cayley Table 8. Then, X is a normal basic implication algebra, and element 1 is not maximal and is not a unit.*

Table 8. Cayley table of operation →.

→	a	b	c	d	1
a	1	1	1	1	1
b	c	1	1	1	1
c	c	c	1	1	1
d	a	c	a	1	c
1	a	b	b	c	1

Definition 13. *A nonempty subset F of basic implication algebra (X, →, ≤) is called a filter of X if it satisfies:*

(1) *F is an upper set of X, that is, $x \in F$ and $x \leq y \in X \Longrightarrow y \in F$;*
(2) *for all $x \in F$, $x \to x \in F$;*
(3) *$x \in F$, $y \in X$, $x \to y \in F \Longrightarrow y \in F$;*
(4) *$x \in X$, $y \to z \in F \Longrightarrow (x \to y) \to (x \to z) \in F$;*
(5) *$x \in X$, $y \to z \in F \Longrightarrow (z \to x) \to (y \to x) \in F$.*

For normal basic implication algebra (X, →, ≤, e), a filter F of X is considered to be regular, if it satisfies:

(6) *$x \in X$, $(x \to y) \to e \in F$ and $(y \to z) \to e \in F \Longrightarrow (x \to z) \to e \in F$.*

Proposition 14. *Let (X, →, ≤, e) be a normal basic implication algebra and $F \subseteq X$. Then, F is a filter of X if and only if it satisfies:*

(1) *$e \in F$;*
(2) *$x \in F$, $y \in X$, $x \to y \in F \Longrightarrow y \in F$;*
(3) *$x \in X$, $y \to z \in F \Longrightarrow (x \to y) \to (x \to z) \in F$;*
(4) *$x \in X$, $y \to z \in F \Longrightarrow (z \to x) \to (y \to x) \in F$.*

Obviously, if *e* is the maximal element of normal basic implication algebra (X, →, ≤, *e*), then any filter of X is regular.

Theorem 3. *Let X be a basic implication algebra and F a filter of X. Define binary $\approx_F$ on X as follows:*

$$x \approx_F y \Longleftrightarrow x \to y \in F \text{ and } y \to x \in F, \text{ where } x, y \in X.$$

Then

(1) *$\approx_F$ is a equivalent relation on X;*
(2) *$\approx_F$ is a congruence relation on X, that is, $x \approx_F y \Longrightarrow (z \to x) \approx_F (z \to y)$, $(x \to z) \approx_F (y \to z)$, for all $z \in X$.*

Proof (1) $\forall x \in X$, from Definition 13 (2), $x \to x \in F$, thus $x \approx_F x$. Moreover, $\forall x, y \in X$, if $x \approx_F y$, then $y \approx_F x$.

If $x \approx_F y$ and $y \approx_F z$. Then $x \to y \in F$, $y \to x \in F$, $y \to z \in F$, and $z \to y \in F$. Applying Definition 13 (4) and (5), we have

$$(x \to y) \to (x \to z) \in F,\ (z \to y) \to (z \to x) \in F.$$

From this and Definition 13 (3), we have $x \to z \in F$, $z \to x \in F$. Thus, $x \approx_F z$.

Hence, $\approx_F$ is a equivalent relation on X.

(2) Assume $x \approx_F y$. By the definition of bianary relation $\approx_F$, we have $x{\rightarrow}y{\in}F$, $y{\rightarrow}x{\in}F$. Using Definition 13 (4),

$$(z{\rightarrow}x){\rightarrow}(z{\rightarrow}y){\in}F,\ (z{\rightarrow}y){\rightarrow}(z{\rightarrow}x){\in}F.$$

This means that $(z{\rightarrow}x) \approx_F (z{\rightarrow}y)$. Moreover, using Definition 13 (5), we have

$$(y{\rightarrow}z){\rightarrow}(x{\rightarrow}z){\in}F,\ (x{\rightarrow}z){\rightarrow}(y{\rightarrow}z){\in}F.$$

Hence, $(x{\rightarrow}z) \approx_F (y{\rightarrow}z)$. □

Theorem 4. *Let $(X, \rightarrow, \leq, e)$ be a normal basic implication algebra and F a regular filter of X. Define quotient operation $\rightarrow$ and binary relation $\leq$ on $X/{\approx_F}$ as follows:*

$$[x]_F{\rightarrow}[y]_F = [x]_F{\rightarrow}[y]_F,\ \forall x, y{\in}X;$$

$$[x]_F{\leq}\ [y]_F \Longleftrightarrow [x]_F \rightarrow[y]_F = [e]_F,\ \forall x, y{\in}X.$$

Then, $(X/{\approx_F}, \rightarrow, \leq, [e]_F)$ is a normal basic implication algebra, and $(X, \rightarrow, \leq, e) \sim (X/{\approx_F}, \rightarrow, \leq, [e]_F)$.

Proof. Firstly, we prove that binary relation $\leq$ on $X/{\approx_F}$ is a partial order.

(1) $\forall x{\in}X$, obviously, $[x]_F \leq [x]_F$.
(2) Assume that $[x]_F \leq [y]_F$ and $[y]_F \leq [x]_F$, then

$$[x]_F{\rightarrow}[y]_F = [x{\rightarrow}y]_F = [e]_F,\ [y]_F{\rightarrow}[x]_F = [y{\rightarrow}x]_F = [e]_F.$$

It follows that $e{\rightarrow}(x{\rightarrow}y){\in}F, e{\rightarrow}(y{\rightarrow}x){\in}F$. Applying Proposition 14 (1) and (2), we get that $(x{\rightarrow}y){\in}F$ and $(y{\rightarrow}x){\in}F$. This means that $[x]_F = [y]_F$.
(3) Assume that $[x]_F \leq [y]_F$ and $[y]_F \leq [z]_F$, then

$$[x]_F{\rightarrow}[y]_F = [x{\rightarrow}y]_F = [e]_F,\ [y]_F{\rightarrow}[z]_F = [y{\rightarrow}z]_F = [e]_F.$$

Using the definition of equivalent relation $\approx_F$, we have

$$e{\rightarrow}(x{\rightarrow}y){\in}F,\ (x{\rightarrow}y){\rightarrow}e{\in}F;\ e{\rightarrow}(y{\rightarrow}z){\in}F,\ (y{\rightarrow}z){\rightarrow}e{\in}F.$$

From $e{\rightarrow}(x{\rightarrow}y){\in}F$ and $e{\rightarrow}(y{\rightarrow}z){\in}F$, applying Proposition 14 (1) and (2), $(x{\rightarrow}y){\in}F$ and $(y{\rightarrow}z){\in}F$. By Proposition 14 (4), $(x{\rightarrow}y){\rightarrow}(x{\rightarrow}z){\in}F$. It follows that $(x{\rightarrow}z){\in}F$. Hence, $(x{\rightarrow}x){\rightarrow}(x{\rightarrow}z){\in}F$, by Proposition 14 (4). Therefore,

$$e{\rightarrow}(x{\rightarrow}z) = (x{\rightarrow}x){\rightarrow}(x{\rightarrow}z){\in}F.$$

Moreover, from $(x{\rightarrow}y){\rightarrow}e{\in}F$ and $(y{\rightarrow}z){\rightarrow}e{\in}F$, applying regularity of F and Definition 13 (6), we get that $(x{\rightarrow}z){\rightarrow}e{\in}F$.

Combining the above $e{\rightarrow}(x{\rightarrow}z){\in}F$ and $(x{\rightarrow}z){\rightarrow}e{\in}F$, we have $x{\rightarrow}z \approx_F e$, that is, $[x{\rightarrow}z]_F = [e]_F$. This means that $[x]_F \leq [z]_F$. It follows that the binary relation $\leq$ on $X/{\approx_F}$ is a partially order.

Therefore, applying Theorem 3, we know that $(X/{\approx_F} \rightarrow, \leq, [e]_F)$ is a normal basic implication algebra, and $(X, \rightarrow, \leq, e) \sim (X/{\approx_F} \rightarrow, \leq, [e]_F)$ in the homomorphism mapping f: $X{\rightarrow}X/{\approx_F}; f(x) = [x]_F$. □

Example 7. *Let $X = \{a, b, c, d, 1\}$. Define operations $\rightarrow$ on X as per the following Cayley Table 9, and the order binary on X is defined as follows: $a \leq b \leq c \leq 1$, $b \leq d \leq 1$. Then $(X, \rightarrow, \leq, 1)$ is a normal basic implication algebra (it is not a quantum B-algebra). Denote $F = \{1\}$, then F is regular filters of X, and the quotient algebras $(X, \rightarrow, \leq, 1)$ is isomorphism to $(X/{\approx_F}, \rightarrow, [1]_F)$.*

Table 9. Cayley table of operation $\rightarrow$.

$\rightarrow$	***a***	***b***	***c***	***d***	**1**
a	1	1	1	1	1
b	d	1	1	1	1
c	b	d	1	d	1
d	a	c	c	1	1
1	a	b	c	d	1

Example 8. *Denote X = {a, b, c, d, 1}. Define operations $\rightarrow$ on X as per the following Cayley Table 10, and the order binary on X is defined as follows: $a \leq b \leq c \leq 1, b \leq d \leq 1$. Then $(X, \rightarrow, \leq, 1)$ is a normal basic implication algebra (it is not a quantum B-algebra). Let F = {1, d}, then F is a regular filters of X, and the quotient algebras $(X/\approx_F, \rightarrow, [1]_F)$ is presented as the following Table 11, where $X/\approx_F = \{\{a\}, \{b, c\}, [1]_F = \{1, d\}\}$. Moreover, $(X, \rightarrow, \leq, 1) \sim (X/\approx_F \rightarrow, [1]_F)$.*

Table 10. Cayley table of operation $\rightarrow$.

$\rightarrow$	***a***	***b***	***c***	***d***	**1**
a	1	1	1	1	1
b	c	1	1	1	1
c	b	d	1	d	1
d	a	c	c	1	1
1	a	b	c	d	1

Table 11. Quotient algebra $(X/\approx_F, \rightarrow, [1]_F)$.

$\rightarrow$	$\{a\}$	$\{b,c\}$	$[1]_F$
$\{a\}$	$[1]_F$	$[1]_F$	$[1]_F$
$\{b,c\}$	$\{b,c\}$	$[1]_F$	$[1]_F$
$[1]_F$	$\{a\}$	$\{b,c\}$	$[1]_F$

5. Conclusions

In this paper, we introduced the notion of a q-filter in quantum B-algebras and investigated quotient structures; by using q-filters as a corollary, we obtained quotient pseudo-BCI algebras by their filters. Moreover, we pointed out that the concept of quantum B-algebra does not apply to non-associative fuzzy logics. From this fact, we proposed the new concept of basic implication algebra, and established the corresponding filter theory and quotient algebra. In the future, we will study in depth the structural characteristics of basic implication algebras and the relationship between other algebraic structures and uncertainty theories (see References [33–36]). Moreover, we will consider the applications of q-filters for Gentzel's sequel calculus.

Author Contributions: X.Z. initiated the research and wrote the draft. B.R.A., Y.B.J., and X.Z. completed the final version.

References

1. Hájek, P. *Metamathematics of Fuzzy Logic*; Kluwer: Dordrecht, The Netherlands, 1998.
2. Klement, E.P.; Mesiar, R.; Pap, E. *Triangular Norm*; Springer: Berlin, Germany, 2000.
3. Esteva, F.; Godo, L. Monoidal t-norm based logic: Towards a logic for left-continuous t-norms. *Fuzzy Sets Syst.* **2001**, *124*, 271–288. [CrossRef]
4. Flaminio, T. Strong non-standard completeness for fuzzy logics. *Soft Comput.* **2008**, *12*, 321–333. [CrossRef]
5. Zhang, X.H. *Fuzzy Logics and Algebraic Analysis*; Science in China Press: Beijing, China, 2008.
6. Zhang, X.H.; Dudek, W.A. BIK+-logic and non-commutative fuzzy logics. *Fuzzy Syst. Math.* **2009**, *23*, 8–20.
7. Iorgulescu, A. *Implicative-Groups vs. Groups and Generalizations*; Matrix Room: Bucuresti, Romania, 2018.

8. Fodor, J.C. Contrapositive symmetry of fuzzy implications. *Fuzzy Sets Syst.* **1995**, *69*, 141–156. [CrossRef]
9. Ruiz-Aguilera, D.; Torrens, J. Distributivity of residual implications over conjunctive and disjunctive uninorms. *Fuzzy Sets Syst.* **2007**, *158*, 23–37. [CrossRef]
10. Yao, O. On fuzzy implications determined by aggregation operators. *Inf. Sci.* **2012**, *193*, 153–162.
11. Baczyński, M.; Beliakov, G.; Bustince Sola, H.; Pradera, A. *Advances in Fuzzy Implication Functions*; Springer: Berlin/Heidelberg, Germany, 2013.
12. Vemuri, N.; Jayaram, B. The ⊗-composition of fuzzy implications: Closures with respect to properties, powers and families. *Fuzzy Sets Syst.* **2015**, *275*, 58–87. [CrossRef]
13. Li, D.C.; Qin, S.J. The quintuple implication principle of fuzzy reasoning based on interval-valued S-implication. *J. Log. Algebraic Meth. Program.* **2018**, *100*, 185–194. [CrossRef]
14. Su, Y.; Liu, H.W.; Pedrycz, W. A method to construct fuzzy implications–rotation construction. *Int. J. Appr. Reason.* **2018**, *92*, 20–31. [CrossRef]
15. Wang, C.Y.; Wan, L.J. Type-2 fuzzy implications and fuzzy-valued approximation reasoning. *Int. J. Appr. Reason.* **2018**, *102*, 108–122. [CrossRef]
16. Luo, M.X.; Zhou, K.Y. Logical foundation of the quintuple implication inference methods. *Int. J. Appr. Reason.* **2018**, *101*, 1–9. [CrossRef]
17. Georgescu, G.; Iorgulescu, A. Pseudo-BCK algebras. In *Combinatorics, Computability and Logic*; Springer: London, UK, 2001; pp. 97–114.
18. Dudek, W.A.; Jun, Y.B. Pseudo-BCI algebras. *East Asian Math. J.* **2008**, *24*, 187–190.
19. Dymek, G. Atoms and ideals of pseudo-BCI-algebras. *Commen. Math.* **2012**, *52*, 73–90.
20. Zhang, X.H.; Jun, Y.B. Anti-grouped pseudo-BCI algebras and anti-grouped filters. *Fuzzy Syst. Math.* **2014**, *28*, 21–33.
21. Zhang, X.H. Fuzzy anti-grouped filters and fuzzy normal filters in pseudo-BCI algebras. *J. Intell. Fuzzy Syst.* **2017**, *33*, 1767–1774. [CrossRef]
22. Emanovský, P.; Kühr, J. Some properties of pseudo-BCK- and pseudo-BCI-algebras. *Fuzzy Sets Syst.* **2018**, *339*, 1–16. [CrossRef]
23. Zhang, X.H.; Park, C.; Wu, S.P. Soft set theoretical approach to pseudo-BCI algebras. *J. Intell. Fuzzy Syst.* **2018**, *34*, 559–568. [CrossRef]
24. Rump, W.; Yang, Y. Non-commutative logical algebras and algebraic quantales. *Ann. Pure Appl. Log.* **2014**, *165*, 759–785. [CrossRef]
25. Rump, W. Quantum B-algebras. *Cent. Eur. J. Math.* **2013**, *11*, 1881–1899. [CrossRef]
26. Han, S.W.; Xu, X.T.; Qin, F. The unitality of quantum B-algebras. *Int. J. Theor. Phys.* **2018**, *57*, 1582–1590. [CrossRef]
27. Han, S.W.; Wang, R.R.; Xu, X.T. On the injective hulls of quantum B-algebras. *Fuzzy Sets Syst.* **2018**. [CrossRef]
28. Rump, W. Quantum B-algebras: Their omnipresence in algebraic logic and beyond. *Soft Comput.* **2017**, *21*, 2521–2529. [CrossRef]
29. Rump, W. L-algebras, self-similarity, and l-groups. *J. Algebra* **2008**, *320*, 2328–2348. [CrossRef]
30. Botur, M.; Paseka, J. Filters on some classes of quantum B-algebras. *Int. J. Theor. Phys.* **2015**, *54*, 4397–4409. [CrossRef]
31. Botur, M.; Halaš, R. Commutative basic algebras and non-associative fuzzy logics. *Arch. Math. Log.* **2009**, *48*, 243–255. [CrossRef]
32. Botur, M. A non-associative generalization of Hájek's BL-algebras. *Fuzzy Sets Syst.* **2011**, *178*, 24–37. [CrossRef]
33. Zhang, X.H.; Smarandache, F.; Liang, X.L. Neutrosophic duplet semi-group and cancellable neutrosophic triplet groups. *Symmetry* **2017**, *9*, 275. [CrossRef]
34. Zhang, X.H.; Bo, C.X.; Smarandache, F.; Park, C. New operations of totally dependent- neutrosophic sets and totally dependent-neutrosophic soft sets. *Symmetry* **2018**, *10*, 187. [CrossRef]
35. Zhang, X.H.; Bo, C.X.; Smarandache, F.; Dai, J.H. New inclusion relation of neutrosophic sets with applications and related lattice structure. *Int. J. Mach. Learn. Cyber.* **2018**, *9*, 1753–1763. [CrossRef]
36. Caponetto, R.; Fortuna, L.; Graziani, S.; Xibilia, M.G. Genetic algorithms and applications in system engineering: A survey. *Trans. Inst. Meas. Control* **1993**, *15*, 143–156. [CrossRef]

15

Determining Crossing Number of Join of the Discrete Graph with Two Symmetric Graphs of Order Five

Michal Staš

Faculty of Electrical Engineering and Informatics, Technical University of Košice, 040 01 Košice, Slovakia; michal.stas@tuke.sk

Abstract: The main aim of the paper is to give the crossing number of the join product $G + D_n$ for the disconnected graph G of order five consisting of one isolated vertex and of one vertex incident with some vertex of the three-cycle, and D_n consists of n isolated vertices. In the proofs, the idea of the new representation of the minimum numbers of crossings between two different subgraphs that do not cross the edges of the graph G by the graph of configurations $\mathcal{G}_D$ in the considered drawing D of $G + D_n$ will be used. Finally, by adding some edges to the graph G, we are able to obtain the crossing numbers of the join product with the discrete graph D_n and with the path P_n on n vertices for three other graphs.

Keywords: graph; good drawing; crossing number; join product; cyclic permutation

---

1. Introduction

The investigation of the crossing number of graphs is a classical and very difficult problem provided that computing of the crossing number of a given graph in general is an NP-complete problem. It is well known that the problem of reducing the number of crossings in the graph has been studied in many areas, and the most prominent area is very large-scale integration technology.

In the paper, we will use notations and definitions of the crossing numbers of graphs like in [1]. We will often use Kleitman's result [2] on crossing numbers of the complete bipartite graphs. More precisely, he proved that:

$$\mathrm{cr}(K_{m,n}) = \left\lfloor \frac{m}{2} \right\rfloor \left\lfloor \frac{m-1}{2} \right\rfloor \left\lfloor \frac{n}{2} \right\rfloor \left\lfloor \frac{n-1}{2} \right\rfloor, \quad \text{if} \quad m \leq 6.$$

Using Kleitman's result [2], the crossing numbers for join of two paths, join of two cycles, and for join of path and cycle were studied in [1]. Moreover, the exact values for crossing numbers of $G + D_n$ and $G + P_n$ for all graphs G of order at most four are given in [3]. Furthermore, the crossing numbers of the graphs $G + D_n$ are known for a few graphs G of order five and six in [4–10]. In all of these cases, the graph G is connected and contains at least one cycle. Further, the exact values for the crossing numbers $G + P_n$ and $G + C_n$ have been also investigated for some graphs G of order five and six in [5,7,11,12].

The methods presented in the paper are new, and they are based on multiple combinatorial properties of the cyclic permutations. It turns out that if the graph of configurations is used like a graphical representation of the minimum numbers of crossings between two different subgraphs, then the proof of the main theorem will be simpler to understand. Similar methods were partially used for the first time in the papers [8,13]. In [4,9,10,14], the properties of cyclic permutations were also verified with the help of software in [15]. In our opinion, the methods used in [3,5,7] do not allow establishing the crossing number of the join product $G + D_n$.

2. Cyclic Permutations and Configurations

Let G be the disconnected graph of order five consisting of one isolated vertex and of one vertex incident with some vertex of the three-cycle. We will consider the join product of the graph G with the discrete graph on n vertices denoted by D_n. The graph $G + D_n$ consists of one copy of the graph G and of n vertices $t_1, \dots, t_n$, where any vertex t_i, $i = 1, \dots, n$, is adjacent to every vertex of G. Let T^i, $1 \leq i \leq n$, denote the subgraph induced by the five edges incident with the vertex t_i. Thus, the graph $T^1 \cup \dots \cup T^n$ is isomorphic with the complete bipartite graph $K_{5,n}$ and:

$$G + D_n = G \cup K_{5,n} = G \cup \left(\bigcup_{i=1}^{n} T^i \right). \tag{1}$$

In the paper, we will use the same notation and definitions for cyclic permutations and the corresponding configurations for a good drawing D of the graph $G + D_n$ like in [9,14]. Let D be a drawing of the graph $G + D_n$. The rotation $\mathrm{rot}_D(t_i)$ of a vertex t_i in the drawing D like the cyclic permutation that records the (cyclic) counter-clockwise order in which the edges leave t_i has been defined by Hernández-Vélez, Medina, and Salazar [13]. We use the notation (12345) if the counter-clockwise order the edges incident with the vertex t_i is t_iv_1, t_iv_2, t_iv_3, t_iv_4, and t_iv_5. We have to emphasize that a rotation is a cyclic permutation. In the paper, each cyclic permutation will be represented by the permutation with one in the first position. Let $\overline{\mathrm{rot}}_D(t_i)$ denote the inverse permutation of $\mathrm{rot}_D(t_i)$. We will deal with the minimal necessary number of crossings between the edges of T^i and the edges of T^j in a subgraph $T^i \cup T^j$ depending on the rotations $\mathrm{rot}_D(t_i)$ and $\overline{\mathrm{rot}}_D(t_j)$.

We will separate all subgraphs T^i, $i = 1, \dots, n$, of the graph $G + D_n$ into three mutually-disjoint subsets depending on how many of the considered T^i cross the edges of G in D. For $i = 1, \dots, n$, let $R_D = \{T^i : \mathrm{cr}_D(G, T^i) = 0\}$ and $S_D = \{T^i : \mathrm{cr}_D(G, T^i) = 1\}$. Every other subgraph T^i crosses the edges of G at least twice in D. Moreover, let F^i denote the subgraph $G \cup T^i$ for $T^i \in R_D$, where $i \in \{1, \dots, n\}$. Thus, for a given subdrawing of G, any subgraph F^i is exactly represented by $\mathrm{rot}_D(t_i)$.

Let us suppose first a good drawing D of the graph $G + D_n$ in which the edges of G do not cross each other. In this case, without loss of generality, we can choose the vertex notation of the graph in such a way as shown in Figure 1a. Our aim is to list all possible rotations $\mathrm{rot}_D(t_i)$ that can appear in D if the edges of T^i do not cross the edges of G. Since there is only one subdrawing of $F^i \setminus \{v_2, v_5\}$ represented by the rotation (143), there are two possibilities for how to obtain the subdrawing of $F^i \setminus v_5$ depending on in which region the edge t_iv_2 is placed. Of course, the vertex v_5 can be placed in one of four regions of the subdrawing $F^i \setminus v_5$ with the vertex t_i on their boundaries. These $2 \times 4 = 8$ possibilities under our consideration will be denoted by A_k and B_l, for $k = 1, 2$ and $l = 1, \dots, 6$. The configuration is of type A or B in the considered drawing D, if the vertex v_5 is placed in the quadrangular or in the triangular region in the subdrawing $D(F^i \setminus v_5)$, respectively. As for our considerations, it does not play a role in which of the regions is unbounded; assume the drawings shown in Figure 2. Thus, the configurations A_1, A_2, B_1, B_2, B_3, B_4, B_5, and B_6 are represented by the cyclic permutations (15432), (12435), (14532), (12453), (14325), (15243), (12543), and (14352), respectively. In a fixed drawing of the graph $G + D_n$, some configurations from $\mathcal{M}$ need not appear. We denote by $\mathcal{M}_D$ the subset of $\mathcal{M} = \{A_1, A_2, B_1, B_2, B_3, B_4, B_5, B_6\}$ consisting of all configurations that exist in the drawing D.

We remark that if two different subgraphs F^i and F^j with their configurations from $\mathcal{M}_D$ cross in a considered drawing D of the graph $G + D_n$, then the edges of T^i are crossed only by the edges of T^j. Let X, Y be the configurations from $\mathcal{M}_D$. We briefly denote by $\mathrm{cr}_D(X, Y)$ the number of crossings in D between T^i and T^j for two different $T^i, T^j \in R_D$ such that F^i, F^j have configurations X, Y, respectively. Finally, let $\mathrm{cr}(X, Y) = \min\{\mathrm{cr}_D(X, Y)\}$ over all good drawings of the graph $G + D_n$ with $X, Y \in \mathcal{M}_D$. Our aim shall be to establish $\mathrm{cr}(X, Y)$ for all pairs $X, Y \in \mathcal{M}$.

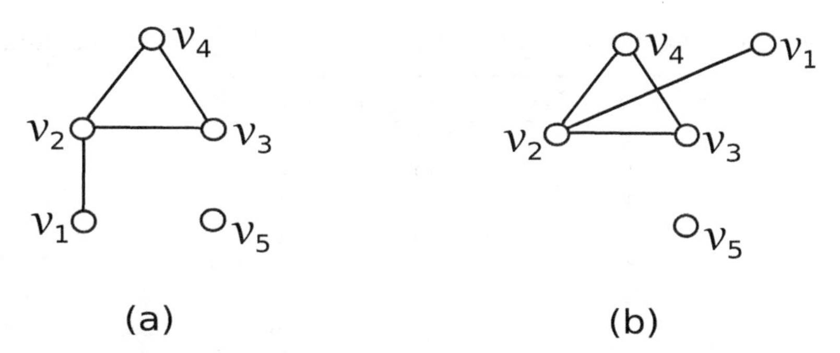

Figure 1. Two good drawings of the graph G. (**a**): the planar drawing of G; (**b**): the drawing of G with $\mathrm{cr}_D(G) = 1$.

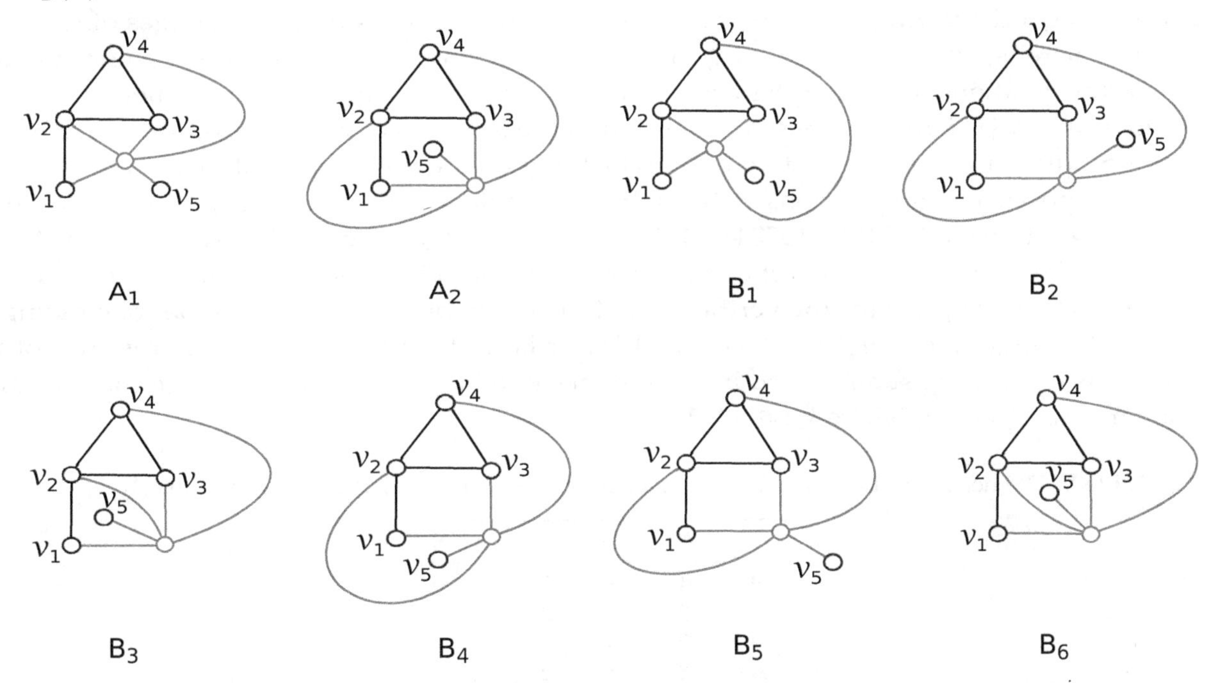

Figure 2. Drawings of eight possible configurations from $\mathcal{M}$ of the subgraph F^i.

The configurations A_1 and A_2 are represented by the cyclic permutations (15432) and (12435), respectively. Since the minimum number of interchanges of adjacent elements of (15432) required to produce cyclic permutation $\overline{(12435)} = (15342)$ is one, any subgraph T^j with the configuration A_2 of F^j crosses the edges of T^i at least once, i.e., $\mathrm{cr}(A_1, A_2) \geq 1$. Details have been worked out by Woodall [16]. The same reason gives $\mathrm{cr}(A_1, B_2) \geq 2$, $\mathrm{cr}(A_1, B_4) \geq 2$, $\mathrm{cr}(A_1, B_6) \geq 2$, $\mathrm{cr}(A_2, B_1) \geq 2$, $\mathrm{cr}(A_2, B_3) \geq 2$, $\mathrm{cr}(A_2, B_5) \geq 2$, $\mathrm{cr}(B_i, B_j) \geq 2$, and $\mathrm{cr}(A_i, B_j) \geq 3$ for $i \equiv j \pmod 2$. Moreover, by a discussion of possible subdrawings, we can verify that $\mathrm{cr}(B_1, B_5) \geq 4$, $\mathrm{cr}(B_3, B_5) \geq 4$, $\mathrm{cr}(B_2, B_6) \geq 4$, and $\mathrm{cr}(B_4, B_6) \geq 4$. Let F^i be the subgraph having the configuration B_5, and let T^j be a subgraph from R_D with $j \neq i$. Using Woodall's result $\mathrm{cr}_D(T^i, T^j) = Q(\mathrm{rot}_D(t_i), \mathrm{rot}_D(t_j)) + 2k$ for some nonnegative integer k, let us also suppose that $Q(\mathrm{rot}_D(t_i), \mathrm{rot}_D(t_j)) = 2$. Of course, any subgraph F^j having the configuration B_1 or B_3 satisfies the mentioned condition. One can easily see that if $t_j \in \omega_{1,2} \cup \omega_{3,4} \cup \omega_{1,2,3}$, then $\mathrm{cr}(T^i, T^j) > 2$. If $t_j \in \omega_{2,4,5}$ and $\mathrm{cr}(T^i, T^j) = 2$, then the subdrawing $D(F^j)$ induced by the edges incident with the vertices v_1 and v_3 crosses the edges of T^i exactly once, and once, respectively. Thus, $\mathrm{rot}_D(t_j) = (12435)$, i.e., the subgraph F^j has the configuration A_2. This forces $\mathrm{cr}(B_5, B_1) \geq 4$ and $\mathrm{cr}(B_5, B_3) \geq 4$. Similar arguments are applied for $\mathrm{cr}(B_6, B_2) \geq 4$ and $\mathrm{cr}(B_6, B_4) \geq 4$. Clearly, also $\mathrm{cr}(A_k, A_k) \geq 4$ and $\mathrm{cr}(B_l, B_l) \geq 4$ for any $k = 1, 2$ and $l = 1, \ldots, 6$. Thus, all lower bounds of the number of crossing of configurations from $\mathcal{M}$ are summarized in the symmetric Table 1 (here, X_k and Y_l are configurations of the subgraphs F^i and F^j, where k, l are integers from $\{1, 2\}$ or $\{1, \ldots, 6\}$, and $X, Y \in \{A, B\}$).

Table 1. The necessary number of crossings between T^i and T^j for the configurations X_k, Y_l.

—	A_1	A_2	B_1	B_2	B_3	B_4	B_5	B_6
A_1	4	1	3	2	3	2	3	2
A_2	1	4	2	3	2	3	2	3
B_1	3	2	4	3	2	3	4	3
B_2	2	3	3	4	3	2	3	4
B_3	3	2	2	3	4	3	4	3
B_4	2	3	3	2	3	4	3	4
B_5	3	2	4	3	4	3	4	3
B_6	2	3	3	4	3	4	3	4

Assume a good drawing D of the graph $G + D_n$ with one crossing among edges of the graph G (in which there is a subgraph $T^i \in R_D$). In this case, without loss of generality, we can choose also the vertex notations of the graph in such a way as shown in Figure 1b. Since there is only one subdrawing of $F^i \setminus \{v_5\}$ represented by the rotation (1324), we have four possibilities for how to obtain the subdrawing of F^i depending on in which region the vertex v_5 is placed. Thus, there are four different possible configurations of the subgraph F^i denoted as A_1, A_2, A_3, and A_4, with the corresponding rotations (13245), (13524), (13254), and (15324), respectively. We denote by $\mathcal{N}_D$ the subset of $\mathcal{N} = \{A_1, A_2, A_3, A_4\}$ consisting of all configurations that exist in the drawing D. The same way as above can be applied for the verification of the lower bounds of the number of crossings of two different configurations from $\mathcal{N}$. Thus, all lower bounds of the numbers of crossings of two configurations from $\mathcal{N}$ are summarized in the symmetric Table 2 (here, A_k and A_l are configurations of the subgraphs F^i and F^j, where $k, l \in \{1, 2, 3, 4\}$).

Table 2. The necessary number of crossings between T^i and T^j for the configurations A_k, A_l.

—	A_1	A_2	A_3	A_4
A_1	4	2	3	3
A_2	2	4	3	3
A_3	3	3	4	2
A_4	3	3	2	4

3. The Graph of Configurations $\mathcal{G}_D$

In general, the low possible number of crossings between two different subgraphs in a good subdrawing of $G + D_n$ is one of the main problems in the proofs on the crossing number of the join of the graph G with the discrete graphs D_n. The lower bounds of the numbers of crossings between two subgraphs, which do not cross the edges of G, were summarized in the symmetric Table 1. Since some configurations from the set $\mathcal{M}$ need not appear in the fixed drawing of $G + D_n$, we will first deal with the smallest possible values in Table 1 as with the worst possible case in the mentioned proofs. Thus, a new graphical representation of Table 1 by the graph of configurations will be useful.

Let us suppose that D is a good drawing of the graph $G + D_n$ with $\mathrm{cr}_D(G) = 0$, and let $\mathcal{M}_D$ be the nonempty set of all configurations that exist in the drawing D belonging to the set $\mathcal{M} = \{A_1, A_2, B_1, B_2, B_3, B_4, B_5, B_6\}$. A graph of configurations $\mathcal{G}_D$ is an ordered triple (V_D, E_D, w_D), where V_D is the set of vertices, E_D is the set of edges, which is formed by all unordered pairs of distinct vertices, and a weight function $w : E_D \to \mathbb{N}$ that associates with each edge of E_D an unordered pair of two vertices of V_D. The vertex $x_k \in V_D$ for some $x \in \{a, b\}$ if the corresponding configuration $X_k \in \mathcal{M}_D$ for some $X \in \{A, B\}$, where $k \in \{1, 2\}$ or $k \in \{1, \ldots, 6\}$. The edge $e = x_k y_l \in E_D$ if x_k and y_l are two different vertices of the graph $\mathcal{G}_D$. Finally, $w_D(e) = m \in \mathbb{N}$ for the edge $e = x_k y_l$, if m is the associated lower bound between two different configurations X_k, and Y_l in Table 1. Of course, $\mathcal{G}_D$ is the simple undirected edge-weighted graph uniquely determined by the drawing D. Moreover, if we define the graph $\mathcal{G} = (V, E, w)$ in the same way over the set $\mathcal{M}$, then $\mathcal{G}_D$ is the subgraph of $\mathcal{G}$

induced by V_D for the considered drawing D. Since the graph $\mathcal{G} = (V, E, w)$ can be represented like the edge-weighted complete graph K_8, it will be more transparent to follow the subcases in the proof of the main theorem; see Figure 3.

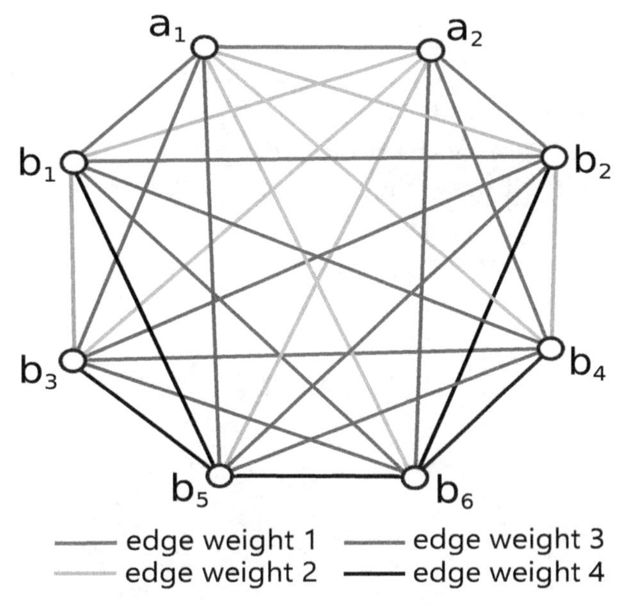

Figure 3. Representation of the lower bounds of Table 1 by the graph $\mathcal{G} = (V, E, w)$.

4. The Crossing Number of $G + D_n$

Two vertices t_i and t_j of $G + D_n$ are antipodal in a drawing of $G + D_n$ if the subgraphs T^i and T^j do not cross. A drawing is antipodal-free if it has no antipodal vertices. In the rest of the paper, each considered drawing of the graph $G + D_n$ will be assumed antipodal-free. In the proof of the main theorem, the following lemma related to some restricted subdrawings of the graph $G + D_n$ is helpful.

Lemma 1. *Let D be a good and antipodal-free drawing of $G + D_n$, $n > 2$. If T^i, $T^j \in R_D$ are different subgraphs such that F^i, F^j have different configurations from any of the sets $\{A_1, B_2\}$, $\{A_1, B_6\}$, $\{A_2, B_1\}$, and $\{A_2, B_5\}$, then:*

$$\mathrm{cr}_D(G \cup T^i \cup T^j, T^k) \geq 4 \qquad \textit{for any } T^k \in S_D.$$

Proof of Lemma 1. Let us suppose the configuration A_1 of the subgraph F^i, and note that it is exactly represented by $\mathrm{rot}_D(t_i) = (15432)$. The unique drawing of the subgraph F^i contains four regions with the vertex t_i on their boundaries (Figure 2). If there is a $T^k \in S_D$ with $\mathrm{cr}_D(T^i, T^k) = 1$, then one can easily see that $t_k \in \omega_{1,2,4,5}$. Of course, the edge $t_k v_3$ must cross one edge of the graph G. If $t_k v_3$ crosses the edge $v_1 v_2$, then the subgraph F^k is represented by $\mathrm{rot}_D(t_k) = (13245)$. If the edge $t_k v_3$ crosses the edge $v_2 v_4$, then there are only three possibilities for the considered subdrawing of F^k, i.e., the subgraph F^k can be represented by three possible cyclic permutations (13452), (15234), or (12354).

For the remaining configurations A_2, B_1, B_2, B_5, and B_6 of F^i, using the same arguments, one can easily verify that the rotations of the vertex t_k are from the sets $\{(15324), (12534), (13425), (13542)\}$, $\{(12345), (14235)\}$, $\{(15342), (15423)\}$, $\{(12345)\}$, and $\{(15342)\}$, respectively. This forces that there is no subgraph $T^k \in S_D$ with $\mathrm{cr}_D(T^i \cup T^j, T^k) = 2$, where the subgraph F^j has the configuration B_2 or B_6. The same reason is given for the case of A_2 with the configurations B_1 and B_5. Finally, $\mathrm{cr}_D(G \cup T^i \cup T^j, T^k) \geq 1 + 3 = 4$ for any $T^k \in S_D$. This completes the proof. □

We have to emphasize that we cannot generalize Lemma 1 for all pairs of different configurations from $\mathcal{M}$. Let us assume the configurations A_1 of F^i and B_4 of F^j. For $T^k \in S_D$, the reader can easily find a subdrawing of $G \cup T^i \cup T^j \cup T^k$ in which $\mathrm{cr}_D(T^i, T^k) = \mathrm{cr}_D(T^j, T^k) = 1$. The same remark holds for pairs A_2 with B_3, B_1 with B_3, and B_2 with B_4.

Theorem 1. $\mathrm{cr}(G + D_n) = 4\left\lfloor \frac{n}{2} \right\rfloor \left\lfloor \frac{n-1}{2} \right\rfloor + \left\lfloor \frac{n}{2} \right\rfloor$ *for* $n \geq 1$.

Proof of Theorem 1. The drawing in Figure 4b shows that $\mathrm{cr}(G + D_n) \leq 4\lfloor \frac{n}{2} \rfloor \lfloor \frac{n-1}{2} \rfloor + \lfloor \frac{n}{2} \rfloor$. We prove the reverse inequality by contradiction. The graph $G + D_1$ is planar; hence, $\mathrm{cr}(G + D_1) = 0$. Since the graph $G + D_2$ contains a subdivision of the complete bipartite graph $K_{3,3}$, we have $\mathrm{cr}(G + D_2) \geq 1$. Thus, $\mathrm{cr}(G + D_2) = 1$ by the good drawing of $G + D_2$ in Figure 4a. Suppose now that for $n \geq 3$, there is a drawing D with:

$$\mathrm{cr}_D(G + D_n) < 4\left\lfloor \frac{n}{2} \right\rfloor \left\lfloor \frac{n-1}{2} \right\rfloor + \left\lfloor \frac{n}{2} \right\rfloor, \tag{2}$$

and let

$$\mathrm{cr}(G + D_m) \geq 4\left\lfloor \frac{m}{2} \right\rfloor \left\lfloor \frac{m-1}{2} \right\rfloor + \left\lfloor \frac{m}{2} \right\rfloor \quad \text{for any integer } m < n. \tag{3}$$

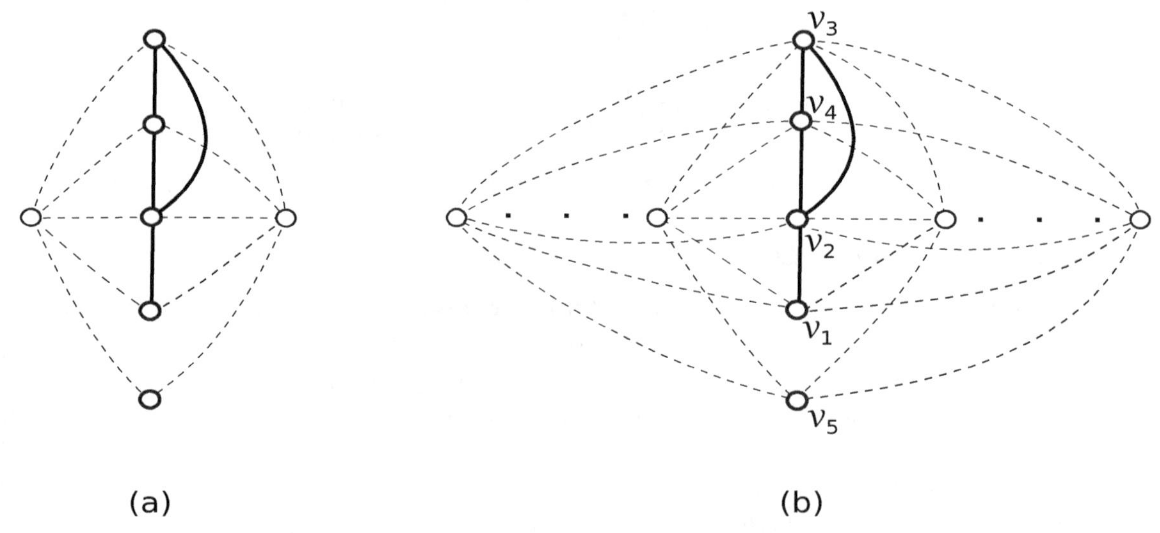

Figure 4. The good drawings of $G + D_2$ and of $G + D_n$. (**a**): the drawing of $G + D_2$ with one crossing; (**b**): the drawing of $G + D_n$ with $4\left\lfloor \frac{n}{2} \right\rfloor \left\lfloor \frac{n-1}{2} \right\rfloor + \left\lfloor \frac{n}{2} \right\rfloor$ crossings.

Let us first show that the considered drawing D must be antipodal-free. As a contradiction, suppose that, without loss of generality, $\mathrm{cr}_D(T^n, T^{n-1}) = 0$. Using positive values in Tables 1 and 2, one can easily verify that both subgraphs T^n and T^{n-1} are not from the set R_D, i.e., $\mathrm{cr}_D(G, T^n \cup T^{n-1}) \geq 1$. The known fact that $\mathrm{cr}(K_{5,3}) = 4$ implies that any T^k, $k = 1, \ldots, n-2$, crosses the edges of the subgraph $T^n \cup T^{n-1}$ at least four times. Therefore, for the number of crossings in the considered drawing D, we have:

$$\mathrm{cr}_D(G + D_n) = \mathrm{cr}_D\,(G + D_{n-2}) + \mathrm{cr}_D(G, T^n \cup T^{n-1}) + \mathrm{cr}_D(T^n \cup T^{n-1}) + \mathrm{cr}_D(K_{5,n-2}, T^n \cup T^{n-1})$$

$$\geq 4\left\lfloor \frac{n-2}{2} \right\rfloor \left\lfloor \frac{n-3}{2} \right\rfloor + \left\lfloor \frac{n-2}{2} \right\rfloor + 1 + 0 + 4(n-2) = 4\left\lfloor \frac{n}{2} \right\rfloor \left\lfloor \frac{n-1}{2} \right\rfloor + \left\lfloor \frac{n}{2} \right\rfloor.$$

This contradiction with the assumption (2) confirms that D must be an antipodal-free drawing. Moreover, if $r = |R_D|$ and $s = |S_D|$, the assumption (3) together with the well-known fact $\mathrm{cr}(K_{5,n}) = 4\left\lfloor \frac{n}{2} \right\rfloor \left\lfloor \frac{n-1}{2} \right\rfloor$ imply that in D, there are at least $\lceil \frac{n}{2} \rceil + 1$ subgraphs T^i, which do not cross the edges of G. More precisely:

$$\mathrm{cr}_D(G) + \mathrm{cr}_D(G, K_{5,n}) \leq \mathrm{cr}_D(G) + 0r + 1s + 2(n - r - s) < \left\lfloor \frac{n}{2} \right\rfloor,$$

i.e.,

$$s + 2(n - r - s) < \left\lfloor \frac{n}{2} \right\rfloor. \tag{4}$$

This forces that $r \geq 2$, and $r \geq \lceil \frac{n}{2} \rceil + 1$. Now, for $T^i \in R_D$, we will discuss the existence of possible configurations of subgraphs $F^i = G \cup T^i$ in the drawing D.

Case 1. $\mathrm{cr}_D(G) = 0$. Without loss of generality, we can choose the vertex notation of the graph G in such a way as shown in Figure 1a. Thus, we will deal with the configurations belonging to the nonempty set $\mathcal{M}_D$. According to the minimum value of the weights of edges in the graph $\mathcal{G}_D = (V_D, E_D, w_D)$, we will fix one, or two, or three subgraphs with a contradiction with the condition (2) in the following subcases:

i. $\{A_1, A_2\} \subseteq \mathcal{M}_D$, i.e., $w_D(a_1a_2) = 1$. Without loss of generality, let us consider two different subgraphs T^n, $T^{n-1} \in R_D$ such that F^n and F^{n-1} have configurations A_1 and A_2, respectively. Then, $\mathrm{cr}_D(G \cup T^n \cup T^{n-1}, T^i) \geq 5$ for any $T^i \in R_D$ with $i \neq n-1, n$ by summing the values in all columns in the considered two rows of Table 1. Moreover, $\mathrm{cr}_D(T^n \cup T^{n-1}, T^i) \geq 3$ for any subgraph T^i with $i \neq n-1, n$ due to the properties of the cyclic permutations. Hence, by fixing the graph $G \cup T^n \cup T^{n-1}$,

$$\mathrm{cr}_D(G + D_n) \geq 4\left\lfloor \frac{n-2}{2} \right\rfloor \left\lfloor \frac{n-3}{2} \right\rfloor + 5(r-2) + 4(n-r) + 1 = 4\left\lfloor \frac{n-2}{2} \right\rfloor \left\lfloor \frac{n-3}{2} \right\rfloor + 4n + r - 9$$

$$\geq 4\left\lfloor \frac{n-2}{2} \right\rfloor \left\lfloor \frac{n-3}{2} \right\rfloor + 4n + \left(\left\lceil \frac{n}{2} \right\rceil + 1 \right) - 9 \geq 4\left\lfloor \frac{n}{2} \right\rfloor \left\lfloor \frac{n-1}{2} \right\rfloor + \left\lfloor \frac{n}{2} \right\rfloor.$$

ii. $\{A_1, A_2\} \not\subseteq \mathcal{M}_D$, i.e., $w_D(e) \geq 2$ for any $e \in E_D$.

Let us assume that $\{A_1, B_2, B_4\} \subseteq \mathcal{M}_D$ or $\{A_2, B_1, B_3\} \subseteq \mathcal{M}_D$, i.e., there is a three-cycle in the graph $\mathcal{G}_D$ with weights of two of all its edges. Without loss of generality, let us consider three different subgraphs T^n, T^{n-1} $T^{n-2} \in R_D$ such that F^n, F^{n-1}m and F^{n-2} have different configurations from $\{A_1, B_2, B_4\}$. Then, $\mathrm{cr}_D(G \cup T^n \cup T^{n-1} \cup T^{n-2}, T^i) \geq 8$ for any $T^i \in R_D$ with $i \neq n-1, n$ by Table 1, and $\mathrm{cr}_D(G \cup T^n \cup T^{n-1} \cup T^{n-2}, T^i) \geq 5$ for any subgraph $T^i \in S_D$ by Lemma 1. Thus, by fixing the graph $G \cup T^n \cup T^{n-1} \cup T^{n-2}$,

$$\mathrm{cr}_D(G + D_n) \geq 4\left\lfloor \frac{n-3}{2} \right\rfloor \left\lfloor \frac{n-4}{2} \right\rfloor + 8(r-3) + 5(n-r) + 6 \geq 4\left\lfloor \frac{n-3}{2} \right\rfloor \left\lfloor \frac{n-4}{2} \right\rfloor + 5n + 3r - 18$$

$$\geq 4\left\lfloor \frac{n-3}{2} \right\rfloor \left\lfloor \frac{n-4}{2} \right\rfloor + 5n + 3\left(\left\lceil \frac{n}{2} \right\rceil + 1 \right) - 18 \geq 4\left\lfloor \frac{n}{2} \right\rfloor \left\lfloor \frac{n-1}{2} \right\rfloor + \left\lfloor \frac{n}{2} \right\rfloor.$$

In the next part, let us suppose that $\{A_1, B_2, B_4\} \not\subseteq \mathcal{M}_D$ and $\{A_2, B_1, B_3\} \not\subseteq \mathcal{M}_D$,

(1) $\{A_j, B_k\} \subseteq \mathcal{M}_D$ for some $k \equiv j + 1 \pmod 2$ or $\{B_j, B_{j+2}\} \subseteq \mathcal{M}_D$, where $j \in \{1, 2\}$. Without loss of generality, let us consider two different subgraphs T^n, $T^{n-1} \in R_D$ such that F^n and F^{n-1} have configurations A_1 and B_2, respectively. Then, $\mathrm{cr}_D(G \cup T^n \cup T^{n-1}, T^i) \geq 6$ for any $T^i \in R_D$ with $i \neq n-1, n$ by Table 1. Moreover, $\mathrm{cr}_D(T^n \cup T^{n-1}, T^i) \geq 2$ for any subgraph T^i with $i \neq n-1, n$ due to properties of the cyclic permutations. Hence, if we fix the graph $G \cup T^n \cup T^{n-1}$,

$$\mathrm{cr}_D(G + D_n) \geq 4\left\lfloor \frac{n-2}{2} \right\rfloor \left\lfloor \frac{n-3}{2} \right\rfloor + 6(r-2) + 3s + 4(n-r-s) + 2 = 4\left\lfloor \frac{n-2}{2} \right\rfloor \left\lfloor \frac{n-3}{2} \right\rfloor$$

$$+4n + r + r - s - 10 \geq 4\left\lfloor \frac{n-2}{2} \right\rfloor \left\lfloor \frac{n-3}{2} \right\rfloor + 4n + \left\lceil \frac{n}{2} \right\rceil + 1 + 1 - 10 \geq 4\left\lfloor \frac{n}{2} \right\rfloor \left\lfloor \frac{n-1}{2} \right\rfloor + \left\lfloor \frac{n}{2} \right\rfloor.$$

(2) $\{A_j, B_k\} \not\subseteq \mathcal{M}_D$ for any $k \equiv j+1 \pmod 2$ and $\{B_j, B_{j+2}\} \not\subseteq \mathcal{M}_D$, where $j = 1, 2$, i.e., $w_D(e) \geq 3$ for any $e \in E_D$. Without loss of generality, we can assume that $T^n \in R_D$. Then, $\mathrm{cr}_D(T^n, T^i) \geq 3$ for any $T^i \in R_D$ with $i \neq n$. Thus, by fixing the graph $G \cup T^n$,

$$\mathrm{cr}_D(G+D_n) \geq 4\left\lfloor\frac{n-1}{2}\right\rfloor\left\lfloor\frac{n-2}{2}\right\rfloor + 3(r-1) + 2(n-r) + 0 = 4\left\lfloor\frac{n-1}{2}\right\rfloor\left\lfloor\frac{n-2}{2}\right\rfloor + 2n + r - 3$$

$$\geq 4\left\lfloor\frac{n-1}{2}\right\rfloor\left\lfloor\frac{n-2}{2}\right\rfloor + 2n + \left(\left\lceil\frac{n}{2}\right\rceil + 1\right) - 3 \geq 4\left\lfloor\frac{n}{2}\right\rfloor\left\lfloor\frac{n-1}{2}\right\rfloor + \left\lfloor\frac{n}{2}\right\rfloor.$$

Case 2. $\mathrm{cr}_D(G) = 1$. Without loss of generality, we can choose the vertex notation of the graph G in such a way as shown in Figure 1b. Thus, we will deal with the configurations belonging to the nonempty set $\mathcal{N}_D$ in the following two cases:

i. $\{A_i, A_{i+1}\} \subseteq \mathcal{N}_D$ for some $i \in \{1, 2\}$. Without loss of generality, let us consider two different subgraphs T^n, $T^{n-1} \in R_D$ such that F^n and F^{n-1} have different configurations from the set $\{A_1, A_2\}$. Then, $\mathrm{cr}_D(G \cup T^n \cup T^{n-1}, T^i) \geq 6$ for any $T^i \in R_D$ with $i \neq n-1, n$ by Table 2. Moreover, $\mathrm{cr}_D(T^n \cup T^{n-1}, T^i) \geq 2$ for any subgraph T^i with $i \neq n-1, n$ due to the properties of the cyclic permutations. Hence, by fixing the graph $G \cup T^n \cup T^{n-1}$,

$$\mathrm{cr}_D(G+D_n) \geq 4\left\lfloor\frac{n-2}{2}\right\rfloor\left\lfloor\frac{n-3}{2}\right\rfloor + 6(r-2) + 3s + 4(n-r-s) + 2 + 1 = 4\left\lfloor\frac{n-2}{2}\right\rfloor\left\lfloor\frac{n-3}{2}\right\rfloor$$

$$+4n + r + r - s - 9 \geq 4\left\lfloor\frac{n-2}{2}\right\rfloor\left\lfloor\frac{n-3}{2}\right\rfloor + 4n + \left\lceil\frac{n}{2}\right\rceil + 1 + 1 - 9 \geq 4\left\lfloor\frac{n}{2}\right\rfloor\left\lfloor\frac{n-1}{2}\right\rfloor + \left\lfloor\frac{n}{2}\right\rfloor.$$

If F^n and F^{n-1} have different configurations from the set $\{A_3, A_4\}$, then the same argument can be applied.

ii. $\{A_i, A_{i+1}\} \not\subseteq \mathcal{N}_D$ for any $i = 1, 2$. Without loss of generality, we can assume that $T^n \in R_D$. Then, $\mathrm{cr}_D(T^n, T^i) \geq 3$ for any $T^i \in R_D$ with $i \neq n$. Thus, by fixing the graph $G \cup T^n$,

$$\mathrm{cr}_D(G+D_n) \geq 4\left\lfloor\frac{n-1}{2}\right\rfloor\left\lfloor\frac{n-2}{2}\right\rfloor + 3(r-1) + 2(n-r) + 1 = 4\left\lfloor\frac{n-1}{2}\right\rfloor\left\lfloor\frac{n-2}{2}\right\rfloor + 2n + r - 2$$

$$\geq 4\left\lfloor\frac{n-1}{2}\right\rfloor\left\lfloor\frac{n-2}{2}\right\rfloor + 2n + \left(\left\lceil\frac{n}{2}\right\rceil + 1\right) - 2 \geq 4\left\lfloor\frac{n}{2}\right\rfloor\left\lfloor\frac{n-1}{2}\right\rfloor + \left\lfloor\frac{n}{2}\right\rfloor.$$

Thus, it was shown that there is no good drawing D of the graph $G + D_n$ with less than $4\lfloor\frac{n}{2}\rfloor\lfloor\frac{n-1}{2}\rfloor + \lfloor\frac{n}{2}\rfloor$ crossings. This completes the proof of Theorem 1. □

5. Three Other Graphs

Finally, in Figure 4b, we are able to add the edges v_3v_5 and v_1v_5 to the graph G without additional crossings, and we obtain three new graphs H_i for $i = 1, 2, 3$ in Figure 5. Therefore, the drawing of the graphs $H_1 + D_n$, $H_2 + D_n$, and $H_3 + D_n$ with $4\lfloor\frac{n}{2}\rfloor\lfloor\frac{n-1}{2}\rfloor + \lfloor\frac{n}{2}\rfloor$ crossings is obtained. On the other hand, $G + D_n$ is a subgraph of each $H_i + D_n$, and therefore, $\mathrm{cr}(H_i + D_n) \geq \mathrm{cr}(G + D_n)$ for any $i = 1, 2, 3$. Thus, the next results are obvious.

Corollary 1. $\mathrm{cr}(H_i + D_n) = 4\left\lfloor\frac{n}{2}\right\rfloor\left\lfloor\frac{n-1}{2}\right\rfloor + \left\lfloor\frac{n}{2}\right\rfloor$ *for* $n \geq 1$, *where* $i = 1, 2, 3$.

We remark that the crossing numbers of the graphs $H_1 + D_n$ and $H_3 + D_n$ were already obtained by Berežný and Staš [4], and Klešč and Schrötter [7], respectively. Moreover, into the drawing in Figure 4b, it is possible to add n edges, which form the path P_n, $n \geq 2$ on the vertices of D_n without another crossing. Thus, the next results are also obvious.

Theorem 2. $\mathrm{cr}(G + P_n) = \mathrm{cr}(H_2 + P_n) = 4\left\lfloor\frac{n}{2}\right\rfloor\left\lfloor\frac{n-1}{2}\right\rfloor + \left\lfloor\frac{n}{2}\right\rfloor$ *for* $n \geq 2$.

The crossing number of the graph $H_1 + P_n$ has been investigated in [12].

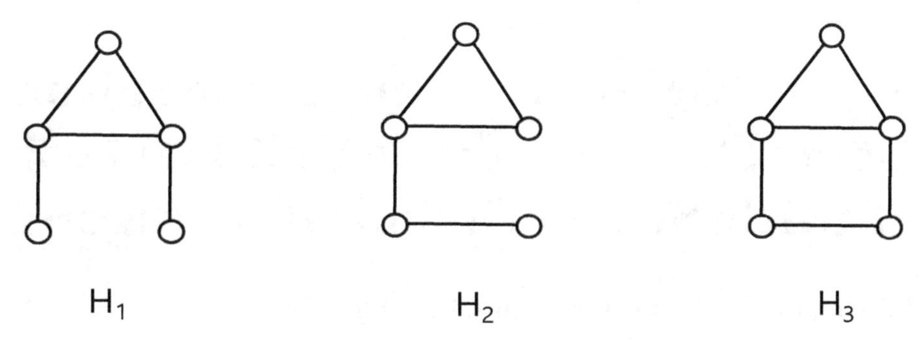

Figure 5. Three graphs H_1, H_2, and H_3 by adding new edges to the graph G.

Acknowledgments: This work was supported by the internal faculty research Project No. FEI-2017-39.

References

1. Klešč, M. The join of graphs and crossing numbers. *Electron. Notes Discret. Math.* **2007**, *28*, 349–355. [CrossRef]
2. Kleitman, D.J. The crossing number of $K_{5,n}$. *J. Comb. Theory* **1970**, *9*, 315–323. [CrossRef]
3. Klešč, M.; Schrötter, Š. The crossing numbers of join products of paths with graphs of order four. *Discuss. Math. Graph Theory* **2011**, *31*, 312–331. [CrossRef]
4. Berežný, Š.; Staš, M. On the crossing number of the join of five vertex graph G with the discrete graph D_n. *Acta Electrotech. Inform.* **2017**, *17*, 27–32. [CrossRef]
5. Klešč, M. The crossing numbers of join of the special graph on six vertices with path and cycle. *Discret. Math.* **2010**, *310*, 1475–1481. [CrossRef]
6. Klešč, M.; Petrillová, J.; Valo, M. On the crossing numbers of Cartesian products of wheels and trees. *Discuss. Math. Graph Theory* **2017**, *37*, 339–413. [CrossRef]
7. Klešč, M.; Schrötter, Š. The crossing numbers of join of paths and cycles with two graphs of order five. In *Lecture Notes in Computer Science: Mathematical Modeling and Computational Science*; Springer: Berlin/Heidelberg, Germany, 2012; Volume 7125, pp. 160–167.
8. Staš, M. On the crossing number of the join of the discrete graph with one graph of order five. *Math. Model. Geom.* **2017**, *5*, 12–19. [CrossRef]
9. Staš, M. Cyclic permutations: Crossing numbers of the join products of graphs. In Proceedings of the Aplimat 2018: 17th Conference on Applied Mathematics, Bratislava, Slovak, 6–8 February 2018; pp. 979–987.
10. Staš, M. Determining crossing numbers of graphs of order six using cyclic permutations. *Bull. Aust. Math. Soc.* **2018**, *98*, 353–362. [CrossRef]
11. Klešč, M.; Valo, M. Minimum crossings in join of graphs with paths and cycles. *Acta Electrotech. Inform.* **2012**, *12*, 32–37. [CrossRef]
12. Staš, M.; Petrillová, J. On the join products of two special graphs on five vertices with the path and the cycle. *Math. Model. Geom.* **2018**, *6*, 1–11.
13. Hernández-Vélez, C.; Medina, C.; Salazar, G. The optimal drawing of $K_{5,n}$. *Electron. J. Comb.* **2014**, *21*, 29.
14. Berežný, Š.; Staš, M. Cyclic permutations and crossing numbers of join products of symmetric graph of order six. *Carpathian J. Math.* **2018**, *34*, 143–155.
15. Berežný, Š.; Buša, J., Jr.; Staš, M. Software solution of the algorithm of the cyclic-order graph. *Acta Electrotech. Inform.* **2018**, *18*, 3–10. [CrossRef]
16. Woodall, D.R. Cyclic-order graphs and Zarankiewicz's crossing number conjecture. *J. Graph Theory* **1993**, *17*, 657–671. [CrossRef]

Optimizing the High-Level Maintenance Planning Problem of the Electric Multiple Unit Train using a Modified Particle Swarm Optimization Algorithm

Jianping Wu [1], Boliang Lin [1,*], Hui Wang [2], Xuhui Zhang [3], Zhongkai Wang [2] and Jiaxi Wang [1]

1 School of Traffic and Transportation, Beijing Jiaotong University, Beijing 100044, China; jianpingwu@bjtu.edu.cn (J.W.); wangjiaxi@bjtu.edu.cn (J.W.)
2 Institute of Computing Technology, China Academy of Railway Sciences, Beijing 100081, China; wanghui2215@163.com (H.W.); winter-light@163.com (Z.W.)
3 Department of Vehicles, China Railway Shanghai Bureau Group Co., Ltd., Shanghai 200071, China; xhzhang121@163.com
* Correspondence: bllin@bjtu.edu.cn.

Abstract: Electric multiple unit (EMU) trains' high-level maintenance planning is a discrete problem in mathematics. The high-level maintenance process of the EMU trains consumes plenty of time. When the process is undertaken during peak periods of the passenger flow, the transportation demand may not be fully satisfied due to the insufficient supply of trains. In contrast, if the process is undergone in advance, extra costs will be incurred. Based on the practical requirements of high-level maintenance, a 0–1 programming model is proposed. To simplify the description of the model, candidate sets of delivery dates, i.e., time windows, are generated according to the historical data and maintenance regulations. The constraints of the model include maintenance regulations, the passenger transportation demand, and capacities of workshop. The objective function is to minimize the mileage losses of all EMU trains. Moreover, a modified particle swarm algorithm is developed for solving the problem. Finally, a real-world case study of Shanghai Railway is conducted to demonstrate the proposed method. Computational results indicate that the (approximate) optimal solution can be obtained successfully by our method and the proposed method significantly reduces the solution time to 500 s.

Keywords: Electric multiple unit trains; high-level maintenance planning; time window; 0–1 programming model; particle swarm algorithm

1. Introduction

In China, high-speed railway has become a priority option for the long trip due to its convenience and comfortableness, and it account for 60% of rail total passenger traffic. It has long been a difficult problem for the China Railway (CR) that supply enough available Electric Multiple Unit (EMU) trains in tourist rush seasons to fulfill heavy transportation tasks. The EMU trains are the unique vehicles running on the China high-speed railway. In addition, they have a high purchase cost and complicated maintenance regulations. Therefore, the problem becomes much worse.

The gigh-level maintenance planning (HMP) is an important part of the operation and maintenance management for the EMU trains, which covers the third-level maintenance, the fourth-level maintenance and the fifth-level maintenance. Due to the complex regular preventive maintenance, uneven distribution of passenger flow, limited maintenance capacity of workshop and several weeks for maintenance service time, the EMU trains' HMP needs to be scheduled in advance, of which the planning horizon lasts for a natural year or more. The HMP's aim is to provide enough available EMU trains for the long peak period of the passenger flow (such as the Spring Festival and the summer holiday) under the conditions of limited maintenance resources and the regular preventive maintenance policy. The HMP is a prerequisite for the EMU operational plan, the second-level maintenance plan, the job shop scheduling at workshop, and the outsourcing plan for high-level maintenance workloads (see Figure 1).

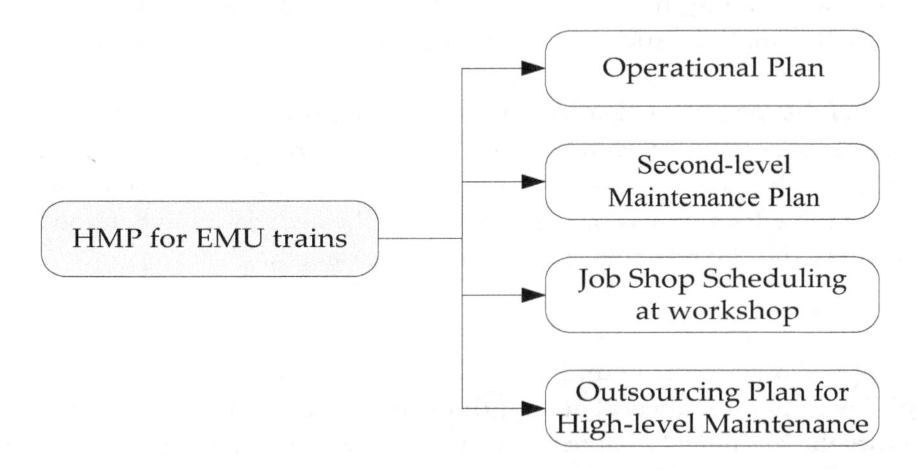

Figure 1. The position of high-level maintenance plan (HMP) in the operation and maintenance management system for the electric multiple unit (EMU) trains.

In the CR system, according to the maintenance regulations [1], the EMU trains that have been put into operation in the first period will undergo the high-level maintenance (HM) procedures together in the next few years. When lots of EMU trains are sent to workshop in tourist rush season, it will lead to the travel needs of passengers cannot be met. Not only the operating income but also the traveler satisfaction level on high-speed railway transportation will decrease. Meanwhile, it will increase the maintenance costs when the HM procedures are undergone in advance. Due to the limited capacities of workshops, the maintenance service time will be prolonged when plenty of EMU trains undergo the maintenance procedures together. Therefore, to scientifically formulate the HMP that meets the travel demands and reduces maintenance cost as much as possible is a complicated combinatorial optimization problem that needs to be solved in the field of high-speed railway operation.

For the maintenance system, Stuchly et al. [2] and Rezvanizaniani et al. [3] developed a condition based maintenance management system, Shimada [4] introduced the accident prevention maintenance system, and Cheng and Tsao [5] proposed a preventive and corrective maintenances system. These maintenance systems can reduce the maintenance costs and improve the utilization efficiency of EMU train.

Many experts and scholars study the first-level and the second-level maintenances. Maróti and Kroon [6,7] proposed that adjust the operation schedule ahead of time to ensure the maintenance procedures can be undertook in time. Tsuji et al. [8] developed a novel approach based on ant colony optimization and Wang et al. [9] designed an algorithm based column generation to solve the EMU train maintenance plan problem. Giacco et al. [10] integrated the rolling stock circulation problem and short-term maintenance planning, and a mixed-integer linear-programming formulation was proposed.

In other fields, scheduled maintenance planning problems also have been researched. Ziarati et al. [11], Lingaya et al. [12], and Wang et al. [13] studied the locomotive operation and maintenance. Moudania and Félix [14] and Mehmet and Bilge [15] developed the aircraft operation and maintenance. Budai et al. [16] researched the long-term planning of railway maintenance works. In addition, Grigoriev et al. [17] tried to find the length of maintenance plan to minimize the total operation costs.

There are a few relevant literatures available for the EMU trains HMP problem. Lin et al. [18] designed a state function to show the state of EMU train on each day during the planning horizon, and a non-linear 0–1 programming model and its solution strategy was proposed. Wu et al. [19] proposed a time-state network to optimize the EMU trains HMP problem. Li et al. [20] presented a forecast method to estimate the maintenance quantity of the EMU trains in future.

Compared to the existing researches, a 0–1 programming model and solution strategy were proposed. In the mathematic model, all necessary regulations and practical constraints were considered.

The remainder of this paper is organized as follows. The problem description of HMP in the CR system is presented in Section 2. In Section 3, a 0–1 programming model is proposed, and then the solution algorithm on the basis of particle swarm algorithm is designed in Section 4. An empirical case is provided to verify the effectiveness of the model and the algorithm in Section 5. The last section gives some conclusions and the possible areas of further research.

2. HMP Problem at CR

The HMP, a typical discrete system, is a tactical plan that determines when the EMU trains to undergo the high-level maintenance. The length of the planning horizon lasts for about one year. Each train undergoes the high-level maintenance at most once during the planning horizon because of the interval between two adjacent HM processes is longer than the span of planning horizon; the order of the maintenance level for a train is the third-level, the fourth-level, the third-level, the fifth level, the third-level and so on until they are scrapped [1]. Therefore, the level of the high-level maintenance can be deduced beforehand according to the records and regulations of maintenance. According to the maintenance regulations, each train has a time window during which the train can be delivered to workshop on any day. The lower bound of the time window is the earliest date on which a train can be delivered to workshop while the upper bound is the latest date. The detailed problem descriptions can be referred to Lin et al. [18] and Wu et al. [19]. Here, we focus on the generation steps of the time window.

The estimated time of arrival (eta) of the HM can be easily calculated [20]. In this process, the average daily operating mileage is used to describe the daily usage of EMU trains before HM procedures. The notations used in the generation process of the time window are listed in Table 1.

Table 1. Notations used in the generation process of the time window.

Notations	Definition
e	The index of EMU trains;
E	The set of all EMU trains;
E'	An empty set;
eta_e	The eta of the HM for the EMU train e;
$e(m)$	The type of the EMU train e;
$e(g)$	The maintenance level of the EMU train e;
R^+_{mg}	The maximum value of the difference between the actual operating mileage before the HM and the eta_e for the train of which the type is m and maintenance level is g;
R^-_{mg}	The maximum value of the difference between the eta_e and the actual operating mileage before the HM for the train of which the type is m and the maintenance level is g;
l_e	The average daily operating mileage for the EMU train e before HM.

The time window of the start time for the HM can be generated as follows.

Step 1. Take an EMU train e from E, and calculate the *eta* of HM [20];

Step 2. According to $e(m)$ and $e(g)$, determine the offset range of the operating mileage for the EMU train e: $[-R^-_{e(m),e(g)}, R^+_{e(m),e(g)}]$;

Step 3. Calculate the offset range of the time window for the EMU train e: $[-R^-_{e(m),e(g)}/l_e, R^+_{e(m),e(g)}/l_e]$;

Step 4. Determine the time window of the EMU train e: $[eta_e - R^-_{e(m),e(g)}/l_e, eta_e + R^+_{e(m),e(g)}/l_e]$, and set $E = E - \{e\}$, $E' = E' + \{e\}$;

Step 5. If $E = \phi$, turn to Step 6, otherwise turn to Step 1;

Step 6. Set $E = E'$, over.

The EMU train's time window is continuous if we set the "day" as the minimum time unit. In this way, the time window can be presented by a time interval.

In addition, the HMP aims to ensure that there are enough well-conditioned trains to meet the passenger transport demand. We set a maximum HM rate to guarantee it. The HM rate is the ratio of the number of trains in HM state to the fleet size.

3. Mathematical Model of the HMP Problem

In this section, we propose a 0–1 programming model for the HMP problem. The constraints of the model include the maintenance interval, the passenger transportation demand, and the capacity of workshop. The objective function is to minimize the mileage losses of all EMU trains.

3.1. Notations

The all notations that used in the model are listed in Table 2.

Table 2. Notations used in the model.

Notations	Definition
Indices	
e	The index of EMU trains;
m	The index of types for the EMU trains;
t	The index of dates during the planning horizon;
g	The index of the maintenance level;
Set	
E	The set of all EMU trains; $e \in E$
M	The set of all types; $m \in M$
T	The set of all dates during the planning horizon; $t \in T$
G	The set of all maintenance levels; $g \in G$
Input parameters	
$e(g)$	The maintenance level of the EMU train e;
$e(m)$	The type of the EMU train e;
c	The unit penalty fee for the unused mileage before the HM;
eta_e	The eta for the EMU train e;
l_e	The average daily operating mileage for the EMU train e before the HM;
WS_e	The first date of the time window for the EMU train e, $WS_e = eta_e - R^-_{e(m),e(g)}/l_e$ (See Section 2);
WE_e	The last date of the time window for the EMU train e, $WE_e = eta_e + R^+_{e(m),e(g)}/l_e$ (See Section 2);
α_m	The conversion coefficient for train of which the type is m, indicates whether the train includes sixteen cars or not, if yes, then $\alpha_m = 2$; otherwise, $\alpha_m = 1$;
θ_t	The maximum HM rate on the t-th day;
Inv	The fleet size (the standard set);
d^t_m	The maximum number of an EMU train in the HM state for the m type on the t-th day;
b_g	The maximum number of an EMU train in the g-th level maintenance state;
N_g	The maximum number of an acceptable EMU train in the g-th level maintenance state at the same time;
Q	A sufficiently large positive number;
H^g_m	The maintenance service time for an EMU train with the m-th type and the g-th level;
J^g_m	The minimum interval time for delivering another train after an EMU train with the m-th type and the g-th level enters the workshop;
$\|T\|$	The length of the planning horizon;
$\|E\|$	The number of the EMU trains which need to be maintained during the planning horizon;

Table 2. *Cont.*

Notations	Definition
Decision Variables	
x_e^t	Binary variable, indicates whether the EMU train e selects the t-th day to start the HM procedures during the planning horizon, $x_e^t = 1$ if yes, $x_e^t = 0$ otherwise;
y_e^t	Binary variable, indicates whether the EMU train e in the $e(g)$-th level maintenance state on the t-th day during the planning horizon, $y_e^t = 1$ if yes, $y_e^t = 0$ otherwise;
z_e^t	Binary variable, indicate whether the EMU train e in the $e(g)$-th level delivery interval on the t-th day during the planning horizon, $z_e^t = 1$ if yes, $z_e^t = 0$ otherwise;

3.2. Optimization Objective

The objective function of the mathematical model is to maximize the service efficiency of the EMU trains, i.e., to minimize the unutilized mileage. In this way, the objective function can be presented as follows.

$$\min Z = c \sum_{t\in[WS_e,WE_e]} \sum_{e\in E} (WE_e - t) l_e x_e^t \tag{1}$$

3.3. Constraints Analysis

According to Section 2, each train e can choose one and only one delivery date during the time window. This is the uniqueness constraint.

$$\sum_{t\in[WS_e,WE_e]} x_e^t = 1 \quad \forall\, e \in E \tag{2}$$

Any time t out of the time window for the train e cannot be selected.

$$x_e^t = 0 \quad \forall\, e \in E, t \notin [WS_e, WE_e] \text{ and } t \in T \tag{3}$$

The θ_t is a variable according to travel demand which should be guaranteed. Theoretically, the value of θ_t is different for each day during the planning horizon. To describe this requirement, a set of constraints established as follows.

$$\sum_{e\in E} \alpha_{e(m)} y_e^t \leq \theta_t \cdot Inv \qquad \forall\, t \in T \tag{4}$$

Because of the various itineraries, the number of the trains with the specific type in the HM state must be less than the given threshold value on the t-th day. Constraints in this respect can be expressed as follows.

$$\sum_{e\in E|e(m)=m} y_e^t \leq d_m^t \qquad \forall m \in M,\ t \in T \tag{5}$$

The number of trains in each level of the HM state should not be exceeded the capacity of workshops. A set of constraints can be listed as follows.

$$\sum_{e\in E|e(g)=g} y_e^t \leq b_g \qquad \forall\, g \in G,\ t \in T \tag{6}$$

Meanwhile, restricted by the limited resources, only a few trains are permitted to enter the workshop over several days. This situation can be described in the form of mathematical inequalities as follows.

$$\sum_{e\in E|e(g)=g} z_e^t \leq N_g \qquad \forall\, g \in G,\ t \in T \tag{7}$$

In addition, the logical relationships between those three sets of decision variables are presented as follows.

$$(x_e^k - 1)Q \leq (\sum_{t=k}^{k+H_{e(m)}^{e(g)}-1} y_e^t - H_{e(m)}^{e(g)}) \leq (1 - x_e^k)Q \qquad k \in [WS_e, WE_e], \forall e \in E \tag{8}$$

$$\sum_{t=1}^{|T|+H_{e(m)}^{e(g)}} y_e^t = H_{e(m)}^{e(g)} \qquad \forall\, e \in E \tag{9}$$

$$(x_e^k - 1)Q \leq (\sum_{t=k}^{k+J_{e(m)}^{e(g)}-1} z_e^t - J_{e(m)}^{e(g)}) \leq (1 - x_e^k)Q \qquad k \in [WS_e, WE_e], \forall e \in E \tag{10}$$

$$\sum_{t=1}^{|T|+J_{e(m)}^{e(g)}} z_e^t = J_{e(m)}^{e(g)} \qquad \forall\, e \in E \tag{11}$$

Finally, all of the decision variables are binary variables.

$$x_e^t,\ y_e^t,\ z_e^t \in \{0,1\} \qquad \forall\, e \in E, t \in T \tag{12}$$

3.4. Model Construction

On the basis of above analysis, a 0–1 programming model for the EMU train HMP problem is proposed as follows.

$$\text{HMP model: } \min Z = c \sum_{t \in [WS_e, WE_e]} \sum_{e \in E} (WE_e - t) l_e x_e^t$$

$$\text{s.t.} \quad \sum_{t \in [WS_e, WE_e]} x_e^t = 1 \qquad \forall\, e \in E$$

$$x_e^t = 0 \qquad \forall\, e \in E, t \notin [\mathrm{WS}_e, WE_e] \text{ and } t \in T$$

$$\sum_{e \in E} \alpha_{e(m)} y_e^t \leq \theta_t \cdot Inv \qquad \forall\, t \in T$$

$$\sum_{e \in E | e(m) = m} y_e^t \leq d_m^t \qquad \forall m \in M,\ t \in T$$

$$\sum_{e \in E | e(g) = g} y_e^t \leq b_g \qquad \forall\, g \in G,\ t \in T$$

$$\sum_{e \in E | e(g) = g} z_e^t \leq N_g \qquad \forall\, g \in G,\ t \in T$$

$$(x_e^k - 1)Q \leq (\sum_{t=k}^{k+H_{e(m)}^{e(g)}-1} y_e^t - H_{e(m)}^{e(g)}) \leq (1 - x_e^k)Q \quad k \in [WS_e, WE_e], \forall e \in E$$

$$\sum_{t=1}^{|T|+H_{e(m)}^{e(g)}} y_e^t = H_{e(m)}^{e(g)} \qquad \forall\, e \in E$$

$$(x_e^k - 1)Q \leq (\sum_{t=k}^{k+J_{e(m)}^{e(g)}-1} z_e^t - J_{e(m)}^{e(g)}) \leq (1 - x_e^k)Q \quad k \in [WS_e, WE_e], \forall e \in E$$

$$\sum_{t=1}^{|T|+J_{e(m)}^{e(g)}} z_e^t = J_{e(m)}^{e(g)} \qquad \forall\, e \in E$$

$$x_e^t,\ y_e^t,\ z_e^t \in \{0,1\} \qquad \forall\, e \in E, t \in T$$

We can see from the HMP model that the number of all decision variables equals to $3 \times |T| \times |E|$ that is the product of two factors: the time span of the planning horizon and the number of trains. But the search space will reach $|T|^{|E|}$ according to the model. It is too complicated to be solved by using CPLEX or Gurobi within a reasonable time. Thus, a meta-heuristic solution strategy based on the particle swarm optimization (PSO) algorithm is designed to address this problem.

4. Modified Particle Swarm Optimization Algorithm

PSO algorithm is a population based stochastic optimization technique motivated by social behavior of organisms. PSO algorithm has the advantage of fast convergence speed and high accuracy solution and it is easy to be applied in most areas [21], which makes it attract great attention from researchers. The algorithm can also be used in solving the combinatorial optimization problem [22,23]. In this section, we present a modified particle swarm optimization (MPSO) algorithm to solve the HMP model based on analysis and preprocess.

4.1. Processing of Model Constraints

The constraint conditions of the HMP model need to be processed before applying the MPSO algorithm. Constraints (2), (3), (8)–(12) are the logical relationships, and they can be observed by the specific encoding rules (see the latter section), while the others can be removed by the penalty function method. For the value of the penalty factor, it is necessary to combine the actual application of the EMU trains and the strength of the constraint. Among the Inequations (4)–(7), the Inequation (7) has the strongest constraint and the Inequation (4) has the weakest constraint. The relationship of the penalty coefficient is $\lambda_4 > \lambda_3 > \lambda_2 > \lambda_1$. The optimization model can be presented as follows.

$$\begin{aligned} \min W = {} & c \sum_{t\in[WS_e,WE_e]} \sum_{e\in E} (WE_e - t) l_e x_e^t + \lambda_1 \sum_{t\in T} \max\{0, \sum_{e\in E} \alpha_{e(m)} y_e^t - \theta_t \cdot Inv\} \\ & + \lambda_2 \sum_{t\in T} \sum_{m\in M} \max\{0, \sum_{e\in E|e(m)=m} y_e^t - d_m^t\} + \lambda_3 \sum_{t\in T} \sum_{g\in G} \max\{0, \sum_{e\in E|e(g)=g} y_e^t - b_g\} + \lambda_4 \sum_{t\in T} \sum_{g\in G} \max\{0, \sum_{e\in E|e(g)=g} z_e^t - N_g\} \end{aligned} \tag{13}$$

s.t. (2), (3), (8)–(12).

4.2. General Particle Swarm Optimization Algorithm

In general PSO algorithm, the basic update equations of the velocity and position of the particles are as follows:

$$V_i(r+1) = \omega V_i(r) + c_1 \xi (HB_i(r) - P_i(r)) + c_2 \eta (GB(r) - P_i(r)) \tag{14}$$

$$P_i(r+1) = P_i(r) + V_i(r+1) \tag{15}$$

where $V_i(r)$ and $P_i(r)$ denote the velocity and the position, respectively, for particle i in the $r-$ th iteration. $HB_i(r)$ denotes the best position in the history for particle i by the end of the $r-$ th iteration; $GB(r)$ denotes the best position in the history for all of the particles by the end of the $r-$ th iteration. ω denotes the inertia weight; c_1 denotes the self-learning factor; c_2 denotes the social learning factor; ξ and η are the random numbers in [0, 1].

4.3. MPSO and Solution Strategy

4.3.1. Inertia Weight

In order to make the particles have a better search ability in the early stage of evolution and have a better development ability in the later stage of the evolution, the linear time-varying inertia weight is adopted in this paper. The inertia weight can be calculated as follows.

$$\omega(r) = \omega_{\max} - (\omega_{\max} - \omega_{\min}) \cdot r / MAXR \tag{16}$$

where $\omega(r)$ denotes the inertia weight at the r-th iteration. $\omega_{\max}$ and $\omega_{\min}$ denotes the maximum inertia weight and the minimum inertia weight, respectively. And $MAXR$ denotes the maximum number of the evolution iterations.

4.3.2. Learning Factor

In the same way, in order to make the particles strengthen the global search ability in the early stage and converge to the global optimum in the later period, we decrease the self-learning factor and increase the social learning factor continuously during the process of optimization. The calculation formulae are as follows.

$$c_1(r) = c'_1 + (c''_1 - c'_1) \cdot r / MAXR \tag{17}$$

$$c_2(r) = c'_2 + (c''_2 - c'_2) \cdot r / MAXR \tag{18}$$

where $c_1(r)$ and $c_2(r)$ denote the self-learning factor and the social learning factor in the $r-$ th iteration. c'_1 and c''_1 denote the initial value and the final value for the self-learning factor; c'_2 and c''_2 denote the initial value and the final value for the social learning factor; they are the constants.

4.3.3. Update Equations

The particle continuously updates its position in the search space at an unfixed speed. The velocity represents the variation of position in magnitude and direction like the definition in classical physics, and it has the same dimension as the position. Let $hbest_i(r)$ denote the corresponding fitness value of $HB_i(r)$. Let $gbest(r)$ denote the corresponding fitness value of $GB(r)$. The fitness value can be calculated by the formula (13). The update equations of the position and velocity for the particle i are as follows.

$$V_i(r+1) = \omega(r) \cdot V_i(r) + c_1(r) \cdot \xi(r) \cdot (HB_i(r) - P_i(r)) + c_2(r) \cdot \eta(r) \cdot (GB(r) - P_i(r)) \tag{19}$$

$$P_i(r+1) = P_i(r) + V_i(r+1) \tag{20}$$

where $\xi(r)$ and $\eta(r)$ are the random numbers in [0,1].

Each dimension of the velocity of a particle is limited to an interval $[-V_{\max}, V_{\max}]$, and if it is out of the interval, we set the boundary value of the interval as the actual velocity component. Similarly, the components of the position vector are limited to the time window for each train. Let $pbest_i(r)$ denote the fitness value for $P_i(r)$.

4.3.4. Encoding Rules and Initial Solution

It is a crucial step to make the particle of the MPSO and the solution of a certain problem correspond with each other. We use a particle to represent an overall HMP for all of the EMU trains. According to Equations (2) and (3), we set the dimensionality of a particle to $|E|$. The value of each dimension represents the start time of the HM for the corresponding EMU train. The detailed description is shown in Figure 2 with the help of schematic diagram.

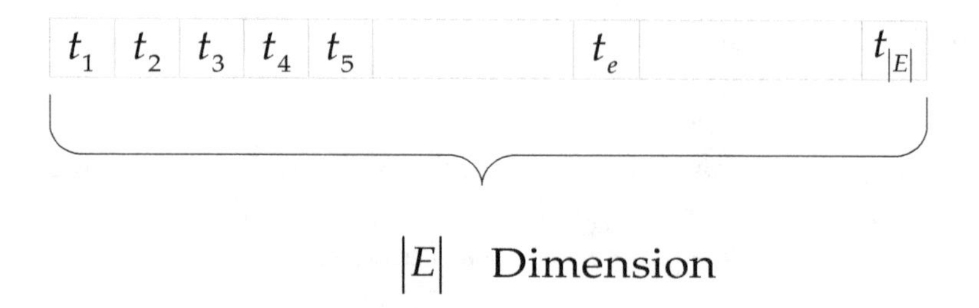

Figure 2. Encoding Rules (a particle).

In Figure 2, t_e denotes the start time of the HM for the $e-$ th EMU train. Therefore, $x_e^t = 1$ when the value of t_e is determined, and the t in x_e^t equals to t_e. Because of the maintenance level and the type of each train can be ensured in advance, the value of $H_{e(m)}^{e(g)}$ and $J_{e(m)}^{e(g)}$ for the train e is known. Then, the value of y_e^t and z_e^t can be determined accordingly, i.e., $y_e^t = 1$ $(t \in [t_e, t_e + H_{e(m)}^{e(g)} - 1])$; $z_e^t = 1$ $(t \in [t_e, t_e + J_{e(m)}^{e(g)} - 1])$. In this way, the Constraints (2), (3), (8)–(12) are well handled.

The initial solution can be generated by selecting the start time of the HM during the time window for each train. The selection is stochastic in this process.

4.4. Algorithm Steps

Step 1. Generate the time window for all of the EMU trains that need to be maintained during the planning horizon, turn to Step 2.

Step 2. Initialization. Assign values to related parameters including the size of particle swarm I, $\omega_{\max}$, $\omega_{\min}$, c'_1, c''_1, c'_2, c''_2, $MAXR$ and $V_{\max}$. Generate the initial solution $P_i(0)$ according to Section 4.3.1, and generate the initial velocity $V_i(0)$ randomly. Set $r = 0$, turn to Step 3.

Step 3. Calculate the fitness value $pbest_i(r)$ of each particle, turn to Step 4.

Step 4. Compare the fitness values of $P_i(r)$ with $HB_i(r)$. If $pbest_i(r) < hbest_i(r)$, then $HB_i(r) = P_i(r)$, $hbest_i(r) = pbest_i(r)$. Turn to Step 5.

Step 5. Compare the fitness values of $HB_i(r)(i \in I)$ with $GB(r)$. If $hbest_i(r) < gbest(r)$, then $GB(r) = HB_i(r)$, $gbest(r) = hbest_i(r)$. Turn to Step 6.

Step 6. Update $\omega(r)$, $c_1(r)$ and $c_2(r)$ according to Formulas (16)–(18), turn to Step 7.

Step 7. Update $V_i(r)$ and $P_i(r)$ according to Formulas (19) and (20). If any dimension in $V_i(r)$ out of $[-V_{\max}, V_{\max}]$, we set the boundary value of the interval as an actual value. If any dimension in $P_i(r)$ out of $[WS_e, WE_e]$, we set the boundary value of the time window as an actual value, turn to Step 8.

Step 8. $r = r + 1$. If $r > MAXR$ or $gbest(r) = gbest(r - 200)$ $(r > 200)$, turn to Step 9; otherwise, turn to Step 3.

Step 9. Make $GB(r)$ feasible according to the HMP model, and output $GB(r)$ and $gbest(r)$. Over.

5. Case Study

In this section, we implement the proposed method to solve a practice problem. The detailed description of the case can be found in the literature [20]. The proposed model is solved by the commercial optimization solver, e.g., Gurobi, as well as the MPSO algorithm. The exact method is coded in Python 2.7 and implemented within Spyder 3.1.4 and the MPSO algorithm is implemented in C++. All the computational experiments are conducted on the computer with Intel Core i5-6200U CPU and 8 GB RAM.

According to the literature [20], some parameters are valued as follows. $Inv = 115$, $|E| = 60$, $|T| = 533$. To protect data confidentiality, we can only generate the time window in advance for each train and use an ID number replace the train. The initial conditions of trains when $|T| = 0$, which include the train ID, type, the average daily operating mileage, the time window, the maintenance level and the maintenance service time, are listed in Table 3.

Table 3. Initial conditions of all trains.

ID	Type	l_e (km)	Time Window	Level	$H_{e(m)}^{e(g)}$ (Day)
1	m1	1600	[64,164]	3	50
2	m1	1600	[64,164]	3	50
3	m1	1600	[90,190]	3	50
4	m1	1600	[124,224]	3	50
5	m1	1600	[144,244]	3	50
6	m1	1600	[188,288]	3	50
7	m1	1600	[200,300]	3	50
8	m1	1600	[274,374]	4	55
9	m1	1600	[370,470]	4	55
10	m1	1600	[379,479]	4	55
11	m1	1600	[429,479]	4	55
12	m2	1600	[72,172]	3	50
13	m2	1600	[158,258]	3	50
14	m2	1600	[216,316]	3	50
15	m2	1600	[264,354]	3	50
16	m2	1600	[387,484]	3	50
17	m2	1600	[396,484]	3	50
18	m2	1600	[409,484]	3	50
19	m2	1600	[443,484]	4	55
21	m3	2000	[134,234]	5	60
21	m3	2000	[149,249]	3	40
22	m3	2000	[150,250]	3	40
23	m3	2000	[153,253]	3	40
24	m3	2000	[158,258]	5	60
25	m3	2000	[158,258]	3	40
26	m3	2000	[159,259]	3	40
27	m3	2000	[167,267]	3	40
28	m3	2000	[172,272]	4	55
29	m3	2000	[172,272]	3	40
30	m3	2000	[182,282]	3	40
31	m3	2000	[184,284]	3	40
32	m3	2000	[185,285]	3	40
33	m3	2000	[190,290]	3	40
34	m3	2000	[190,290]	3	40
35	m3	2000	[191,291]	3	40
36	m3	2000	[192,292]	3	40
37	m3	2000	[193,293]	3	40
38	m3	2000	[206,306]	3	40
39	m3	2000	[209,309]	3	40
40	m3	2000	[211,311]	3	40
41	m3	2000	[211,311]	3	40
42	m3	2000	[249,349]	3	40
43	m3	2000	[276,376]	3	40
44	m3	2000	[280,380]	3	40
45	m3	2000	[281,381]	3	40
46	m3	2000	[289,389]	3	40
47	m3	2000	[290,390]	3	40
48	m3	2000	[299,399]	3	40
49	m3	2000	[300,400]	3	40
50	m3	2000	[309,409]	3	40
51	m3	2000	[309,409]	3	40
52	m3	2000	[320,420]	3	40
53	m3	2000	[321,421]	3	40
54	m3	2000	[325,425]	3	40
55	m3	2000	[395,494]	3	40
56	m3	2000	[398,494]	3	40
57	m3	2000	[425,494]	3	40
58	m3	2000	[429,494]	3	40
59	m3	2000	[448,494]	3	40
60	m3	2000	[459,494]	3	40

From Table 3, it can be seen that there are 60 EMU trains will undergo HM procedures during the planning period. Among them, there are 101 candidate dates at most, and the minimum values are only 36 candidate dates.

During the planning horizon, the HM rate is valued as follows:

$$\theta_t = \begin{cases} 0 & if\ t \in [149, 189) \\ 7\% & if\ t \in [189, 318) \\ 6\% & if\ t \in [318, 380) \\ 0 & if\ t = 533 \\ 9\% & otherwise \end{cases} \tag{21}$$

All of these trains undergo the HM procedures in the factory. The maximal capacity is ten, i.e., $b_1 + b_2 + b_3 = 10$. Meanwhile, the receiving capability per day for a factory is limited, e.g., $N_3+N_4+N_5 = 1$; $J^3_{m1} = J^4_{m1} = J^5_{m1} = J^3_{m2} = J^4_{m2} = J^5_{m2} = J^3_{m3} = J^4_{m3} = J^5_{m3} = 1$. According to the demand, $d^t_{m1} = d^t_{m2} = 4$, $d^t_{m3} = 10$ $(t \in T)$. Therefore, the Formula (13) can be converted to the following form.

$$\begin{aligned} \min \mathrm{W} = c \sum_{t\in[WS_e, WE_e]} \sum_{e\in E} (WE_e - t) l_e x^t_e + \lambda_1 \sum_{t\in T} \max\{0, \sum_{e\in E} \alpha_{e(m)} y^t_e - \theta_t Inv\} \\ + \lambda_2 \sum_{t\in T} \sum_{m\in M} \max\{0, \sum_{e\in E|e(m)=m} y^t_e - d^t_m\} + \lambda_3 \sum_{t\in T} \max\{0, -10 + \sum_{g\in G} \sum_{e\in E|e(g)=g} y^t_e\} + \lambda_4 \sum_{t\in T} \max\{0, -1 + \sum_{g\in G} \sum_{e\in E|e(g)=g} z^t_e\} \end{aligned} \tag{22}$$

In addition, the values of other parameters are as follows. $c = 0.001$, $\lambda_1 = 100$, $\lambda_2 = 500$, $\lambda_3 = 800$, $\lambda_4 = 1000$; $I = 1000$, $\omega_{\max} = 1.2$, $\omega_{\min} = 0.8$, $c'_1 = 2.5$, $c''_1 = 0.5$, $c'_2 = 0.5$, $c''_2 = 2.5$, $V_{\max} = (WE_e\text{-}WS_e+1)/2$ $MAXR = 1000$.

Based on the data given above, the HMP model is solved by the proposed algorithm. The program runs for 500 s. The returned optimal fitness value is 3,213,121. The curve of the optimal fitness value in the iterative process is depicted in Figure 3.

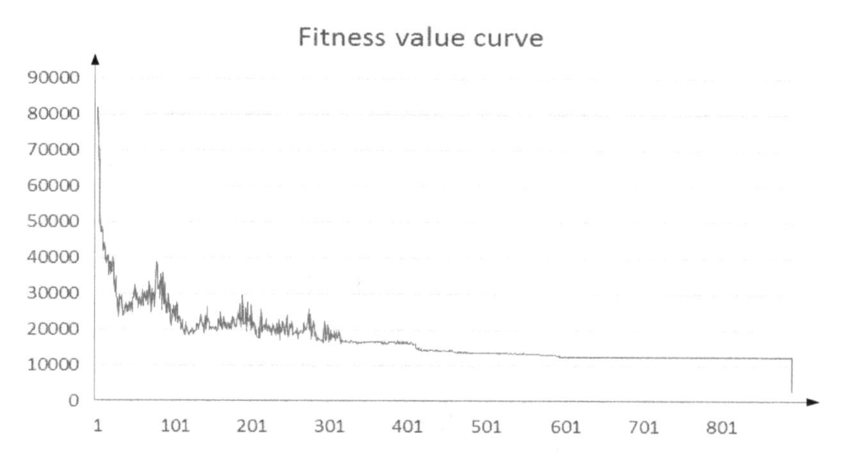

Figure 3. The optimal fitness value curve.

As can be seen in Figure 3, in the first 300 iterations, the algorithm has a strong search capability, which can effectively avoid the occurrence of the premature phenomena; and from about the 300th to the 690th iteration, the development ability of the algorithm is strengthened, which is helpful to search the optimal solution; the fitness value remains the same in the last 200 iterations, indicating that the (approximate) optimal solution for the HMP problem has been generated. The (approximate) optimal solution is listed in Table 4, and the first column is the train's ID; the second column is the start date of HM procedures denoted by t_e.

Table 4. Optimal solution.

e	t_e	e	t_e	e	t_e	e	t_e	e	t_e	e	t_e
1	108	11	481	21	182	31	280	41	310	51	409
2	103	12	142	22	220	32	252	42	349	52	419
3	153	13	212	23	250	33	278	43	376	53	421
4	193	14	333	24	240	34	273	44	379	54	425
5	198	15	360	25	248	35	282	45	381	55	472
6	300	16	460	26	188	36	276	46	388	56	498
7	312	17	492	27	218	37	238	47	390	57	500
8	387	18	458	28	268	38	303	48	398	58	502
9	482	19	490	29	270	39	243	49	400	59	496
10	480	20	190	30	246	40	308	50	407	60	494

We present the daily HM rates from the (approximate) optimal solution, and compare those with the predefined maintenance rate thresholds (see Figure 4).

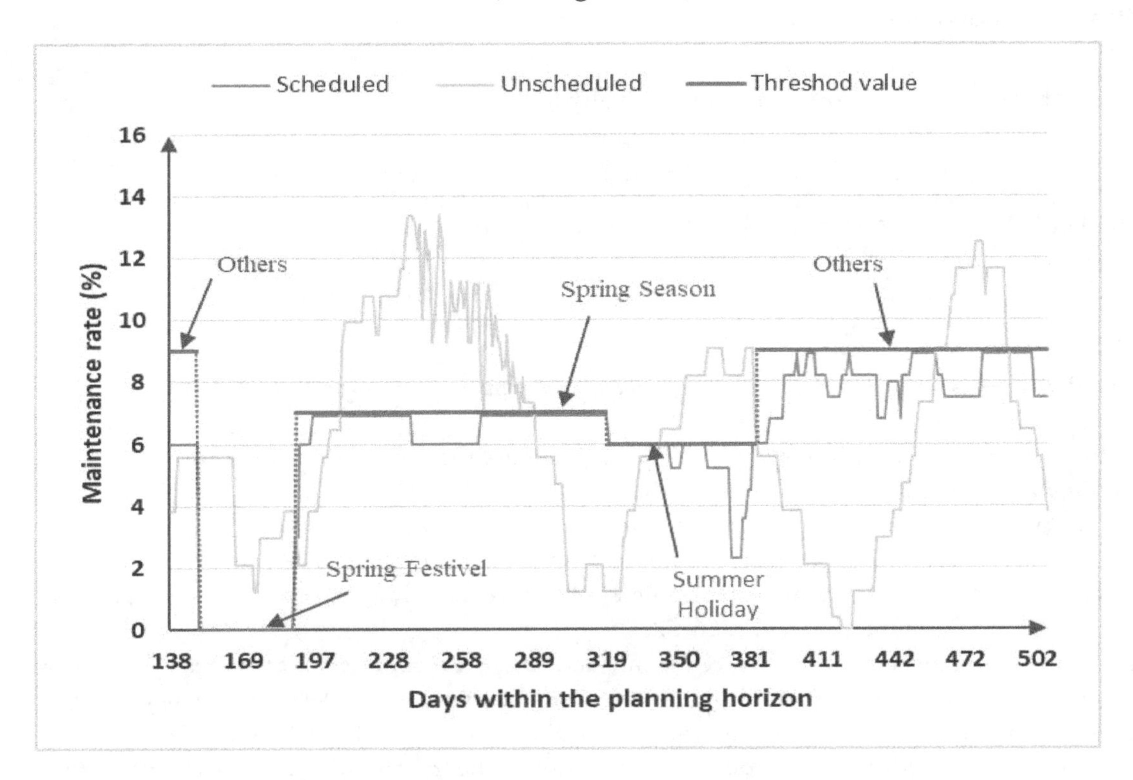

Figure 4. The distribution of the high-level maintenance rate in 2017.

From Figure 4, we can see that the HM rate of the optimal solution remains below the threshold in each period. Therefore, the proposed HMP model and the algorithm can meet the travel demands. However, the HM rate from the unscheduled solution fluctuates sharply.

In addition, to compare the performance of the proposed algorithm, we solve the HMP model by the Gurobi solver because the HMP model is linear. The detailed numerical comparison results of two solution methods, i.e., the Gurobi and the MPSO algorithm, are shown in Table 5.

Table 5. Numerical comparison results.

Method	z (km)	Time Consumption (s)
Gurobi	3,213,121	3186 s
MPSO	3,213,121	500 s

We can conclude that our proposed approach is efficient and effective from Table 5. This result demonstrates that the proposed algorithm is more efficient than the commercial optimization solver with reducing 84.31% in the solution time consumption.

6. Conclusions

This paper researches the electric multiple unit train's HMP problem. A 0–1 integer programming model and a modified particle swarm optimization algorithm are proposed. The objective function of the model is to minimize the unutilized mileage for all trains, and the model considers the necessary regulations and practical constraints, including passenger transport demand, workshop maintenance capacity, and maintenance regulations. A real-world instance demonstrates that the proposed method can efficaciously obtain the (approximate) optimal solution (see Table 4), and the solution strategy significantly reduces the solution time to 500 s (see Table 5). This result also demonstrates that the proposed algorithm is more efficient than the commercial optimization solver with reducing 84.31% in the solution time consumption. Optimize the workshop's overhaul process to shorten the maintenance service time is needed in future research.

Author Contributions: The authors contributed equally to this work.

Acknowledgments: This work was supported by the National Railway Administration of the People's Republic of China under the Grant Number KF2017-015, and the China Railway (formerly Ministry of Railways) under the Grant Number 2015X004-C. We would like to thank Lirong Diao of State Intellectual Property Office of the P.R.C. and Yinan Zhao of Beijing Jiaotong University for her valuable comments and suggestions.

References

1. China Railway. *User Manual for EMU Operation and Maintenance Procedures*, 1st ed.; Railway Publishing House: Beijing, China, 2013.
2. Stuchly, V.; Grencik, J.; Poprocky, R. Railway vehicle maintenance and information systems. *WIT Trans. Built Environ.* **2000**, *50*, 885–894.
3. Rezvanizanianil, S.M.; Valibeiglool, M.; Asgharil, M.; Barabady, J.; Kumar, U. Reliability centered maintenance for rolling stock: A case study in coaches' wheel sets of passenger trains of Iranian railway. In Proceedings of the IEEE International Conference on Industrial Engineering and Engineering Management, Singapore, 8–11 December 2008; pp. 516–520.
4. Shimada, N. Rolling stock maintenance for safe and stable transport. *Jpn. Railw. Eng.* **2006**, *46*, 4–7.
5. Cheng, Y.H.; Tsao, H.L. Rolling stock maintenance strategy selection, spares parts' estimation, and replacements' interval calculation. *Int. J. Prod. Econ.* **2010**, *128*, 404–412. [CrossRef]
6. Maróti, G.; Kroon, L. Maintenance routing for train units: The transition model. *Transp. Sci.* **2005**, *39*, 518–525. [CrossRef]
7. Maróti, G.; Kroon, L. Maintenance routing for train units: The interchange model. *Comput. Oper. Res.* **2007**, *34*, 1121–1140. [CrossRef]
8. Tsuji, Y.; Kuroda, M.; Imoto, Y. Rolling stock planning for passenger trains based on ant colony optimization. *Trans. Jpn. Soc. Mech. Eng. Part C* **2010**, *76*, 397–406. [CrossRef]
9. Wang, Y.; Liu, J.; Miao, J.R. Column generation algorithms based optimization method for maintenance scheduling of multiple units. *China Railw. Sci.* **2010**, *31*, 115–120.
10. Giacco, G.L.; D'Ariano, A.; Pacciarelli, D. Rolling stock rostering optimization under maintenance constraints. *J. Intell. Transp. Syst.* **2014**, *18*, 95–105. [CrossRef]
11. Ziaratia, K.; Soumisa, F.; Desrosiers, J.; Gélinas, S.; Saintonge, A. Locomotive assignment with heterogeneous consists at CN North America. *Eur. J. Oper. Res.* **1997**, *97*, 281–292. [CrossRef]
12. Lingaya, N.; Cordeau, J.F.; Desaulniers, G.; Desrosiers, J.; Soumis, F. Operational car assignment at VIA Rail Canada. *Transp. Res. B Meth.* **2002**, *36*, 755–778. [CrossRef]
13. Wang, L.; Ma, J.J.; Lin, B.L.; Chen, L.; Wen, X.H. Method for optimization of freight locomotive scheduling and routing problem. *J. China Railw. Soc.* **2014**, *36*, 7–15. [CrossRef]

14. Moudania, W.E.; Félix, M.C. A dynamic approach for aircraft assignment and maintenance scheduling by airlines. *J. Air Transp. Manag.* **2000**, *6*, 233–237. [CrossRef]
15. Mehmet, B.; Bilge, Ü. Operational aircraft maintenance routing problem with remaining time consideration. *Eur. J. Oper. Res.* **2014**, *235*, 315–328. [CrossRef]
16. Budai, G.; Huisman, D.; Dekker, R. Scheduling preventive railway maintenance activities. *J. Oper. Res. Soc.* **2006**, *57*, 1035–1044. [CrossRef]
17. Grigoriev, A.; Klundert, J.V.D.; Spieksma, F.C.R. Modeling and solving the periodic maintenance problem. *Eur. J. Oper. Res.* **2006**, *172*, 783–797. [CrossRef]
18. Lin, B.L.; Lin, R.X. An Approach to the high-level maintenance planning for EMU trains based on simulated annealing. *arXiv* **2017**, arXiv:1704.02752v1. Available online: https://arxiv.org/abs/1704.02752 (accessed on 10 April 2017).
19. Wu, J.P.; Lin, B.L.; Wang, J.X.; Liu, S.Q. A network-based method for the EMU train high-level maintenance planning problem. *Appl. Sci.* **2018**, *8*, 2. [CrossRef]
20. Li, Y.; Zhang, W.J.; Jia, Z.K. Forecast method of annual senior over haul amount for EMU. *J. Transp. Eng.* **2013**, *13*, 102–107.
21. Wang, D.W.; Wang, J.W.; Wang, H.F.; Zhang, R.Y; Guo, Z. *Intelligent Optimization Method*; High Education Press: Beijing, China, 2007; pp. 217–259.
22. Li, J.; Lin, B.L.; Wang, Z.K.; Chen, L.; Wang, J.X. A pragmatic optimization method for motor train set assignment and maintenance scheduling problem. *Discrete Dyn. Nat. Soc.* **2016**, *3*, 1–13. [CrossRef]
23. Wang, J.X.; Lin, B.L.; Jin, J.C. Optimizing the shunting schedule of electric multiple units depot using an enhanced particle swarm optimization algorithm. *Comput. Intell. Neurosci.* **2016**, *1*, 1–11. [CrossRef] [PubMed]

Permissions

All chapters in this book were first published in MDPI; hereby published with permission under the Creative Commons Attribution License or equivalent. Every chapter published in this book has been scrutinized by our experts. Their significance has been extensively debated. The topics covered herein carry significant findings which will fuel the growth of the discipline. They may even be implemented as practical applications or may be referred to as a beginning point for another development.

The contributors of this book come from diverse backgrounds, making this book a truly international effort. This book will bring forth new frontiers with its revolutionizing research information and detailed analysis of the nascent developments around the world.

We would like to thank all the contributing authors for lending their expertise to make the book truly unique. They have played a crucial role in the development of this book. Without their invaluable contributions this book wouldn't have been possible. They have made vital efforts to compile up to date information on the varied aspects of this subject to make this book a valuable addition to the collection of many professionals and students.

This book was conceptualized with the vision of imparting up-to-date information and advanced data in this field. To ensure the same, a matchless editorial board was set up. Every individual on the board went through rigorous rounds of assessment to prove their worth. After which they invested a large part of their time researching and compiling the most relevant data for our readers.

The editorial board has been involved in producing this book since its inception. They have spent rigorous hours researching and exploring the diverse topics which have resulted in the successful publishing of this book. They have passed on their knowledge of decades through this book. To expedite this challenging task, the publisher supported the team at every step. A small team of assistant editors was also appointed to further simplify the editing procedure and attain best results for the readers.

Apart from the editorial board, the designing team has also invested a significant amount of their time in understanding the subject and creating the most relevant covers. They scrutinized every image to scout for the most suitable representation of the subject and create an appropriate cover for the book.

The publishing team has been an ardent support to the editorial, designing and production team. Their endless efforts to recruit the best for this project, has resulted in the accomplishment of this book. They are a veteran in the field of academics and their pool of knowledge is as vast as their experience in printing. Their expertise and guidance has proved useful at every step. Their uncompromising quality standards have made this book an exceptional effort. Their encouragement from time to time has been an inspiration for everyone.

The publisher and the editorial board hope that this book will prove to be a valuable piece of knowledge for researchers, students, practitioners and scholars across the globe.

List of Contributors

Abeer Al-Siyabi and Nazife Ozdes Koca
Department of Physics, College of Science, Sultan Qaboos University, Al-Khoud, Muscat 123, Oman

Mehmet Koca
Department of Physics, Cukurova University, Adana 1380, Turkey

Jia-Bao Liu
School of Mathematics and Physics, Anhui Jianzhu University, Hefei 230601, China

Haidar Ali, Muhammad Kashif Shafiq and Usman Munir
Department of Mathematics, Government College University, Faisalabad 38000, Pakistan

Shunyi Liu
School of Science, Chang'an University, Xi'an 710064, China

Xiujun Zhang
Key Laboratory of Pattern Recognition and Intelligent Information Processing, Institutions of Higher Education of Sichuan Province, Chengdu University, Chengdu 610106, China

Xinling Wu
South China Business College, Guang Dong University of Foreign Studies, Guangzhou 510545, China

Shehnaz Akhter
Department of Mathematics, School of Natural Sciences (SNS), National University of Sciences and Technology (NUST), Sector H-12, Islamabad 44000, Pakistan

Muhammad Kamran Jamil
Department of Mathematics, Riphah Institute of Computing and Applied Sciences, Riphah International University Lahore, Lahore 54660, Pakistan

Mohammad Reza Farahani
Department of Applied Mathematics, Iran University of Science and Technology, Narmak, Tehran 16844, Iran

Zhan-Ao Xue, Dan-Jie Han, Min-Jie Lv and Min Zhang
College of Computer and Information Engineering, Henan Normal University, Xinxiang 453007, China
Engineering Lab of Henan Province for Intelligence Business & Internet of Things, Henan Normal University, Xinxiang 453007, China

Michal Staš
Faculty of Electrical Engineering and Informatics, Technical University of Košice, 042 00 Košice, Slovakia

Fabian Ball and Andreas Geyer-Schulz
Karlsruhe Institute of Technology, Institute of Information Systems and Marketing, Kaiserstr. 12, 76131 Karlsruhe, Germany

Xian-wei Xin
College of Computer and Information Engineering, Henan Normal University, Xinxiang 453007, China
Engineering Lab of Henan Province for Intelligence Business & Internet of Things, Henan Normal University, Xinxiang 453007, China

Jelena Dakić and Sanja Jančić-Rašović
Department of Mathematics, Faculty of Natural Science and Mathematics, University of Montenegro, 81000 Podgorica, Montenegro

Irina Cristea
Centre for Information Technologies and Applied Mathematics, University of Nova Gorica, 5000 Nova Gorica, Slovenia

Marija Maksimović
Department of Mathematics, University of Rijeka, Rijeka 51000, Croatia

Huilin Xu and Yuhui Xiao
College of Mathematics and Computer Science, Gannan Normal University, Ganzhou 341000, China

Xiaoying Wu
Department of Mathematics, Shaanxi University of Science & Technology, Xi'an 710021, China

Hu Zhao
School of Science, Xi'an Polytechnic University, Xi'an 710048, China

Hong-Ying Zhang
School of Mathematics and Statistics, Xi'an Jiaotong University, Xi'an 710049, China

Xiaohong Zhang
Department of Mathematics, Shaanxi University of Science and Technology, Xi'an 710021, China
Department of Mathematics, Shanghai Maritime University, Shanghai 201306, China

Rajab Ali Borzooei
Department of Mathematics, Shahid Beheshti University, Tehran 1983963113, Iran

Young Bae Jun
Department of Mathematics, Shahid Beheshti University, Tehran 1983963113, Iran
Department of Mathematics Education, Gyeongsang National University, Jinju 52828, Korea

Jianping Wu, Boliang Lin and Jiaxi Wang
School of Traffic and Transportation, Beijing Jiaotong University, Beijing 100044, China

Hui Wang and Zhongkai Wang
Institute of Computing Technology, China Academy of Railway Sciences, Beijing 100081, China

Xuhui Zhang
Department of Vehicles, China Railway Shanghai Bureau Group Co., Ltd., Shanghai 200071, China

Index

Printed in the USA
CPSIA information can be obtained
at www.ICGtesting.com
JSHW061909131123
51979JS00006B/55